Soil Science and Environment

Soil Science and Environment

Edited by **Henry Wang**

New York

Published by Callisto Reference,
106 Park Avenue, Suite 200,
New York, NY 10016, USA
www.callistoreference.com

Soil Science and Environment
Edited by Henry Wang

International Standard Book Number: 978-1-63239-625-9 (Hardback)

Printed in the United States of America.

Contents

Preface

The purpose of the book is to provide a glimpse into the dynamics and to present opinions and studies of some of the scientists engaged in the development of new ideas in the field from very different standpoints. This book will prove useful to students and researchers owing to its high content quality.

Soil science studies in detail all the properties of soil like biological, physical, and chemical properties, as well as soil formation. Different soils have their own unique properties which affect the type of agricultural as well as other activities that can be performed on them. This book consists of contributions made by international experts in this area. Also included in it is a detailed explanation of the various uses and management techniques of different types of soils. It will serve as a valuable source of reference for agronomists, geologists, ecologists, scientists and students. Those in search of information to further their knowledge will be greatly assisted by this book.

At the end, I would like to appreciate all the efforts made by the authors in completing their chapters professionally. I express my deepest gratitude to all of them for contributing to this book by sharing their valuable works. A special thanks to my family and friends for their constant support in this journey.

Editor

Organo-mineral interactions in contrasting soils under natural vegetation

Edward Jones and Balwant Singh*

Department of Environmental Sciences, Faculty of Agriculture and Environment, The University of Sydney, Sydney, NSW, Australia

Edited by:
Xinhua Peng, Chinese Academy of Sciences, China

Reviewed by:
Nabeel K. Niazi, University of Agriculture Faisalabad, Pakistan
Geertje J. Pronk, Technische Universität München, Germany

*Correspondence:
Balwant Singh, Department of Environmental Sciences, Faculty of Agriculture and Environment, The University of Sydney, 1 Central Ave., Australian Technology Park, Eveleigh, NSW 2015, Australia
e-mail: balwant.singh@sydney.edu.au

Organo-mineral interactions are important for the cycling and preservation of organic carbon (OC) in soils. To understand the role of soil mineral surfaces in organo-mineral interactions, we used a sequential density fractionation procedure to isolate <1.6, 1.6–1.8, 1.8–2.0, 2.0–2.2, 2.2–2.6, and >2.6 g cm^{-3} density fractions from topsoils (0–10 cm) of contrasting mineralogies. These soils were under natural vegetation of four major Australian soil types - Chromosol, Ferrosol, Sodosol, and Vertosol. The soils and their organic matter (OM) contents were found to be partitioned in four distinct pools: (i) particulate organic matter <1.6 g cm^{-3}; (ii) phyllosilicate dominant 1.8–2.2 g cm^{-3}; (iii) quartz and feldspar dominant >2.6 g cm^{-3}; and (iv) Fe oxides dominant >2.0 g cm^{-3} (in the Ferrosol). X-ray photoelectron spectroscopy was used to investigate organic C and N bonding environments associated within each density fraction. Mineral pools were shown to be enriched in distinct organic functional groups: phyllosilicate dominant fractions were enriched with oxidized OC species (C-O, C=O, O=C-O) and protonated amide forms; quartz and feldspar dominated fractions were enriched in aliphatic C and protonated amide forms; Fe oxides dominant fractions had the greatest proportions of oxidized OC species and were low in protonated amide forms. The enrichment of different C species was related to the interaction of functional groups with the mineral surfaces. These results demonstrate the potential of mineral surfaces in influencing the chemical composition of OM bound in surfaces reactions and subsequently the stability of OM in organo-mineral interactions.

Keywords: sequential density fractionation, X-ray photoelectron spectroscopy, organic matter composition, mineralogy, clay minerals, soil organic matter

INTRODUCTION

Soil organic matter (SOM) describes a diverse continuum of organic compounds with highly variable composition, as well as different inherent stabilities and turnover rates (Kölbl and Kögel-Knabner, 2004). It has been shown that the stability of organic matter (OM) in soils can be enhanced by the formation of various organo-mineral associations between domains of SOM particles and the surfaces of soil minerals (Baldock and Skjemstad, 2000; Kaiser and Guggenberger, 2003; Kögel-Knabner et al., 2008). This increased stability is hypothesized to be gained from the protection of OM from mineralizing agents, thus increasing C mean residence time, up to millennial time periods (Baldock and Skjemstad, 2000). The amount of OM stabilized by soil minerals is known to be influenced by the mineral size, surface functional groups, specific surface area, and porosity of the minerals involved (Anderson, 1988; Mayer, 1994; Baldock and Skjemstad, 2000; Kaiser and Guggenberger, 2000). However, there remains a lack of understanding between mineralogy and chemistry of mineral associated SOM and the stabilizing mechanisms involved in the protection of SOM *in-situ* (von Lützow et al., 2006; Kögel-Knabner et al., 2008). Improved knowledge of these relationships is essential to developing management strategies in order to increase the quantity and stability of OM in soils. To model the turnover dynamics of SOC under changing climatic conditions or land management practices, the factors influencing organo-mineral formation, surface coverage, and organic C loading need to be parameterized. This research is of vital importance in Australia, a land with inherently low soil organic carbon (SOC) levels.

The chemical composition of SOM is largely driven by the quality of OM inputs, the stage of decomposition, soil management practices and the level of physical and chemical protection provided by soil minerals (Baldock and Skjemstad, 2000). However, the biotic and/or abiotic processes and modifications involved in transforming particulate OM into a material deposited on the surface of a soil mineral remains unclear. The process of unraveling this mystery is made difficult by the variety and structural complexity of the mineral and organic species involved, as well as the simultaneous operation of a range of bonding mechanisms (Kögel-Knabner et al., 2008). Sequential density fractionation (SDF) is a procedure that can minimize these complications by isolating distinct pools of SOM allowing the structure and function of organic molecules to be modeled (Christensen, 1992). When applying SDF fractions <1.60 or <1.85 g cm^{-3} contain mostly particulate OM forms, while OM found in higher densities is associated in some way with the mineral phase. The use of SDF has highlighted the importance of organo-mineral associations in controlling C turnover dynamics with up to >90 % of total OM found associated with mineral phases in soils and sediments (Mayer, 1994; Basile-Doelsch et al., 2007).

Sequential density fractionation has also shown that in general C and N contents, and C:N ratios of fractions decrease with increasing particle densities, while δ^{13}C ratios and C turnover times increase (e.g., Baisden et al., 2002; Sollins et al., 2006). Decreased C:N ratios indicate increasing proportions of well-decomposed, microbial processed OM with increasing fraction density (Baldock et al., 1992; Golchin et al., 1994). Meanwhile, enrichment in δ^{13}C of soil samples may be attributed to disproportionate retention of δ^{13}C rich organic species or a result of continued microbial processing, a phenomenon that is well documented though not yet fully understood (Gleixner et al., 1993; Fernandez et al., 2003). These observations have led to the development of a layered mode of organo-mineral interactions on aluminosilicate minerals, with higher densities isolating organo-mineral associations with thinner deposits of OM that reveal microbial processed, proteinaceous substances at the mineral surface (Sollins et al., 2006). Sollins et al. (2009) observed an enrichment in δ^{13}C within the density range of ~1.8–2.6 g cm^{-3} of four soils with contrasting mineralogies; however, above this density increases in C:N may be observed as well as decreases in both δ^{13}C ratios and lignin phenol oxidation. This trend also coincides with shift in mineralogy, from phyllosilicates to primary minerals and Fe oxides, indicating that mineral surfaces may be driving these differences in OM composition. However, because the density of organo-mineral particles is determined by both the amount of OM associated with the particle as well the density of the sorbent mineral, it is difficult to discern to what extent mineral properties play in these relationships and to what extent these observations can be attributed to a reduction in the thickness of associated OM.

X-ray photoelectron spectroscopy (XPS) is able to quantitatively investigate the surface composition of materials (~10 nm) and is thus useful in investigating the composition of OM bound to mineral surfaces. The technique has been used, with or without Ar surface etching, to identify that OM is accumulated primarily on the surfaces of soil particles and that enrichment at mineral surfaces can reach levels up to 1000 times that found for bulk soil samples (Amelung et al., 2002; Gerin et al., 2003). Lombardi et al. (2006) used XPS analysis of chemically treated kaolinite clay deposits to identify that OM, probably humin, was strongly bound to the surface of the mineral. Meanwhile, Mikutta et al.

(2009) provided evidence for mineral influence on both the quantity and composition of associated OM when investigating the formation of organo-mineral associations over a mineralogical timescale (0.3–4100 kyr) on Hawaiian Island soils derived from basaltic tephra. This study found that mineralogical development from low specific surface area primary minerals to high specific surface area and poorly crystalline minerals was accompanied with a large increase in mineral associated OM and an increase in lignin phenols over carbohydrates. Further mineralogical development to secondary Fe and Al (hydr)oxides and kaolin minerals then saw a relative decrease in associated OM and a depletion in lignin and carbohydrates when compared to particulate OM. These results remain cryptic without given further context and investigation in other environments and there remains a need to expand the knowledge on the effect of mineralogy on associated OM especially in Australian soils that possess a very different mineralogy and climatic conditions to those studied.

In this study we hypothesize that mineral properties have the potential to determine not only the content but also the composition of associated OM in major Australian soil groups. To investigate this we used XPS analysis of sequential density fractionated soils to identify: (i) if OM pools in different soil density fractions show discrete chemical composition; and (ii) if the chemical composition of OM in different soil density fractions varies with the mineral composition of soils.

MATERIALS AND METHODS
SITE SELECTION AND SAMPLE COLLECTION
Four soil types, known to possess a diverse range of clay minerals, were selected for the study. All selected soil sites were situated in the state of New South Wales, Australia (Table 1). The sites were under native or naturalized vegetation and had never been cultivated for agricultural purposes, thus they provided a good representation of undisturbed native conditions for investigating organo-mineral associations. Bulk samples were taken from the topsoil (0–10 cm) at each of the four sites. Samples were air-dried and ground to obtain <2 mm for laboratory analyses.

GENERAL CHARACTERIZATION OF BULK SOILS
Routine soil procedures were used to determine pH (1:5 H$_2$O), electrical conductivity (1:5 H$_2$O), cation exchange capacity (ammonium-acetate at pH 7) and exchangeable cations (Rayment and Higginson, 1992). Particle size analysis was performed using the pipette method (Gee and Bauder, 1986). Total C and N in the

Table 1 | Location, elevation and climatic data of the four sites used in the study.

ASC soil order	Latitude	Longitude	Nearest town	Average annual precipitation (mm)	Average annual temperature (°C)	Elevation (m)
Chromosol[a]	−31° 5′ 48.4″	150° 43′ 50.7″	Tamworth	673	17.3	404
Ferrosol[b]	−28° 48′ 55.8″	153° 24′ 0.7″	Wollongbar	1800	19.2	166
Sodosol[a]	−31° 21′ 11.0″	150° 4′ 46.1″	Gunnedah	622	18.5	285
Vertosol[a]	−31° 43′ 13.1″	150° 40′ 54.6″	Quirindi	683	16.8	390

[a] Climate data sourced from Australian Bureau of Meteorology (BOM, 2013).

[b] Climate data sourced from the New South Wales Department of Primary Industries (NSWDPI, 2013).

soils were determined by dry combustion method using a Vario Max CNS analyzer (Elementar Analysensysteme GmbH, Hanau, Germany).

Speciation of Fe (as well as Mn, Al, and Si) was determined by three methods. Total free Fe oxides were quantified using the dithionite-citrate-bicarbonate (DCB) method (Mehra and Jackson, 1960); poorly crystalline Fe and Al species were extracted by the acid ammonium oxalate (pH = 3.0) in the dark (Schwertmann, 1964); and organically bound Fe was estimated using the Na-pyrophosphate (pH = 10.0) extractable Fe procedure (McKeague, 1967). Alkali dissolution with 0.5 M NaOH was used to investigate the presence of poorly ordered aluminosilicate species in soils (Hashimoto and Jackson, 1960). The soil extracts were analyzed for Fe, Al, Mn, and Si using a Varian 720-ES inductively coupled plasma optical atomic emission spectrometer.

MINERALOGICAL ANALYSIS

Semi-quantitative mineralogical analysis of sand (2000–20 μm), silt (20–2 μm) and clay (<2 μm) sized particles was performed using X-ray diffraction (XRD) following the isolation of these particle size fractions using a sedimentation method based on Stoke's Law (Gee and Bauder, 1986). The size and density fraction samples were ground to a fine powder and randomly oriented samples were analyzed using monochromatic CuKα radiation at 30 kV and 28.5 mA (GBC MMA diffractometer). The samples were scanned from 4 to 65° 2θ at a speed of 1° 2θ min^{-1} and using a step size of 0.01° 2θ. Oriented samples from the clay fraction was also analyzed after various pre-treatments (i.e., Mg-saturated, Mg-saturated and ethylene glycol solvated, K-saturated and K-saturated and heated to 550°C) for the identification of phyllosilicate species (Brown and Brindley, 1980).

SEQUENTIAL DENSITY FRACTIONATION

Soils were isolated into <1.6, 1.6–1.8, 1.8–2.0, 2.0–2.2, 2.2–2.6, and >2.6 g cm^{-3} fractions using the procedure adopted from Sollins et al. (2006, 2009). The density fractions were decided based on the clay mineral analysis of bulk soils to isolate SOM associated with various suites of soil minerals. Briefly, 25 g of air-dry soil was added to a 250 ml polycarbonate centrifuge bottle with 125 ml of 1.6 g cm^{-3} sodium polytungstate (SPT) solution. The centrifuge tube was shaken vigorously by hand to ensure the soil was well wetted with SPT solution and then shaken for 3 h on a horizontal shaker (300 rpm). The tubes were then centrifuged at 970g using a Spintron 175 for 30 min. Floating material (<1.6 g cm^{-3} fraction) was then aspirated under suction. SPT was recovered by filtering the aspirated liquid through a 0.7 μm glass fiber filter, readjusted to the target density and returned to the centrifuge tube. The tube was again shaken for a further 1 h and centrifuged. Following the second aspiration of floating material the density of the supernatant was recorded using a 20 ml pipette (all densities fell within ±0.05 g cm^{-3} of the target density). The supernatant was aspirated down to the pellet and filtered through a 0.7 μm glass fiber filter to recover the floating fraction. The two recovered soil fractions were rinsed on a 0.7 μm glass fiber filter multiple times with RO water to remove residual SPT to the point where the filtrate had an EC <50 μS cm^{-1}. The two yielded fractions were then dried at 60°C to obtain the <1.6 g cm^{-3} fraction.

125 ml of SPT solution of the next target density was then added to the centrifuge tube and the process repeated. The process was repeated for all fractions with the exception that removal of residual SPT in clay rich fractions was achieved by repeated washing with RO water in the centrifuge tubes as filtration became unviable. All recovered fractions were hand ground to a fine powder for further analyses.

TOTAL C AND N AND δ^{13}C

Density fractions were analyzed for total C, total N and δ^{13}C by isotope ratio mass spectrometry (Delta V Thermo Finnigan). The δ^{13}C values were measured against standards of beet sucrose, cane sucrose, IVA soil and BW algae and the analytical precision of the δ^{13}C analysis was <0.1%.

X-RAY PHOTOELECTRON SPECTROSCOPY (XPS)

Photoelectron spectra were recorded on finely ground density fractions using an X-ray photoelectron spectrometer (ESCALAB250Xi) at the University of New South Wales. Samples were excited using monochromatic AlKα radiation (1486.68 eV) with a 90° photoelectron take off angle and vacuum conditions remaining below 9.1×10^{-4} mbar throughout the analysis. A 500 μm spot was analyzed with pass energies of 100 eV for survey scans and 20 eV for regions scans over C1s and N1s spectral lines. Atomic quantification was achieved by a process of linear background subtraction and fitting lines with a set of Gaussian curves with conversion of intensities to atomic concentrations through sensitivity factors (Moulder et al., 1992). Carbon peaks at 285.0, 286.5, 288.0, and 289.5 eV were assigned to aromatic and aliphatic functional groups (C-C/C-H), alcoholic and phenolic (C-O), carbonyl of amide/carboxylic (C=O) and carboxylate (O=C-O) bonding environments, respectively (Proctor and Sherwood, 1982). Nitrogen peaks at 400.9 and 402.5 eV were assigned to amine/amide (C-N) and protonated amine (C-N$^+$), respectively (Abe and Watanabe, 2004).

STATISTICAL ANALYSIS

Analysis of variance was used to investigate significant differences between OC (g g^{-1}), N content (g g^{-1}), C:N ratio and δ^{13}C across density fractions of a soil and corresponding density fractions of different soils. Data were log transformed where appropriate to achieve normality. Tukey's HSD was used for multiple comparison of mean values of three replicates for statistically significant results at $p < 0.05$. Linear regression was used to investigate the relationship of Fe concentration (% w/w) as determined by XPS and the relative concentration of O=C-O moieties for all density fractions of all soils.

RESULTS AND DISCUSSION
GENERAL CHARACTERIZATION OF BULK SOILS

The clay fraction of the four soils showed contrasting mineral compositions (**Figure 1**). The clay fraction of the Chromosol was dominated by illite with lesser amount of kaolinite, and the opposite trend was seen in the Sodosol with the dominance of kaolinite and illite being present in smaller amount (**Figures 1A,C**). The Vertosol clay fraction was dominated by smectite with illite and kaolinite being the accessory minerals (**Figure 1D**). In the

Ferrosol, only kaolinite was identified in the oriented clay samples (**Figure 1B**), the random powder of the clay fraction also showed the presence of hematite, goethite and gibbsite (XRD pattern not shown). The silt and sand sized fractions were characterized by increasing proportions of primary minerals (quartz, feldspars and/or Fe bearing minerals) with reduced levels of phyllosilicates. Traces of phyllosilicates were still identifiable in all sand sized

fractions except for the Sodosol. Small amounts of hematite were also identified in the sand and silt sized fractions of the Vertosol (data not shown). The Ferrosol was found to have the highest total C content of 6.11%, while the other three soils had very similar C levels that ranged from 1.82 to 2.06% (**Table 2**). As all of the soils investigated had acidic pH (≤ 6.5; **Table 2**), the total C thus represented OC in these soils. The Vertosol exhibited the highest

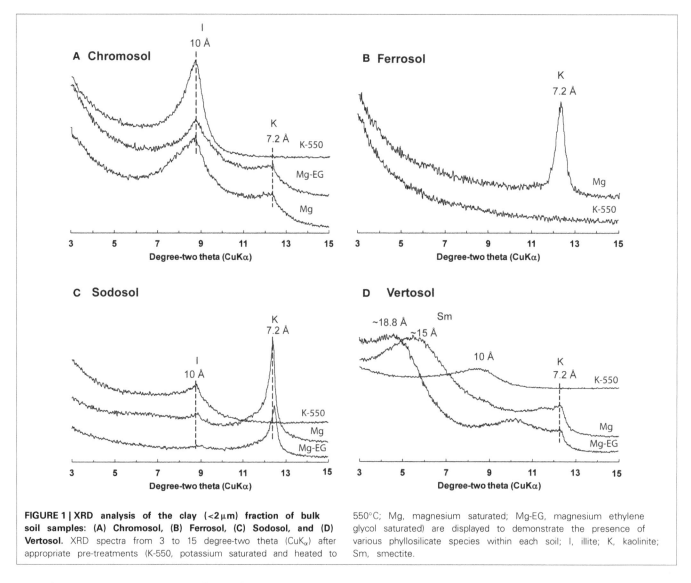

FIGURE 1 | XRD analysis of the clay (<2 μm) fraction of bulk soil samples: (A) Chromosol, (B) Ferrosol, (C) Sodosol, and (D) Vertosol. XRD spectra from 3 to 15 degree-two theta (CuK$_\alpha$) after appropriate pre-treatments (K-550, potassium saturated and heated to 550°C; Mg, magnesium saturated; Mg-EG, magnesium ethylene glycol saturated) are displayed to demonstrate the presence of various phyllosilicate species within each soil; I, illite; K, kaolinite; Sm, smectite.

Table 2 | Properties of bulk soil samples (< 2 mm) used in this study.

ASC soil order	pH (1:5 water)	EC (1:5, dS m^{-1})	Total C* (%)	Total N* (%)	CEC (mmol$_c$ kg^{-1})	Exchangeable cations (mmol$_c$ kg^{-1})				Sand (%)	Silt (%)	Clay (%)
						Ca	Mg	Na	K			
Chromosol	5.9	0.22	2.06	0.19	183	130	44	2	7	40	28	32
Ferrosol	5.6	0.19	6.11	0.55	64	32	10	0.7	4	26	43	31
Sodosol	6.0	0.13	1.82	0.11	6.5	2.9	1.8	<0.2	0.5	91	4	6
Vertosol	6.5	0.08	2.01	0.12	403	210	180	4	9	16	15	69

*Total N and C contents are mean of three replicates, SE for C = 0.02–0.05, and N = <0.01%.

CEC ($403\,mmol_c\,kg^{-1}$), which is consistent with the dominance of smectite in the clay fraction of this soil. The Chromosol had the second highest CEC ($183\,mmol_c\,kg^{-1}$) in accordance with its illite dominated clay fraction. Conversely the Sodosol had an extremely low CEC of $6.5\,mmol_c\,kg^{-1}$, which is in agreement with a texture comprising of 91% sand sized particles and kaolinite dominated clay fraction of the soil. Although the Ferrosol had high clay content, this was composed of variable charge minerals that conveyed a low CEC ($64\,mmol_c\,kg^{-1}$) under the acidic conditions of the soil. The high organic carbon content has probably contributed most of the CEC in the Ferrosol.

Extractable cation analyses showed that the Ferrosol had the highest amount of free Fe and Al oxides, with Fe accounting for approximately 15.0% of the total soil mass (**Table 3**). The Sodosol had the lowest concentration of all soils with only 0.3% Fe. The Vertosol had the highest concentration of oxalate extractable Fe, Mn and Si, which was up to half of that extracted using DCB suggesting a relatively high proportion of these elements exist in amorphous or poorly crystalline forms. In contrast, the Ferrosol had a small (4.0% of the DCB) proportion of oxalate extractable Fe, indicating the presence of mainly crystalline form of Fe (hematite and goethite) in the soil, as confirmed by XRD analysis. The ratio of oxalate: DCB Fe in the Chromosol and Sodosol was 0.15 and 0.27, respectively. Pyrophosphate extractable elements were also the highest in the Ferrosol indicating a relatively large amount of organically bound Fe (0.9%) and Al (0.4%) compared to the other soils. Alkali dissolution with NaOH gave very low Si and Al yields (data not presented) indicating negligible amounts of poorly crystalline aluminosilicates in these soils.

MINERALOGY OF THE DENSITY FRACTIONS
The inherent diversity of minerals, varying content of associated OM, micro-aggregation of minerals and coatings of

phyllosilicates on primary mineral crystals or particulate OM species inhibits complete isolation of individual mineral species. For example, quartz was identified, at least in trace quantities, in all density fractions of all soil samples (**Table 4**). Nevertheless, suites of soil minerals were found to be enriched in certain fractions as expected from their densities.

The lightest two fractions (<1.6 and 1.6–$1.8\,g\,cm^{-3}$) of all soils were dominated by organic matter and/or other amorphous material, visual observation identified large proportions of fresh or slightly decomposed plant debris as well as significant quantities of charcoal (**Table 4**). Varying quantities of mineral species were also identified in the $<1.8\,g\,cm^{-3}$ fractions, which could be due to mineral coatings on particulate OM surfaces or entrainment of mineral species during density fractionation.

The proportion of minerals then increased with increasing fraction density. The pattern of mineral isolation in fractions of the Chromosol, Sodosol and Vertosol density fractions were found to be similar and distinct to the mineralogy identified for the Ferrosol as expected from the preliminary mineralogical analysis. For the Chromosol, Sodosol and Vertosol, the mineralogy of the 1.8–2.0 and 2.0–$2.2\,g\,cm^{-3}$ density fractions were dominated by phyllosilicates with some inclusions from quartz and feldspars. The mineralogy of the 2.2–$2.6\,g\,cm^{-3}$ in these soils was generally dominated by quartz and feldspar except for the Chromosol where illite was present in about the same proportion. The increased proportion of illite in the 2.2–$2.6\,g\,cm^{-3}$ fraction of the Chromosol is likely due the higher specific gravity, 2.6–2.9 of illite than smectite (2.0–$3.0\,g\,cm^{-3}$) and kaolinite ($2.65\,g\,cm^{-3}$) (Lide, 2007) and the larger crystal size of illite in this soil compared to phyllosilicate minerals in the Sodosol and Vertosol. The heaviest fraction ($>2.6\,g\,cm^{-3}$) of the Chromosol, Sodosol and Vertosol was dominated by quartz and feldspars with only traces of hematite identified in the Vertosol (**Table 4**).

In contrast, hematite, goethite and gibbsite were identified in all fractions of the Ferrosol. Some contributions of kaolinite were identified in the 1.8–2.0, 2.0–2.2 and 2.2–$2.6\,g\,cm^{-3}$ fractions, however, all heavier fractions ($>2.0\,g\,cm^{-3}$) were dominated by hematite along with gibbsite and goethite with only traces of quartz identified.

C AND N CONCENTRATIONS AND C/N RATIO OF THE DENSITY FRACTIONS
Mass spectrometry results showed that total C and N contents in the density fractions decreased significantly with increasing density in the soils with phyllosilicates dominated clay fractions (i.e., Chromosol, Sodosol and Vertosol), except for increased total C in the 1.6–$1.8\,g\,cm^{-3}$ (cf. $<1.6\,g\,cm^{-3}$) in the Chromosol and increased total N content in the 1.8–$2.0\,g\,cm^{-3}$ (cf. 1.6–$1.8\,g\,cm^{-3}$) in the Vertosol (**Figures 2A,B; Table 5**). The OC content of the phyllosilicate dominant 1.8–$2.0\,g\,cm^{-3}$ fraction was not statistically significant between the three soils even though the content and proportion of phyllosilicate minerals did vary. However, in the smectite rich, 2.0–$2.2\,g\,cm^{-3}$ fraction of the Vertosol both C and N were significantly lower than the corresponding fractions in the Chromosol and Sodosol, which were dominated by illite and kaolinite. The difference in the total C and N contents in

Table 3 | Mean ($n = 3$) values and standard errors ($mg\,kg^{-1}$) of iron, aluminium, manganese, and silicon extracted from bulk soil samples ($<2\,mm$) using dithionite-citrate-bicarbonate (DCB), acid oxalate and sodium pyrophosphate (Na-pyro) extraction procedures.

Extraction procedure	ASC soil order	Extractable cation (mean $\pm$ *SE*, $mg\,kg^{-1}$)			
		Fe	Al	Mn	Si
DCB	Chromosol	16896 ± 776	1476 ± 71	528 ± 25	724 ± 29
	Ferrosol	149970 ± 1741	12757 ± 124	697 ± 7	1027 ± 3
	Sodosol	3394 ± 9	549 ± 2	85 ± 0.3	747 ± 6
	Vertosol	15072 ± 247	1341 ± 24	820 ± 13	1641 ± 44
Acid oxalate	Chromosol	2475 ± 11	1070 ± 13	790 ± 7	323 ± 17
	Ferrosol	6040 ± 35	3227 ± 32	1094 ± 17	143 ± 3
	Sodosol	910 ± 17	290 ± 5	127 ± 3	36 ± 0.1
	Vertosol	7947 ± 174	2014 ± 27	1629 ± 29	761 ± 19
Na-pyro	Chromosol	2475 ± 11	1070 ± 13	790 ± 7	323 ± 17
	Ferrosol	6040 ± 35	3227 ± 32	1094 ± 17	143 ± 3
	Sodosol	910 ± 17	290 ± 5	127 ± 3	36 ± 0.1
	Vertosol	7947 ± 174	2014 ± 27	1629 ± 29	761 ± 19

Table 4 | Semi-quantitative X-ray diffraction mineralogical analysis of the six density fractions of the four soils.

ASC soil order	Density fraction (g cm^{-3})	Phyllosilicates			Oxide minerals					Feldspars				Quartz
		Ka	Il	Sm	Go	He	Gi	Ru	An	Mi	Or/K	Pl	Al	
Chromosol	<1.6	x	x	–	–	–	–	–	–	–	–	–	x	x
	1.6–1.8	x	x	–	–	–	–	–	–	–	–	–	tr	x
	1.8–2.0	xx	xx	–	–	–	–	–	–	–	–	–	tr	x
	2.0–2.2	x	xx	–	–	–	–	–	–	–	–	–	x	x
	2.2–2.6	–	xx	–	–	–	–	–	–	–	–	–	xx	xx
	>2.6	–	x	–	–	–	–	–	–	–	–	–	xx	xxx
Ferrosol	<1.6	tr	–	–	tr	x	x	–	–	–	–	–	–	tr
	1.6–1.8	x	–	–	tr	x	x	–	–	–	–	–	–	tr
	1.8–2.0	xx	x	–	x	x	xx	–	tr	–	–	–	–	tr
	2.0–2.2	xx	–	–	–	xx	xx	–	tr	–	–	–	–	tr
	2.2–2.6	xx	–	–	–	xx	xx	–	–	–	–	–	–	tr
	>2.6	tr	–	–	x	xxx	x	–	–	–	x	–	–	x
Sodosol	< 1.6	x	x	–	–	–	–	–	–	–	–	–	x	x
	1.6–1.8	xx	x	–	–	–	–	–	–	–	–	–	x	x
	1.8–2.0	xxx	–	–	–	–	–	tr	tr	–	–	–	xx	x
	2.0–2.2	xxx	–	–	–	–	–	–	–	–	–	–	–	xx
	2.2–2.6	x	–	–	–	–	–	–	–	–	x	–	–	xxx
	>2.6	–	–	–	–	–	–	–	–	–	–	–	–	xxxx
Vertosol	<1.6	–	x	–	–	–	–	–	–	–	–	–	–	x
	1.6–1.8	–	xx	–	–	–	–	–	–	–	–	–	–	x
	1.8–2.0	x	x	xxx	–	–	–	–	–	–	–	–	–	x
	2.0–2.2	xx	–	xx	–	–	–	–	–	–	–	–	–	xx
	2.2–2.6	x	–	x	–	–	–	–	–	xx	–	–	–	xxx
	>2.6	x	–	–	–	tr	–	–	–	xx	–	xx	–	xxx

Mineral abbreviations used: Ka, kaolinite; Il, illite; Sm, smectite; Go, goethite; He, hematite; Gi, gibbsite; Ru, rutile; An, anatase; Mi, microcline; Or/K, orthoclase/K-feldspar; Pl, plagioclase; Al, albite; Estimated proportion of mineral: xxxx, dominant (>60%), xxx, large (40–60%), xx, moderate (20–40%), x, small (5–20%); tr, trace (< 5%); –, non-detectable.

the 2.0–2.2 g cm^{-3} fraction of the three phyllosilicates dominated soils (i.e., Chromosol, Sodosol and Vertosol) could be attributed to the fraction of the soil recovered in this density fraction. In the Vertosol (4.4%) a substantially greater soil mass was recovered than the Chromosol (1.1%) and the Sodosol (0.5%), hence significantly lower C and N contents (i.e., dilution effect) than the other two soils. The trend in total C and N contents in the density fractions was reinforced by the X-ray photoelectron spectroscopy analysis, which also showed the mass percentage of C and N decreased with increasing density (**Figures 3A,B**). The decrease was most substantial from <1.6 g cm^{-3} fraction to the 1.6–1.8 g cm^{-3} fraction, and then it decreased steadily in the subsequent heavier density fractions, with few exceptions. These results are in accordance with previous studies covering a wide range of mineralogy and climatic conditions (e.g., Sollins et al., 2009; Plante et al., 2010; Bonnard et al., 2012) and are intuitive as the association of OM acts to reduce net organo-mineral particle density meaning that fractions with a higher OM content are generally lighter. The Ferrosol had similar C and N concentrations in the first two fractions as compared to the other soils, but significantly

higher total C and N contents in the fractions heavier than 1.8 g cm^{-3}.

The C:N ratio also decreased with increasing density fractions with few, yet notable exceptions (**Figure 2C**). The increase in C:N ratio to the 1.6–1.8 g cm^{-3} fraction was attributed to charcoal observed in this fraction, although the high porosity of charcoal gives it a low specific gravity, the skeletal density of produced chars have been shown to have maximum densities of up to 2.0–2.1 g cm^{-3} (Brown et al., 2006). The C:N increase in the Vertosol for the heaviest fraction was intriguing, although it has been observed in other studies (Sollins et al., 2009; Throop et al., 2013). The presence of hematite and the high Fe content of this fraction as observed by XPS (**Table 6**) suggest a role of Fe oxides in this observation, which is reinforced by the fact that the C:N ratio converges toward that of the Ferrosol. Similarly, in a study by Sollins et al. (2009) two soils displayed similar trend with a shift in mineralogy from phyllosilicate dominated (chlorite, vermiculite, kaolinite) lighter fractions to inclusions of hematite or other Fe bearing primary minerals in the fractions.

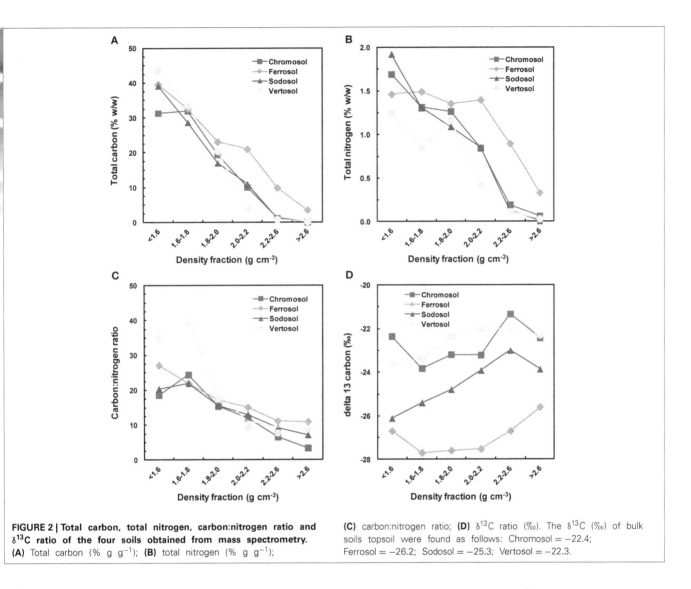

FIGURE 2 | Total carbon, total nitrogen, carbon:nitrogen ratio and δ^{13}C ratio of the four soils obtained from mass spectrometry. **(A)** Total carbon (% g g^{-1}); **(B)** total nitrogen (% g g^{-1});

(C) carbon:nitrogen ratio; **(D)** δ^{13}C ratio (‰). The δ^{13}C (‰) of bulk soils topsoil were found as follows: Chromosol = −22.4; Ferrosol = −26.2; Sodosol = −25.3; Vertosol = −22.3.

δ^{13}C OF DENSITY FRACTIONS

All soils showed a progressive increase in δ^{13}C ratios across between the 1.6–1.8 g cm^{-3} fraction up to the 2.2–2.6 g cm^{-3} (**Figure 2D**) reflecting many other studies (Basile-Doelsch et al., 2009; Sollins et al., 2009; Throop et al., 2013). Following this the Chromosol, Sodosol and Vertosol showed a decline to the heaviest fraction (>2.6 g cm^{-3}) while the δ^{13}C of the Ferrosol showed a substantial increase in the heaviest fraction (**Figure 2D**).

SOIL MASS, TOTAL C AND N DISTRIBUTION ACROSS THE DENSITY FRACTIONS

Total C and N was calculated as the product of C or N content of the fractions and the total mass yielded for each fraction. The distribution of total C and N in the density fractions showed similar trends in the soils with a phyllosilicate dominated clay fraction, i.e., Chromosol, Sodosol, Vertosol (**Figures 4A,C,D**). For these soils the first density fraction (<1.6 g cm^{-3}) contained either a local or absolute maximum value for the proportion of total soil C. The proportion of total C then decreased immediately in the 1.6–1.8 g cm^{-3} fraction before rising toward another local or absolute maximum value in the phyllosilicate dominated

fractions (1.8–2.0 or 2.0–2.2 g cm^{-3}). The proportion of total C then fell sharply in the heaviest fraction (>2.6 g cm^{-3}). The heaviest fraction contained very little of the total C even though it often accounted for a large proportion of the soil mass (e.g., 84% for the Sodosol). In sharp contrast to other soils, the total C in the Ferrosol was found to be concentrated in the two heaviest fractions (>2.2 g cm^{-3}) dominated by Fe oxides (**Figure 4B**). The proportions of soil mass, total C and total N were found to be 98, 85, and 85%, respectively in the heaviest two fractions (>2.2 g cm^{-3}) of the Ferrosol compared to average values of 72, 18, and 36%, respectively for the corresponding fractions in the other three soils. The proportion of total N in the density fractions of the four soils followed a trend similar to total C except the occurrence of substantially higher total N content in the 2.2–2.6 g cm^{-3} fraction of the Chromosol and in the 2.0–2.2 and 2.2–2.6 g cm^{-3} fractions of the Vertosol.

XPS COMPOSITION OF THE SOIL DENSITY FRACTIONS

XPS derived data show that mass percentage of mineral structural components such as O, Si, Al, Fe, Mg, Na, and K increased with the increasing density of fractions (**Table 6**). The majority

Table 5 | Statistical analysis between the differences in mean values of carbon and nitrogen contents, C:N ratios and δ^{13}C ratios between different soils and density fractions.

Soil property	ASC soil order	Density fraction (g cm^{-3})					
		<1.6	1.6–1.8	1.8–2.0	2.0–2.2	2.2–2.6	>2.6
Carbon (g g^{-1})	Chromosol	aA	nsA	abB	aC	aD	aE
	Ferrosol	abA	nsA	aB	bB	bC	bD
	Sodosol	abA	nsB	bC	aD	acE	cF
	Vertosol	aA	nsA	abB	cC	cD	aE
Nitrogen (g g^{-1})	Chromosol	aA	aB	abB	aC	aD	aE
	Ferrosol	bA	bA	bB	bAB	bC	bD
	Sodosol	cA	aB	cC	aD	aE	aF
	Vertosol	dA	cB	acA	cC	aD	aD
C:N ratio	Chromosol	aA	aB	aC	aD	aE	aF
	Ferrosol	bA	bB	abC	bD	bE	bE
	Sodosol	cA	bA	aB	aC	cD	cE
	Vertosol	dA	cB	bC	cD	acD	dD
δ^{13}C (‰)	Chromosol	aA	aB	aC	aC	aD	aA
	Ferrosol	bA	bB	bB	bB	bA	bC
	Sodosol	cA	cB	cC	cD	cE	cD
	Vertosol	dA	aA	dB	dB	dB	dB

Uppercase letters indicate significant differences, p < 0.05, between density fractions of a given soil, while lowercase letters indicate significant differences between soils in corresponding density fractions.

of this increase generally occurred in the 1.8–2.0 g cm^{-3} density fraction after which concentrations plateaued or continued to increase slowly with few exceptions. Conversely, C and N concentrations were found to decrease with increasing density. The C:N ratio decreased with increasing density up to 2.0–2.2 g cm^{-3} fraction for all soils, and after this it remained similar or increased substantially in the heaviest fraction of Vertosol and Sodosol (**Figure 3C**). XPS elemental data are consistent with the XRD mineralogical analysis, for example, high contents of K in the Chromosol signifying the presence of illite and albite in the soil; the highest level of Si was identified in the >2.6 g cm^{-3} fraction of the Sodosol which is primarily quartz; very high levels of Fe in the Vertosol in the >2.6 g cm^{-3} fraction reaffirming the presence of hematite observed in the XRD analysis. The presence of Fe in all samples suggests that Fe was intimately associated with OM whether in crystalline (Ferrosol) or amorphous forms (other soils).

For the soils with phyllosilicate dominated clay fraction (Sodosol, Chromosol, and Vertosol), the proportion of aliphatic to total C decreased with increasing density reaching to a minimum value in the 2.0–2.2 g cm^{-3} fraction before rising again in the quartz and other primary mineral dominated fractions (**Table 7**). Conversely, the proportions of C-O and C=O moieties were seen to increase in the phyllosilicate dominated fractions, reaching maximum values in the 2.0–2.2 g cm^{-3} fraction before decreasing in the >2.6 g cm^{-3} fraction, where levels were smaller than those observed in the POM fractions. The proportion of carboxylic compounds were also elevated in the phyllosilicate

dominated fractions, however no clear trend was seen in the >2.6 g cm^{-3} fraction. However, a strong positive relationship was observed between the log-transformed values of Fe (**Table 6**) and the proportion of O=C-O functional group (**Table 7**) in all density fractions of all soils ($p < 0.001$, $R^2 = 0.556$) (**Figure 5**).

The proportions of quaternary bonded protonated amide form of N increased with increasing fraction density for all soils with a complementary decrease in amide/amine N forms (**Table 7**). The Vertosol showed greater increases in the proportion of protonated amide form in the 2.0–2.2 g cm^{-3} fraction in comparison to the Chromosol and Sodosol.

A comparison of the C contents determined using mass spectrometry and XPS shows that XPS derived C content is lower in lighter fractions (1.6–1.8 and 1.8–2.0 g cm^{-3}) and higher in heavier fractions (2.2–2.6 and > 2.6 g cm^{-3}) than the bulk C analysis in the corresponding fractions obtained by mass spectrometry (**Figure 3D**). Although, the samples were finely ground for the XPS analysis, these results reflect the nature of XPS as a surface technique, probing ~10 nm of the surface layer; the variability in the C data suggest possible existence of OM in the micropores in lighter fractions (<2.0 g cm^{-3}) and variable or incomplete coating of larger crystal size of minerals by OM in heavier fractions (>2.2 g cm^{-3}). A similar result was also obtained for N (data not presented).

Variation in the proportions of organic structures across the density fractions was less pronounced in the Ferrosol as compared to the other soils (**Table 7**). The proportion of aliphatic to oxidized C groups was much smaller in the Ferrosol compared to the other three soils across all density fractions and there was no increase in the proportion of aliphatic C in the >2.6 g cm^{-3} fraction as observed for the other three soils. The proportion of C-O moieties was constant across all density fractions, and the proportion of C=O moieties was also constant across Fe oxides dominant fractions (>2.0 g cm^{-3}). Only the proportion of O=C-O was found to increase with increasing density in the Fe oxides dominated fractions. The Ferrosol also showed the highest proportion of amide/amine N to protonated amide forms in all fractions and did not display the strong decreasing trend as observed in the other soils (**Table 7**).

GENERAL DISCUSSION
BONDING ENVIRONMENTS OF ORGANIC SPECIES WITHIN DENSITY FRACTIONS IN RELATION TO MINERAL SURFACE PROPERTIES

All fractions yielded heterogeneous C and N bonding environments, representing diverse assemblies of organic compounds. However, clear enrichment or depletion in particular organic functional groups were observed across density fractions and soil types. The composition of OM was found to be associated with (i) phyllosilicates, (ii) quartz and feldspars, and (iii) Fe oxides, and how they may be influenced by mineral surface properties are discussed below with focus on bonding mechanisms, including: ligand exchange, H-bonding, polyvalent cation bridges, direct electrostatic attraction and weak interactions.

Phyllosilicate dominated fractions

The chemical composition of OC associated in the phyllosilicate dominated fractions was characterized by the highest proportions of oxidized C species (C-O, C=O, O=C-O) of all fractions and

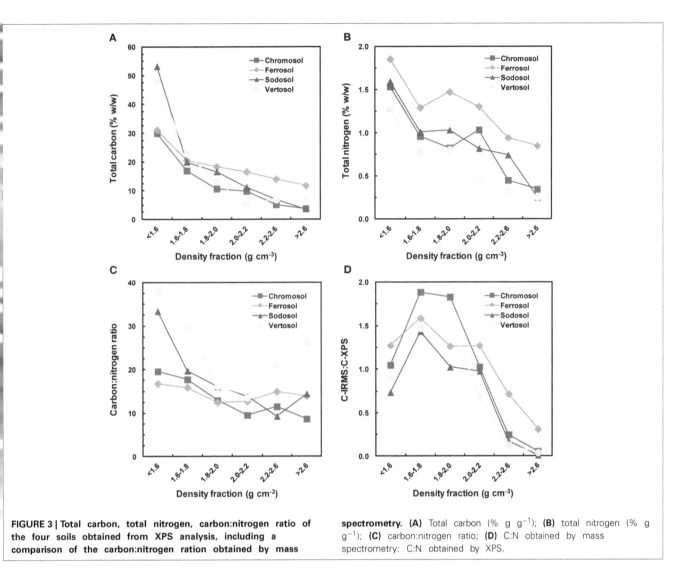

FIGURE 3 | Total carbon, total nitrogen, carbon:nitrogen ratio of the four soils obtained from XPS analysis, including a comparison of the carbon:nitrogen ration obtained by mass spectrometry. **(A)** Total carbon (% g g^{-1}); **(B)** total nitrogen (% g g^{-1}); **(C)** carbon:nitrogen ratio; **(D)** C:N obtained by mass spectrometry: C:N obtained by XPS.

an increase in protonated amide forms relative to the particulate organic matter (**Table 7**). These results reflect those of Sollins et al. (2009), who found that OM associated with phyllosilicate fractions had an increased oxidation state as determined using alkaline cupric-oxide oxidation. The enrichment of O-C and O=C-O moieties provides support for the importance of ligand exchange reactions and/or H-bonding in these fractions. Ligand exchange reactions produce a strong bond between singly coordinate hydroxyl groups at the edges of phyllosilicates, and carboxyl and phenolic groups of the OM (von Lützow et al., 2006; Kögel-Knabner et al., 2008). While the delocalization of electrons produced by the presence of O in O-C and O=C-O, as well as C=O moieties, may also encourage H-bonding processes with the mineral surface (von Lützow et al., 2006). Mortland (1986) suggested that the basal O planes of expansible minerals, such as smectite, are weaker electron donors and subsequently these minerals would form weaker H-bonds with the oxidized OM species. Our study provided evidence for this as the proportions of oxidized C species were lower in smectite rich fractions of the Vertosol compared to the Chromosol and Sodosol, however the increases in oxidized C species were similar to the other

soils when compared relative to POM and could be indicative of differences in OM inputs. Thus it remains difficult to elucidate to what extent each of the bonding mechanisms play in stabilizing OM at the mineral surface, however both appear to be important. Another bonding mechanism proposed for expansible minerals is polyvalent cation bridges (Kleber et al., 2007). In these reactions negatively charged organic functional groups, such as disassociated carboxyls and hydroxyls, form bonds with delocalized charges on phyllosilicates through intermediary polyvalent cations that act to neutralize the repulsive force between the two entities. We suggest that polyvalent cation bridging plays a minor role in OM stabilization in these soils as the kaolinite rich fraction of the Sodosol, which has the lowest potential to form these bonds due its low CEC, has the highest proportion of O=C-O and O-C moieties compared to the corresponding fractions in the Chromosol and Vertosol. Conversely the increase in protonated amide forms, especially in the 2.0–2.2 g cm^{-3} fraction of the Vertosol, indicates that stabilization by direct ionic attraction between positively charged OM species and delocalized negative charges at the mineral surface may be important, especially in the bonding of N rich proteinaceous species at the mineral surface.

Table 6 | Surface elemental composition of the six density fractions of the four soils obtained from the survey scans of X-ray photoelectron spectra (% w/w).

ASC soil order	Density fraction (g cm^{-3})	C1s	N1s	O1s	Si2p	Al2p	Mg1s	Na1s	Fe2p	Ca2p	K2s	W4d5	Cl2p	Ti2p
Chromosol	<1.6	30.0	1.5	38.3	16.3	9.3	0.5	0.8	2.4	0.0	1.0	0.0	0.0	0.0
	1.6–1.8	17.0	1.0	48.4	24.0	13.6	0.8	0.8	3.1	0.0	1.6	0.0	0.0	0.0
	1.8–2.0	10.6	0.8	44.6	23.6	12.0	0.7	1.3	2.8	0.0	1.4	2.1	0.0	0.1
	2.0–2.2	9.8	1.0	53.7	32.6	12.0	0.8	1.3	3.5	0.0	1.3	0.0	0.0	0.0
	2.2–2.6	5.2	0.4	56.6	33.3	15.2	0.9	2.4	5.3	0.0	2.0	0.0	0.0	0.3
	>2.6	3.0	0.3	45.6	24.6	13.6	1.4	2.0	5.9	0.0	2.1	0.9	0.0	0.4
Ferrosol	<1.6	31.1	1.9	40.3	9.2	10.5	0.3	0.9	3.5	0.6	0.0	1.5	0.4	0.0
	1.6–1.8	20.5	1.3	45.2	12.0	14.1	0.4	0.6	5.4	0.0	0.0	0.0	0.0	0.5
	1.8–2.0	18.3	1.5	43.0	13.0	13.2	1.3	0.0	4.8	0.0	0.0	4.5	0.0	0.4
	2.0–2.2	16.6	1.3	45.8	13.1	15.4	0.2	0.8	6.4	0.0	0.0	0.0	0.0	0.6
	2.2–2.6	14.0	0.9	45.3	13.7	16.0	0.2	1.1	5.7	0.0	0.0	2.7	0.0	0.4
	>2.6	11.8	0.8	46.0	12.2	15.6	0.0	0.9	9.3	0.0	0.0	2.8	0.0	0.5
Sodosol	<1.6	53.3	1.6	30.8	8.3	4.5	0.0	0.7	0.8	0.0	0.0	0.0	0.0	0.0
	1.6–1.8	20.0	1.0	43.1	19.5	11.9	0.7	1.0	2.4	0.0	0.2	0.0	0.0	0.0
	1.8–2.0	16.6	1.0	49.2	24.3	13.1	0.8	1.4	3.4	0.0	0.3	0.0	0.0	0.0
	2.0–2.2	11.3	0.8	46.5	23.4	12.7	0.6	1.2	2.5	0.0	0.3	0.5	0.0	0.1
	2.2–2.6	6.9	0.7	46.6	25.2	12.9	0.7	1.3	3.4	0.0	0.6	1.6	0.0	0.1
	>2.6	3.6	0.3	47.4	39.3	5.3	0.1	0.8	2.7	0.0	0.2	0.0	0.0	0.2
Vertosol	<1.6	48.2	1.3	30.0	9.9	4.8	0.2	0.4	1.9	1.6	0.0	1.6	0.0	0.0
	1.6–1.8	22.8	0.8	35.9	15.5	6.0	0.6	1.0	3.2	1.7	0.0	12.5	0.0	0.0
	1.8–2.0	13.1	0.8	43.6	23.9	9.8	0.7	1.2	5.0	0.5	0.0	1.0	0.0	0.3
	2.0–2.2	5.8	0.4	46.5	25.2	12.1	1.0	1.9	5.7	0.0	0.2	0.7	0.0	0.4
	2.2–2.6	6.8	0.3	44.8	23.9	10.7	1.0	1.8	7.6	0.2	0.4	2.1	0.0	0.4
hjgjhg	>2.6	6.7	0.3	44.6	21.0	8.2	0.9	1.4	12.1	0.4	0.2	2.1	0.0	2.1

High tungsten values are most likely due to residual SPT used in the density fractionation procedure. Na levels may also have been influenced by the fractionation procedure.

Quartz and feldspar dominated fractions

The OM associated with quartz and feldspar dominated fractions (>2.6 g cm^{-3}) was characterized by the highest proportions of aliphatic C and protonated amide N for all densities (**Table 7**). This increase in the proportion of aliphatic C was mirrored by a strong decrease in C-O and C=O functional groups, however, the proportion of carboxylic groups in oxidized C species varied between the soils; a strong increase was shown for the Vertosol, a slight decrease for the Chromosol and a strong decrease was seen for the Sodosol. Quartz and feldspar surfaces are typified by low specific surface areas, low surface reactivity and consist of some unsatisfied siloxane bonds. These surfaces generally have a lower potential to protect SOM as compared to phyllosilicate surfaces; however, coating with Fe oxides or phyllosilicate species may increase their surface chemistry and adsorption capacity, which present difficulties in the interpretation of results. Many other studies have demonstrated a strong correlation between quartz and feldspar minerals and aliphatic C, for example, aliphatic C was found to be enriched in bulk samples of coarse textured soils (Capriel et al., 1995) as well as the fine-fraction of coarse-textured soils (Galantini et al., 2004; Mao et al., 2007); and that the relative proportions of mineral associated aliphatic C was increased significantly with increasing sand content of soils (Jindaluang et al., 2013). The high C content of aliphatic species in this fraction could explain the increase in C:N observed in the Vertosol by mass spectrometry, while the high concentration of aliphatic C and O=C-O moieties we observed in these fractions reinforce the importance of lipids and ligand bonded waxes in this fraction, with lipids apparently more important in the Sodosol which showed reduced proportions of the O=C-O group. Thus we conclude that the bonding of OM to quartz and feldspar surfaces in these soils appears to be characterized by H-bonding of non-ionic lipids and polysaccharides and ligand bonding of waxes with the small amount of unsatisfied siloxane bonds. However, the increase in protonated amide forms and low C:N ratios seen in the two soils with purer quartz and feldspar surfaces suggest that electrostatic attraction of proteinaceous species to unsatisfied siloxane bonds is also important.

Fe oxides dominated fractions

Iron oxides dominant fractions (>2.0 g cm^{-3}) in the Ferrosol are characterized by higher proportion of oxidized C bonding environments as well as a higher proportion of amide/amine N to protonated amide species compared to corresponding densities

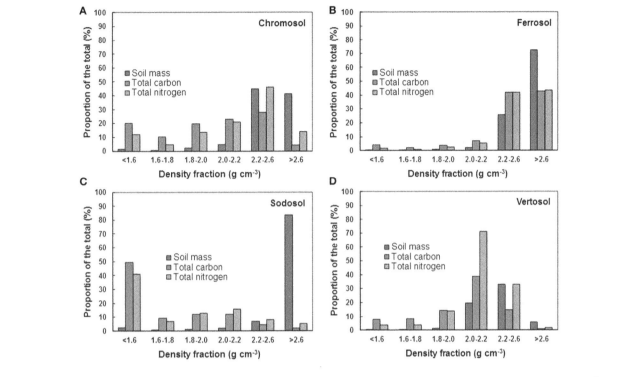

FIGURE 4 | Proportion of soil mass, total carbon, and total nitrogen distributed in density fraction of the four soils. (A) Chromosol—recovery of soil mass, total carbon, and total nitrogen = 94.8, 105.0, and 110.9%; **(B)** ferrosol—recovery of soil mass, total carbon, and total nitrogen = 102.2, 102.1, and 95.8%; **(C)** sodosol—recovery of soil mass, total carbon, and total nitrogen = 96.8, 89.5, and 89.6%; **(D)** vertosol—recovery of soil mass, total carbon, and total nitrogen = 60.9, 84.2, and 127.5%. High yields of nitrogen observed by the Chromosol and Vertosol are likely due to adsorption of nitrogen from the sodium-polytungstate solution by smectite and illite during sequential density fractionation. Low mass and carbon yields may be attributable to the small size of smectite particles most likely lost during the washing process.

in other soils (**Table 7**). These findings are consistent with the observations that Fe oxides exhibit a strong affinity for highly oxidized, lignin derived organic species (Kaiser et al., 1997; Mikutta et al., 2007). The low observed levels of protonated amide species compared to non-Fe oxides soils are intuitive as Fe/Al oxides carry a net positive charge at the existing soil pH, thus protonated amide forms would be repelled by the mineral surfaces. We suggest that the association of SOM with mineral surfaces is occurring primarily through ligand exchange reactions in these fractions. In these reactions negatively charged organic domains, such as hydrolyzed carboxylic acids or phenolic compounds, form strong monodentate and bidentate covalent bonds with unsatisfied hydroxyl groups of Fe or Al oxides (Sollins et al., 1996). We suggest carboxyl groups are relatively more important in this interaction as the proportion of O=C-O moieties was seen to increase with density, while C=O moieties remained relatively stable. We believe that this finding is justified as carboxylic acids have a lower pK_a compared to phenols (e.g., pK_a of methoxyacetic acid is 3.54 compared to pK_a of 9.65 of methoxyphenol) and are thus more likely to disassociate in the acidic conditions of these soils to provide the anionic form necessary for ligand bond formation (Essington, 2004).

The role of Fe in organo-mineral formation may also be seen in the other three soils with a positive correlation observed between Fe concentration and O=C-O moieties across all fractions of all soils (**Figure 5**). The effect of Fe coatings on other soils minerals may be seen in the $>2.6 \, \text{g cm}^{-3}$ fraction of the Vertosol, which is dominated by quartz and feldspars but contains traces of hematite and other amorphous Fe forms (**Tables 4, 6**). We observed that this fraction behaves more like the corresponding fraction in the Ferrosol, compared to the Chromosol or Sodosol, in three ways. Firstly the proportion of carboxylic functional groups is seen to increase, secondly the decrease in $\delta^{13}C$ was not as well pronounced, thirdly the C:N ratio as measured by mass spectrometry converges toward that of the Ferrosol. We thus suggest that even small amounts of Fe can have significant influence on the composition of associated OM.

MINERAL CONTROL OVER C CONTENT AND TOTAL C WITHIN DENSITY FRACTIONS

The analysis uncovered general decreases in C and N contents across successively dense fractions. The Ferrosol was found to store significantly ($p < 0.05$) higher C and N concentrations in heavier density fractions ($>2.0 \, \text{g cm}^{-3}$) compared to the other soils, reinforcing the greater OM adsorption capacity of Fe oxides compared to phyllosilicates (Kahle et al., 2004; Kaiser et al., 1997), with hematite and goethite possessing large and reactive surface areas and sorption capacities of up to 110–140 mg OC per g oxide (Tipping, 1981; Kaiser and Guggenberger, 2007).

Table 7 | The relative atomic concentrations of different C and N bonding environments of the six density fractions of the four soils determined by deconvoluted XPS spectra.

ASC soil order	Density fraction (g cm^{-3})	C1sA (C-C/C-H)	C1sB (C-O)	C1sC (C=O)	C1sD (O=C-O)	N1sA (C-N)	N1sB (C-N$^+$)
Chromosol	<1.6	65.4	23.5	7.2	3.9	86.4	13.6
	1.6–1.8	63.2	21.9	8.5	6.5	77.2	22.8
	1.8–2.0	57.9	26.3	8.9	6.9	75.4	24.6
	2.0–2.2	52.7	31.0	9.7	6.7	78.2	21.8
	2.2–2.6	60.9	24.8	5.3	9.0	69.4	30.6
	>2.6	70.2	18.5	2.4	8.8	51.3	48.7
Ferrosol	<1.6	54.3	31.3	8.1	6.3	93.5	6.5
	1.6–1.8	50.7	30.4	9.9	9.0	89.2	10.8
	1.8–2.0	48.1	32.5	10.8	8.6	89.5	10.5
	2.0–2.2	47.2	32.9	10.2	9.6	87.1	12.9
	2.2–2.6	47.4	32.3	10.9	9.3	87.9	12.1
	>2.6	46.6	32.3	10.3	10.8	88.6	11.4
Sodosol	<1.6	69.9	21.0	5.7	3.4	88.6	11.4
	1.6–1.8	60.8	25.7	7.1	6.5	84.2	15.8
	1.8–2.0	55.4	28.6	8.7	7.2	83.8	16.2
	2.0–2.2	52.6	30.4	8.8	8.2	80.3	19.7
	2.2–2.6	53.8	28.9	8.8	8.6	76.3	23.7
	>2.6	74.7	16.6	3.5	5.2	69.9	30.1
Vertosol	<1.6	75.5	15.0	4.5	5.1	89.4	10.6
	1.6–1.8	70.4	16.4	6.0	7.2	79.4	20.6
	1.8–2.0	61.2	24.3	7.3	7.2	72.3	27.7
	2.0–2.2	58.9	25.2	8.0	7.8	58.1	41.9
	2.2–2.6	69.0	18.9	5.8	6.3	56.1	43.9
	>2.6	69.9	17.8	4.2	8.0	50.0	50.0

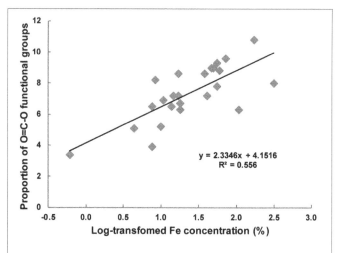

FIGURE 5 | The relationship between log transformed Fe concentration (% w/w) and the proportion of O=C-O functional groups derived from the XPS analysis of various density fractions of the four soils.

Another interesting result was the lower C content in the smectite-rich 2.0–2.2 g cm^{-3} fraction of the Vertosol compared to the Chromosol and Sodosol, as laboratory experiments have shown that OC is adsorbed faster and in greater quantities to smectite compared to other phyllosilicates (Varadachari et al., 1995; Dontsova and Bigham, 2005), and studies have shown lower mineralization rates when dissolved OM is incubated with smectite minerals compared to illite and significantly lower when compared to kaolinite (Saidy et al., 2012). However, investigations of OC and mineralogy *in situ* suggest that smectite rich soils do not hold more OC compared to soils containing other phyllosilicates, including Ferrosols (Dalal and Mayer, 1986; Wattel-Koekkoek et al., 2001; Krull and Skjemstad, 2003; Wattel-Koekkoek and Buurman, 2004). In our study, a substantially greater proportion of soil mass was recovered in the 2.0–2.2 g cm^{-3} fraction of the Vertosol (i.e., lower C loading) compared to the Chromosol and the Sodosol, hence the variable C (and N) contents in the density fraction of these soils cannot be attributed to the adsorption capacity of the clay minerals.

When considering the total C stored in each fraction large variations were seen between the soils, which appeared to be influenced by mineral size, surface area and mineral type

(**Figures 4A–D**). For instance the Sodosol, which is comprised primarily of sand sized quartz particles and has little reactive surface area, stored 65.6% of total recovered C in the POM fraction (<1.8 g cm^{-3}) (**Figure 4C**). Meanwhile, the Vertosol and the Chromosol stored large proportions, 63.8 and 66.8% respectively, of total recovered C in the phyllosilicate dominated fractions (**Figures 4A,D**), with the Ferrosol having 83.2% of total C recovered in the two heaviest fractions (>2.2 g cm^{-3}) (**Figure 4B**). This demonstrates that the soils have different inherent capacities to form organo-mineral associations. Jindaluang et al. (2013) demonstrated that, within soil groups, the amount of mineral associated C and N in the 1.85–2.60 g cm^{-3} fractions was significantly increased with increasing clay content. Our results reflect this relationship, with the proportion of total C held in phyllosilicate dominated fractions of the soils correlated to total clay content independent of phyllosilicate type: i.e., Vertosol > Chromosol > Sodosol. The presence of most of the OC in the free POM form in the Sodosol offers a contrast to other studies that have found that OM is primarily associated with mineral phases in soils and sediments (Mayer, 1994; Basile-Doelsch et al., 2007).

MINERAL CONTROL OVER SOM STABILITY WITHIN DENSITY FRACTIONS AND MANAGEMENT IMPLICATIONS

The different levels of mineral associated OM in each soil will confer different levels of OM stability under disturbances, e.g., cultivation or increased atmospheric temperatures, which may disrupt soil aggregates and increase turnover OM rates in soils (Six et al., 2002; Chivenge et al., 2007). von Lützow et al. (2006) described the <1.8 g cm^{-3} fraction as an active SOM pool due to its high turnover rates. X-ray photoelectron spectroscopy spectra support this idea showing that C in these fractions is enriched in aliphatic C, which is less stable than aromatic bonded C and more easily metabolized by decomposing agents (Demyan et al., 2012) (**Table 7**). The fact that most of the Sodosol's C is stored in this relatively less stable due to its reduced capacity to form organo-mineral interactions may explain why the Sodosol contains 20% less C compared to the Chromosol and Vertosol (**Table 2**), even though it receives similar rainfall (**Table 1**). Meanwhile the ligand exchange reactions between oxidized C species and the surface of Fe oxides are known as the strongest organo-mineral interactions, with incubation studies show that OM sorbed to Fe oxides are resistant to desorption by chemical reagents (Kaiser and Guggenberger, 2007; Mikutta et al., 2007). This suggests that SOM in the Ferrosol, which is predominately associated with mineral fractions, may be the most chemically stable of all soils, a theory that has grounding in another study reporting a much longer turnover time of SOM in clayey Oxisols (400–700 years) than coarse textured Ultisols (~300 years) and Spodosols (<100 years) in the Brazilian Amazon (Telles et al., 2003). The mineralogical composition of the Oxisol was similar to the Ferrosol used in our study, with a mixture of kaolinite, sesquioxides, and quartz. Organic matter associated with mineral surfaces may limit C losses during agricultural production as demonstrated by a study of paired sites, under natural conditions versus continuous cultivation, showed that the relative proportions of OM associated with the 2.1–2.4 g cm^{-3} fractions increased from 21 to 32%, and from 6 to 16% at the two sites studied (Plante et al., 2010). Thus the potential of organo-mineral interactions to protect SOM over long time scales and under cultivation reinforces the need for further research to improve our understanding of the complex factors driving formation of organo-mineral interactions.

SUPPORTING EVIDENCE FOR THE LAYERED MODE OF OM ACCUMULATION ON MINERAL SURFACES (Sollins et al., 2006, 2009)

Our trends in C, N, C:N ratios and δ^{13}C over phyllisilicate fractions closely resemble those of previous studies and give weight to the argument of a layered mode of OM accumulation (Sollins et al., 2006, 2009). A similar trend was observed across successively dense Fe oxide dominated fractions in the Ferrosol, indicating a similar effect in these soils. Further, evidence provided by the XPS analysis in our study provides further evidence for the model. When investigating the ratio of C concentration observed from mass spectrometry (which measure total C in a sample) compared to those of XPS (a surface only technique, ~10 nm) we observed that the XPS technique increasingly over-estimated C content with successively dense mineral fractions (>1.8 g cm^{-3}) (**Figure 3D**). This indicates that decreases in C observed with increasing density of fractions are derived from a greater proportion from decreases in the thickness of the OM deposit as compared to reductions in the coverage area of OM deposits. Another interpretation of this observation may be due to increased mineral size at higher densities increasing the amount of C observed by XPS, however we believe that our process of hand grinding all samples to a fine powder would have reduced this effect.

Our analysis also identified some other trends not fully covered by the layered mode including the decrease in δ^{13}C observed in each of the >2.6 cm^{-3} fractions for the Chromosol, Sodosol, and Vertosol, as well as the increase in C:N ratio observed in the same fraction for some soils. It is difficult to elucidate whether the observed decreases in δ^{13}C ratios is a reflection of a reduced

level of microbial processing of this OM prior to attachment or a different source material of the attached OM. However, as this change also coincides with a shift in mineralogy from phyllosilicate to quartz and feldspars species there appears to be an underlying effect of the associated minerals behind this observation. The influence of mineral surfaces on microbial communities within the soil environment offers a possible explanation, as fungal colonies are known to be concentrated in the sand sized fractions which account for the majority of the >2.6 g cm^{-3} fraction, while bacteria predominate in the clay and silt sized fractions (Sessitsch et al., 2001). Furthermore, Sylvia et al. (2005) identified higher C:N ratios for fungal tissues, ~10:1, compared to bacteria, ~4:1, which could explain the increase in C:N ration observed in the >2.6 g cm^{-3} fraction of the Vertosol. As much of the OM associated with mineral surfaces has been processed by micro-organisms, further investigation is warranted into the complex interactions between minerals and micro-organisms in the formation of organo-mineral associations.

CONCLUSION

A distinct chemical fingerprint of SOM was isolated in the density fractions of soils displaying similar mineralogies. The sequential density fractionation procedure was thus found effective in isolating functional pools of discrete mineralogy and OM composition. The relative enrichment or depletion of C and N functional groups was controlled by mineral surface properties with the chemical composition of OM becoming enriched in domains that are able to most effectively associate with mineral surfaces. For example Fe oxides were enriched in O=C-O moieties that are able to form strong ligand bonds with hydroxyls groups at the mineral surface; meanwhile phyllosilicate minerals were enriched in O-C, O=O, and O=C-O forms, which are able to associate via ligand bonding to singly coordinated hydroxyls at crystal edges and/or induce H-bonding across basal O planes of the minerals. Differences in OM composition were also seen between different phyllosilicate species, for example smectite dominated fractions were enriched in protonated amide forms that may form electrostatic bonds with delocalized surface charges of the mineral. This study provides insight into the interactions between mineral surfaces and SOM, and how mineral surfaces may influence not only the content but also the composition of associated OM. This knowledge is useful in modeling SOM dynamics and C flows between labile and stable pools when attempting to optimize SOM levels.

ACKNOWLEDGMENTS

The authors would like to thank Mr. Jim Harris, Mr. Philip Pearson, Mr. Noel Ticehurst and Dr. Lukas Van Zwieten for their assistance in soil sampling; Drs. Bill Gong and Claudia Keitel for XPS and IRMS analysis, respectively; and Ms Lorie Watson for providing various laboratory equipment.

REFERENCES

Abe, T., and Watanabe, A. (2004). X-ray photoelectron spectroscopy of nitrogen functional groups in soil humic acids. *Soil Sci.* 169, 35–43. doi: 10.1097/01.ss.0000112016.97541.28

Amelung, W., Kaiser, K., Kammerer, G., and Sauer, G. (2002). Organic carbon at soil particle surfaces—evidence from x-ray photoelectron spectroscopy and

surface abrasion. *Soil Sci. Soc. Am. J.* 66, 1526–1530. doi: 10.2136/sssaj200
2.1526

Anderson, D. W. (1988). The effect of parent material and soil development
on nutrient cycling in temperate ecosystems. *Biogeochemistry* 5, 71–97. doi:
10.1007/BF02180318

Baisden, W. T., Amundson, R., Cook, A. C., and Brenner, D. L. (2002). Turnover
and storage of C and N in five density fractions from California annual grassland
surface soils. *Global Biogeochem. Cycles* 16, 1117. doi: 10.1029/2001GB001822

Baldock, J. A., Oades, J. M., Waters, A. G., Peng, X., Vassallo, A. M., and Wilson,
M. A. (1992). Aspects of the chemical structure of soil organic materials as
revealed by solid-state13C NMR spectroscopy. *Biogeochemistry* 16, 1–42. doi:
10.1007/BF00024251

Baldock, J., and Skjemstad, J. (2000). Role of the soil matrix and minerals in protecting natural organic materials against biological attack. *Org. Geochem.* 31,
697–710. doi: 10.1016/S0146-6380(00)00049-8

Basile-Doelsch, I., Amundson, R., Stone, W. E. E., Borschneck, D., Bottero, J. Y.,
Moustier, S., et al. (2007). Mineral control of carbon pools in a volcanic soil
horizon. *Geoderma* 137, 477–489. doi: 10.1016/j.geoderma.2006.10.006

Basile-Doelsch, I., Brun, T., Borschneck, D., Masion, A., Marol, C., and Balesdent, J.
(2009). Effect of landuse on organic matter stabilized in organomineral
complexes: a study combining density fractionation, mineralogy and δ13C.
Geoderma 151, 77–86. doi: 10.1016/j.geoderma.2009.03.008

Bonnard, P., Basile-Doelsch, I., Balesdent, J., Masion, A., Borschneck, D., and
Arrouays, D. (2012). Organic matter content and features related to associated mineral fractions in an acid, loamy soil. *Eur. J. Soil Sci.* 63, 625–636. doi:
10.1111/j.1365-2389.2012.01485.x

Brown, G., and Brindley, G. W. (1980). "X-ray identification procedures for clay
mineral identification," in *Crystal Structures of Clay Minerals and Their X-
Ray Identification*, eds G. W. Brindley and G. Brown (London: Mineralogical
Society), 305–359.

Brown, R. A., Kercher, A. K., Nguyen, T. H., Nagle, D. C., and Ball,
W. P. (2006). Production and characterization of synthetic wood chars for
use as surrogates for natural sorbents. *Org. Geochem.* 37, 321–333. doi:
10.1016/j.orggeochem.2005.10.008

Bureau of Meteorology. (2013). Climate and past weather. Available online at:
http://www.bom.gov.au/climate/.

Capriel, P., Beck, T., Borchert, H., Gronholz, J., and Zachmann, G. (1995).
Hydrophobicity of the organic matter in arable soils. *Soil Biol. Biochem.* 27,
1453–1458. doi: 10.1016/0038-0717(95)00068-P

Chivenge, P. P., Murwira, H. K., Giller, K. E., Mapfumo, P., and Six, J. (2007).
Long-term impact of reduced tillage and residue management on soil carbon
stabilization: implications for conservation agriculture on contrasting soil. *Soil
Tillage Res.* 94, 328–337. doi: 10.1016/j.still.2006.08.006

Christensen, B. (1992). Physical fractionation of soil and organic matter in primary
particle size and density separates. *Adv. Soil Sci.* 20, 1–90. doi: 10.1007/978-1-
4612-2930-8_1

Dalal, R. C., and Mayer, R. J. (1986). Long term trends in fertility of soils under
continuous cultivation and cereal cropping in southern Queensland. II. Total
organic carbon and its rate of loss from the soil profile. *Aust. J. Soil Res.* 24,
281–292. doi: 10.1071/SR9860281

Demyan, M. S., Rasche, F., Schulz, E., Breulmann, M., Müller, T., and Cadisch, G.
(2012). Use of specific peaks obtained by diffuse reflectance fourier trans-
form mid-infrared spectroscopy to study the composition of organic matter
in a Haplic Chernozem. *Eur. J. Soil Sci.* 63, 189–199. doi: 10.1111/j.1365-
2389.2011.01420.x

Dontsova, K. M., and Bigham, J. M. (2005). Anionic polysaccharide sorption by
clay minerals. *Soil Sci. Soc. Am. J.* 69, 1026–1035. doi: 10.2136/sssaj2004.0203

Essington, M. E. (2004). *Soil and Water Chemistry: An Integrative Approach*. Boca
Raton, FL: CRC press.

Fernandez, I., Mahieu, N., and Cadisch, G. (2003). Carbon isotopic fractionation
during decomposition of plant materials of different quality. *Global Biogeochem.
Cycles* 17, 1075.

Galantini, J. A., Senesi, N., Brunetti, G., and Rosell, R. (2004). Influence of texture
on organic matter distribution and quality and nitrogen and sulphur status in
semiarid Pampean grassland soils of Argentina. *Geoderma* 123, 143–152. doi:
10.1016/j.geoderma.2004.02.008

Gee, G. W., and Bauder, J. W. (1986). "Particle-size analysis," in *Methods of Soil
Analysis: Part 1. Physical and Mineralogical Methods II*, ed A. Klute (Madison:
Soil Science Society of America), 383–411.

Gerin, P. A., Genet, M. J., Herbillon, A. J., and Delvaux, B. (2003). Surface analysis of soil material by X-ray photoelectron spectroscopy. *Eur. J. Soil Sci.* 54,
589–604. doi: 10.1046/j.1365-2389.2003.00537.x

Gleixner, G., Danier, H. J., Werner, R. A., and Schmidt, H. L. (1993). Correlations
between the 13C content of primary and secondary plant products in different
cell compartments and that in decomposing basidiomycetes. *Plant Physiol.* 102,
1287–1290. doi: 10.1104/pp.102.4.1287

Golchin, A., Oades, J., Skjemstad, J., and Clarke, P. (1994). Study of free and
occluded particulate organic matter in soils by solid state 13 C Cp/MAS NMR
spectroscopy and scanning electron microscopy. *Aust. J. Soil Res.* 32, 285–309.
doi: 10.1071/SR9940285

Hashimoto, M. L., and Jackson, I. (1960). Rapid dissolution of allophane
and kaolinite after dehydration. *Clays Clay Miner.* 7, 102–113. doi:
10.1346/CCMN.1958.0070104

Jindaluang, W., Kheoruenromne, I., Suddhiprakarn, A., Singh, B. P., and Singh, B.
(2013). Influence of soil texture and mineralogy on organic matter content and
composition in physically separated fractions soils of Thailand. *Geoderma* 195,
207–219. doi: 10.1016/j.geoderma.2012.12.003

Kahle, M., Kleber, M., and Jahn, R. (2004). Retention of dissolved organic matter
by phyllosilicate and soil clay fractions in relation to mineral properties. *Org.
Geochem.* 35, 269–276. doi: 10.1016/j.orggeochem.2003.11.008

Kaiser, K., and Guggenberger, G. (2000). The role of DOM sorption to mineral surfaces in the preservation of organic matter in soils. *Org. Geochem.* 31, 711–725.
doi: 10.1016/S0146-6380(00)00046-2

Kaiser, K., and Guggenberger, G. (2003). Mineral surfaces and soil organic matter.
Eur. J. Soil Sci. 54, 219–236. doi: 10.1046/j.1365-2389.2003.00544.x

Kaiser, K., and Guggenberger, G. (2007). Sorptive stabilization of organic matter by
microporous goethite: sorption into small pores vs. surface complexation. *Eur.
J. Soil Sci.* 58, 45–59. doi: 10.1111/j.1365-2389.2006.00799.x

Kaiser, K., Guggenberger, G., Haumaier, L., and Zech, W. (1997). Dissolved organic
matter sorption on subsoil and minerals studied by 13C-NMR and DRIFT spectroscopy. *Eur. J. Soil Sci.* 48, 301–310. doi: 10.1111/j.1365-2389.1997.tb00550.x

Kleber, M., Sollins, P., and Sutton, R. (2007). A conceptual model of organo-
mineral interactions in soils: self-assembly of organic molecular fragments
into zonal structures on mineral surfaces. *Biogeochemistry* 85, 9–24. doi:
10.1007/s10533-007-9103-5

Kögel-Knabner, I., Guggenberger, G., Kleber, M., Kandeler, E., Kalbitz, K., Scheu,
S., et al. (2008). Organo-mineral associations in temperate soils: integrating
biology, mineralogy, and organic matter chemistry. *J. Plant Nutr. Soil Sci.* 171,
61–82. doi: 10.1002/jpln.200700048

Kölbl, A., and Kögel-Knabner, I. (2004). Content and composition of free and
occluded particulate organic matter in a differently textured arable Cambisol as
revealed by solid-state13C NMR spectroscopy. *J. Plant Nutr. Soil Sci.* 167, 45–53.
doi: 10.1002/jpln.200321185

Krull, E. S., and Skjemstad, J. O. (2003). δ13C and δ15N profiles in 14C-dated
Oxisol and Vertisols as a function of soil chemistry and mineralogy. *Geoderma*
112, 1–29. doi: 10.1016/S0016-7061(02)00291-4

Lide, D. R. (2007). *CRC Handbook of Chemistry and Physics*. Boca Raton, FL: CRC
Press.

Lombardi, K. C., Mangrich, A. S., Wypych, F., Rodrigues-Filho, U. P., Guimarães, J.
L., and Schreiner, W. H. (2006). Sequestered carbon on clay mineral probed
by electron paramagnetic resonance and X-ray photoelectron spectroscopy.
J. Colloid Interface Sci. 295, 135–140. doi: 10.1016/j.jcis.2005.08.015

Mao, J., Fang, X., Schmidt-Rohr, K., Carmo, A. M., Hundal, L. S., and
Thompson, M. L. (2007). Molecular-scale heterogeneity of humic acid
in particle-size fractions of two Iowa soils. *Geoderma* 140, 17–29. doi:
10.1016/j.geoderma.2007.03.014

Mayer, L. M. (1994). Relationships between mineral surfaces and organic carbon concentrations in soils and sediments. *Chem. Geol.* 114, 347–363. doi:
10.1016/0009-2541(94)90063-9

McKeague, J. A. (1967). An evaluation of 0.1 M pyrophosphate and pyrophosphate – dithionite in comparison with oxalate as extractants of the accumulation
products in podzols and some other soils. *Can. J. Soil Sci.* 47, 95–99. doi:
10.4141/cjss67-017

Mehra, O. P., and Jackson, M. L. (1960). Iron oxide removal from soils and clays by
a dithionite-citrate system buffered with sodium bicarbonate. *Clays Clay Miner.*
7, 317–327. doi: 10.1346/ccmn.1958.0070122

Mikutta, R., Mikutta, C., Kalbitz, K., Scheel, T., Kaiser, K., and Jahn, R.
(2007). Biodegradation of forest floor organic matter bound to minerals via

different binding mechanisms. *Geochim. Cosmochim. Acta* 71, 2569–2590. doi: 10.1016/j.gca.2007.03.002

Mikutta, R., Schaumann, G. E., Gildemeister, D., Bonneville, S., Kramer, M. G., Chorover, J., et al. (2009). Biogeochemistry of mineral–organic associations across a long-term mineralogical soil gradient (0.3–4100kyr), Hawaiian Islands. *Geochim. Cosmochim. Acta* 73, 2034–2060. doi: 10.1016/j.gca.2008.12.028

Mortland, M. (1986). "Mechanisms of adsorption of nonhumic organic species by clays," in *Interactions of Soil Minerals with Natural Organics and Microbes*, eds P. M. Huang and M. Schnitzer (Madison: Soil Science Society of America), 59–76.

Moulder, J. F., Stickle, W. F., Sobol, P. E., and Bomben, K. D. (1992). *Handbook of X-ray Photoelectron Spectroscopy*. Stanford, CA: Perin-Elmer.

New South Wales Department of Primary Industries. (2013). Climate at the wollongbar primary industries institute. Available online at: http://www.dpi.nsw.gov.au/research/centres/wollongbar/climate.

Plante, A. F., Virto, I., and Malhi, S. S. (2010). Pedogenic, mineralogical and land-use controls on organic carbon stabilization in two contrasting soils. *Can. J. Soil Sci.* 90, 15–26. doi: 10.4141/CJSS09052

Proctor, A., and Sherwood, P. (1982). X-ray photoelectron spectroscopic studies of carbon fiber surface. *Surf. Interface Anal.* 4, 212–219. doi: 10.1002/sia.740040508

Rayment, G. E., and Higginson, F. R. (1992). *Australian Laboratory Handbook for Soil and Water Chemical Methods*. Melbourne, VIC: Inkata.

Saidy, A. R., Smernik, R. J., Baldock, J. A., Kaiser, K., Sanderman, J., and Macdonald, L. M. (2012). Effects of clay mineralogy and hydrous iron oxides on labile organic carbon stabilisation. *Geoderma* 173, 104–110. doi: 10.1016/j.geoderma.2011.12.030

Schwertmann, U. (1964). The differentiation of iron oxides in soils by a photochemical extraction with acid ammonium oxalate. *Z. Pflanz. Bodenkunde* 105, 194–202. doi: 10.1002/jpln.3591050303

Sessitsch, A., Weilharter, A., Gerzabek, M. H., Kirchmann, H., and Kandeler, E. (2001). Microbial population structures in soil particle size fractions of a long-term fertilizer field experiment. *Appl. Environ. Microbiol.* 67, 4215–4224. doi: 10.1128/AEM.67.9.4215-4224.2001

Six, J., Conant, R. T., Paul, E. A., and Paustian, K. (2002). Stabilization mechanisms of soil organic matter: implications for C-saturation of soils. *Plant Soil* 241, 155–176. doi: 10.1023/A:1016125726789

Sollins, P., Homann, P., and Caldwell, B. A. (1996). Stabilization and destabilization of soil organic matter: mechanisms and controls. *Geoderma* 74, 65–105. doi: 10.1016/S0016-7061(96)00036-5

Sollins, P., Kramer, M. G., Swanston, C., Lajtha, K., Filley, T., Aufdenkampe, A. K., et al. (2009). Sequential density fractionation across soils of contrasting mineralogy: evidence for both microbial- and mineral-controlled soil organic matter stabilization. *Biogeochemistry* 96, 209–231. doi: 10.1007/s10533-009-9359-z

Sollins, P., Swanston, C., Kleber, M., Filley, T., Kramer, M., Crow, S., et al. (2006). Organic C and N stabilization in a forest soil: evidence from sequential den-

sity fractionation. *Soil Biol. Biochem.* 38, 3313–3324. doi: 10.1016/j.soilbio.2006.04.014

Sylvia, D., Fuhrmannm, J. J., Hartel, P., and Zuberer, D. A. (2005). *Principles and Applications of Soil Microbiology*. New Jersey, NJ: Pearson Prentice Hall.

Telles, E. C. C., Camargo, P. B., Martinelli, L. A., Trumbore, S. E., Costa, E. S., Santos, J., et al. (2003). Influence of soil texture on carbon dynamics and storage potential in tropical forest soils of Amazonia. *Global Biogeochem. Cycles* 17, 1040. doi: 10.1029/2002GB001953

Throop, H. L., Lajtha, K., and Kramer, M. (2013). Density fractionation and 13C reveal changes in soil carbon following woody encroachment in a desert ecosystem. *Biogeochemistry* 112, 409–422. doi: 10.1007/s10533-012-9735-y

Tipping, E. (1981). The adsorption of aquatic humic substances by iron oxides. *Geochim. Cosmochim. Acta* 45, 191–199. doi: 10.1016/0016-7037(81)90162-9

Varadachari, C., Mondal, A. H., and Ghosh, K. (1995). The influence of crystal edges on clay-humus complexation. *Soil Sci.* 159, 185–190. doi: 10.1097/00010694-199515930-00005

von Lützow, M., Kogel-Knabner, I., Ekschmitt, K., Matzner, E., Guggenberger, G., Marschner, B., et al. (2006). Stabilization of organic matter in temperate soils: mechanisms and their relevance under different soil conditions - a review. *Eur. J. Soil Sci.* 57, 426–445. doi: 10.1111/j.1365-2389.2006.00809.x

Wattel-Koekkoek, E. J. W., and Buurman, P. (2004). Mean residence time of kaolinite and smectite-bound organic matter in Mozambiquan Soils. *Soil Sci. Soc. Am. J.* 68, 154–161. doi: 10.2136/sssaj2004.1540

Wattel-Koekkoek, E. J. W., van Genuchten, P. P., Buurman, P., and van Lagen, B. (2001). Amount and composition of clay-associated soil organic matter in a range of kaolinitic and smectitic soils. *Geoderma* 99, 27–49. doi: 10.1016/S0016-7061(00)00062-8

Conflict of Interest Statement: The authors declare that the research was conducted in the absence of any commercial or financial relationships that could be construed as a potential conflict of interest.

Extreme climatic events: impacts of drought and high temperature on physiological processes in agronomically important plants

Urs Feller[1]* and Irina I. Vaseva[1,2]

[1] Institute of Plant Sciences and Oeschger Centre for Climate Change Research, University of Bern, Bern, Switzerland
[2] Plant Stress Molecular Biology Department, Institute of Plant Physiology and Genetics, Bulgarian Academy of Sciences, Sofia, Bulgaria

Edited by:
Pankaj Kumar Arora, Yeungnam University, South Korea

Reviewed by:
Claudio Lovisolo, University of Turin, Italy
Martin Zimmer, Leibniz Center for Tropical Marine Ecology, Germany

*Correspondence:
Urs Feller, Institute of Plant Sciences and Oeschger Centre for Climate Change Research, University of Bern, Altenbergrain 21, CH-3013 Bern, Switzerland
e-mail: urs.feller@ips.unibe.ch

Climate models predict more frequent and more severe extreme events (e.g., heat waves, extended drought periods, flooding) in many regions for the next decades. The impact of adverse environmental conditions on crop plants is ecologically and economically relevant. This review is focused on drought and heat effects on physiological status and productivity of agronomically important plants. Stomatal opening represents an important regulatory mechanism during drought and heat stress since it influences simultaneously water loss via transpiration and CO_2 diffusion into the leaf apoplast which further is utilized in photosynthesis. Along with the reversible short-term control of stomatal opening, stomata and leaf epidermis may produce waxy deposits and irreversibly down-regulate the stomatal conductance and non-stomatal transpiration. As a consequence photosynthesis will be negatively affected. Rubisco activase—a key enzyme in keeping the Calvin cycle functional—is heat-sensitive and may become a limiting factor at elevated temperature. The accumulated reactive oxygen species (ROS) during stress represent an additional challenge under unfavorable conditions. Drought and heat cause accumulation of free amino acids which are partially converted into compatible solutes such as proline. This is accompanied by lower rates of both nitrate reduction and *de novo* amino acid biosynthesis. Protective proteins (e.g., dehydrins, chaperones, antioxidant enzymes or the key enzyme for proline biosynthesis) play an important role in leaves and may be present at higher levels under water deprivation or high temperatures. On the whole plant level, effects on long-distance translocation of solutes via xylem and phloem and on leaf senescence (e.g., anticipated, accelerated or delayed senescence) are important. The factors mentioned above are relevant for the overall performance of crops under drought and heat and must be considered for genotype selection and breeding programs.

Keywords: drought, heat, abiotic stress, stomates, protein pattern, leaf senescence, xylem, phloem

INTRODUCTION

Besides the general temperature increase global change models predict more frequent and more severe extreme events such as drought periods, heat waves or flooding (Easterling et al., 2000; Schar et al., 2004; Fuhrer et al., 2006; Wehner et al., 2011; Mittal et al., 2014). These regional climatic extremes (Gilgen and Buchmann, 2009) are ecologically and economically relevant for agriculture and forestry (IPCC, 2012; Smith and Gregory, 2013; Nair, 2014). The susceptibility to abiotic stresses may differ considerably among species or varieties of a crop (Yordanov et al., 2000; Simova-Stoilova et al., 2009; Vassileva et al., 2011; Chen et al., 2012; Wishart et al., 2014). Therefore, the selection of suitable genotypes and breeding of less susceptible varieties could reduce negative effects of extreme climate events on plant productivity (Neumann, 2008; Mir et al., 2012; Jogaiah et al., 2013), which is particularly important for the annual crops.

The apparent significance of stress period for the crop productivity does not rule out the fact that subsequent recovery stages are equally crucial for a proper evaluation of the overall performance (Subramanian and Charest, 1998; Gallé and Feller, 2007; Gallé et al., 2007; Vassileva et al., 2011). The progression and duration of stress, plant developmental stage and other biotic and abiotic factors may influence the stress response. For example certain species may be affected at early developmental stage, but still be capable to recover and finally to survive. Others could cope with suboptimal conditions comparatively well at the beginning of the stress period remaining still quite productive. Later on their surviving potential could be exhausted leaving the plants irreversibly damaged. A comprehensive evaluation of plant stress response includes the overall characterization of plant physiological behavior and survival. Here we summarize some of the major physiological parameters which characterize stress response reactions and which could be implemented as tools for evaluation of stress effects.

The impact of drought and heat on physiological status and productivity of agronomically important plants will become even more relevant during the next decades since these two major stress factors are associated with the predicted extreme events in the course of the global climate change. Assimilatory processes in leaves, long-distance translocation of solutes via xylem and phloem, changes in protein patterns and free amino acids, as well as the physiological phenomena associated with induced leaf senescence are addressed.

REGULATION OF STOMATAL OPENING BY DROUGHT AND HEAT

Together with internal CO_2 concentration, light and hormone levels, leaf temperature is one of the important factors for the regulation of stomatal opening. The three parameters: leaf temperature, water status and stomatal conductance represent a so-called « magic triangle» (Valladares and Pearcy, 1997; Reynolds-Henne et al., 2010). Leaf temperature may increase throughout the day reaching values above 40°C during the late afternoon in a sunny day in summer (**Figure 1**). Temperature sensors which monitor leaf temperature are integrated in modern equipment for measuring CO_2-assimilation, fluorescence or stomatal conductance. However, the measuring equipment itself influences leaf temperature by affecting external conditions (e.g., air convection, local air temperature, local humidity or photon flux density) therefore the detected values can differ considerably from the real temperature on the surface of undisturbed leaves. Ergo such leaf temperature data must be interpreted with certain precaution. Additional measurements from undisturbed leaves taken with an infrared thermometer which does not enter in contact and does not shadow the leaf are therefore recommended in this context.

Temperature of fully sun-exposed leaves is often 5–10°C higher than the one of shady leaves from the same plant. The interactions between leaf temperature and stomatal conductance are illustrated for a series of plants in **Figure 2**.

CO_2 is a major player in the regulation of stomatal opening (Medlyn et al., 2001). Opened stomata facilitate CO_2 diffusion from the ambient air into the leaf, but at the same time this is accompanied with additional water loss via enhanced transpiration. Therefore, the continuous increase in CO_2 partial pressure in the context of Global Change should be regarded as an important environmental factor capable to influence stomatal regulation. Although the relevance of stomatal opening for CO_2 assimilation is obvious, it must be considered that non-stomatal limitations such as changes in mesophyll conductance for CO_2 or in metabolic processes can also occur under drought and/or elevated temperature (Rosati et al., 2006; Signarbieux and Feller, 2011). Oscillations of leaf temperature after transition from darkness to high light intensity were reported recently (Feller, 2006; Reynolds-Henne et al., 2010) and are illustrated in **Figure 3**. After the transfer from shadow to strong light leaf temperature rises immediately, while stomates react within several minutes which explains the delay in cooling via transpiration. Stomatal opening and transpiration result in decreased leaf temperature which may lead again to a partial closure of stomates.

The water status of crop plants strongly depends on rainfall patterns and soil properties. Furthermore, agronomic practices

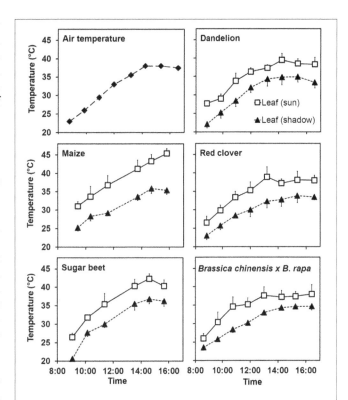

FIGURE 1 | Leaf and air temperature during a sunny day (August 13) of the exceptionally hot and dry summer 2003. Air temperature was measured with an electronic thermometer at the level of the top leaves. Leaf temperatures for various plant species were monitored in a field near Bern (Switzerland) with an infrared thermometer avoiding leaf contact and shadowing. Means and standard deviations (in one direction only for clarity) of 6 replicates are shown for leaf temperatures.

influence soil water availability which affects plant water status (Lenssen et al., 2007; Sturny et al., 2007; Gan et al., 2010). Abscisic acid (ABA) produced in roots exposed to soil with a low water potential, reaches the leaves via transpiration stream and causes stomatal closure. It also has been observed that ABA shifts the heat-induced stomatal opening toward a higher temperature (Feller, 2006; Reynolds-Henne et al., 2010; **Figure 4**). Thus, heat and drought act in an opposite manner on stomates. Sustainable agronomic techniques focused on good soil structure may contribute to a better productivity under abiotic stress. This is documented by a comparison of till and no-till plots at the same location during a dry and hot summer (**Figure 5**).

Another physiological phenomenon which may affect stomatal conductance is the deposition of waxy substances on the leaf surface. The cuticle is situated at the interface between the plant and its atmospheric environment. It is continuously exposed to natural and anthropogenic influences (Percy and Baker, 1987). Air pollutants and other environmental stresses may induce deposition of cuticular waxes which results in morphological changes to epicuticular wax layers. This could provoke reduced transpiration (Sanchez et al., 2001; Gallé and Feller, 2007; Seo et al., 2011; Yang et al., 2011; Zhu et al., 2014). Such effects become relevant immediately, but are not (or are

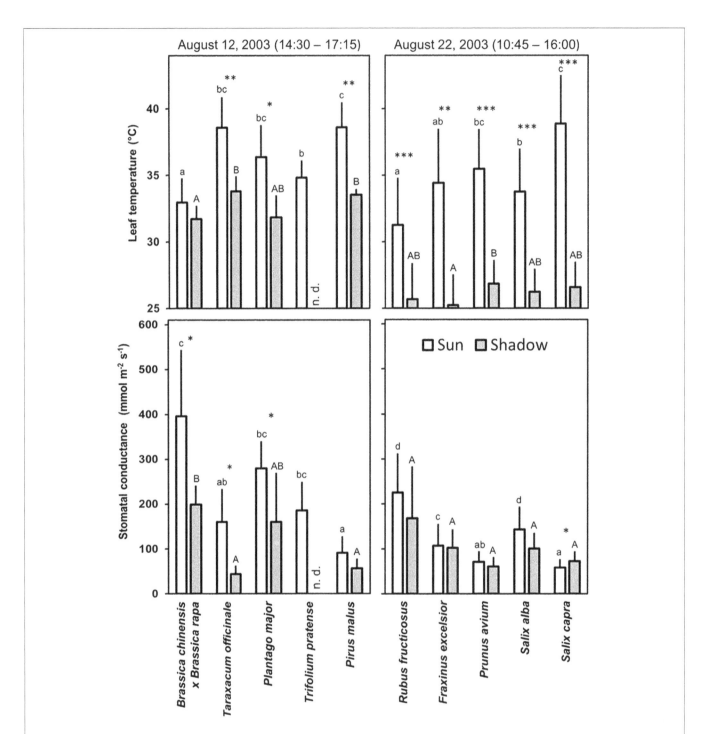

FIGURE 2 | Leaf temperature (measured at the undisturbed leaf with an infrared thermometer) and stomatal conductance in plant species grown at the same farm. Total stomatal conductance of lower and upper leaf surfaces are shown as means + SD of 5 measurements during the time intervals indicated. The photon flux densities were 1300–1900 μmol m^{-2} s^{-1} for sun-exposed leaves and around 100 μmol m^{-2} s^{-1} for shadowed leaves.

Values for leaves of *Trifolium pratense* could not be determined (n. d.) in the shadow. Sun-exposed leaves of different species with the same letter (a,b,c,d) and shadowed leaves with the same letter (A,B) in the same diagram are not significantly different at $P = 0.05$. Significant differences between sun-exposed and shadowed leaves of the same species at the *$P = 0.05$, **$P = 0.01$, and ***$P = 0.001$ are indicated above the column pair.

only partially) reversible, since the deposits remain after drought period.

On the cellular level, aquaporins—channels involved in water and CO_2 transport across membranes—are also integrated in

drought and heat stress response by influencing the water flux from the xylem to the leaf surface and may indirectly influence stomatal opening (Prado and Maurel, 2013). Aquaporins serve in a double function facilitating water and CO_2 fluxes across

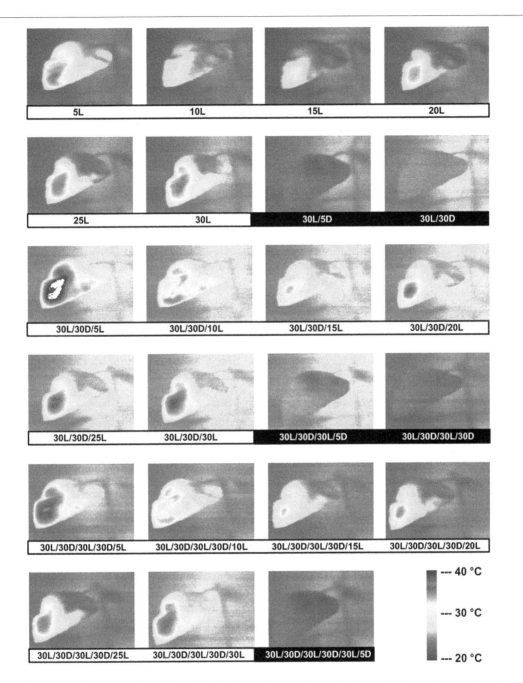

FIGURE 3 | Changes in sunflower leaf temperature during dark/light cycles. A dark-adapted sunflower plant was illuminated with a strong halogen light source for 30 min and then kept in darkness for 30 min before starting two other cycles with 30 min light (30L) followed by 30 min dark (30D). Leaf temperature was visualized in regular intervals with an infrared camera. The numbers below each picture indicate the pretreatment with the number of min in light (L) and dark (D). The white pixels at 30L/30D/5L were caused by a leaf temperature above 30°C.

membranes and must be considered as important players in the response of plants to abiotic stresses (Uehlein et al., 2003; Katsuhara and Hanba, 2008).

PHOTOSYNTHETIC CAPACITY DURING AND AFTER EXTREME EVENTS

Some drought and heat effects on photosynthesis are reversible and may even change repeatedly during 1 day, while other processes lead to irreversible damages. It is important to consider the reversibility of such effects on the organ and on the whole plant level when evaluating overall impacts. A reversible decrease of CO_2 fixation was observed in tree leaves and in grassland species (although less pronounced) at midday or in the afternoon under moderate drought (Haldimann et al., 2008; Bollig and Feller, 2014). An extended drought period may irreversibly damage leaves causing an anticipated and often atypical senescence

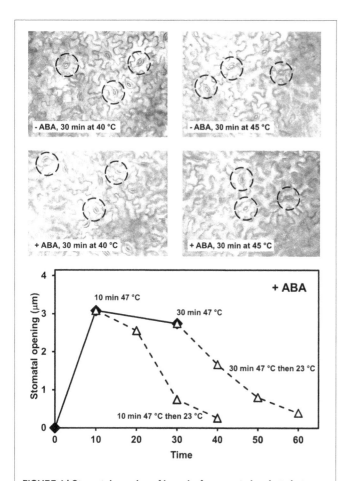

FIGURE 4 | Stomatal opening of bean leaf segments incubated at elevated temperature in dark in the presence (+ABA) and absence (−ABA) of 0.1 μM abscisic acid. The leaf segments were preincubated with ABA for 30 min before starting the temperature treatments. The pictures at the top were taken with a microsope camera from a pre-incubated leaf segment without any further preparation to avoid changes in stomatal opening. Some easily visible stomates are encircled in each picture. The diagram at the bottom illustrates the opening of the stomates in the presence of 0.1 μM abscisic acid at 47°C and the closure during a subsequent recovery phase at 23°C.

characterized by an incomplete nitrogen remobilization as a consequence of altered source/sink pattern (Feller and Fischer, 1994). The early loss of leaves reduces plant assimilatory capacity and prolonged drought period leads to plant death (Haldimann et al., 2008).

Photosynthesis and plant productivity can be reversibly or irreversibly affected by extreme environmental conditions such as drought or heat (Haldimann and Feller, 2005; Sharkey, 2005; Signarbieux and Feller, 2012). Stomatal opening as well as nonstomatal limitations (e.g., effects on mesophyll conductance for CO_2 or on metabolic processes) may influence CO_2 assimilation in drought-stressed leaves (Signarbieux and Feller, 2011). Since photon flux density is often very high during drought periods or heat waves and the demand for ATP and reduction equivalents for assimilatory processes is decreased, the channeling of absorbed light energy becomes crucial to avoid detrimental effects of reactive oxygen species (ROS) often accumulating under

abiotic stresses (Velikova and Loreto, 2005; Vickers et al., 2009). Particularly important in this regard is the antioxidant capacity of the plants comprising a system of enzymatic reactions as well as biosynthesis and accumulation of non-enzymatic low molecular metabolites, such as ascorbate, reduced glutathione, α-tocopherol, carotenoids, flavonoids and proline (reviewed by Gill and Tuteja, 2010).

ELECTRON TRANSPORT

Plant ecophysiology under adverse environmental conditions such as reduced water availability or heat can be investigated by non-destructive ≪ *in situ*≫ analyses of photosystem II functionality based on chlorophyll fluorescence measurements (Maxwell and Johnson, 2000). The ratio of variable fluorescence F_v to maximal fluorescence F_m in dark-adapted leaves is a measure of the maximum efficiency of photosystem II and in healthy leaves it is around 0.8 (Maxwell and Johnson, 2000). A decrease in this value is an indicator for irreversible damages and may be used to evaluate impacts of extreme events in field conditions. The different leaves of one and the same plant may be unequally affected by abiotic stress as demonstrated on **Figure 6**. Changes in non-photochemical quenching (which increases during abiotic stress) and in ϕ_{PSII} (PSII quantum yield which decreases during abiotic stress) are at least initially reversible and serve as indicators for the actual status of the photosynthetic apparatus. More sophisticated analyses indicate that the thermostability of photosystem II is improved under drought stress (Oukarroum et al., 2009) and as well as after growth at moderately elevated temperature (Haldimann and Feller, 2005). Some studies have indicated that photosystem II and the thylakoid membrane can be considered as comparatively thermotolerant components of the photosynthetic apparatus (Sharkey, 2005).

RUBISCO ACTIVASE

Rubisco—the key enzyme for CO_2 assimilation—is the most abundant protein on earth and it is quite heat-tolerant (Crafts-Brandner and Salvucci, 2000). Rubisco remains functional at temperatures above 50°C. However, high temperature causes a more rapid inactivation which is reverted in an ATP-dependent reaction (carbamylation) catalyzed by Rubisco activase (Crafts-Brandner and Salvucci, 2004; Kim and Portis, 2006). Since Rubisco activase is highly heat-sensitive, this enzyme becomes a key player for the rate of photosynthesis at elevated temperature (Feller et al., 1998; Salvucci et al., 2001; Yamori et al., 2012). Depending on the plant species, Rubisco activase activity is negatively affected by temperatures above 30°C (Salvucci and Crafts-Brandner, 2004). Rubisco activase is present under two forms which may be encoded by only one gene (the two forms originate by alternative splicing of the pre-mRNA) or by different genes (Salvucci et al., 2003). The slightly larger form contains two cysteine residues in the C-terminal extension allowing a redox regulation via the thioredoxin system (Portis et al., 2008). The heat sensitivity of photosynthesis was found to be due to thermal denaturation of Rubisco activase and not to the oxidation of the cysteine residues in the larger form (Salvucci et al., 2006). The complex regulation of Rubisco activase (and as a consequence of Rubisco) and CO_2 fixation is not yet fully explored for all

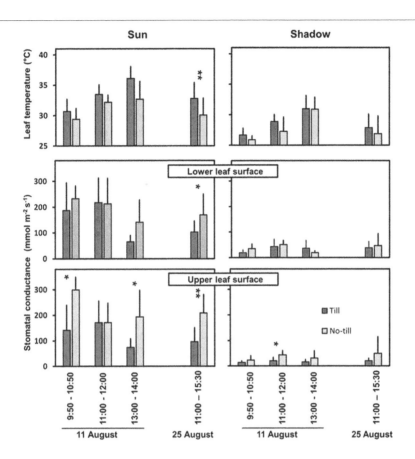

FIGURE 5 | Leaf temperature and stomatal conductance in sugar beet leaves of till and no-till plots in the same field during sunny days of the exceptionally hot and dry summer 2003. Temperature of the undisturbed leaves were measured with an infrared thermometer in field plots near Bern (Switzerland). Stomatal conductances of the upper and lower leaf surface are shown separately. The columns represent means + SD of 5 replicates. Significant differences between till and no-till at $^*P = 0.05$ and $^{**}P = 0.01$ are indicated.

major crop plants and will remain a subject of research during the next years.

Considerable differences in the heat tolerance of Rubisco activase in various plant species were reported (Salvucci and Crafts-Brandner, 2004). Rubisco activase has been identified as a possible target for novel breeding practices of crop plants which are still productive during a heat phase (Kim and Portis, 2005; Kurek et al., 2007; Kumar et al., 2009; Parry et al., 2011). Furthermore, Rubisco may be regulated via inhibitor levels making the evaluation of its functionality under stress even more complex (Parry et al., 2008).

ACCUMULATION AND DETOXIFICATION OF REACTIVE OXYGEN SPECIES

Plants which are exposed to stress cannot properly use ATP and reduction equivalents for biosynthetic processes and accumulate ROS. ROS are very reactive compounds with an obvious destructive potential, but they must be also regarded as signaling molecules (Suzuki and Mittler, 2006; Miller et al., 2007). ROS like superoxide anion radical, hydroxyl radical, and hydrogen peroxide are recognized to act as initiators and signals in programmed cell death (Mittler et al., 1999; Apel and Hirt, 2004; Locato et al., 2008; Van Breusegem et al., 2008). The promotion of ROS production (Lee et al., 2012) and the loss of antioxidant defenses (Munne-Bosch et al., 2001) may induce or accelerate senescence in plants subjected to abiotic stress.

The accumulation and detoxification of ROS become more important during drought (Miller et al., 2010) and during growth stages characterized with elevated ambient temperature (Wahid et al., 2007). A rapid removal of ROS is necessary to avoid deleterious effects such as lipid peroxidation and their negative influence over plant metabolism (Oberschall et al., 2000; Locato et al., 2009). The production/detoxification of ROS is important for several subcellular compartments and it is not restricted to chloroplasts (Noctor et al., 2002; Pastore et al., 2007). Antioxidant enzyme activities such as catalases, peroxidases and superoxide dismutases play important role in the detoxification of ROS (Selote et al., 2004; Pastore et al., 2007; Bian and Jiang, 2009). A study on cotton varieties differing in thermotolerance suggests that there is a potential to incorporate the knowledge regarding the role of antioxidant enzymes in stress response for breeding of tolerant varieties (Snider et al., 2010) by the enhancement of *in vivo* levels of antioxidant enzymes. The relevance of high constitutive activities of ROS-detoxifying enzymes and of their on-going increase during abiotic stress was reported by Turkan et al. (2005) for bean plants.

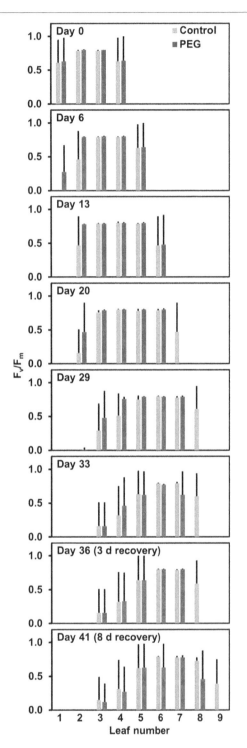

FIGURE 6 | Intactness of photosystem II in different leaves of drought-stressed and control plants of *Lolium perenne*. The water potential in nutrient medium was artificially decreased by addition of polyethyleneglycol 6000. Leaves were numbered from 1 (oldest) to 9 (youngest). F_v/F_m in healthy and fully expanded leaves is close to 0.8. In young and not yet expanded leaves, the mean value may be lower and increase during further expansion, while a decrease in old leaves indicates irreversible damages (e.g., senescence). Means + SD of 5 replicates are shown. A value of 0.0 was entered for missing leaves (relevant only for the youngest leaves).

In addition to enzymatic ROS detoxification, hydrophilic and lipophilic antioxidant compounds contribute to the antioxidant response and may serve as radical scavengers (Fryer, 1992; Loreto et al., 2001; Larkindale and Huang, 2004; Pose et al., 2009). Increased levels of such compounds assist for a rapid detoxification of ROS and aid the protection of subcellular structures. Enzymes involved in the biosynthesis of antioxidant compounds, their expression before and during abiotic stress, their subcellular compartmentalization, as well as the regulation of their activity must be considered in the context of ROS detoxification.

PHOTORESPIRATION

As mentioned above, photosynthesis decreases under drought or heat, but the leaves are often exposed to a high photon flux density and a low CO_2 partial pressure in the leaf apoplast. Oxygenase activity (the starting point of the photorespiratory metabolism) is an inherent property of Rubisco and depends on CO_2 and O_2 partial pressure (Osmond and Grace, 1995). Modifications in the large subunit of Rubisco can alter the relative oxygenase/carboxylase activities (Whitney et al., 1999). Therefore, the large subunit of Rubisco which is encoded in the chloroplast DNA is considered for breeding strategies in the future in order to improve the assimilatory capacity of crops (Parry et al., 2011). Stomatal closure during drought periods may decrease the CO_2 partial pressure in the leaves and alter the relative oxygenase/carboxylase activities of Rubisco in favor of oxygenase. Protective effects of photorespiration in drought-exposed C_3 plants under high irradiance were studied by various research teams (Wingler et al., 1999; Haupt-Herting et al., 2001; Noctor et al., 2002; Guan et al., 2004; Bai et al., 2008). Increased transcript levels of enzymes involved in the photorespiratory carbon cycle were detected in tobacco under drought (Rivero et al., 2009). Detailed studies with *Phaseolus vulgaris* brought to a conclusion that photorespiration, although stimulated under water deficit, does not play a major role in photoprotection of leaf cells under drought (Brestic et al., 1995). In contrast to C_3 plants, the rate of photorespiration remains low in C_4 plants exposed to drought (Carmo-Silva et al., 2008). Photorespiration and monoterpene production were considered as mechanisms involved in the thermotolerance of oak (Penuelas and Llusia, 2002). To summarize: heat and drought increase the rate of photorespiration in leaves of C_3 plants, but the question to which extent photorespiration plays a protective role in different crop species remains still open.

NITROGEN METABOLISM

Several stages of nitrogen metabolism could be affected by abiotic stress. One important step is the assimilation of nitrate into organic compounds. The activity of the first enzyme involved (nitrate reductase) is negatively influenced by abiotic stresses (Ferrario-Mery et al., 1998; Xu and Zhou, 2006). The adverse drought effect may be decreased by the improved availability of inorganic nitrogen (Krcek et al., 2008; Zhang et al., 2012). Nitrogen fixation in legume nodules is also severely reduced during drought periods (Larrainzar et al., 2009; Aranjuelo et al., 2011; Gil-Quintana et al., 2013). A negative effect of accumulated free amino acids on nitrogen fixation (N-feedback inhibition) and increased oxygen resistance in the nodules were among the proposed mechanisms for this below-ground drought impact (Aranjuelo et al., 2011; Gil-Quintana et al., 2013).

The balance between free and protein-bound amino acids is also affected by abiotic stresses. Under drought, the quantity of proteins usually declines, while free amino acids tend to accumulate being partially converted into compatible solutes (e.g., proline) as reported by several groups during the past decades (Yoshiba et al., 1997; Su and Wu, 2004; Gruszka Vendruscolo et al., 2007; Parida et al., 2008; Bowne et al., 2012). Proline accumulation under abiotic stresses was reviewed in detail by Verbruggen and Hermans (2008). The reversible accumulation of proline in drought-stressed clover is illustrated in **Figure 7**. A 10- to 100-fold increase in proline content can be observed during a stress phase. During a subsequent recovery proline levels in leaves decrease again and reach values similar to those of unstressed control plants.

LEAF SURVIVAL AND ALTERED TIMING OF SENESCENCE

Senescence is a complex process (Hörtensteiner and Feller, 2002). The number and the area of active leaves per plant is important for the overall performance of a plant (Munne-Bosch and Alegre, 2004). The formation and expansion of young leaves and senescence of old leaves are equally important in this context (Lefi et al., 2004; Simova-Stoilova et al., 2010; Mahdid et al., 2011; Gilgen and Feller, 2014). The catabolism of proteins in older leaves allows a redistribution of nitrogen from senescing tissues to other plant parts (Feller and Fischer, 1994), while the small percentage of nitrogen present in chlorophyll remains in modified form in the vacuoles of senescing or senesced cells (Hörtensteiner, 2006). Chlorophyll in intact chloroplasts is present in photosystems I and II together with chlorophyll-binding proteins in well-organized structures. During senescence the photosystems are degraded. Chlorophyll outside these structures would have detrimental physiological consequences (Hörtensteiner, 2006). Chlorophyll

catabolism prevents such negative effects on one hand and allows the remobilization of chlorophyll-binding proteins on the other (Hörtensteiner and Feller, 2002).

Besides phytohormones and ROS source/sink interactions and C/N ratios must be also considered as endogenous senescence-regulating factors (Feller and Fischer, 1994; Thoenen et al., 2007; Luquet et al., 2008). Sink capacities may be strongly reduced under drought and heat. This may lead to an abnormal type of senescence accompanied by accumulation of free amino acids which could be partially converted into osmoprotectants in source leaves (Bowne et al., 2012). This process is initially reversible, but when prolonged it may turn into senescence finally leading to organ death.

An interesting observation concerning interactions between leaf senescence and drought tolerance was reported by Rivero et al. (2007). In their studies they compared wild-type plants and transgenic plants with a delayed drought-induced senescence. The latter were characterized with an excellent drought tolerance and maintained a high physiological potential. Considerable differences in the drought response in relation to senescence were reported also for various genotypes of maize (Messmer et al., 2011), millet (Dai et al., 2011), wheat (Hafsi et al., 2000; Verma et al., 2004), and alfalfa (Erice et al., 2011). The relevance of the recovery phase after an extended drought period was emphasized by several groups (Merewitz et al., 2010; Vassileva et al., 2011; Yao et al., 2012).

ACCUMULATION OF STRESS-RELATED PROTEINS

Drought and high temperatures, together with other environmental factors like chemical pollutants, cold and high salt concentrations have similar effects on plants. They damage plant cell and lead to osmotic and oxidative stress (Reddy et al., 2004; Foyer and Noctor, 2009). Changes in expression and post-translational modification of proteins are an important part of perception and response to abiotic stress (Hashiguchi et al., 2010). Drought and high temperature involve, as a common feature, increased numbers of inactive proteins—denatured, aggregated or oxidatively damaged. Protein homeostasis under stress is maintained via different biochemical mechanisms that regulates their biosynthesis, folding, trafficking and degradation (Gottesman et al., 1997; Chen et al., 2011). Plants respond to dehydration stress by synthesis of protective proteins such as dehydrins and chaperones and by degradation of irreversibly damaged proteins by proteases (reviewed in Vaseva et al., 2011). Protein breakdown has been recognized as one of the important mechanisms for the adaptation of plants to environmental conditions (Vierstra, 1996). Proteolysis is performed by an impressive number of proteases—approximately 2% of the genes code for proteolytic enzymes (Rawlings et al., 2004). Proteases vary significantly in size and molecular structure and could be composed of single molecules with small size of approximately 20 kDa as well as they could be represented by big proteolytic complexes with molecular mass around 6 MDa. Some proteases are able to act as chaperones under specific conditions. They are called chaperonines and comprise complex elements of regulated proteolysis participating in the fine-tuning of gene expression (Sakamoto, 2006).

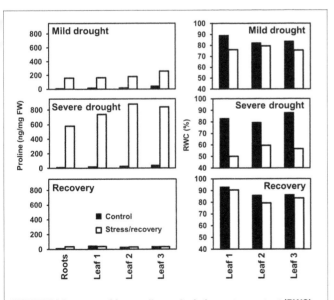

FIGURE 7 | Contents of free proline and relative water content (RWC) in leaves of soil-grwon *Trifolium repens* during drought stress and recovery.

Chaperones have essential function in protein homeostasis under normal condition and are highly responsive to various stresses (Wang et al., 2004). Their main physiological role is to maintain proteins in a functional conformation and to prevent aggregation of non-native proteins. Chaperones also participate in refolding of denatured proteins to their native conformation and in removal of non-functional and potentially harmful polypeptides. Heat-shock proteins (HSPs) belong to the group of stress-related proteins with chaperone function. Plant HSPs comprise five classes according to their approximate molecular weight: Hsp100, Hsp90, Hsp70, Hsp60, and small heat-shock proteins (sHsps) (Kotak et al., 2007). Transcription of heat-shock protein genes is controlled by regulatory proteins called heat stress transcription factors (Hsfs). *Arabidopsis* genome contains 21 genes encoding Hsfs (Scharf et al., 2012).

ENZYMES INVOLVED IN THE DETOXIFICATION OF REACTIVE OXYGEN SPECIES (ROS)

Plants have developed efficient non-enzymatic and enzymatic detoxification mechanisms to scavenge ROS. Superoxide dismutase (EC 1.15.1.1), catalase (CAT; EC 1.11.1.6), ascorbate peroxidase (APX; EC 1.11.1.11), and glutathione peroxidase (EC 1.11.1.7) are the major enzymes involved in oxidative stress response in plants (Mittler, 2002; Apel and Hirt, 2004). The regulation of ROS levels and fine-tuning of ROS homeostasis is performed at several biochemical steps. The three types of plant superoxide dismutases have different functional metals and subcellular localization (Bowler et al., 1994; Alscher et al., 2002). Cu/Zn-superoxide dismutases localized mainly in the cytosol, but have also been detected in peroxisomes and chloroplasts. Fe-superoxide dismutase is a chloroplast enzyme, while Mn- superoxide dismutases has been found in the mitochondrial matrix and peroxisomes (Bowler et al., 1994). Initially superoxide dismutase converts superoxide to H_2O_2 which can be further metabolized by catalase or ascorbate peroxidase to oxygen and water—processes mainly localized in peroxisomes. Most probably the better tolerance toward oxidative stress, often assigned to higher superoxide dismutase, ascorbate peroxidase or catalase levels, is a result of a complex interplay between these antioxidant enzymes (Xu et al., 2013).

D-1-PYRROLINE-5-CARBOXYLATE SYNTHETASE (P5CS)

Proline acts as an osmoprotectant in response to osmotic stress and its accumulation has been recognized as a marker for tolerance toward drought and high salt concentrations (Hmida-Sayari et al., 2005; Kishor et al., 2005; Deng et al., 2013). It has been proved to be a very effective singlet oxygen quencher (Alia et al., 2001). The first two steps of proline biosynthesis in plants are catalyzed by the bifunctional enzyme D-1-pyrroline-5-carboxylate synthetase (P5CS, EC not assigned) that encompasses both γ-glutamyl kinase and glutamic-γ-semialdehyde dehydrogenase activities (Pérez-Arellano et al., 2010). P5CS plays a key role in plant intracellular accumulation of proline and is subjected to feedback inhibition by proline, controlling the level of the free imino acid under both normal and stress conditions (Hong et al., 2000). It has been confirmed that D-1-pyrroline-5-carboxylate synthetase is encoded by two differentially regulated genes in

different plant species (Turchetto-Zolet et al., 2009). Usually one of the P5CS isoforms is osmo-regulated and the other is associated with developmentally governed processes (Hur et al., 2004; Székely et al., 2008; Pérez-Arellano et al., 2010).

DEHYDRINS

Dehydrins belong to the group of Late Embryogenesis-Abundant (LEA) proteins which are expressed in late stages of seed maturation and/or upon water stress conditions in plants (Rorat, 2006). They constitute a highly divergent group of thermostable intrinsically disordered proteins that can be classified into different types according to the presence of distinct, short sequence motifs. All dehydrins have at least one conserved, lysine-rich 15-amino acid domain, EKKGIMDKIKEKLPG, named the K-segment (Close, 1997). In addition dehydrin molecule could contain a track of serine residues (the S-segment) and/or a consensus motif, T/VDEYGNP (the Y-segment) which is usually located near the N-terminus. The less conserved regions of dehydrins are characterized by a high polar amino acid content and usually are referred to as Φ-segments. The number and order of the Y-, S-, and K-segments define the different dehydrin sub-classes: YnSKn, YnKn, SKn, Kn, and KnS, which may possess a specific function and tissue distribution (Close, 1997). As intrinsically disordered proteins dehydrins are characterized by high flexibility, structural adaptability, and extended conformational states (Tompa, 2009) which most probably contributes to conferring plant desiccation stress tolerance via various possible biochemical mechanisms—sequestering ions, stabilizing membranes, or acting as chaperones (Danyluk et al., 1998; Rorat, 2006; Tompa, 2009). *Arabidopsis* dehydrins ERD10 and ERD14 fulfill protective functions acting as potent chaperones of broad substrate specificity and they also have membrane-binding capacity (Kovacs et al., 2008). It was also reported that both ERD10 and ERD14 can be phosphorylated at various sites, which promotes the binding of divalent metal ions, and this might be related to their ion-sequestering activity (Rorat, 2006).

Drought tolerance is assessed as the ability of plants to maintain a certain level of production under water shortage, which is relevant for most economically important crops (Volaire and Lelievre, 2001). Accumulation of dehdrins in leaves under drought is a quite general phenomenon, but the dehydrin patterns may differ considerably between species subjected to the same drought treatment which makes them suitable as diagnostic tools (Close, 1997; Vaseva et al., 2014). Immunodetection of strong dehydrin accumulation in four plant species (*Trifolium repens, Helianthus uniflorus, Dactylis glomerata,* and *Lolium perenne*) subjected to uniform dehydration is represented on **Figure 8**. Immunosignals are revealed with antibodies against both the K- and the Y-dehydrin segments. The analyzed plants are important forage crops, often used in pasture seed mixes. The considerable differences among immunosignal spectra of the tested species (**Figure 8**) indicate that these drought-stress markers are highly specific for the different plants and a universal assessment approach is not applicable for dehydrins.

A recent study on *Trifolium repens* dehydrins revealed complex structure of dehydrin-coding sequences, which could be a prerequisite for high variability among the transcripts originating

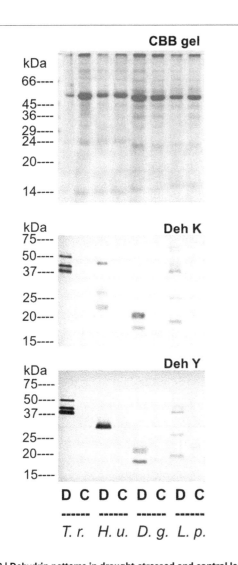

FIGURE 8 | Dehydrin patterns in drought-stressed and control leaves of *Trifolium repens* (*T. r.*), *Helianthus uniflorus* (*H. u.*), *Dactylis glomerata* (*D. g.*), and *Lolium perenne* (*L. p.*). The various plant species were grown in the same containers to ensure identical conditions. Controls (C) were incubated on standard nutrient medium, while polyethyleneglycol 6000 was added to this medium for incubations under artificial drought (D). Crude extract was analyzed by SDS-PAGE electrophoresis followed by staining with Coomassie Brilliant Blue (loading control). The supernatant of heat-treated and then centrifuged crude extract was used for immunoblots with specific antibodies against the well conserved dehydrin K- (Deh K) and Y-segments (Deh Y).

from a single gene (Vaseva et al., 2014). For some dehydrins, natural antisense transcripts have been identified (Vaseva and Feller, 2013). It has been suggested that natural antisense RNAs hold potential to regulate the expression of their sense partner(s) at either transcriptional or post-transcriptional level (Sunkar et al., 2007), which remains to be experimentally verified for dehydrins.

AQUAPORINS

Aquaporins represent a group of membrane proteins facilitating the transport of water across a membrane (Lovisolo et al., 2007; Prado and Maurel, 2013; Li et al., 2014). Although aquaporins

were initially identified as membrane intrinsic proteins facilitating water transport, it is well accepted now that they play also an important role in CO_2 transport across plant menbranes (Uehlein et al., 2003; Katsuhara and Hanba, 2008; Secchi and Zwieniecki, 2013; Kaldenhoff et al., 2014). Both functions are highly relevant under abiotic stresses, especially for the regulation of leaf hydraulics under drought stress (Prado and Maurel, 2013). The expression of aquaporins under various environmental conditions is well regulated (Chaumont and Tyerman, 2014). Previous studies have reported that environmental stresses, among which these with dehydration element, regulate the expression of aquaporins (Maurel et al., 2002; Suga et al., 2002; Vera-Estrella et al., 2004; Ayadi et al., 2011; Mirzaei et al., 2012).

Aquaporins are present in plants under various forms and they may differ considerably in their properties (Lovisolo et al., 2007). Aquaporins are not only important in various shoot parts, but may also play a key role in regulating the hydraulic conductance in roots (Perrone et al., 2012). However, these authors concluded that a root-specific aquaporin is more important in the regulation of water flow from the roots to the shoot in well-watered than in drought-stressed plants.

LONG-DISTANCE TRANSPORT VIA XYLEM AND PHLOEM

Solute transport via the two long-distance transport systems xylem and phloem are highly important for the supply of various organs with nutrients and assimilates (Bahrun et al., 2002; Sevanto, 2014). The transport network is strongly affected by abiotic stresses. This provokes changes in the translocation of nutrients and assimilates (including phytohormones) via the xylem from the roots to the shoot under adverse conditions. Redistribution processes via the phloem within the shoot or from the shoot to the roots is also strongly affected by stress.

SOLUTE ALLOCATION VIA THE XYLEM

Root development and root metabolism are both influenced by drought (Mori and Inagaki, 2012; Comas et al., 2013). The transport in the xylem is driven by the water potential difference between the soil and the atmosphere (transpiration) and strongly depends on stomatal conductance (Miyashita et al., 2005; Bollig and Feller, 2014). The relative transpiration rates of various shoot organs determine the distribution of solutes present in the xylem sap. Besides the quantity of xylem sap transported from the roots to the shoot, the composition of the xylem sap may be affected by drought as a consequence of altered root physiology (Bahrun et al., 2002; Comas et al., 2013). For example abscisic acid which is involved in decreasing stomatal conductance is synthesized in roots and it is a well-known signaling molecule in the xylem sap of drought-stressed plants (Ismail et al., 1994; Hansen and Dorffling, 1999; Alvarez et al., 2008). Air embolism caused by a fall in hydraulic conductivity in the xylem of vascular plants may become an issue under severe drought (Kolb and Davis, 1994; Cochard, 2002; Kaufmann et al., 2009). A partial repair of embolism during the recovery was reported for grapevine (Lovisolo et al., 2008). Abscisic acid may accumulate in the roots during the drought period, reach after rehydration the leaves via the xylem, cause stomatal closure and improve as a consequence the water potential in various shoot parts facilitating the repair of embolism (Lovisolo et al., 2008). More recently Secchi and Zwieniecki (2014) reported a strong up-regulation of aquaporin gene expression when xylem embolism was formed. Furthermore, these authors concluded from experiments with transgenic poplar plants that the expression of aquaporin genes is important for the recovery from embolism.

REDISTRIBUTION PROCESSES VIA THE PHLOEM

The export of nutrients and assimilates from source leaves to sink organs is important for the development of vegetative and reproductive organs and for the overall performance of crop plants (Van Bel, 2003). Possible mechanisms for drought effects on phloem transport were reviewed recently by Sevanto (2014). The accumulation of dehydrins in the phloem of *Solanaceae* plants under drought stress were reported and discussed in the context of protecting sieve tubes and companion cells under abiotic stresses (Szabala et al., 2014). Besides the mass flow in the sieve tubes, the composition of the phloem sap represents another key issue. Strong influences of soil drought on the source/sink network must be expected (Gilgen and Feller, 2013). The allocation of solutes, traced by ^{134}Cs label of control and drought-stressed

plants from leaf 3 (third-oldest leaf) to roots, older leaves (leaves 1 and 2) and younger leaves of wheat during vegetative growth is illustrated in **Figure 9**. Lower solute content, as evident from the measured label signal, was transported under drought from leaf 3 to the roots, while the supply of the other plant parts was not significantly influenced by artificial drought (polyethylene glycol 6000). Root development and productivity under drought may differ between wheat genotypes indicating that there might be some potential for novel breeding strategies in the future (Mori and Inagaki, 2012). Soil may not suffer water deprivation uniformly during a drought period. Such conditions were simulated in an experiment with a split root system of white clover (**Figure 10**) where the allocation of ^{134}Cs from a leaf to other plant parts was monitored. The low water potential in the environment of some roots caused a decreased solute supply via the phloem presumably as a consequence of a reduced sink capacity in these roots. This demonstrates that root growth and development in rapidly drying soil regions are more severely affected. This could result in highly asymmetrical root systems which on the other hand may obstruct the recovery after re-watering. The unequal root distribution in the soil would not allow an optimal use of resources at the beginning of recovery phase. Later, after re-watering new roots may be formed and this will allow the exploration of previously inaccessible soil regions.

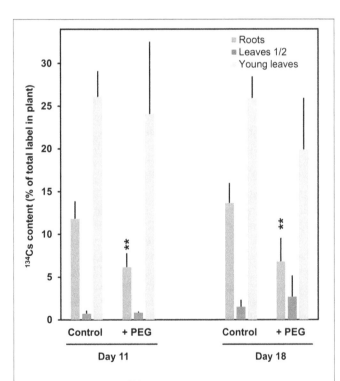

FIGURE 9 | Allocation of ^{134}Cs from leaf 3 to other parts of control and drought-stressed wheat. The plants were grown on standard nutrient medium for 17 d before starting the experiment (day 0). The water potential in the nutrient medium was decreased by addition of polyethylene glycol 6000 (PEG; 100 g PEG plus 1 liter standard nutrient medium at the beginning). The label was introduced via a flap into the lamina of leaf 3 at day 4 according to Schenk and Feller (1990) (collected at day 11) and at day 11 (collected at day 18). The transfer of the label to roots, two oldest (leaves 1/2) and younger leaves (leaf 4 and younger) was detected by gamma spectrometry. Means + SD of 4 replicates are shown. Significant differences between roots of drought-stressed and control plants of the same age at **$P = 0.01$ are indicated. No significant differences were detected in leaves.

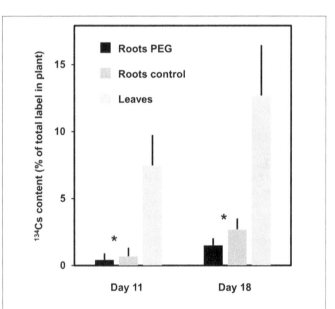

FIGURE 10 | Allocation of ^{134}Cs from a leaf to other plant parts of white clover in a split-root system with one part of the root system in standard medium and the other part in medium containing polyethylene glycol 6000 (PEG; 100 g PEG plus 1 liter standard nutrient medium to lower the water potential). The plants were grown with both parts of the split roots in standard nutrient medium for 52 day before starting the experiment (day 0), then the nutrient medium was replaced by new standard medium for one part of the root system and with standard medium containing PEG for the other part. The label was introduced at day 4 (collected at day 11) and at day 11 (collected at day 18) via a petiole flap into the largest fully expanded trifoliate. "Leaves" represent all leaves with petioles except the labeled leaf. Means + SD of 6 replicates are shown. Significant differences between roots in PEG and control roots at *$P = 0.05$ are indicated.

CONCLUSIONS

Climate change is a challenge for plant breeders, physiologists, agronomists and decision makers (Ingram et al., 2008). Various species differ in their drought and heat tolerance. Furthermore, a moderate temperature increase may be beneficial for certain crops (e.g., maize) which are cold-sensitive (Klein et al., 2013). The identification of key processes on the whole plant level is important for genotype selection and organizing breeding programs in the future (Gornall et al., 2010). Organ development, assimilatory processes, morphological adaptations, long-distance transport, senescence and seed maturation may contribute to the overall response. Our knowledge in this field is still quite limited. Not only species, but also genotypes of the same species may differ considerably in their tolerance to abiotic stresses such as drought or heat. The performance during stress and subsequent recovery phases must be considered in this context (Walter et al., 2011). A plant with a poor performance during the stress phase may survive longer and recover more efficiently than a plant which initially remains productive under unfavorable conditions. In monocultures breeding, genotype selection and agronomic practices represent challenges to cope with climatic changes including more frequent extreme events during the next decades as predicted from regional climate models.

In mixed cultures (e.g., grasslands) the competition between species must be taken additionally into account. Stress periods may affect various plants differently and cause a shift in the species spectrum (Jentsch et al., 2011) negatively influencing the competition between the cultivated plants and certain weeds (Gilgen et al., 2010). Accordingly such interactions not only result in decrease yield but they require extra measures in weed control management.

ACKNOWLEDGMENTS

We thank Iwona Anders, Regula Blösch, Klimentina Demirevska, Ninetta Graf, Anelia Kostadinova, Anita Langenegger, Jan Mani, Roza Nenkova, Valya Vassileva, Thomas von Känel, Bistra Yuperlieva-Mateeva and Anita Zumsteg for stimulating discussions and for providing data for the illustrations. The investigations were partially supported by NCCR "Climate" (Project "Plant Soil"), by Sciex-NMS (Project No. 11.113: "Identification of Dehydrin Types involved in Abiotic Stress Responses in *Trifolium repens*"—IDAST) and by SCOPES program of the Swiss National Science Foundation (project DILPA—JRP—IB73AO-111142/1).

REFERENCES

Alia, Mohanty, P., and Matysik, J. (2001). Effect of proline on the production of singlet oxygen. *Amino Acids* 21, 195–200. doi: 10.1007/s007260170026

Alscher, R. G., Erturk, N., and Heath, L. S. (2002). Role of superoxide dismutases (SODs) in controlling oxidative stress in plants. *J. Exp. Bot.* 53, 1331–1341. doi: 10.1093/jexbot/53.372.1331

Alvarez, S., Marsh, E. L., Schroeder, S. G., and Schachtman, D. P. (2008). Metabolomic and proteomic changes in the xylem sap of maize under drought. *Plant Cell Environ.* 31, 325–340. doi: 10.1111/j.1365-3040.2007.01770.x

Apel, K., and Hirt, H. (2004). Reactive oxygen species: metabolism, oxidative stress, and signal transduction. *Annu. Rev. Plant Biol.* 55, 373–399. doi: 10.1146/annurev.arplant.55.031903.141701

Aranjuelo, I., Molero, G., Erice, G., Christophe Avice, J., and Nogues, S. (2011). Plant physiology and proteomics reveals the leaf response to drought in alfalfa (*Medicago sativa* L.). *J. Exp. Bot.* 62, 111–123. doi: 10.1093/jxb/erq249

Ayadi, M., Cavez, D., Miled, N., Chaumont, F., and Masmoudi, K. (2011). Identification and characterization of two plasma membrane aquaporins in durum wheat (*Triticum turgidum* L. subsp. durum) and their role in abiotic stress tolerance. *Plant Physiol. Biochem.* 49, 1029–1039. doi: 10.1016/j.plaphy.2011.06.002

Bahrun, A., Jensen, C. R., Asch, F., and Mogensen, V. O. (2002). Drought-induced changes in xylem pH, ionic composition, and aba concentration act as early signals in field-grown maize (*Zea mays* L.). *J. Exp. Bot.* 53, 251–263. doi: 10.1093/jexbot/53.367.251

Bai, J., Xu, D. H., Kang, H. M., Chen, K., and Wang, G. (2008). Photoprotective function of photorespiration in *Reaumuria soongorica* during different levels of drought stress in natural high irradiance. *Photosynthetica* 46, 232–237. doi: 10.1007/s11099-008-0037-5

Bian, S. M., and Jiang, Y. W. (2009). Reactive oxygen species, antioxidant enzyme activities and gene expression patterns in leaves and roots of Kentucky bluegrass in response to drought stress and recovery. *Sci. Hortic.* 120, 264–270. doi: 10.1016/j.scienta.2008.10.014

Bollig, C., and Feller, U. (2014). Impacts of drought stress on water relations and carbon assimilation in grassland species at different altitudes. *Agric. Ecosyst. Environ.* 188, 212–220. doi: 10.1016/j.agee.2014.02.034

Bowler, C., Vancamp, W., Vanmontagu, M., and Inze, D. (1994). Superoxide-dismutase in plants. *Crit. Rev. Plant Sci.* 13, 199–218. doi: 10.1080/07352689409701914

Bowne, J. B., Erwin, T. A., Juttner, J., Schnurbusch, T., Langridge, P., Bacic, A., et al. (2012). Drought responses of leaf tissues from wheat cultivars of differing drought tolerance at the metabolite level. *Mol. Plant* 5, 418–429. doi: 10.1093/mp/ssr114

Brestic, M., Cornic, G., Fryer, M. J., and Baker, N. R. (1995). Does photorespiration protect the photosynthetic apparatus in French bean-leaves from photoinhibition during drought stress. *Planta* 196, 450–457. doi: 10.1007/BF00203643

Carmo-Silva, A. E., Powers, S. J., Keys, A. J., Arrabaca, M. C., and Parry, M. A. J. (2008). Photorespiration in C(4) grasses remains slow under drought conditions. *Plant Cell Environ.* 31, 925–940. doi: 10.1111/j.1365-3040.2008.01805.x

Chaumont, F., and Tyerman, S. D. (2014). Aquaporins: highly regulated channels controlling plant water relations. *Plant Physiol.* 164, 1600–1618. doi: 10.1104/pp.113.233791

Chen, B., Retzlaff, M., Roos, T., and Frydman, J. (2011). Cellular strategies of protein quality control. *Cold Spring Harb. Perspect. Biol.* 3:a004374. doi: 10.1101/cshperspect.a004374

Chen, J., Xu, W., Velten, J., Xin, Z., and Stout, J. (2012). Characterization of maize inbred lines for drought and heat tolerance. *J. Soil Water Conserv.* 67, 354–364. doi: 10.2489/jswc.67.5.354

Close, T. J. (1997). Dehydrins: a commonality in the response of plants to dehydration and low temperature. *Physiol. Plant.* 100, 291–296. doi: 10.1111/j.1399-3054.1997.tb04785.x

Cochard, H. (2002). Xylem embolism and drought-induced stomatal closure in maize. *Planta* 215, 466–471. doi: 10.1007/s00425-002-0766-9

Comas, L. H., Becker, S. R., Cruz, V. V., Byrne, P. F., and Dierig, D. A. (2013). Root traits contributing to plant productivity under drought. *Front. Plant Sci.* 4:442. doi: 10.3389/fpls.2013.00442

Crafts-Brandner, S. J., and Salvucci, M. E. (2000). Rubisco activase constrains the photosynthetic potential of leaves at high temperature and CO_2. *Proc. Natl. Acad. Sci. U.S.A.* 97, 13430–13435. doi: 10.1073/pnas.230451497

Crafts-Brandner, S. J., and Salvucci, M. E. (2004). Analyzing the impact of high temperature and CO_2 on net photosynthesis: biochemical mechanisms, models and genomics. *Field Crops Res.* 90, 75–85. doi: 10.1016/j.fcr.2004.07.006

Dai, H. P., Zhang, P. P., Lu, C., Jia, G. L., Song, H., Ren, X. M., et al. (2011). Leaf senescence and reactive oxygen species metabolism of broomcorn millet (*Panicum miliaceum* L.) under drought condition. *Aust. J. Crop Sci.* 5, 1655–1660. Available online at: http://www.cropj.com/dai_5_12_2011_1655_1660.pdf

Danyluk, J., Perron, A., Houde, M., Limin, A., Fowler, B., Benhamou, N., et al. (1998). Accumulation of an acidic dehydrin in the vicinity of the plasma membrane during cold acclimation of wheat. *Plant Cell* 10, 623–638. doi: 10.1105/tpc.10.4.623

Deng, G., Liang, J., Xu, D., Long, H., Pan, Z., and Yu, M. (2013). The relationship between proline content, the expression level of P5CS (Delta(1)-pyrroline-5-carboxylate synthetase), and drought tolerance in Tibetan hulless barley (*Hordeum vulgare* var. Nudum). *Russ. J. Plant Physl.* 60, 693–700. doi: 10.1134/S1021443713050038

Easterling, D. R., Meehl, G. A., Parmesan, C., Changnon, S. A., Karl, T. R., and Mearns, L. O. (2000). Climate extremes: observations, modeling, and impacts. *Science* 289, 2068–2074. doi: 10.1126/science.289.5487.2068

Erice, G., Louahlia, S., Irigoyen, J. J., Sanchez-Diaz, M., Alami, I. T., and Avice, J. C. (2011). Water use efficiency, transpiration and net CO_2 exchange of four alfalfa genotypes submitted to progressive drought and subsequent recovery. *Environ. Exp. Bot.* 72, 123–130. doi: 10.1016/j.envexpbot.2011.02.013

Feller, U. (2006). Stomatal opening at elevated temperature: an underestimated regulatory mechanism. *Gen. Appl. Plant Physiol.* 32, 19–31. Available online at: http://www.bio21.bas.bg/ipp/gapbfiles/pisa-06/06_pisa_19-31.pdf

Feller, U., Crafts-Brandner, S. J., and Salvucci, M. E. (1998). Moderately high temperatures inhibit ribulose-1,5-bisphosphate carboxylase/oxygenase (Rubisco) activase-mediated activation of Rubisco. *Plant Physiol.* 116, 539–546. doi: 10.1104/pp.116.2.539

Feller, U., and Fischer, A. (1994). Nitrogen-metabolism in senescing leaves. *Crit. Rev. Plant Sci.* 13, 241–273. doi: 10.1080/07352689409701916

Ferrario-Mery, S., Valadier, M. H., and Foyer, C. H. (1998). Overexpression of nitrate reductase in tobacco delays drought-induced decreases in nitrate reductase activity and mRNA. *Plant Physiol.* 117, 293–302. doi: 10.1104/pp.117.1.293

Foyer, C. H., and Noctor, G. (2009). Redox regulation in photosynthetic organisms: signaling, acclimation, and practical implications. *Antioxid. Redox Signal.* 11, 861–905. doi: 10.1089/ars.2008.2177

Fryer, M. J. (1992). The antioxidant effects of thylakoid vitamin E (alpha-tocopherol). *Plant Cell Environ.* 15, 381–392. doi: 10.1111/j.1365-3040.1992.tb00988.x

Fuhrer, J., Beniston, M., Fischlin, A., Frei, C., Goyette, S., Jasper, K., et al. (2006). Climate risks and their impact on agriculture and forests in Switzerland. *Clim. Change* 79, 79–102. doi: 10.1007/s10584-006-9106-6

Gallé, A., and Feller, U. (2007). Changes of photosynthetic traits in beech saplings (*Fagus sylvatica*) under severe drought stress and during recovery. *Physiol. Plant.* 131, 412–421. doi: 10.1111/j.1399-3054.2007.00972.x

Gallé, A., Haldimann, P., and Feller, U. (2007). Photosynthetic performance and water relations in young pubescent oak (*Quercus pubescens*) trees during drought stress and recovery. *New Phytol.* 174, 799–810. doi: 10.1111/j.1469-8137.2007.02047.x

Gan, Y. T., Warkentin, T. D., Bing, D. J., Stevenson, F. C., and McDonald, C. L. (2010). Chickpea water use efficiency in relation to cropping system, cultivar, soil nitrogen and rhizobial inoculation in semiarid environments. *Agric. Water Manag.* 97, 1375–1381. doi: 10.1016/j.agwat.2010.04.003

Gilgen, A. K., and Buchmann, N. (2009). Response of temperate grasslands at different altitudes to simulated summer drought differed but scaled with annual precipitation. *Biogeosciences* 6, 2525–2539. doi: 10.5194/bg-6-2525-2009

Gilgen, A. K., and Feller, U. (2013). Drought stress alters solute allocation in broadleaf dock (*Rumex obtusifolius*). *Weed Sci.* 61, 104–108. doi: 10.1614/WS-D-12-00074

Gilgen, A. K., and Feller, U. (2014). Effects of drought and subsequent rewatering on Rumex obtusifolius leaves of different ages: reversible and irreversible damages. *J. Plant Interact.* 9, 75–81. doi: 10.1080/17429145.2013.765043

Gilgen, A. K., Signarbieux, C., Feller, U., and Buchmann, N. (2010). Competitive advantage of *Rumex obtusifolius* L. Might increase in intensively managed temperate grasslands under drier climate. *Agric. Ecosyst. Environ.* 135, 15–23. doi: 10.1016/j.agee.2009.08.004

Gill, S. S., and Tuteja, N. (2010). Reactive oxygen species and antioxidant machinery in abiotic stress tolerance in crop plants. *Plant Physiol. Biochem.* 48, 909–930. doi: 10.1016/j.plaphy.2010.08.016

Gil-Quintana, E., Larrainzar, E., Arrese-Igor, C., and Gonzalez, E. M. (2013). Is N-feedback involved in the inhibition of nitrogen fixation in drought-stressed *Medicago truncatula*? *J. Exp. Bot.* 64, 281–292. doi: 10.1093/jxb/ers334

Gornall, J., Betts, R., Burke, E., Clark, R., Camp, J., Willett, K., et al. (2010). Implications of climate change for agricultural productivity in the early twenty-first century. *Philos. Trans. R. Soc. B Biol. Sci.* 365, 2973–2989. doi: 10.1098/rstb.2010.0158

Gottesman, S., Wickner, S., and Maurizi, M. R. (1997). Protein quality control: triage by chaperones and proteases. *Genes Dev.* 11, 815–823. doi: 10.1101/gad.11.7.815

Gruszka Vendruscolo, E. C., Schuster, I., Pileggi, M., Scapim, C. A., Correa Molinari, H. B., Marur, C. J., et al. (2007). Stress-induced synthesis of proline confers tolerance to water deficit in transgenic wheat. *J. Plant Physiol.* 164, 1367–1376. doi: 10.1016/j.jplph.2007.05.001

Guan, X. Q., Zhao, S. J., Li, D. Q., and Shu, H. R. (2004). Photoprotective function of photorespiration in several grapevine cultivars under drought stress. *Photosynthetica* 42, 31–36. doi: 10.1023/B:PHOT.0000040566.55149.52

Hafsi, M., Mechmeche, W., Bouamama, L., Djekoune, A., Zaharieva, M., and Monneveux, P. (2000). Flag leaf senescence, as evaluated by numerical image analysis, and its relationship with yield under drought in durum wheat. *J. Agron. Crop Sci.* 185, 275–280. doi: 10.1046/j.1439-037x.2000.00436.x

Haldimann, P., and Feller, U. (2005). Growth at moderately elevated temperature alters the physiological response of the photosynthetic apparatus to heat stress in pea (*Pisum sativum* L.) leaves. *Plant Cell Environ.* 28, 302–317. doi: 10.1111/j.1365-3040.2005.01289.x

Haldimann, P., Gallé, A., and Feller, U. (2008). Impact of an exceptionally hot dry summer on photosynthetic traits in oak (*Quercus pubescens*) leaves. *Tree Physiol.* 28, 785–795. doi: 10.1093/treephys/28.5.785

Hansen, H., and Dorffling, K. (1999). Changes of free and conjugated abscisic acid and phaseic acid in xylem sap of drought-stressed sunflower plants. *J. Exp. Bot.* 50, 1599–1605. doi: 10.1093/jxb/50.339.1599

Hashiguchi, A., Ahsan, N., and Komatsu, S. (2010). Proteomics application of crops in the context of climatic changes. *Food Res. Int.* 43, 1803–1813. doi: 10.1016/j.foodres.2009.07.033

Haupt-Herting, S., Klug, K., and Fock, H. P. (2001). A new approach to measure gross CO_2 fluxes in leaves. Gross CO_2 assimilation, photorespiration, and mitochondrial respiration in the light in tomato under drought stress. *Plant Physiol.* 126, 388–396. doi: 10.1104/pp.126.1.388

Hmida-Sayari, A., Gargouri-Bouzid, R., Bidani, A., Jaoua, L., Savoure, A., and Jaoua, S. (2005). Overexpression of Delta(1)-pyrroline-5-carboxylate synthetase increases proline production and confers salt tolerance in transgenic potato plants. *Plant Sci.* 169, 746–752. doi: 10.1016/j.plantsci.2005.05.025

Hong, Z. L., Lakkineni, K., Zhang, Z. M., and Verma, D. P. S. (2000). Removal of feedback inhibition of Delta(1)-pyrroline-5-carboxylate synthetase results in increased proline accumulation and protection of plants from osmotic stress. *Plant Physiol.* 122, 1129–1136. doi: 10.1104/pp.122.4.1129

Hörtensteiner, S. (2006). Chlorophyll degradation during senescence. *Annu. Rev. Plant Biol.* 57, 55–77. doi: 10.1146/annurev.arplant.57.032905.105212

Hörtensteiner, S., and Feller, U. (2002). Nitrogen metabolism and remobilization during senescence. *J. Exp. Bot.* 53, 927–937. doi: 10.1093/jexbot/53.370.927

Hur, J., Jung, K. H., Lee, C. H., and An, G. H. (2004). Stress-inducible OsP5CS2 gene is essential for salt and cold tolerance in rice. *Plant Sci.* 167, 417–426. doi: 10.1016/j.plantsci.2004.04.009

Ingram, J. S. I., Gregory, P. J., and Izac, A. M. (2008). The role of agronomic research in climate change and food security policy. *Agric. Ecosyst. Environ.* 126, 4–12. doi: 10.1016/j.agee.2008.01.009

IPCC. (2012). "Summary for policymakers," in *Managing the Risks of Extreme Events and Disasters to Advance Climate Change Adaptation. A Special Report of Working Groups I and II of the Intergovernmental Panel on Climate Change*, eds C. B. Field, V. Barros, T. F. Stocker, D. Qin, D. J. Dokken, K. L. Ebi, et al. (Cambridge, MA: Cambridge University Press), 1–19.

Ismail, A. M., Hall, A. E., and Bray, E. A. (1994). Drought and pot size effects on transpiration efficiency and carbon-isotope discrimination of cowpea accessions and hybrids. *Aust. J. Plant Physiol.* 21, 23–35. doi: 10.1071/PP9940023

Jentsch, A., Kreyling, J., Elmer, M., Gellesch, E., Glaser, B., Grant, K., et al. (2011). Climate extremes initiate ecosystem-regulating functions while maintaining productivity. *J. Ecol.* 99, 689–702. doi: 10.1111/j.1365-2745.2011.01817.x

Jogaiah, S., Govind, S. R., and Tran, L. S. P. (2013). Systems biology-based approaches toward understanding drought tolerance in food crops. *Crit. Rev. Biotechnol.* 33, 23–39. doi: 10.3109/07388551.2012.659174

Kaldenhoff, R., Kai, L., and Uehlein, N. (2014). Aquaporins and membrane diffusion of CO_2 in living organisms. *Biochim. Biophys. Acta* 1840, 1592–1595. doi: 10.1016/j.bbagen.2013.09.037

Katsuhara, M., and Hanba, Y. T. (2008). Barley plasma membrane intrinsic proteins (pip aquaporins) as water and CO_2 transporters. *Pflugers Arch.* 456, 687–691. doi: 10.1007/s00424-007-0434-9

Kaufmann, I., Schulze-Till, T., Schneider, H. U., Zimmermann, U., Jakob, P., and Wegner, L. H. (2009). Functional repair of embolized vessels in maize roots after temporal drought stress, as demonstrated by magnetic resonance imaging. *New Phytol.* 184, 245–256. doi: 10.1111/j.1469-8137.2009.02919.x

Kim, K., and Portis, A. R. (2005). Temperature dependence of photosynthesis in *Arabidopsis* plants with modifications in Rubisco activase and membrane fluidity. *Plant Cell Physiol.* 46, 522–530. doi: 10.1093/pcp/pci052

Kim, K. M., and Portis, A. R. (2006). Kinetic analysis of the slow inactivation of Rubisco during catalysis: effects of temperature, O_2 and Mg^{++}. *Photosynth. Res.* 87, 195–204. doi: 10.1007/s11120-005-8386-4

Kishor, P. B. K., Sangam, S., Amrutha, R. N., Laxmi, P. S., Naidu, K. R., Rao, K., et al. (2005). Regulation of proline biosynthesis, degradation, uptake and transport in higher plants: its implications in plant growth and abiotic stress tolerance. *Curr. Sci.* 88, 424–438. Available online at: http://www.iisc.ernet.in/currsci/feb102005/424.pdf

Klein, T., Holzkamper, A., Calanca, P., Seppelt, R., and Fuhrer, J. (2013). Adapting agricultural land management to climate change: a regional multi-objective optimization approach. *Landsc. Ecol.* 28, 2029–2047. doi: 10.1007/s10980-013-9939-0

Kolb, K. J., and Davis, S. D. (1994). Drought tolerance and xylem embolism in cooccurring species of coastal sage and chaparral. *Ecology* 75, 648–659. doi: 10.2307/1941723

Kotak, S., Larkindale, J., Lee, U., von Koskull-Doring, P., Vierling, E., and Scharf, K. D. (2007). Complexity of the heat stress response in plants. *Curr. Opin. Plant Biol.* 10, 310–316. doi: 10.1016/j.pbi.2007.04.011

Kovacs, D., Kalmar, E., Torok, Z., and Tompa, P. (2008). Chaperone activity of ERD10 and ERD14, two disordered stress-related plant proteins. *Plant Physiol.* 147, 381–390. doi: 10.1104/pp.108.118208

Krcek, M., Slamka, P., Olsovska, K., Brestic, M., and Bencikova, M. (2008). Reduction of drought stress effect in spring barley (*Hordeum vulgare* L.) by nitrogen fertilization. *Plant Soil Environ.* 54, 7–13. Available online at: http://www.agriculturejournals.cz/publicFiles/00567.pdf

Kumar, A., Li, C. S., and Portis, A. R. (2009). *Arabidopsis thaliana* expressing a thermostable chimeric Rubisco activase exhibits enhanced growth and higher rates of photosynthesis at moderately high temperatures. *Photosynth. Res.* 100, 143–153. doi: 10.1007/s11120-009-9438-y

Kurek, I., Chang, T. K., Bertain, S. M., Madrigal, A., Liu, L., Lassner, M. W., et al. (2007). Enhanced thermostability of *Arabidopsis* Rubisco activase improves photosynthesis and growth rates under moderate heat stress. *Plant Cell* 19, 3230–3241. doi: 10.1105/tpc.107.054171

Larkindale, J., and Huang, B. (2004). Thermotolerance and antioxidant systems in *Agrostis stolonifera*: involvement of salicylic acid, abscisic acid, calcium, hydrogen peroxide, and ethylene. *J. Plant Physiol.* 161, 405–413. doi: 10.1078/0176-1617-01239

Larrainzar, E., Wienkoop, S., Scherling, C., Kempa, S., Ladrera, R., Arrese-Igor, C., et al. (2009). Carbon metabolism and bacteroid functioning are involved in the regulation of nitrogen fixation in *Medicago truncatula* under drought and recovery. *Mol. Plant Microbe Interact.* 22, 1565–1576. doi: 10.1094/MPMI-22-12-1565

Lee, S., Seo, P. J., Lee, H. J., and Park, C. M. (2012). A NAC transcription factor NTL4 promotes reactive oxygen species production during drought-induced leaf senescence in *Arabidopsis*. *Plant J.* 70, 831–844. doi: 10.1111/j.1365-313X.2012.04932.x

Lefi, E., Gulias, J., Cifre, J., Ben Younes, M., and Medrano, H. (2004). Drought effects on the dynamics of leaf production and senescence in field-grown *Medicago arborea* and *Medicago citrina*. *Ann. Appl. Biol.* 144, 169–176. doi: 10.1111/j.1744-7348.2004.tb00330.x

Lenssen, A. W., Johnson, G. D., and Carlson, G. R. (2007). Cropping sequence and tillage system influences annual crop production and water use in semi-arid Montana, USA. *Field Crops Res.* 100, 32–43. doi: 10.1016/j.fcr.2006.05.004

Li, G. W., Santoni, V., and Maurel, C. (2014). Plant aquaporins: roles in plant physiology. *Biochim. Biophys. Acta* 1840, 1574–1582. doi: 10.1016/j.bbagen.2013.11.004

Locato, V., de Pinto, M. C., and De Gara, L. (2009). Different involvement of the mitochondrial, plastidial and cytosolic ascorbate-glutathione redox enzymes in heat shock responses. *Physiol. Plant.* 135, 296–306. doi: 10.1111/j.1399-3054.2008.01195.x

Locato, V., Gadaleta, C., De Gara, L., and De Pinto, M. C. (2008). Production of reactive species and modulation of antioxidant network in response to heat

shock: a critical balance for cell fate. *Plant Cell Environ.* 31, 1606–1619. doi: 10.1111/j.1365-3040.2008.01867.x

Loreto, F., Mannozzi, M., Maris, C., Nascetti, P., Ferranti, F., and Pasqualini, S. (2001). Ozone quenching properties of isoprene and its antioxidant role in leaves. *Plant Physiol.* 126, 993–1000. doi: 10.1104/pp.126.3.993

Lovisolo, C., Perrone, I., Hartung, W., and Schubert, A. (2008). An abscisic acid-related reduced transpiration promotes gradual embolism repair when grapevines are rehydrated after drought. *New Phytol.* 180, 642–651. doi: 10.1111/j.1469-8137.2008.02592.x

Lovisolo, C., Secchi, F., Nardini, A., Salleo, S., Buffa, R., and Schubert, A. (2007). Expression of pip1 and pip2 aquaporins is enhanced in olive dwarf genotypes and is related to root and leaf hydraulic conductance. *Physiol. Plant.* 130, 543–551. doi: 10.1111/j.1399-3054.2007.00902.x

Luquet, D., Clement-Vidal, A., Fabre, D., This, D., Sonderegger, N., and Dingkuhn, M. (2008). Orchestration of transpiration, growth and carbohydrate dynamics in rice during a dry-down cycle. *Funct. Plant Biol.* 35, 689–704. doi: 10.1071/FP08027

Mahdid, M., Kameli, A., Ehlert, C., and Simonneau, T. (2011). Rapid changes in leaf elongation, aba and water status during the recovery phase following application of water stress in two durum wheat varieties differing in drought tolerance. *Plant Physiol. Biochem.* 49, 1077–1083. doi: 10.1016/j.plaphy.2011.08.002

Maurel, C., Javot, H., Lauvergeat, V., Gerbeau, P., Tournaire, C., Santoni, V., et al. (2002). Molecular physiology of aquaporins in plants. *Int. Rev. Cytol.* 215, 105–148. doi: 10.1016/S0074-7696(02)15007-8

Maxwell, K., and Johnson, G. N. (2000). Chlorophyll fluorescence—a practical guide. *J. Exp. Bot.* 51, 659–668. doi: 10.1093/jexbot/51.345.659

Medlyn, B. E., Barton, C. V. M., Broadmeadow, M. S. J., Ceulemans, R., De Angelis, P., Forstreuter, M., et al. (2001). Stomatal conductance of forest species after long-term exposure to elevated CO_2 concentration: a synthesis. *New Phytol.* 149, 247–264. doi: 10.1046/j.1469-8137.2001.00028.x

Merewitz, E. B., Gianfagna, T., and Huang, B. R. (2010). Effects of SAG12-ipt and HSP18.2-ipt expression on cytokinin production, root growth, and leaf senescence in creeping bentgrass exposed to drought stress. *J. Am. Soc. Hort. Sci.* 135, 230–239. Available online at: http://journal.ashspublications.org/content/135/3/230.full.pdf

Messmer, R., Fracheboud, Y., Banziger, M., Stamp, P., and Ribaut, J. M. (2011). Drought stress and tropical maize: QTLs for leaf greenness, plant senescence, and root capacitance. *Field Crops Res.* 124, 93–103. doi: 10.1016/j.fcr.2011.06.010

Miller, G., Suzuki, N., Ciftci-Yilmaz, S., and Mittler, R. (2010). Reactive oxygen species homeostasis and signalling during drought and salinity stresses. *Plant Cell Environ.* 33, 453–467. doi: 10.1111/j.1365-3040.2009.02041.x

Miller, G., Suzuki, N., Rizhsky, L., Hegie, A., Koussevitzky, S., and Mittler, R. (2007). Double mutants deficient in cytosolic and thylakoid ascorbate peroxidase reveal a complex mode of interaction between reactive oxygen species, plant development, and response to abiotic stresses. *Plant Physiol.* 144, 1777–1785. doi: 10.1104/pp.107.101436

Mir, R. R., Zaman-Allah, M., Sreenivasulu, N., Trethowan, R., and Varshney, R. K. (2012). Integrated genomics, physiology and breeding approaches for improving drought tolerance in crops. *Theor. Appl. Genet.* 125, 625–645. doi: 10.1007/s00122-012-1904-9

Mirzaei, M., Pascovici, D., Atwell, B. J., and Haynes, P. A. (2012). Differential regulation of aquaporins, small GTPases and V-ATPases proteins in rice leaves subjected to drought stress and recovery. *Proteomics* 12, 864–877. doi: 10.1002/pmic.201100389

Mittal, N., Mishra, A., Singh, R., and Kumar, P. (2014). Assessing future changes in seasonal climatic extremes in the Ganges river basin using an ensemble of regional climate models. *Clim. Change* 123, 273–286. doi: 10.1007/s10584-014-1056-9

Mittler, R. (2002). Oxidative stress, antioxidants and stress tolerance. *Trends Plant Sci.* 7, 405–410. doi: 10.1016/S1360-1385(02)02312-9

Mittler, R., Herr, E. H., Orvar, B. L., van Camp, W., Willekens, H., Inze, D., et al. (1999). Transgenic tobacco plants with reduced capability to detoxify reactive oxygen intermediates are hyperresponsive to pathogen infection. *Proc. Natl. Acad. Sci. U.S.A.* 96, 14165–14170. doi: 10.1073/pnas.96.24.14165

Miyashita, K., Tanakamaru, S., Maitani, T., and Kimura, K. (2005). Recovery responses of photosynthesis, transpiration, and stomatal conductance in kidney bean following drought stress. *Environ. Exp. Bot.* 53, 205–214. doi: 10.1016/j.envexpbot.2004.03.015

Mori, M., and Inagaki, M. N. (2012). Root development and water-uptake under water deficit stress in drought-adaptive wheat genotypes. *Cereal Res. Commun.* 40, 44–52. doi: 10.1556/CRC.40.2012.1.6

Munne-Bosch, S., and Alegre, L. (2004). Die and let live: leaf senescence contributes to plant survival under drought stress. *Funct. Plant Biol.* 31, 203–216. doi: 10.1071/FP03236

Munne-Bosch, S., Jubany-Mari, T., and Alegre, L. (2001). Drought-induced senescence is characterized by a loss of antioxidant defenses in chloroplasts. *Plant Cell Environ.* 24, 1319–1327. doi: 10.1046/j.1365-3040.2001.00794.x

Nair, P. K. R. (2014). Grand challenges in agroecology and land use systems. *Front. Environ. Sci.* 2:1. doi: 10.3389/fenvs.2014.00001

Neumann, J. (2008). Regional climate change and variability: impacts and responses. *J. Reg. Sci.* 48, 460–462. doi: 10.1111/j.1467-9787.2008.00559_3.x

Noctor, G., Veljovic-Jovanovic, S., Driscoll, S., Novitskaya, L., and Foyer, C. H. (2002). Drought and oxidative load in the leaves of c-3 plants: a predominant role for photorespiration? *Ann. Bot.* 89, 841–850. doi: 10.1093/aob/mcf096

Oberschall, A., Deak, M., Torok, K., Sass, L., Vass, I., Kovacs, I., et al. (2000). A novel aldose/aldehyde reductase protects transgenic plants against lipid peroxidation under chemical and drought stresses. *Plant J.* 24, 437–446. doi: 10.1046/j.1365-313x.2000.00885.x

Osmond, C. B., and Grace, S. C. (1995). Perspectives on photoinhibition and photorespiration in the field - quintessential inefficiencies of the light and dark reactions of photosynthesis. *J. Exp. Bot.* 46, 1351–1362. doi: 10.1093/jxb/46.special_issue.1351

Oukarroum, A., Schansker, G., and Strasser, R. J. (2009). Drought stress effects on photosystem I content and photosystem II thermotolerance analyzed using Chl a fluorescence kinetics in barley varieties differing in their drought tolerance. *Physiol. Plant.* 137, 188–199. doi: 10.1111/j.1399-3054.2009.01273.x

Parida, A. K., Dagaonkar, V. S., Phalak, M. S., and Aurangabadkar, L. P. (2008). Differential responses of the enzymes involved in proline biosynthesis and degradation in drought tolerant and sensitive cotton genotypes during drought stress and recovery. *Acta Physiol. Plant.* 30, 619–627. doi: 10.1007/s11738-008-0157-3

Parry, M. A. J., Keys, A. J., Madgwick, P. J., Carmo-Silva, A. E., and Andralojc, P. J. (2008). Rubisco regulation: a role for inhibitors. *J. Exp. Bot.* 59, 1569–1580. doi: 10.1093/jxb/ern084

Parry, M. A. J., Reynolds, M., Salvucci, M. E., Raines, C., Andralojc, P. J., Zhu, X. G., et al. (2011). Raising yield potential of wheat. II. Increasing photosynthetic capacity and efficiency. *J. Exp. Bot.* 62, 453–467. doi: 10.1093/jxb/erq304

Pastore, D., Trono, D., Laus, M. N., Di Fonzo, N., and Flagella, Z. (2007). Possible plant mitochondria involvement in cell adaptation to drought stress—a case study: durum wheat mitochondria. *J. Exp. Bot.* 58, 195–210. doi: 10.1093/jxb/erl273

Penuelas, J., and Llusia, J. (2002). Linking photorespiration, monoterpenes and thermotolerance in quercus. *New Phytol.* 155, 227–237. doi: 10.1046/j.1469-8137.2002.00457.x

Percy, K. E., and Baker, E. A. (1987). Effects of simulated acid rain on production, morphology and composition of epicuticular wax and on cuticular membrane development. *New Phytol.* 107, 577–589 doi: 10.1111/j.1469-8137.1987.tb02928.x

Pérez-Arellano, I., Carmona-Alvarez, F., Martinez, A. I., Rodriguez-Diaz, J., and Cervera, J. (2010). Pyrroline-5-carboxylate synthase and proline biosynthesis: from osmotolerance to rare metabolic disease. *Protein Sci.* 19, 372–382. doi: 10.1002/pro.340

Perrone, I., Gambino, G., Chitarra, W., Vitali, M., Pagliarani, C., Riccomagno, N., et al. (2012). The grapevine root-specific aquaporin VvPIP2;4N controls root hydraulic conductance and leaf gas exchange under well-watered conditions but not under water stress. *Plant Physiol.* 160, 965–977. doi: 10.1104/pp.112.203455

Portis, A. R., Li, C. S., Wang, D. F., and Salvucci, M. E. (2008). Regulation of Rubisco activase and its interaction with Rubisco. *J. Exp. Bot.* 59, 1597–1604. doi: 10.1093/jxb/erm240

Pose, D., Castanedo, I., Borsani, O., Nieto, B., Rosado, A., Taconnat, L., et al. (2009). Identification of the *Arabidopsis* dry2/sqe1-5 mutant reveals a central role for sterols in drought tolerance and regulation of reactive oxygen species. *Plant J.* 59, 63–76. doi: 10.1111/j.1365-313X.2009.03849.x

Prado, K., and Maurel, C. (2013). Regulation of leaf hydraulics: from molecular to whole plant levels. *Front. Plant Sci.* 4:255. doi: 10.3389/fpls.2013.00255

Rawlings, N. D., Tolle, D. P., and Barrett, A. J. (2004). Evolutionary families of peptidase inhibitors. *Biochem. J.* 378, 705–716. doi: 10.1042/BJ20031825

Reddy, A. R., Chaitanya, K. V., and Vivekanandan, M. (2004). Drought-induced responses of photosynthesis and antioxidant metabolism in higher plants. *J. Plant Physiol.* 161, 1189–1202. doi: 10.1016/j.jplph.2004.01.013

Reynolds-Henne, C. E., Langenegger, A., Mani, J., Schenk, N., Zumsteg, A., and Feller, U. (2010). Interactions between temperature, drought and stomatal opening in legumes. *Environ. Exp. Bot.* 68, 37–43. doi: 10.1016/j.envexpbot.2009.11.002

Rivero, R. M., Kojima, M., Gepstein, A., Sakakibara, H., Mittler, R., Gepstein, S., et al. (2007). Delayed leaf senescence induces extreme drought tolerance in a flowering plant. *P. Natl. Acad. Sci. U.S.A.* 104, 19631–19636. doi: 10.1073/pnas.0709453104

Rivero, R. M., Shulaev, V., and Blumwald, E. (2009). Cytokinin-dependent photorespiration and the protection of photosynthesis during water deficit. *Plant Physiol.* 150, 1530–1540. doi: 10.1104/pp.109.139378

Rorat, T. (2006). Plant dehydrins - tissue location, structure and function. *Cell. Mol. Biol. Lett.* 11, 536–556. doi: 10.2478/s11658-006-0044-0

Rosati, A., Metcalf, S., Buchner, R., Fulton, A., and Lampinen, B. (2006). Tree water status and gas exchange in walnut under drought, high temperature and vapour pressure deficit. *J. Horticult. Sci. Biotechnol.* 81, 415–420. Available online at: http://ucanr.edu/sites/LampinenLab/files/80447.pdf

Sakamoto, W. (2006). Protein degradation machineries in plastids. *Annu. Rev. Plant Biol.* 57, 599–621. doi: 10.1146/annurev.arplant.57.032905.105401

Salvucci, M. E., and Crafts-Brandner, S. J. (2004). Relationship between the heat tolerance of photosynthesis and the thermal stability of Rubisco activase in plants from contrasting thermal environments. *Plant Physiol.* 134, 1460–1470. doi: 10.1104/pp.103.038323

Salvucci, M. E., DeRidder, B. P., and Portis, A. R. (2006). Effect of activase level and isoform on the thermotolerance of photosynthesis in *Arabidopsis*. *J. Exp. Bot.* 57, 3793–3799. doi: 10.1093/jxb/erl140

Salvucci, M. E., Osteryoung, K. W., Crafts-Brandner, S. J., and Vierling, E. (2001). Exceptional sensitivity of Rubisco activase to thermal denaturation *in vitro* and *in vivo*. *Plant Physiol.* 127, 1053–1064. doi: 10.1104/pp.010357

Salvucci, M. E., van de Loo, F. J., and Stecher, D. (2003). Two isoforms of Rubisco activase in cotton, the products of separate genes not alternative splicing. *Planta* 216, 736–744. doi: 10.1007/s00425-002-0923-1

Sanchez, F. J., Manzanares, M., de Andres, E. F., Tenorio, J. L., and Ayerbe, L. (2001). Residual transpiration rate, epicuticular wax load and leaf colour of pea plants in drought conditions. Influence on harvest index and canopy temperature. *Eur. J. Agron.* 15, 57–70. doi: 10.1016/S1161-0301(01)00094-6

Schar, C., Vidale, P. L., Luthi, D., Frei, C., Haberli, C., Liniger, M. A., et al. (2004). The role of increasing temperature variability in European summer heatwaves. *Nature* 427, 332–336. doi: 10.1038/nature02300

Scharf, K. D., Berberich, T., Ebersberger, I., and Nover, L. (2012). The plant heat stress transcription factor (Hsf) family: structure, function and evolution. *BBA-Gene Regul. Mech.* 1819, 104–119. doi: 10.1016/j.bbagrm.2011.10.002

Schenk, D., and Feller, U. (1990). Rubidium export from individual leaves of maturing wheat. *J. Plant Physiol.* 137, 175–179. doi: 10.1016/S0176-1617(11)80077-5

Secchi, F., and Zwieniecki, M. A. (2013). The physiological response of *Populus tremula x alba* leaves to the down-regulation of PIP1 aquaporin gene expression under no water stress. *Front. Plant Sci.* 4:507. doi: 10.3389/fpls.2013.00507

Secchi, F., and Zwieniecki, M. A. (2014). Down-regulation of plasma intrinsic protein1 aquaporin in poplar trees is detrimental to recovery from embolism. *Plant Physiol.* 164, 1789–1799. doi: 10.1104/pp.114.237511

Selote, D. S., Bharti, S., and Khanna-Chopra, R. (2004). Drought acclimation reduces O_2. accumulation and lipid peroxidation in wheat seedlings. *Biochem. Biophys. Res. Commun.* 314, 724–729. doi: 10.1016/j.bbrc.2003.12.157

Seo, P. J., Lee, S. B., Suh, M. C., Park, M. J., Go, Y. S., and Park, C. M. (2011). The MYB96 transcription factor regulates cuticular wax biosynthesis under drought conditions in *Arabidopsis*. *Plant Cell* 23, 1138–1152. doi: 10.1105/tpc.111.083485

Sevanto, S. (2014). Phloem transport and drought. *J. Exp. Bot.* 65, 1751–1759. doi: 10.1093/jxb/ert467

Sharkey, T. D. (2005). Effects of moderate heat stress on photosynthesis: importance of thylakoid reactions, Rubisco deactivation, reactive oxygen species, and thermotolerance provided by isoprene. *Plant Cell Environ.* 28, 269–277. doi: 10.1111/j.1365-3040.2005.01324.x

Signarbieux, C., and Feller, U. (2011). Non-stomatal limitations of photosynthesis in grassland species under artificial drought in the field. *Environ. Exp. Bot.* 71, 192–197. doi: 10.1016/j.envexpbot.2010.12.003

Signarbieux, C., and Feller, U. (2012). Effects of an extended drought period on physiological properties of grassland species in the field. *J. Plant Res.* 125, 251–261. doi: 10.1007/s10265-011-0427-9

Simova-Stoilova, L., Demirevska, K., Petrova, T., Tsenov, N., and Feller, U. (2009). Antioxidative protection and proteolytic activity in tolerant and sensitive wheat (*Triticum aestivum* L.) varieties subjected to long-term field drought. *Plant Growth Regul.* 58, 107–117. doi: 10.1007/s10725-008-9356-6

Simova-Stoilova, L., Vaseva, I., Grigorova, B., Demirevska, K., and Feller, U. (2010). Proteolytic activity and cysteine protease expression in wheat leaves under severe soil drought and recovery. *Plant Physiol. Biochem.* 48, 200–206. doi: 10.1016/j.plaphy.2009.11.003

Smith, P., and Gregory, P. J. (2013). Climate change and sustainable food production. *P. Nutr. Soc.* 72, 21–28. doi: 10.1017/S0029665112002832

Snider, J. L., Oosterhuis, D. M., and Kawakami, E. M. (2010). Genotypic differences in thermotolerance are dependent upon prestress capacity for antioxidant protection of the photosynthetic apparatus in *Gossypium hirsutum*. *Physiol. Plant.* 138, 268–277. doi: 10.1111/j.1399-3054.2009.01325.x

Sturny, W. G., Chervet, A., Maurer-Troxler, C., Ramseier, L., Muller, M., Schafflutzel, R., et al. (2007). Comparison of no-tillage and conventional plough tillage—a synthesis. *Agrarforschung* 14, 350–357. Available online at: http://www.agrarforschungschweiz.ch/archiv_11en.php?id_artikel=1297

Su, J., and Wu, R. (2004). Stress-inducible synthesis of proline in transgenic rice confers faster growth under stress conditions than that with constitutive synthesis. *Plant Sci.* 166, 941–948. doi: 10.1016/j.plantsci.2003.12.004

Subramanian, K. S., and Charest, C. (1998). Arbuscular mycorrhizae and nitrogen assimilation in maize after drought and recovery. *Physiol. Plant.* 102, 285–296. doi: 10.1034/j.1399-3054.1998.1020217.x

Suga, S., Komatsu, S., and Maeshima, M. (2002). Aquaporin isoforms responsive to salt and water stresses and phytohormones in radish seedlings. *Plant Cell Physiol.* 43, 1229–1237. doi: 10.1093/pcp/pcf148

Sunkar, R., Chinnusamy, V., Zhu, J. H., and Zhu, J. K. (2007). Small RNAs as big players in plant abiotic stress responses and nutrient deprivation. *Trends Plant Sci.* 12, 301–309. doi: 10.1016/j.tplants.2007.05.001

Suzuki, N., and Mittler, R. (2006). Reactive oxygen species and temperature stresses: a delicate balance between signaling and destruction. *Physiol. Plant.* 126, 45–51. doi: 10.1111/j.0031-9317.2005.00582.x

Szabala, B. M., Fudali, S., and Rorat, T. (2014). Accumulation of acidic SK3 dehydrins in phloem cells of cold- and drought-stressed plants of the *Solanaceae*. *Planta* 239, 847–863. doi: 10.1007/s00425-013-2018-6

Székely, G., Abraham, E., Cselo, A., Rigo, G., Zsigmond, L., Csiszar, J., et al. (2008). Duplicated P5CS genes of *Arabidopsis* play distinct roles in stress regulation and developmental control of proline biosynthesis. *Plant J.* 53, 11–28. doi: 10.1111/j.1365-313X.2007.03318.x

Thoenen, M., Herrmann, B., and Feller, U. (2007). Senescence in wheat leaves: Is a cysteine endopeptidase involved in the degradation of the large subunit of Rubisco? *Acta Physiol. Plant.* 29, 339–350. doi: 10.1007/s11738-007-0043-4

Tompa, P. (2009). *Structure and Function of Intrinsically Disordered Proteins*. Boca Raton, FL: Chapman and Hall/CRC Press.

Turchetto-Zolet, A. C., Margis-Pinheiro, M., and Margis, R. (2009). The evolution of pyrroline-5-carboxylate synthase in plants: a key enzyme in proline synthesis. *Mol. Genet. Genomics* 281, 87–97. doi: 10.1007/s00438-008-0396-4

Turkan, I., Bor, M., Ozdemir, F., and Koca, H. (2005). Differential responses of lipid peroxidation and antioxidants in the leaves of drought-tolerant *P. acutifolius* Gray and drought-sensitive *P. vulgaris* L. subjected to polyethylene glycol mediated water stress. *Plant Sci.* 168, 223–231. doi: 10.1016/j.plantsci.2004.07.032

Uehlein, N., Lovisolo, C., Siefritz, F., and Kaldenhoff, R. (2003). The tobacco aquaporin NtAQP1 is a membrane CO_2 pore with physiological functions. *Nature* 425, 734–737. doi: 10.1038/nature02027

Valladares, F., and Pearcy, R. W. (1997). Interactions between water stress, sun-shade acclimation, heat tolerance and photoinhibition in the sclerophyll *Heteromeles arbutifolia*. *Plant Cell Environ.* 20, 25–36. doi: 10.1046/j.1365-3040.1997.d01-8.x

Van Bel, A. J. E. (2003). The phloem, a miracle of ingenuity. *Plant Cell Environ.* 26, 125–149. doi: 10.1046/j.1365-3040.2003.00963.x

Van Breusegem, F., Bailey-Serres, J., and Mittler, R. (2008). Unraveling the tapestry of networks involving reactive oxygen species in plants. *Plant Physiol.* 147, 978–984. doi: 10.1104/pp.108.122325

Vaseva, I. I., Anders, I., and Feller, U. (2014). Identification and expression of different dehydrin subclasses involved in the drought response of *Trifolium repens*. *J. Plant Physiol.* 171, 213–224. doi: 10.1016/j.jplph.2013.07.013

Vaseva, I. I., and Feller, U. (2013). Natural antisense transcripts of *Trifolium repens* dehydrins. *Plant Signal. Behav.* 8:e27674. doi: 10.4161/psb.27674

Vaseva, I., Sabotic, J., Sustar-Vozlic, J., Meglic, V., Kidric, M., Demirevska, K., et al. (2011). "The response of plants to drought stress—the role of dehydrins, chaperones, proteases and protease inhibitors in maintaining cellular protein function," in *Droughts: New Research*, eds D. F. Neves and J. D. Sanz (New York, NY: Nova Science Publishers), 1–45.

Vassileva, V., Signarbieux, C., Anders, I., and Feller, U. (2011). Genotypic variation in drought stress response and subsequent recovery of wheat (*Triticum aestivum* L.). *J. Plant Res.* 124, 147–154. doi: 10.1007/s10265-010-0340-7

Velikova, V., and Loreto, F. (2005). On the relationship between isoprene emission and thermotolerance in phragmites australis leaves exposed to high temperatures and during the recovery from a heat stress. *Plant Cell Environ.* 28, 318–327. doi: 10.1111/j.1365-3040.2004.01314.x

Vera-Estrella, R., Barkla, B. J., Bohnert, H. J., and Pantoja, O. (2004). Novel regulation of aquaporins during osmotic stress. *Plant Physiol.* 135, 2318–2329. doi: 10.1104/pp.104.044891

Verbruggen, N., and Hermans, C. (2008). Proline accumulation in plants: a review. *Amino Acids* 35, 753–759. doi: 10.1007/s00726-008-0061-6

Verma, V., Foulkes, M. J., Worland, A. J., Sylvester-Bradley, R., Caligari, P. D. S., and Snape, J. W. (2004). Mapping quantitative trait loci for flag leaf senescence as a yield determinant in winter wheat under optimal and drought-stressed environments. *Euphytica* 135, 255–263. doi: 10.1023/B:EUPH.0000013255.31618.14

Vickers, C. E., Possell, M., Cojocariu, C. I., Velikova, V. B., Laothawornkitkul, J., Ryan, A., et al. (2009). Isoprene synthesis protects transgenic tobacco plants from oxidative stress. *Plant Cell Environ.* 32, 520–531. doi: 10.1111/j.1365-3040.2009.01946.x

Vierstra, R. D. (1996). Proteolysis in plants: mechanisms and functions. *Plant Mol. Biol.* 32, 275–302. doi: 10.1007/BF00039386

Volaire, F., and Lelievre, F. (2001). Drought survival in *Dactylis glomerata* and *Festuca arundinacea* under similar rooting conditions in tubes. *Plant Soil* 229, 225–234. doi: 10.1023/A:1004835116453

Wahid, A., Gelani, S., Ashraf, M., and Foolad, M. R. (2007). Heat tolerance in plants: an overview. *Environ. Exp. Bot.* 61, 199–223. doi: 10.1016/j.envexpbot.2007.05.011

Walter, J., Nagy, L., Hein, R., Rascher, U., Beierkuhnlein, C., Willner, E., et al. (2011). Do plants remember drought? Hints towards a drought-memory in grasses. *Environ. Exp. Bot.* 71, 34–40. doi: 10.1016/j.envexpbot.2010.10.020

Wang, W. X., Vinocur, B., Shoseyov, O., and Altman, A. (2004). Role of plant heat-shock proteins and molecular chaperones in the abiotic stress response. *Trends Plant Sci.* 9, 244–252. doi: 10.1016/j.tplants.2004.03.006

Wehner, M., Easterling, D. R., Lawrimore, J. H., Heim, R. R., Vose, R. S., and Santer, B. D. (2011). Projections of future drought in the continental United States and Mexico. *J. Hydrometeorol.* 12, 1359–1377. doi: 10.1175/2011JHM1351.1

Whitney, S. M., von Caemmerer, S., Hudson, G. S., and Andrews, T. J. (1999). Directed mutation of the Rubisco large subunit of tobacco influences photorespiration and growth. *Plant Physiol.* 121, 579–588. doi: 10.1104/pp.121.2.579

Wingler, A., Quick, W. P., Bungard, R. A., Bailey, K. J., Lea, P. J., and Leegood, R. C. (1999). The role of photorespiration during drought stress: an analysis utilizing barley mutants with reduced activities of photorespiratory enzymes. *Plant Cell Environ.* 22, 361–373. doi: 10.1046/j.1365-3040.1999.00410.x

Wishart, J., George, T. S., Brown, L. K., White, P. J., Ramsay, G., Jones, H., et al. (2014). Field phenotyping of potato to assess root and shoot characteristics associated with drought tolerance. *Plant Soil* 378, 351–363. doi: 10.1007/s11104-014-2029-5

Xu, J., Duan, X. G., Yang, J., Beeching, J. R., and Zhang, P. (2013). Enhanced reactive oxygen species scavenging by overproduction of superoxide dismutase and catalase delays postharvest physiological deterioration of cassava storage roots. *Plant Physiol.* 161, 1517–1528. doi: 10.1104/pp.112.212803

Xu, Z. Z., and Zhou, G. S. (2006). Combined effects of water stress and high temperature on photosynthesis, nitrogen metabolism and lipid peroxidation of a perennial grass *Leymus chinensis*. *Planta* 224, 1080–1090. doi: 10.1007/s00425-006-0281-5

Yamori, W., Masumoto, C., Fukayama, H., and Makino, A. (2012). Rubisco activase is a key regulator of non-steady-state photosynthesis at any leaf temperature and, to a lesser extent, of steady-state photosynthesis at high temperature. *Plant J.* 71, 871–880. doi: 10.1111/j.1365-313X.2012.05041.x

Yang, J., Ordiz, M. I., Jaworski, J. G., and Beachy, R. N. (2011). Induced accumulation of cuticular waxes enhances drought tolerance in *Arabidopsis* by changes in development of stomata. *Plant Physiol. Biochem.* 49, 1448–1455. doi: 10.1016/j.plaphy.2011.09.006

Yao, X., Chu, J., Liang, L., Geng, W., Li, J., and Hou, G. (2012). Selenium improves recovery of wheat seedlings at rewatering after drought stress. *Russ. J. Plant Physl.* 59, 701–707. doi: 10.1134/S1021443712060192

Yordanov, I., Velikova, V., and Tsonev, T. (2000). Plant responses to drought, acclimation, and stress tolerance. *Photosynthetica* 38, 171–186. doi: 10.1023/A:1007201411474

Yoshiba, Y., Kiyosue, T., Nakashima, K., Yamaguchi-Shinozaki, K., and Shinozaki, K. (1997). Regulation of levels of proline as an osmolyte in plants under water stress. *Plant Cell Physiol.* 38, 1095–1102. doi: 10.1093/oxfordjournals.pcp.a029093

Zhang, L. X., Li, S. Q., Li, S. X., and Liang, Z. S. (2012). How does nitrogen application ameliorate negative effects of long-term drought in two maize cultivars in relation to plant growth, water status, and nitrogen metabolism? *Commun. Soil Sci. Plant Anal.* 43, 1632–1646. doi: 10.1080/00103624.2012.681735

Zhu, L., Guo, J. S., Zhu, J., and Zhou, C. (2014). Enhanced expression of EsWAX1 improves drought tolerance with increased accumulation of cuticular wax and ascorbic acid in transgenic *Arabidopsis*. *Plant Physiol. Biochem.* 75, 24–35. doi: 10.1016/j.plaphy.2013.11.028

Conflict of Interest Statement: The authors declare that the research was conducted in the absence of any commercial or financial relationships that could be construed as a potential conflict of interest.

3

Soil fauna and soil functions: a jigsaw puzzle

María Jesús I. Briones *

Departamento de Ecología y Biología Animal, Facultad de Biología, Universidad de Vigo, Vigo, Spain

Edited by:
Wilfred Otten, Abertay University, UK

Reviewed by:
Victor Satler Pylro, Federal University of Viçosa, Brazil
Raymon Shange, Tuskegee University, USA

*Correspondence:
María Jesús I. Briones,
Departamento de Ecología y Biología Animal, Facultad de Biología, Universidad de Vigo, 36310 Vigo, Spain
e-mail: mbriones@uvigo.es

Terrestrial ecologists and soil modelers have traditionally portrayed the inhabitants of soil as a black box labeled as "soil fauna" or "decomposers or detritivores" assuming that they just merely recycle the deposited dead plant material. Soil is one of the most diverse habitats on Earth and contains one of the most diverse assemblages of living organisms; however, the opacity of this world has severely limited our understanding of their functional contributions to soil processes and to ecosystem resilience. Traditional taxonomy, based on morphological and anatomical aspects, is becoming replaced by rapid processing molecular techniques (e.g., with marker gene-based approaches). However, this may be impracticable in many ecological studies and consequently, the majority of the current knowledge, still contributes little to our understanding of their role in ecosystem functioning. Over the years, different workers have produced several "functional classifications" based on the body width, feeding regime, certain behavioral and reproductive aspects and ecological niches of soil organisms. Unfortunately, the information available is severely restricted to "major" groups. A better physiological and metabolic understanding of when and how a complex community of soil organisms access nutrients, alter their environment and in turn, affect soil processes, will allow a more realistic quantitative evaluation of their ecological roles in the biogeochemical cycles. Here, I review the applicability of the available approaches, highlight future research challenges and propose a dynamic conceptual framework that could improve our ability to solve this functional puzzle.

Keywords: feeding ecology, functional classifications, resilience, soil organisms, soil processes, ecosystem functioning

INTRODUCTION

According to the United Nations, we are in a "decade of biodiversity" and one of the decisions adopted during the 10th meeting of Conference of the Parties held in Nagoya in October 2010 was the implementation of the "Strategic Plan for Biodiversity 2011–2020" (http://www.cbd.int/decision/cop/?id=12268). The rationale for the new plan is that "*biological diversity underpins ecosystem functioning and the provision of ecosystem services essential for human well-being.*" One important reason for attracting interest in functional diversity is that there will be more chances for the society to support biodiversity conservation if there is an economical value associated to it.

Soils are multicomponent and multifunction systems which provide a series of "ecosystem goods" (biomass production) to humans, but also many "regulating services" (e.g., SOM decomposition, soil structure maintenance, nutrient cycling, etc.) which ensure ecosystem sustainability. All these functions are provided by the variety of organisms which live in the soil, i.e., they are functional outputs of biological processes (Kibblewhite et al., 2008). Scientific advances in soils research have increased significantly since the publication of a special issue in *Science* [Soils—The Final Frontier (Science 11 June 2004, Vol. 304, Issue 5677)] and the appearance of some key papers in *Nature* (e.g., Copley, 2000). However, whereas significant progress have been achieved regarding its physical and chemical characterization, its soil communities remain practically underexplored, with very low

percentages of described species being available (Hawksworth and Mound, 1991; Torsvik et al., 1994; Walter and Proctor, 1999).

Until 1960 the role of soil fauna in ecosystems mainly concerned earthworms (reviewed by Huhta, 2007). In the following two decades, research on the contribution of soil organisms to soil processes revealed the disappointing result that their direct effects on soil organic matter (SOM) decomposition (via litter consumption) were negligible due to their low assimilation efficiency (less than 10%; Petersen and Luxton, 1982). However, their indirect effects (through grazing on microbial communities) seemed to exceed, by several times, their metabolic contribution (MacFadyen, 1963). This is the result of soils being packed with microorganisms and thus, soil animals are believed to either rely on microorganisms as food or to use them as an "external rumen" to compensate for their poor enzymatic capabilities (Swift et al., 1979).

Consequently, SOM decomposition has been traditionally interpreted by considering climatic factors and chemical composition of the plant litter as the main drivers of this process at global scale, whereas the influence of soil biota has been relegated to local scale and for this reason, they are consistently excluded from global decomposition models (Wall et al., 2008). One interesting result from a large field scale experiment (Wall et al., 2008) was the suggestion that taxonomic richness, and not abundance or biomass (Cole et al., 2004), was the primarily driver of the observed responses (Wall et al., 2008). The main cause for this discrepancy was the organisms investigated in this study not being identified to species level, but to Order or even to higher taxonomical levels, which possibly led to a more functional diverse assemblage (Heemsbergen et al., 2004). This contrasts with the opinion of Mulder (2006) who argued that numerical abundance of each taxa is the most reliable method to describe ecosystem functioning, since it is a more "flexible" parameter in reflecting ecosystem state than average body mass. In support of this view, there are some other claims that functional importance of soil organisms does not match their numerical abundance (Anderson, 1988) and that functional complexity does not always influence soil processes (Liiri et al., 2002).

However, if we take into account that dominant species are likely to affect their habitat and resources availability and that keystone species or functional groups could determine the abundance, diversity and activities of other soil invertebrates (Lavelle, 1996), then the need for a functional classification is clear. Yet reducing the huge species diversity present in soils into a smaller number of functional groups will inevitably result in grouping "redundant" species regarding their function, with this being more likely to happen in species-rich faunal groups (Bardgett, 2002). This is also supported by the general assumption of a generalist feeding behavior being predominant amongst soil biota under field conditions (Luxton, 1972; Ponsard and Arditi, 2000; Scheu and Falca, 2000).

On the other hand, if higher diversity implies more species performing the same functions or services, perhaps the role of diversity is as a reserve of "natural insurance capital" (Folke et al., 1996; Yachi and Loreau, 1999) ready to enter in action in response to future environmental conditions/perturbations or to exotic invasions. The idea that *organisms have evolved through selection to maximize their contribution to future generations.... not to serve functions in the ecosystem* has long being coined (Andrén et al., 1995, 1999). Furthermore, some species which are "functionally redundant" for a given ecosystem process at a given time might nevertheless no longer be redundant under changed conditions; similarly, they could be initially inactive but become functionally active under a changing environment (Hodkinson and Wookey, 1999). This resembles the "stability vs. complexity theory" (McCann, 2000) which has been severely criticized because the concept "stability" is a vague term which implies that there are not redundant species, i.e., each one has a place in the ecosystem (Andrén and Balandreau, 1999) and refers to a wide range of different properties, which could change across different organization levels. According to this theory, processes that are carried out by a higher number of species are "more stable," whereas those carried out by a small number are most vulnerable to biodiversity losses (Hooper et al., 1995). In other words, increased species richness can lead to decreases in temporal variability in ecosystem properties (Pfisterer and Schmid, 2002; Pfisterer et al., 2004); in contrast, species losses in functional groups consisting of just a few species could result in habitat degradation. However, if all species differ in their contribution to soil processes, the loss of one species might not be functionally compensated and due to species interactions, species losses could also derive in unprecedented cascade effects (Freckman et al., 1997).

In view of this, it becomes clear that soil ecologists are in dire need to resolve the dilemma of which, "functional dissimilarity" (Heemsbergen et al., 2004) or "functional redundancy" (Andrén and Balandreau, 1999), is dominating in soils and which level of soil biodiversity (population, community or ecosystem level) should be estimated to underpin the biological regulation of soil processes. However, this is difficult to assess at large scales in the field due to the high percentage of species still waiting to be identified (ca. 75% of the total figure; Decaëns et al., 2006, 2008), the huge range of expertise required (Wall et al., 2008) and the unavailability of well developed techniques for accurate taxa differentiation (Freckman et al., 1997).

Then, the next questions are whether the presence of one species belonging to a particular functional group is enough to maintain a particular soil process, or if having more than one species within that functional group will improve the rate of it, or whether having several functional groups within a particular animal group (e.g., earthworms) is more beneficial for ecosystem functioning. In other words, should we be promoting diversity of functional groups or diversity within the functional groups?

Interestingly, hierarchical models, in which certain taxa may be determining the abundance and diversity of other groups (Lavelle, 1996; Wardle and Lavelle, 1996) or modifying the environment so that another organism finds its niche ("metabiosis" *sensu* Waid, 1999), has also highlighted the importance of functional classifications that not only identify taxa with significant functional roles in the ecosystem, but also "redundant taxa" that could have a significant existence by acting as indicators of soil resilience, if under the eventual action of disturbance the soil community composition becomes altered (Lavelle, 1996; Fitter et al., 2005). Therefore, knowing the minimum number of functional groups and species within a functional group to ensure soil resilience should also be a research priority (Brussaard et al., 1997).

This review tries to synthesize the existing knowledge on functional approaches to classify soil organisms and to help in reducing the profuse use in the literature of various terms, which are not always clearly defined: diversity, species richness, functional diversity, feeding guild, functional response trait, functional effect trait, functional domains, spheres of influence, ecosystem/soil functioning (also related to soil quality, soil health and soil fertility), ecosystem properties, ecosystem goods, ecosystem services, etc. The overall aim is to seek for a common framework to better link the diversity of soil organisms to their function in the ecosystems that could improve predictions and policies.

SPECIES RICHNESS: THE THIRD BIOTIC FRONTIER

Soil biota includes bacteria, fungi, protozoans, nematodes, mites (Acari), collembolans (springtails), annelids (enchytraeids and earthworms) and macroarthropods (e.g., spiders, myriapoda, insects, woodlice). It also includes plant roots and their exudates attracting a variety of organisms, which either feed directly on these secretions or graze on the microorganisms concentrated near the roots and hence, receiving the name of "rhizosphere" (first coined by Hiltner, 1904; see also the review of his work by Hartmann et al., 2008). Soil biodiversity is often used as a synonym for the number of heterotrophic species below-ground

(Hooper et al., 2005); however, taxonomic deficit increases with decreasing body size and in the case of soil biota is usually above 90% for those organisms smaller than 100 μm (Decaëns, 2010). Consequently, research has been limited by their immense diversity, their small size and by technical problems. For these reasons, it has been described as the "third biotic frontier" after oceanic abysses and tropical forest canopies (André et al., 1994; Hågvar, 1998).

Species' lists provide a practical metric for assessing biological diversity and soil quality, distinguishing habitats and managing resources (Costello et al., 2013). However, trying to link species diversity with ecosystem function is often hampered by the assumption that all species are potentially equal with respect to function, when there is not mechanistic relationship between these two concepts (Bengtsson, 1998). Indeed, Heemsbergen et al. (2004) concluded that "species richness" has very little influence on soil processes and consequently, communities containing the same number of species but different species combinations have different effects on ecosystem functioning. Furthermore, according to this study, saturation in processes' rates occurred with just adding more than one species and the nature of the interspecific interactions [e.g., positive (facilitation, complementary), negative (inhibition due to competition) or neutral] play the most significant role in the direction and magnitude of their response (Heemsbergen et al., 2004; see also Hooper et al., 2005). This partly confirms previous experimental work, which showed that, for example, more nitrate was mineralized in the presence of solely one collembolan species than when more species were added (Faber and Verhoef, 1991). This can be explained in terms of species' functional attributes and thus, the greater differences in functional characteristics between species, the greater strength of the interspecific interactions (Stevens and Carson, 2001; Chesson et al., 2002). In addition, environmental factors (abiotic and biotic) could also change the strength of these interactions (e.g., Hooper and Dukes, 2004).

SPECIES' FUNCTIONAL ATTRIBUTES: KEY PLAYERS IN REGULATING ECOSYSTEM FUNCTIONING

Advances in molecular techniques have revolutionized the way we describe species. Over the last 15 years the Natural History Museum in London (UK) has seen a 12% replacement of "traditional" taxonomists with molecular biologists (Boxshall and Self, 2011). Molecular markers appear to be the promising tool in assessing the diversity of soil organisms when morphological taxonomy is unachievable and when trying to solve some nomenclature problems derived from the existence of an elevated number of synonyms and species complexes (e.g., Coomans, 2002; Briones et al., 2009; Emerson et al., 2011; Porco et al., 2013). However, at least for certain soil organisms, much work remains to be done in this field to reconcile species synonyms, which might take long time as result of the low accessibility to many species descriptions (Costello et al., 2013).

One added complication is that even the species' definition is not without problems since many soil organisms contain species that are parthenogenetic (e.g., earthworms, plant-parasitic nematodes, mites, and collembolans) or that reproduce asexually by fragmentation (e.g., some enchytraeids).

Finally, the existence of cryptic species further complicates the application of molecular markers-based approaches (Huang et al., 2007; King et al., 2008; Richard, 2008; Rougerie et al., 2009; Porco et al., 2012). For example, very recent work trying to assess the genetic diversity of the genus *Cognettia* (Annelida, Oligochaeta, Enchytraeidae) has revealed that the most relevant species in soil ecological studies (*C. sphagnetorum*) comprises at least four different lineages which can co-ocurr at the same site, and with potential physiological and ecological differences (Martinsson and Erséus, 2014). Importantly, not only the identity of this species and other species within this genus were questioned, but also the phylogenetic relationships among these species varied between the gene trees as a result of "phylogenetic errors" not corrected by the model. Therefore, despite their great potential, these modern tools are in dire need for improving their mathematical procedures to analyze the species trees and for overcoming other methodological shortcomings in relation to nucleic acids extraction procedures, standardization of the methods to allow qualitative and quantitative comparisons, sequencing error rates, missing confident databases, etc. (Chang et al., 2009).

According to Brussaard et al. (1997), the goal of molecular taxonomy is not only to link molecular data to species, rather than broad taxonomic groups, but also to performance in the field. One way of reducing this major challenge is perhaps, to concentrate our efforts in "key stone species," those species whose effects on their communities or ecosystems are larger than expected from their abundance or biomass (Power and Mills, 1995). If only a few species are more relevant than others, because their functional attributes could have a strong impact on soil processes, it becomes obvious that determining each species-specific contribution to each soil function should be a research priority. For example, the enchytraeid species *C. sphagnetorum* has been classified as one of this type in boreal coniferous forest soils, due to its relevant role in the decomposition processes in these particular systems (Laakso and Setälä, 1999). However, it has been argued that a dominant species could also act as a keystone species (Bengtsson, 1998) and indeed, in these boreal systems, *C. sphagnetorum* could make up more than 95% of total enchytraeid numbers (Huhta et al., 1986). By focusing on the causal effect, "keystone-process species" (Folke et al., 1996), it would be possible to obtain a functional attribute irrespective of their number or size.

If we accept that not every species in the soil has the same importance for ecosystem functioning, we could also argue that perhaps the members of the same species might also not be identical and hence, they might not have the same function in the soil (Lavelle et al., 1997; Wolters, 2001). Therefore, identifying genotypes with functional attributes could be our next research aim if we wish to be able to manage soil biodiversity and ecosystems and to face environmental changes. Accordingly, "metagenomes" [i.e., genomic, evolutionary, and functional information recovered directly from soil samples and not from artificial systems (clonal cultivations)] represent the new era to gain new insights into the relationships between genetic and functional diversity (Torsvik and Øvreås, 2002). These new approaches have been successfully applied to microbial communities from distinct biomes (Fierer et al., 2012) and showed, for the first time, that desert communities could be linked to both higher osmoregulation and dormancy capabilities and to lower influences on nutrient cycling and the catabolism of plant-derived organic compounds, whereas non-desert ones seem to have more competitive skills in terms of greater antibiotic resistance. Despite the authors' acknowledgement of some methodological, spatial and temporal limitations in their conclusions, it is the most comprehensive attempt to relate functional gene diversity to large-scale geographical gradients. Since the number of species or genotypes necessary to preserve ecosystem functioning increases with increasing spatial and temporal scales (Hooper et al., 2005), the research task ahead is overwhelming, but as sequencing capacities and analyzing procedures become more effective, coupling genetic and functional diversity with environmental gradients across time and space seems plausible in the near future (Fierer et al., 2012).

FUNCTIONAL GROUPS: THE IMPORTANCE OF TEAM WORK

Although certain soil functions can be performed by one key species, it could also be hypothesized that a battery of similarly equipped players (rich taxonomical diversity but functional redundancy) could be responsible for the same soil function or at least, contribute to that particular function in some ways. Indeed, the term "functional group" was first coined by Cummins (1974) and defined as *a group of organisms which affects a process in a similar way.* The concept immediately attracted scientific attention and consequently, over the years, several classifications, in which several taxa with a potential similar role in ecosystem functioning were fitted into major broad groups, have been produced.

BODY SIZE

Perhaps, the first attempt to provide a functional classification of soil organisms was using their body width as the main classificatory criterion: (i) <100 μm including microbiota (bacteria and fungi) and micro-fauna (protozoa and nematodes), (ii) >100 μm and <2 mm referring to mesofauna (acari, springtails, diplura, symphylans, enchytraeids) and (iii) >2 mm which includes the macrofauna (e.g., mollusc, spiders, insects, earthworms) (Wallwork, 1970; Swift et al., 1979). Unfortunately, the ranges that determine each group size are not exact for all the members of each group, often leading to considerable confusion as to whether a particular organism should be considered macro, meso or micro. Furthermore, in order to get a better understanding of the quantitative role of soil animals in ecosystems live biomass data is required (Abrahamsen, 1973), which is not always feasible. Therefore, for comparison purposes, body length (**Figure 1**) might represent a better parameter since a number of length-weight conversion models are available (e.g., Abrahamsen, 1973; Petersen, 1975).

Nonetheless, the size classification based on body width has been successfully applied in addressing several ecological questions (primarily their effects on SOM decomposition) and it is still in practical use today. For example, Huhta (2007) ranked the contribution of soil fauna to ecosystem functions so that macrofauna (earthworms followed by isopods and diplopods) was ranked first and mesofauna (enchytraeids) second, with the latter only having a predominant role when those larger bodied animals were less numerous or totally absent. This ranking

was based on previous experimental work in which the absence of macrofauna (earthworms) significantly increased N-uptake of pine and birch seedlings (Setälä and Huhta, 1991; Setälä, 1995, 2000). Similarly, Bradford et al. (2007) also concluded that smaller-bodied fauna do not always compensate for the absence of macrofauna in terms of regulating the dynamics of recently photosynthesized carbon.

However, a number of laboratory and field studies failed to show the pivotal role of macrofauna. Thus, Setälä et al. (1996) found that the reduction of macrofauna population numbers, through increased predation by coleopterans and centipedes, did not affect growth of poplar seedlings or the nitrogen content of their leaves, whereas the removal of soil mesofauna led to reduced nitrogen uptake by the seedlings. Similarly, despite the greater dominance of macrofauna (earthworms) in an upland organic grassland, was the mesofauna (mites and collembolans), representing only 3% of the total biomass, the group responsible for processing the majority of the recent photosynthesized carbon (Ostle et al., 2007). Finally, other laboratory studies (Schulz and Scheu, 1994; Filser, 2003) concluded that while macrofauna (earthworms, Diplopoda, Isopoda), mesofauna (enchytraeids) and bacterial feeding microfauna all enhanced SOM decomposition, other mesofaunal groups (collembolans and oribatid mites) showed highly variable results depending on the species considered and incubation conditions.

Among the smaller-sized organisms, microbiota deserves special consideration since they seem to have limitless capacities to access and digest any substrate in the soil. However, our understanding of the functional capabilities of soil microbial communities is still very limited and often focused on particular processes (e.g., N_2 fixation) and requires further assessment to resolve large-scale biogeographical patterns in microbial diversity, community composition and functional attributes (Fierer et al., 2012).

Therefore, although body size/biomass is relatively easy to measure and could provide a good indication of the potential key players driving a particular soil function, they only represent coarse groupings which do not take into account the fact that different soil organisms within the same size group could exhibit different physiological capabilities and that their relative importance in performing a specific function could change across temporal and spatial scales.

REPRODUCTIVE STRATEGIES

The way animals reproduce and their developing time or "life-history traits" also reflect the species' responses to environmental conditions (Moore et al., 1998) and therefore, provide important information on soil processes (Bongers and Bongers, 1998; Brussaard, 1998). For example, root feeding insects in temperate climates have often long life cycles because they have to cope with low quality plant material or its limited availability, whereas in tropical climates they tend to be multivoltine (Brown and Gange, 1990).

The most comprehensive classification using this criterion was provided by Siepel (1994) who described up to 12-life story tactics for soil microarthropods (primarily for mites, but Collembola were also used to validate his classification) based on four traits:

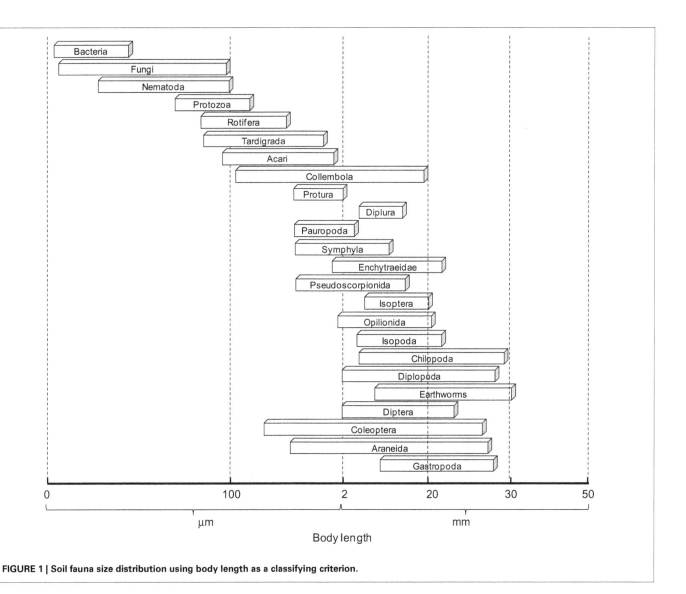

FIGURE 1 | Soil fauna size distribution using body length as a classifying criterion.

(i) Reproduction:

  (a) Sexual or parthenogenesis [thelytoky (automictic or apomictic), arrhenotoky, amphitoky]

  (b) Ovoposition timing: semelparity (on one occasion) and iteroparity (spread over time: continuous or seasonal)

(ii) Development (developmental stages and factors that control development): slow, moderate, fast, quite fast

(iii) Synchronization: diapause (obligate or facultative), aestivation and quiescence

(iv) Dispersal: phoresy (facultative and obligate, carrier-specific and carrier unspecific) and anemocory

Unfortunately, a similar detailed reproductive classification for other soil animal groups is not yet available. In addition, the fact that some species can exhibit more than one of these life-tactics and that some are either seasonal or facultative and therefore, could change under different environmental and food availability conditions, makes its practical use for linking reproductive diversity to a particular soil function rather limited.

FEEDING

Feeding perfectly describes the consumer-resource interaction and represents the fundamental basis of population-community ecology. In fact, the "guild" concept (Root, 1967) refers to *"groups of species having similar exploitation patterns"* and was later modified (Root, 1973) by adding "mode of feeding" as a secondary criterion. Both terms appear to be profusely used as synonyms in the literature (Hawkins and MacMahon, 1989), although the term "guild" appears to refer to a broader species classification than the term "functional group" (Brussaard, 1998).

The importance of feeding niches for describing different functional levels in soil communities is easily corroborated by the success of soil food-webs in modeling energy and nutrient transfers between the different compartments (e.g., Persson et al., 1980; Hunt et al., 1987; Moore and De Ruiter, 1991; De Ruiter et al., 1993, 1995; Bengtsson et al., 1995). The triumph of these models is derived from the causal-effect relationship between two of the parameters which are often included in these relationships: size/biomass and feeding/physiological rates, so that any alterations in their patterns would have a direct effect on processes rates (Bengtsson, 1998). However, one important drawback in these models (e.g., Hunt et al., 1987) is the assumption of soil fauna having a constant biomass, which is not realistic.

A good number of soil fauna classifications, based on feeding habits, have been proposed for different soil biota. They are in continuous revision and over the years, they have been either updated or completely renovated. The majority of them are "organism centered," but a few include the complete soil community. Some of these most influential schemes in ecological research are described below (see also **Table 1**).

A preliminary classification of soil Protozoa based on their preferred diets was proposed by Coûteaux and Darbyshire (1998) by adapting that previously developed by Pratt and Cairns (1985) for freshwater species. This was achieved by just adding the "mycophagous group" to the six ones already established (**Table 1**).

Yeates et al. (1993) identified eight feeding groups for nematodes in plant and soil systems according to their feeding habit or the food source (**Table 1**). Earlier on, Ferris (1982) described four different feeding groups within plant-parasitic nematodes, depending on their location and movement patterns and thus, (i) sedentary endoparasites were those that feed entirely within the root system in specialized feeding sites, (ii) migratory endoparasites feed along the root tissues, (iii) ectoparasites remain on the outside of the root and they can feed either superficially or on deeper tissues and finally, (iv) one intermediate category, the sedentary semi-endoparasitic nematodes which feed with its head deep inside in the vascular root tissues but with most of the body outside the root. Next, in order to get a better understanding of nematode roles in soil functioning, Bongers and Bongers (1998) combined this information and integrated nematode's life stories (from "colonizers" to "persisters") within the trophic classification, which resulted in further subdivisions of the plant feeders and animal predators (**Table 1**). In practice, a simplified system is applied, consisting of primary consumers (plant feeders), secondary consumers (bacterivores and fungivores), and tertiary consumers (predators and omnivores).

In the case of oribatid mites, Luxton (1972) summarized the available knowledge on nutritional biology of these organisms and developed a framework on the basis of their function being primarily litter fragmentation. Accordingly, he defined three major feeding habits: macrophytophages, microphytophages, panphytophages (**Table 1**), with the large majority of mites possibly belonging to the third group and hence, enabling them to colonize a wide range of habitats. In addition, because some oribatid mites were also seen to feed occasionally on other food sources, an additional category of feeding strategies was established (including zoophages, necrophages, coprophages; **Table 1**), but without being able to decide whether these alternative diets obeyed to a general rule, a random occurrence or were imposed by specific conditions (e.g., starvation). Interestingly, according to this study, the three major feeding groups exhibited

different assimilation efficiencies which were likely to be in the order: macrophytophages (10–15%) <panphytophages (40–50%) <microphytophages (50–65%) and hence, with potential different effects on soil processes.

However, due to increased awareness of the close interactions between oribatid mites and microbial populations (through grazing) and of their different digestive enzymatic capabilities, it became obvious that their ecological role could be two-fold: as primary decomposers by egesting a more readily decomposable substrate for microbial colonies and thus, serving as true "catalysts" (MacFadyen, 1964; Luxton, 1972), and as secondary decomposers by gaining their nutrition once it has been made more palatable by the microbiota (Luxton, 1972). Following this, Siepel and De Ruiter-Dijkman (1993) used the activities of three carbohydrases, able to degrade three important food components: cellulose (cell-walls of green plants), chitin (fungal cell-walls) and threalose (fungal cell-contents), as the main criteria for defining the feeding guilds of this particular group of mites. In their classification (**Table 1**), grazers (able to digest both cell-walls and contents) are clearly differentiated from "browsers" (capable of digesting cell-contents only). However, 11 years later, Schneider et al. (2004) simplified this framework and differentiated a total number of four feeding strategies for oribatid mites: phycophages/fungivores, primary decomposers (comminuting litter), secondary decomposers (consuming litter and fungi), and carnivores/scavengers/omnivores.

Because Collembola are typically considered to be fungivores, classifications based on their feeding habits are very scarce. For example, Addison et al. (2003) identified four collembolan feeding groups in a succession of Canadian coastal temperate forests on the basis of their gut contents and the most surprising finding was that, out of the four groups identified, one was not a fungal feeder but comprised species that ingested particulate organic matter. This is in agreement with other studies which also concluded that many species of Collembola are not fungal-feeders, but consumers of living or dead plant tissue or liquids, algae, pollen grains and even other soil biota such as Protozoa, Rotatoria, enchytraeids and nematodes (e.g., Rusek, 1998; Chamberlain et al., 2005, 2006).

The possibility that previous assumptions on soil animal's diets might not reflect reality was also confirmed by other studies which suggested the possibility that earthworms may also digest protozoans, nematodes and even enchytraeids (Gorny, 1984; Roesner, 1986; Bonkowski and Schaefer, 1997). This seems to suggest that predation could be a wider spread feeding strategy in soils than previously assumed, also complicated by the co-existence of cannibalism and intra-guild predation (Polis, 1991). Scheu and Setälä (2002) concluded that the difficulties in finding preys in the opaque and porous soil environment and the inability of potential preys to acquire evolutionary traits to avoid predation by specific predators might explain why soil predators have been usually considered as "generalist feeders." In relation to this, the "trophic cascade theory" or "top-down regulation" has been profusely applied when describing soil food-webs and according to it, prey populations sizes are closely related to that of their prey. However, the effect of increased predation and subsequent reduction in population numbers of potential preys on soil processes

Table 1 | Functional classifications of different soil invertebrates based on different criteria and including the definition of the proposed groups.

Author	Main criteria	Soil animal group	Classification proposed	Definition
Pratt and Cairns, 1985, but adapted by Coûteaux and Darbyshire, 1998	Feeding regime	Protozoa	Photosynthetic autotrophs	Primarily photosynthetic
			Bactivores-detritivores	Feeding on bacteria and detrital particles attached to bacteria
			Mycophagous	Fungal feeders
			Saprotrophs	Feeding on humus particles
			Algivores	Feeding primarily on algae (specially diatoms and small filaments)
			Non-selective omnivores	Feeding on a variety of food materials
			Predators	Feeding on other protozoa or higher taxa (e.g., rotifers)
Yeates et al., 1993, updated by Bongers and Bongers (1998)	Feeding habit and food source	Nematodes	Plant feeders	Feeding on vascular plants. They could be subdivided into six groups: (i) sedentary parasites, (ii) migratory endoparasites, (iii) semi-endoparasite, (iv) ectoparasites, epidermal cell and root hair feeders, (v) algal, lichen or moss feeders, (vi) feeders on above-ground plant tissue
			Fungal feeders	Feeding on hyphae of saprophytic fungi
			Bacterial feeders	Feeding on any prokaryote food source
			Substrate ingester	More than a pure food source is ingested, possibly incidentally
			Animal predators	Feeding on invertebrates (e.g., protozoa, nematode, rotifer and enchytraeids) either as (i) ingesters or (ii) piercers
			Unicellular eucaryote feeders	Feeding on diatoms or other algae, fungal spores and whole yeast cells
			Dispersal or infective stages of animal parasites	Entomogenous species
			Omnivores	Feeding on a wide range of foods
Luxton, 1972	Feeding regime	Oribatid mites	Macrophytophages	Feeding on plant material: (phyllophages) or wood (xylophages) tissues
			Microphytophages	Feeding on microbiota: fungi and yeasts (mycophages), bacteria (bacteriophages) or algae (phycophages)
			Panphytophages	Feeding on all kinds of plant or fungal tissues
			Zoophages	Feeding on living animal material
			Necrophages	Feeding on carrion
			Coprophages	Feeding on fecal material
Siepel and De Ruiter-Dijkman, 1993	Carbohydrase activity	Oribatid mites	Herbivorous grazers	Cellulase activity: feeding on both cell-walls and cell-contents of plants (including algae)
			Herbivorous browsers	Lacking carbohydrase activity: carrion feeders and bacteria feeders
			Fungivorous grazers	Chitinase and threolase activities: feeding on both cell-walls and cell-contents of fungi
			Fungivorous browsers	Threalose activity: feeding on lichens and cell-contents of fungi
			Herbo-fungivorous grazers	Cellulase, chitinase and threalose activities: feeding on all kind of green plants and fungi
			Oportunistic herbo-fungivorous	Cellulase and threolase activities: feeding on green plants but taking advantage of periodic increases in fungal growth in their biotopes (the semi-aquatic group could also feed on cyanophyta and those living on moss on trees or stones, dropped cones, etc., might also feed on lichens)

(Continued)

Table 1 | Continued

Author	Main criteria	Soil animal group	Classification proposed	Definition
			Omnivores	Cellulase and chitinase activities: feeding on green plants and on chitin containing food source (arthropods)
Faber, 1991	Feeding regimes Habitat microes-tratification Specific impacts on soil processes	Fungal grazers	Epigeic fungus engulfers	Surface dwelling species that feed on fungi growing on fresh leaf substrates (L-layer)
			Hemiedaphic fungus engulfers	Partially feeding on saprophytic fungi in the F-layer
			Euedaphic fungus engulfers	Feeding on mycorrhizal hyphae or saprophytic fungi growing in the rhizosphere
Gisin, 1943	Size and morphology according to their vertical stratification	Collembola	Atmobios Hemiedaphic hydrophiles Hemiedaphic mesophiles Hemiedaphic xerophiles Euedaphic	Living on macrophytes Living on the water surface Occupying the top soil layers Living in tree barks, lichens, mosses in dry areas True soil inhabitants
Bouché, 1977	Body size Feeding regime Pigmentation Burrowing abilities Mobility Longevity Generation time Drought survival Predation	Earthworms	Epigeic	Small-medium bodied, heavily pigmented, non or some burrowing confined to the upper few cm by intermediate species, feeding on decomposing residues at the top layers (little or no soil ingested), rapid movement response to disturbance, short-lived with short generation times, drought survival in cocoon stage and often predated by arthropods, birds and mammals
			Anecic	Large bodied, darkly dorsally pigmented and forming large, permanent vertical burrows which use to emerge during the night to collect decomposing litter from the soil surface, rapid withdrawal when disturbed, long-lived with long generation times, survival to drought in a quiescent stage, predated when they are at the surface
			Endogeic	Medium sized worms, usually pale in color, building extensive subhorizontal burrows in the mineral horizons and feeding on material rich in organic matter, slow movements, intermediate longevity with short generation times, survival to drought by entering in diapauses, some predation from time to time by ground-dwelling fauna
Lavelle, 1983	Quality of organic matter ingested	Endogeic Earthworms	Polyhumic endogeics	Fairly pigmented, large bodied and form burrows, feeding on soil with high organic content
			Mesohumic endogeics	Unpigmented and otherwise are intermediate, feeding on mineral and organic particles
			Oligohumic endogeics	Unpigmented, large bodied, have no escape behavior and feed on deeper horizons with low organic matter content
Grassé, 1984; Abe, 1987; Donovan et al., 2001; Eggleton and Tayasu, 2001 Rückamp et al., 2010	Feeding regime Associated microorganisms	Termites	Feeding Group I-single piece nesters	Feeding on undecayed substrates: wood (wet and dry), grass and detritus in the same discrete substrate where they also nest; having mutuallistic flagellates in their guts
			Feeding Group I-intermediate nesters	Feeding on undecayed substrates: wood (wet and dry), grass and detritus in the same discrete

Table 1 | Continued

Author	Main criteria	Soil animal group	Classification proposed	Definition
				substrate where they also nest, but also forage in other patches away from the colony center; having mutuallistic flagellates in their guts
			Feeding Group I-separate piece nesters	Feeding on undecayed substrates: wood (wet and dry), grass and detritus away from their nest; having mutuallistic flagellates in their guts
			Feeding Group II-intermediate nesters	Feeding on undecayed substrates: wood, fungus, grass, detritus, litter and microepiphytes in the same discrete substrate where they also nest, but also forage in other patches away from the colony center; no flagellates in their guts
			Feeding Group II-separate nesters	Feeding on undecayed substrates: wood, fungus, grass, detritus, litter and microepiphytes, away from their nest; no flagellates in their guts
			Feeding Group III	"Humus feeders": feeding on the soil-wood interface and soil (some visible plant fragments present in their guts)
			Feeding Group IV	Feeding on soil only (no visible plant remains present in their guts)
Greenslade and Halliday, 1983; Andersen, 1990, 1995, 1997; Folgarait, 1998; Brown, 2000	Biogeography and habitat requirements Relative behavioral dominance	Ants	Dominant Dolichoderinae	Abundant, highly active and aggressive species, very strong competitors with other ants; prefer hot and open habitats
			Subordinate Camponotini	Co-occurring with, and behaviorally submissive to, Dominant dolichoderines; large body size, often foraging at night
			Hot climate specialists	Arid-adapted taxa with morphological, physiological or behavioral specializations to reduce interactions with Dominant dolichoderines
			Cold climate specialists	Cold and temperate-adapted taxa, occurring in habitats where Dominant dolichoderines are generally not abundant
			Tropical climate specialists	Distribution centered in the humid tropics, in particular in those habitats where Dominant dolichoderines are generally not abundant
			Cryptic species	Taxa foraging predominantly in the soil and litter layers, having relatively little interaction with epigenic ants
			Opportunists	Unspecialized, ruderal taxa, characteristic of disturbed sites or other habitats with low ant diversity
			Generalized Myrmicinae	Cosmopolitan genera occurring in most habitats; sub-dominant ants with the ability for rapid recruitment and successful defense of clumped food resources
			Specialist predators	Large body size but small colony size; little interaction with other ants due to their specialist diet

appears to be either a major controlling agent (Fitter et al., 2005), undetectable (e.g., Laakso and Setälä, 1999) or even inhibitory (Cortet et al., 2002).

To date, a tentative classification of the feeding habits enchytraeid does not exist and much of the research effort is still being placed in trying to determine their preferential diet. For example, while some studies suggest they are fungivorous rather than bacteriovorous (Standen and Latter, 1977; Krištůfek et al., 2001), others showed that they are also saprovores (Briones and Ineson, 2002).

A possible reason for this gap in knowledge in soil animals' feeding ecology is the difficulties associated with accurate assessing of the feeding preferences of soil organisms. Advances in describing soil animal diets and trophic interactions have been gained from stable isotope techniques, which provide an accurate estimate of their assimilated diet (i.e., the isotopic composition of animal tissues reflects that of the animals' diet). In research on soil animals, isotope ratios of carbon (expressed as $\delta^{13}C$) and nitrogen ($\delta^{15}N$), either as bulk or compound specific, have provided valuable information on the dietary preferences for certain groups (e.g., Briones et al., 1999; Tayasu et al., 2002; Ruess et al., 2004; Schneider et al., 2004; Chamberlain et al., 2005, 2006). However, despite being promising tools, several ontogenetic, physiological and biochemical factors can affect the isotopic composition of the animal tissues (reviewed by Briones and Schmidt, 2004). As a result, species from putative different trophic groups, such as collembolans, oribatid mites, diplopods and earthworms, could show similar range of isotopic values and thus, suggesting similar food resource preferences (Scheu and Falca, 2000).

COMBINING FEEDING PREFERENCES WITH OTHER ECO-BEHAVIORAL ASPECTS

Because the same functional group could exploit different feeding habitats as a result of their vertical stratification along the soil profile (Walter et al., 1988), "feeding guild" provides a coarse classification and a new concept, "league" was coined instead (Faber, 1991), which incorporates microhabitat distributions together with feeding preferences. Accordingly, *"a league is a group of organisms, not necessarily taxonomically related, that exploit and process more than one habitat resource in a homologous manner."* This implies that soil organisms should be classified according to two axes, food/feeding strategies and microhabitat preference, which will result in a stratified soil community. For example, the three leagues comprising the fungus grazing fauna (i.e., nematodes, mites and some collembolans and enchytraeids) have different feeding preferences along the fresh, fragmented and humus layers (**Table 1**).

One of the earliest efforts to integrate the vertical distribution of soil organisms within a functional classification was proposed by Gisin (1943) who differentiated several "life forms" of Collembola according to their habitat adaptations and soil stratification (**Table 1**). However, not all collembolan species show a life strategy reflecting their morphology (e.g., Takeda, 1995) and for example, smaller-sized animals tend to prefer deeper layers, possibly due to juveniles and tiny animals being more sensitive to desiccation than their adults (Verhoef and Witteveen, 1980). Furthermore, down-ground migration could be just a survival strategy in response to lowered soil moisture levels during dry periods and different sized invertebrates such as collembolans (Detsis, 2000), enchytraeids (Briones et al., 1997) and termites (Lavelle et al., 1997) find refuge in the deeper layers.

Bouché (1977) used the differences in vertical distribution of earthworms along with other physiological and behavioral characteristics (body size, food, pigmentation, burrowing, mobility, longevity and generation time, drought survival, and susceptibility to predation) to establish three "ecological" groupings among European lumbricids (**Table 1**). Interestingly, the differences in

burrowing abilities exhibited by earthworms led Keudel and Schrader (1999) to measure the axial and radial pressures exerted when penetrating the soil and found that they were actually different: 14–25 and 39–63 kPa for epigeics, 46–65 and 72–93 kPa for anecic and 26–39 and 59–195 kPa for endogeics. However, Lavelle (1979) argued that not all earthworms fit within these categories since that, for example, in tropical savannahs the majority of the earthworms are either endogeic or intermediate between epigeic and endogeic and thereby, he subdivided the endogeic group into "epiendogeic," living in the upper soil horizons and "hypoendogeic," living in the deeper ones. A few years later, Lavelle (1983) completed his previous work and defined three endogeic categories based on the degree of humification of the organic matter ingested (**Table 1**). Interestingly, he also found that the relative abundance of these functional groups varied along a thermo-latitudinal gradient, with cold climates being dominated by both epigeic and polyhumic endogeic worms and as temperature increases they become gradually integrated by anecics and mesohumic and oligohumics endogeics able to digest lower quality resources (Lavelle, 1983).

The first functional classification of termites was proposed by Abe (1987) on the basis of their overlap in feeding and nesting strategies and consisted of four different groups: (i) single piece nesters (those that feed and nest in the same substrate), (ii) intermediate nesters (although feeding and nesting in a specific location, they also forage outside their colony center), (iii) separate-piece nesters (actively seek for their nutrition away from the nest), (iv) soil feeders (those that feed and nest in the soil). Some years later, Donovan et al. (2001) matched the anatomical aspects of the workers with their gut content analyses and defined four feeding groups distributed along a humification gradient: (i) feeding on wood, litter and grass (only in non-Termitidae), (ii) feeding wood, litter and grass (only in Termitidae), (iii) feeding on very decayed wood or high organic content soil (only in Termitidae) and (iv) true soil feeders. Finally, Eggleton and Tayasu (2001) combined these two existing schemes into a "two-way lifeway" classification (**Table 1**), after considering that the substrate's humification and position in relation to the nest center have more ecological importance than the substrate itself. A link with soil processes was provided by Rückamp et al. (2010), who concluded that organic P dominates the P fraction in the xilophagous (feeding group II) termite nests, whereas the inorganic forms tend to dominate in the humivores (feeding group III) and grass-feeder (feeding group II) ones. Conversely, in an earlier study (Wood, 1988), it was shown that the nature of the original materials used by termites to build their structures, rather than their feeding habits, has a more crucial role in N fixation and thus, in those nests made of carton (a mixture of faces and macerated wood fiber) N is actively fixed and C:N ratios can range from 20:1 to 100:1, whereas an absence of N fixation appears to characterize those structures made using highly decayed wood or SOM.

Ants are largely considered to be omnivores and opportunistic feeders, although some subfamilies and genera also include general predators as well as specialized predators (e.g., preferring collembolans, termites or ants) and herbivorous (feeding on seeds, honeydew, plant nectar, and leaves, etc.). Consequently, their communities are better described as a continuum from predominantly vegetarian taxa to purely predators with a high degree of omnivory (Blüthgen et al., 2003; Gibb and Cunningham, 2011). Furthermore, their relative trophic positions are considered to be relatively conservative but their community assemblage (species composition) rather flexible in its use of the available resources, which may result in higher resilience to land use changes (Gibb and Cunningham, 2011). However, other studies suggest that ant assemblages are very sensitive to human impacts and can act as indicators of habitat perturbations, successional stages and land use and climate changes (e.g., Andersen, 1997; Folgarait, 1998; Gómez et al., 2003). This is also the consequence of their overlapping foraging requirements and hence, feeding, as a single criterion, does not adequately describe the huge diversity and behavioral adaptations within the ant communities. As a result, the current functional classification of ants is based in the successive additions of other classificatory factors such as demography, climate and soil type (Greenslade and Halliday, 1983; Andersen, 1990, 1995, 1997, 2000; Brown, 2000) and includes seven different functional groups, one of them with three subdivisions (**Table 1**). This scheme has been considered to be advantageous in reducing the complexity of the ecological systems and in enabling a basis for evaluating environmental change in relation to ant community structure (Vineesh et al., 2007).

ESTABLISHING THE PECKING ORDER: CHOOSING THE RULING TEAM

Lavelle (1996) was the first one to produce a "hierarchical" model of soil biota by combining, on one hand, the effects of certain biological structures on the biodiversity of smaller organisms and on soil processes and on the other, their feeding regime (**Table 2**). Accordingly, at the lowest level, the "microbiota" act upon organic matter and nutrient cycles, root and rhizosphere processes and plant production (with both positive and negative effects). Next, the "micropredators," being primarily microfauna, such as nematodes and protozoa, do not produce any physical structures and survive by predation on microbiota and other organisms and thus, they stimulate mineralization of organic matter and plant nutrient availability. At a higher level, "litter transformers," including many macro- (Diplopoda, Chilopoda, Isopoda or insects) and micro-arthropods (mites and collembolans), enchytraeid worms and other detritus feeders, stimulate the breakdown and decomposition of surface litter and organic matter, producing small, rather fragile and primarily organic fecal pellets. Finally, "ecosystem engineers," comprising big-sized organisms (termites, ants and earthworms) and whose bioturbating activities produce structures that can last long periods of time (outlasting the organisms that produced them), affect SOM dynamics and soil physical processes. The most remarkable aspect of this classification is the fact that it takes into account the potential top-down regulatory controls of larger organisms (e.g., the ecosystem engineers) over smaller ones. The greater importance given to the biogenic structures produced by ecosystem engineers is two-fold: (i) their effects on soil processes spread along spatial and temporal scales: nutrient mineralization (microsites for microorganisms), physical stabilization (hydrological properties, resistance to erosion) and chemical stabilization (humification and nutrient retention) of organic matter (Lavelle et al., 1997)

and (ii) they can be differentiated from the surrounding bulk soil and constitute true functional domains (Lavelle, 2002): "termitosphere" (termites), "myrmecosphere" (ants) and "drilosphere" (earthworms).

Surface mixing by soil invertebrates have been suggested to be one of the most important soil processes (Mulder, 2006) and in particular, casting activities of annelids (including the families Lumbricidae and Enchytraeidae) have long been recognized as one of the main determinants of soil structure. For example, in tropical soils, under some circumstances, the top layers merely consist of earthworm casts of different ages (Lavelle, 1988) and in organic grasslands 90% of the SOM is processed by a few species of earthworms and enchytraeids (Davidson et al., 2002). However, one aspect of Lavelle's classification which deserves some attention is the importance given to the life-span of these cast materials and which allows the differentiation between ecosystem engineers and litter transformers. In Lavelle's study it is stressed that, in the case of the latter group, their casts being predominantly organic, makes them highly susceptible to be ingested by other soil invertebrates and thus, very unlikely to have significant effects in the longer term (Lavelle et al., 1997). Thus, it has been observed that the endogeic earthworm species *Octolasion lacteum* can obtain additional nutritional value from the fecal material produced by millipedes (Bonkowski et al., 1998). In contrast, other studies have shown that mite fecal pellets can persist in the soil profile for a long time and release nitrogen very slowly (Pawluk, 1987), and Heisler et al. (1996) found increased aggregate stability of casts from Collembola compared with soil aggregates. Furthermore, different species of endogeic earthworms can produce different cast material with different stability, and this aspect was used by Blanchart et al. (1999) to classify them into two functional groups: "compacting" earthworm species that produce large and stable "globular" casts and de-compacting worms that produce fragile "granular" casts and partially feed on large compact casts (**Table 2**). Consequently, the life-spans of these structures and their impact on soil structure and ecological processes highly depend on the organic content of the soil and of the food consumed as well as on the intensity of the faunal activities (Lavelle et al., 1997).

The concept of "ecosystem engineering" has been widely accepted by the scientific community, possibly because this eye-catching term perfectly summarizes the long known fact that living organisms can create and structure habitats and modulate resources' availability. However, it is no without problems, mostly associated with deciding by which process (physical, chemical, assimilatory/dissimilatory) soil organisms modify their environment and hence, removing the possibility that nearly every organism could be included under this term. For example, Berke (2010) defined four subcategories, with only three of them actually applying to terrestrial systems (i.e., "structural," "bioturbators," and "chemical" engineers; **Table 2**). However, according to this classification, and in the case of the soil biota, these subdivisions tend to overlap and for example, plants are both structural and chemical engineers, termites and ants are structural engineers as well as bioturbators; the only exceptions are earthworms which are only classified as burrowers and mycorrizal fungi as chemical engineers (**Table 2**).

Table 2 | Functional classifications based on the pivotal role of ecosystem engineers.

Author	Main criteria	Soil animal group	Classification proposed	Definition	Implications for ecosystem functioning
Lavelle, 1996	Physical structures Effects on biodiversity Effects on function	Roots and soil invertebrates	Roots	Creating rhizosphere structures; effects on biodiversity by secreting polysaccharides, selecting microbiota, associating food-webs and attracting root+root litter feeders	Enhanced mineralization
			Micropredators	No physical structures, effects on biodiversity by microbial grazing	Enhanced mineralization
			Litter transformers	Producing organic fecal pellets; effects on biodiversity by selecting microbiota, creating microhabitats for smaller invertebrates and providing food for other invertebrates	Enhance mineralization and SOM sequestration (depending on time scale)
			Ecosystem engineers	Building large compact organo-mineral structures, smaller aggregates and a large variety of pores (galleries, burrows and chambers); effect on biodiversity by selecting microbiota and/or litter transformers and promoting root development	Bioturbation, dissemination of fungal spores, regulation of structural porosity and aeration, aggregation (compaction or decompaction), infiltration rates, water storage capacity, root growth, SOM transformations and nutrient cycling
Blanchart et al., 1999	Casts	Endogeic earthworms	Compacting	Producing large (>2 mm) "globular" casts and large macropores	Increased soil compaction and water retention capacity, decreased infiltration rate
			Decompacting	Producing small (<2 mm) "granular" casts and small macropores	Increased soil porosity, breakdown of the large aggregates, infiltration rate and decreased water retention capacity
Berke, 2010		Most plants and mound-building insects	Structural engineers		Create living space, reduce disturbance, alter hydrodinamics, sedimentation and diversity (usually by enhancing it)
		Earthworms, ants and termites	Bioturbators		Enhance disturbance, mix sediment, alter biogeochemistry and diversity (usually by reducing it)
		Most plants, mycorrhizal fungi	Chemical engineers		Create biogeochemical gradients (physically or physiologically)
Hedde et al., 2005	Casts Nests Mounds	Earthworms Ants Termites	Accumulators of protected organic matter	Carton termite mounds showing high protection of organic matter and high potential mineralization rates	They might represent a pool of protected organic matter
			Soil compactors	Organo-mineral termite mounds and earthworm casts with low concentrations of organic matter and mineralization rates	Higher efficiency and longer stability of the protected organic matter
			Soil decompactors	Ant mounds and termite sheathings with loose structure and low organic content and mineralization rates	Limited action on organic matter dynamics but greater influence on physical properties by disaggregating organo-mineral complexes

Hedde et al. (2005) used a different approach, the molecular composition of the organic matter present in the different biogenic structures produced by ecosystem engineers and its relationships with selected biological and chemical characteristics, to differentiate between three functional groups of engineers (**Table 2**). In this classification, groups were established according to: (i) whether the structures were produced after gut transit (earthworm casts, mounds built by humivorous termites) or by displacement of soil particles (ant nests and termite sheathings), (ii) whether intestinal mucus or saliva have been added (earthworm casts and termite mounds and sheathings) or they were merely an aggregation of separated soil particles (ant nests) and (iii) whether the effect of microbial activities (either by mutualistic association with the gut of the engineer or by colonization of the fresh biostructures) changed the molecular composition of SOM.

There are others who claim that the ecosystem engineering concept should couple the abiotic environment with the population dynamics feedback (Cuddington et al., 2009). For example, in the case of the burrowing fauna (e.g., earthworms), they should be considered as "obligate engineers" since they have to build their burrows in order to survive; however, if worm densities reach high values, the resulting intensive burrowing activity could negatively affect the stability of the substrate and in turn, species population dynamics. One illustrative example can be found in tropical soils, where the activities of the earthworm *Amynthas hawayanus* reduce the SOM content at the top layers and could result in water runoff and soil erosion (Burtelow et al., 1998). Similarly, ants could build two types of mounds with different pedological impacts (Paton et al., 1995): (i) type I, being small in size and crater-shaped and consisting of sand particles linked by a clay matrix and very susceptible to erosion and (ii) type II, being large size and elliptical shape and made of reworked material and highly resistant to erosion.

Another interesting example is the one provided by the peregrine invasive earthworm species *Pontoscolex corethrurus*, which has successfully colonized pastures after forest conversion in the humid tropics and, during the rainy season, its engineering activities result in the formation of a surface soil crust (compact and asphyxiating horizon of 2–5 cm thick and produced by coalescence of a large amounts of casts with a high water content and a muddy structure) through which it cannot feed (Blanchart et al., 1999; Chauvel et al., 1999; Jouquet et al., 2006). This state also results in profound alterations in water infiltration and promotes methane production; however, it is reversible and if decompacting species, with their antagonistic effects, are present or organic residues are added, the soil structure can be conserved (Blanchart et al., 1999). This contrasts with its role in other ecosystems, where this same species improves soil structure and fertility (Marichal et al., 2010).

The direction of the feedback effects of these engineering activities (positive or negative) was used by Jones et al. (1994, 1997) to distinguish between "extended phenotype engineers" (recalling the work by Dawkins, 1982), whose structures have a positive effect on the organism growth, and "accidental engineers," whose activities do not have a positive effect on themselves. The former typically include those social species (e.g., termites and ants) which build mounds and nests that clearly benefit their fitness, whereas the biogenic structures of the earthworms could or could not have a direct positive effect on themselves and hence,

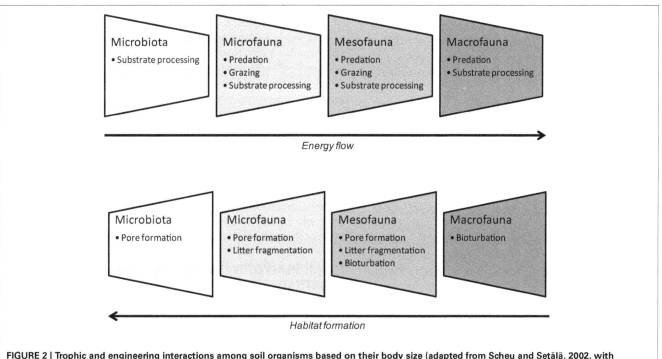

FIGURE 2 | Trophic and engineering interactions among soil organisms based on their body size (adapted from Scheu and Setälä, 2002, with permission from Cambridge University Press).

depending on the species they could be included in either group (Jouquet et al., 2006).

Another way of integrating the influences of both engineers and smaller organisms on soil processes is by converting them into complementary transferring actions over a time scale. Accordingly, because the engineer activities regulate the provision of nutrients to other organisms, they have an indirect effect on trophic interactions and hence, on energy flow. Consequently, the action of small organisms can be viewed as "energy transfer agents" in the short term, whereas engineering activities act as "habitat agents" in the longer term (Scheu and Setälä, 2002; **Figure 2**).

INCORPORATING THE ABOVE-GROUND ENGINEERS: ROLE OF PLANTS IN TEAM PERFORMANCE

Linking the interactions between the above- and below-ground biota, the so-called "holistic view" of the ecosystem is becoming increasingly accepted (e.g., De Deyn and Van der Putten, 2005; Huhta, 2007). The role of plants in soil processes is so relevant that they can also be considered as ecosystem engineers since they also create suitable habitats for other organisms (Lavelle et al., 1993; Brussaard, 1998; Lavelle, 2002; Berke, 2010). Plants influence below-ground processes (e.g., SOM decomposition and nutrient cycling) in several ways, through litter quality, root exudation and mycorrizal association, all of which have been considered to be "functional attributes" or "functional traits." As a consequence, the functional characteristics of the above-ground of dominant plant species are more important than the species richness (e.g., Hooper et al., 2005). For example, different plants support different bacterial communities, both beneficial and pathogenic, and hence, plant performance and resistance (Hartmann et al., 2008) and root exudates are known to attract soil fauna activities. The importance of understanding soil organisms in the context of their microhabitat has led to include the "rhizosphere" as a distinct "functional domain" from the "litter system" (Hiltner, 1904; Lavelle, 2002). These functional domains recall two of the "spheres of influence" of soil biota defined by Brussaard et al. (1997): "rhizosphere" or "root biota," and "decomposers" (also called "litter transformers" or "shredders").

The fact that bacterial communities close to the roots are very different from those in the bulk soil (Marilley and Aragno, 1999) confirms this functional dissimilarity. Similarly, the "mycorrhizosphere" (the volume immediately surrounding the mycorrhizas associated to the roots) also concentrates specific bacterial communities with contrasting functional characteristics from the surrounding soil (Uroz et al., 2013). By influencing the root recognition and receptivity of the mycobiont, the fungal growth and propagation as well as modifying the surrounding soil they can improve the quality of the mycorrhizal association and thus, deserving the consideration of "mycorrhization helper bacteria" (Rigamonte et al., 2010).

Interestingly, these two domains or spheres are considered to be both mutualistic and complementary and thus, while active microbial grazers in the rhizosphere (such as protozoans and nematodes) function as "bacteria-mediated mutualists" by facilitating the liberation of essential nutrients for plant growth (e.g., nitrogen), in the mycorrhizal mutualism the plant transfers

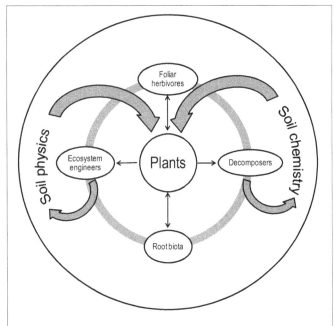

FIGURE 3 | "Plant-centric framework" (redrawn from Brussaard, 1998, with permission from Elsevier).

carbon to fungi in exchange for phosphorous and therefore, complementing each other (Bonkowski et al., 2001).

In view of all these important plant influences, Brussaard (1998) concluded that plants should take a central place in soil ecosystem functioning because they govern the role that soil fauna plays in soil processes. Therefore, in his model plants interact, directly or indirectly, with the physical and chemical environment and with three biotic components or "guilds" of soil organisms influencing decomposition processes: root herbivores (living in association with or feeding directly on roots), decomposers (microbiota and micro-/mesofauna occupying either the bulk soil or "hotspots" where they graze microorganisms or comminute the litter entering into the soil) and "bioturbators" or ecosystem engineers (meso-/macrofauna which create favorable environments for other colonizing organisms) (**Figure 3**). Each of these guilds could show some degree of resource partitioning, which allows coexistence, and for example, among that of root herbivores, different insect species can coexist by having different feeding niches, such as the root collar, central vascular tissue, root cortex or externally on the root (Brown and Gange, 1990). Species spatial aggregation is an influential factor of species coexistence and in turn, ecosystem processes (e.g., Freckleton and Watkinson, 2000; Stoll and Prati, 2001).

FUTURE PERSPECTIVES TO ENABLE A MORE REALISTIC QUANTITATIVE EVALUATION OF SOIL FAUNA ROLES IN ECOSYSTEM FUNCTIONING

Current "functional classifications" can only be considered as "major groups" connected to "major functions" (Lavelle, 2002) and therefore, there has been a claim to be more rigorous when applying their definition criteria (Brussaard et al., 1997). Although numbers and their weights and sizes are relatively easy

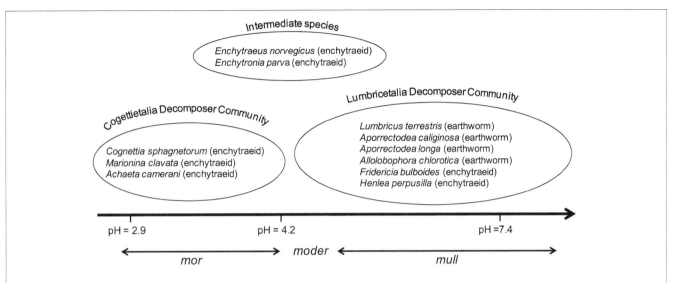

FIGURE 4 | Distribution of 4 earthworm and 5 enchytraeid species along a soil pH gradient (redrawn from Graefe and Beylich, 2003, with permission from Wageningen University).

to measure, quantifying the contribution of soil fauna to soil processes would require incorporating additional information on animal activities, their life styles, feeding habits, reproduction rates and reproductive strategies. In other words, defining "functional groups" would not only imply assigning specific functions to certain taxa (or mixed groups), but also integrating the physical, chemical and climatic environment of the soil with the interactive activities of the organisms living in it (both above- and below-ground). More specifically, a further refinement of soil faunal functional classifications will involve including the following aspects:

(i) Because these groupings are aggregated units whose definition is associated with some degree of arbitrariness (Bengtsson, 1998) and they include a variable number of taxa, the definition of a functional group with respect to one particular function might not fit with respect to another.

(ii) Some groups are associated with more than one function; therefore, we should be changing the focus from "organism-centered" to "processes-centered" (Hodkinson and Wookey, 1999).

(iii) Their functional activities could be temporally and spatially separated and different groups could act at different stages of a particular process (e.g., decomposition) or at different depths along the soil profile (Clapp et al., 1995).

(iv) Some organisms can perform different functions during their life cycle (Lavelle, 1996). In addition, immature stages (juveniles, larvae) could be dominant in numbers in particular habitats or in specific seasons (Vineesh et al., 2007), and have different feeding requirements, different assimilation efficiencies and egesting rates from their adults (Luxton, 1972).

(v) Many soil processes are intimately coupled: for example, SOM decomposition, nutrient cycling and primary productivity.

(vi) "Sanitation" effects (*sensu* Van der Drift, 1965) resulting, for example, from mites feeding on antibiotic-producing fungi (Luxton, 1972) could favor other microorganisms to grow. Similarly, selective consumption of pathogenic fungi by earthworms could enhance or reduce their incidence (Moody et al., 1995).

(vii) Soil characteristics (e.g., soil texture, moisture conditions, pH) could alter the hierarchical role of soil biota on soil processes. For example, in sandy soils endogeic earthworms play a major role in soil structure formation and maintenance, whereas in clayed soils the earthworms had a secondary role and the effect of roots and organic materials becomes predominant (Blanchart et al., 1999). Similarly, the effects of microarthropod communities on N mineralization appear to be greater under drier conditions (Persson, 1989), possibly due to the regulating effect they exert on enchytraeid populations which does not operate at higher moisture contents (Huhta et al., 1998).

The influence of soil pH on shaping soil biota communities is best exemplified by Graefe and Beylich (2003), who established a threshold value of 4.2 (pH $CaCl_2$) below which, these communities completely change their species composition and functionality, resulting in the development of different types of humus in soils (**Figure 4**). This could explain why in mineral soils macrofauna and earthworms, in particular, are expected to be the key faunal component, whereas in organic rich soils other smaller sized organisms (e.g., oribatid mites and enchytraeids) may play a more determinant role in soil processes (e.g., Luxton, 1972; Laakso and Setälä, 1999). Following this, liming, a common practice to raise soil pH in arable soils, is expected to benefit SOM turnover and nutrient cycling; however, this is not always the case and for example, Gray et al. (2003) found that, despite of increasing plant productivity and changing soil biota composition, liming did not have any detectable effect on soil respiration. This

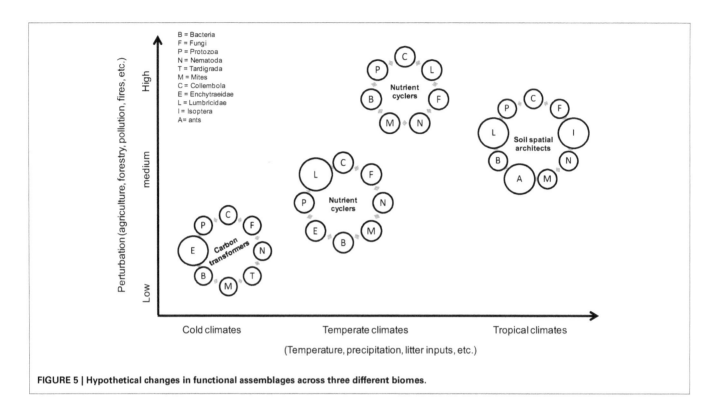

FIGURE 5 | Hypothetical changes in functional assemblages across three different biomes.

discrepancy was explained as a result of soil pH not increasing beyond the turning point of 4.2 when lime is applied to mineral soils, which favors intermediate species (**Figure 4**; Graefe and Beylich, 2003).

(viii) Another issue that should be considered is whether the results obtained for any given soil process in a given ecosystem could be extrapolated to another process and ecosystem (Loreau et al., 2001). For example, in temperate grasslands above-ground processes (e.g., primary production and nutrient retention) appear to be under direct plant control and it does not seem to be linked to below-ground processes (e.g., SOM turnover and nutrient cycling) which are under microbial control.

(ix) Because climate, which operates at large scales of time and space, is the main factor determining soil fauna contribution to soil processes at global scale (Swift et al., 1979; Lavelle et al., 1993), functional classifications could be structured in a different way in different biomes. Lavelle et al. (1997) give some interesting examples of disappearance and functional substitution of different soil animal groups in response to latitudinal changes: (i) high acidity and waterlogging conditions which prevent ecosystem engineers to succeed can result in other invertebrates (isopods and coleopterans, gastropods, diplopods, centipedes, dipteral larvae, and enchytraeids) assuming similar, although limited, roles; (ii) earthworms tend to be better represented in grasslands situated in humid areas than in forested and dry areas, whereas litter arthropods seem to be predominant in ecosystems where sufficient litter is available; (iii) in cold and temperate climates earthworms feed on more organic material and the proportion of endogeic worms is lower, whereas warmer

climates are likely to change earthworm communities with a relative increase of endogeic populations and hence, higher activity in the deeper layers; (iv) in tropical soils there seems to exist an inverse relationship between earthworms and termites, with dominance of termites during the dry seasons and earthworms during the wet ones.

Therefore, as highlighted by Lavelle (1996) there is a need for functional classifications that: (1) group redundant species, (2) are based on well identified functions in soils, (3) give prominent status to species or groups with key functions, but (4) recognize that some groups that do not significantly affect the rates of any ecosystem process may serve as indicators of ongoing processes. This represents a major scientific challenge which will require a great number of multidisciplinary scientific efforts.

Among possible solutions to achieve this enormous task, it has been suggested to adopt a continuous approach, in which a set of relevant biological traits are used as predictors of their function in soil (Hedde et al., 2005). This recalls the distinction made by Hodgson et al. (1999) between "soft biological traits," easy to measure, but not directly related to a specific functional mechanism (e.g., size, growth form, maturation age, life span, etc.) and "hard biological traits," difficult (or impossible) to measure but having direct functional role (e.g., growth rate in response to environmental factors, ecosystem engineers biostructures).

Another alternative approach is to categorize each functional group according to its relative position along several axes determining species' interactions with the abiotic and biotic environment. For example, Hodkinson and Wookey (1999) proposed four axes: population responsiveness (i.e., the speed and magnitude of the response to changing conditions to maintain its position within the community: life cycles, K and r strategies, generation times), dispersability (horizontal dispersal and colonization abilities), ecophysiological flexibility (stress tolerance, growth and survival strategies) and resource use flexibility (ability to cope with spatial and temporal variations in resource availability).

Because species respond to climate, edaphic factors, resource quality and availability from the primary produces and perturbations, integrated knowledge is needed. At the same time, unidirectional causality approaches, in which diversity is either a cause or an effect, should be avoided and the focus should be placed on finding interactive relationships and feedbacks among biodiversity and environmental changes and ecosystem functioning (Loreau et al., 2001). In addition, diversity, ecosystem properties and environmental conditions should be better viewed as dynamic variables (Hooper et al., 2005), in terms of their responses and influences on soil processes.

New mathematical techniques are emerging, such as the "competitivity graph," developed by Criado et al. (2013) to compare the competitiveness of a family of rankings. Although competition does not appear to have a strong influence on below-ground communities (Wardle, 2002; Bardgett et al., 2005; Decaëns, 2010), one may assume that biomes/ecosystems are dynamic systems in which different functional assemblages compete for resources leading to competitive ranks, so that the functional contribution of a particular keystone species or functional group to a given soil process is more pivotal than that of the remaining soil organisms (hierarchy). The number of interactive players and the relevance of a particular group as a "ruling player" or a "ruling team" (e.g., plants, litter fragmenters, ecosystem engineers) would be subjected to changes in climatic, edaphic conditions and to perturbation (e.g., agricultural practices, fires, pollution) operating in those systems and hence, the magnitude and direction of the regulated soil function. But, even in the absence of perturbation, these rankings could change with season and nutrient inputs (via plants) and therefore, the "dynamic" model is superimposed to the hierarchical model.

One illustrative example is shown in **Figure 5** in which a hypothetical change in functional structure of soil biota across three different biomes and different degrees of perturbation, together with their main contributing roles, are depicted. Because the factors that shape biomes (prevailing climate) also drive functional composition of plants and soil organisms, it could be hypothesized that one axis of functional variation could be represented by the correlation between environmental and biotic factors and the geographical spatial patterns, whereas any parameter altering both the ecological and physiological responses and reproduction and dispersability rates of biota could conform one or more perturbation axes (i.e., a "multidimensional" approach). Therefore, for example, in cold climates (**Figure 5**), where environmental conditions restrict the presence of macrofauna and the C storage function dominates, enchytraeids become the pivotal organisms in performing soil mixing and C transformations (i.e., they are "C-transformers"). Although, these small sized-worms have been traditionally classified as "litter transformers" (Lavelle, 1996), at least in C rich soils, their bioturbating (which increase hydraulic conductivity and hence nutrient leaching) and casts activities

(which are hotspots for microbial activities, etc.) would entitle them to be considered the "ecosystem engineers of the organic layers."

Under milder climatic conditions (e.g., temperate and tropical climates), ecosystem engineers (or "soil architects," if we accept that enchytraeids, and possibly other biota are also engineers) take over mesofauna's role and become the key players in the mineralization of the SOM (**Figure 5**). However, if the system is perturbed (e.g., land use change), their functional contribution could be either greatly reduced (e.g., as a result of higher mortality rates within a specific feeding or ecological group) or completely missed (e.g., as a result of heavy use of pesticides and herbicides). The outcome of those species/groups' losses will depend on the functional abilities of the remaining players (which could change their role from functionally redundant to functionally active) and the duration of the perturbation (i.e., resistance or resilience).

This conceptual scheme could also be useful for studying intra- and inter-annual differences in the functional assemblages and whether, at any particular time, the interactions among a selective number of players are stronger than those occurring in a different season. At smaller scale, it could also be possible to investigate whether these groupings change with microhabitat stratification or under different plant species and, in this case, the axes of variation would be specific soil properties and plant traits.

This new overview agrees with the conclusions from the work by Hodkinson and Wookey (1999), who indicated that, at least for Arctic ecosystems, the community resilience to environmental changes largely depends on the long term fluctuations in environmental factors (which provides the "environmental inertia" and "biotic inertia") and on the short term environmental variation [which provides "amplitude" (ecosystem stress tolerance) and "elasticity" (speed to return to the equilibrium state)].

Biological evolution is full of examples of over-specialization as well as wasteful structures; perhaps, the huge diversity present in soils allows "biotic plasticity" and its practical value will only become evident under different circumstances and this, itself, represents a good reason to preserve it. In a near future we will need to know the extent of this property to be able to determine the minimum number of species to produce an ecosystem good, to maintain an ecosystem function or to mitigate climate change.

ACKNOWLEDGMENTS

I would like to thank Mrs. J. Coward (librarian at CEH Lancaster, UK) for providing some of the articles used in this review and to Dr. F. Ramil (Universidad de Vigo, Spain) for helping with the German translation of some of these papers. I gratefully acknowledge the valuable information provided by Dr. P. Eggleton (Natural History Museum, London, UK) and Dr. C. Gómez (Universidad de Gerona, Spain) regarding the functional classifications of termites and ants, respectively.

REFERENCES

Abe, T. (1987). "Evolution of life types in termites," in *Evolution and Coadaptation in Biotic Communities*, eds S. Kawano, J. H. Connell, and T. Hidaka (Tokyo: University of Tokyo Press), 125–148.

Abrahamsen, G. (1973). Studies on body-volume, body-surface area, density and live weight on Enchytraeidae (*Oligochaeta*). *Pedobiologia* 13, 6–15.

Addison, J. A., Trofymow, J. A., and Marshall, V. G. (2003). Functional role of Collembola in succesional coastal temperate forests on Vancouver Island, Canada. *Appl. Soil Ecol.* 24, 247–261. doi: 10.1016/S0929-1393(03)00089-1

Andersen, A. N. (1990). The use of ant communities to evaluate change in Australian terrestrial ecosystems: a review and a recipe. *Proc. Ecol. Soc. Aust.* 16, 347–357.

Andersen, A. N. (1995). A classification of Australian ant communities, based on functional-groups which parallel plant life-forms in relation to stress and disturbance. *J. Biogeogr.* 22, 15–29. doi: 10.2307/2846070

Andersen, A. N. (1997). Functional groups and patterns of organization in North American ant communities: a comparison with Australia. *J. Biogeogr.* 24, 433–460. doi: 10.1111/j.1365-2699.1997.00137.x

Andersen, A. N. (2000). "A global ecology of rainforest ants: functional groups in relation to environmental stress and disturbance. Chapter 3," in *Standard Methods for Measuring and Monitoring Biodiversity*, eds D. Agosti, J. D. Majer, L. Alonso, and R. Schultz (Washington, DC: Smithsonian Institution Press), 25–34.

Anderson, J. M. (1988). Spatio-temporal effects of invertebrates on soil processes. *Biol. Fert. Soils* 6, 216–227. doi: 10.1007/BF00260818

André, H. M., Noti, M. I., and Lebrun, P. (1994). The soil fauna: the other last biotic frontier. *Biodivers. Conserv.* 3, 45–56. doi: 10.1007/BF00115332

Andrén, O., and Balandreau, J. (1999). Biodiversity and soil functioning—from black box to can of worms? *Appl. Soil Ecol.* 13, 105–108. doi: 10.1016/S0929-1393(99)00025-6

Andrén, O., Bengtsson, J., and Clarholm, M. (1995). "Biodiversity and species redundancy among litter decomposers," in *The Significance and Regulation of Biodiversity*, eds H. P. Collins, G. P. Robertson, and M. J. Klug (Dordrecht: Kluwer), 141–151.

Andrén, O., Brussaard, L., and Clarholm, M. (1999). Soil organism influence on ecosystem-level processes—bypassing the ecological hierarchy? *Appl. Soil Ecol.* 11, 177–188. doi: 10.1016/S0929-1393(98)00144-9

Bardgett, R. D. (2002). Causes and consequences of biological diversity in soil. *Zoology* 105, 367–374. doi: 10.1078/0944-2006-00072

Bardgett, R. D., Yeates, G. W., and Anderson, J. M. (2005). "Patterns and determinants of soil biological diversity," in *Biological Diversity and Function in Soils*, eds R. D. Bardgett, M. B. Usher, and D. W. Hopkins (Cambridge: Cambridge University Press), 100–118. doi: 10.1017/CBO9780511541926.007

Bengtsson, J. (1998). Which species? What kind of diversity? Which ecosystem function? Some problems in studies of relations between biodiversity and ecosystem function. *Appl. Soil Ecol.* 10, 191–199. doi: 10.1016/S0929-1393(98)00120-6

Bengtsson, J., Setälä, H., and Zheng, D. W. (1995). "Food webs and nutrient cycling in soils: interactions and positive feedbacks," in *Food Webs: Patterns and Processes*, eds G. Polis and K. Winemiler (London: Chapman and Hall), 30–38.

Berke, S. K. (2010). Functional groups of ecosystem engineers: a proposed classification with comments current issues. *Integr. Comp. Biol.* 50, 147–157. doi: 10.1093/icb/icq077

Blanchart, E., Albrecht, A., Alegre, J., Duboisset, A., Gilot, C., Pashanasi, B., et al. (1999). "Effects of earthworms on soil structure and physical properties," in *Earthworm Management in Tropical Agroecosystems*, eds P. Lavelle, L. Brussaard, and P. Hendrix (Wallingford, WA: CAB International), 149–172.

Blüthgen, N., Gebauer, G., and Fiedler, K. (2003). Disentagling a rainforest food-web using stable isotopes: dietary diversity in a species-rich ant community. *Oecologia* 137, 426–435. doi: 10.1007/s00442-003-1347-8

Bongers, T., and Bongers, M. (1998). Functional diversity of nematodes. *Appl. Soil Ecol.* 10, 239–251. doi: 10.1016/S0929-1393(98)00123-1

Bonkowski, M., Jentsche, G., and Scheu, S. (2001). Contrasting interests in the rhizosphere: interactions between Norway spruce seedlings (*Picea abies* Karst.), mycorrhiza (*Paxillus involutus* (Bartsch) Fr.) and naked amoeba (Protozoa). *Appl. Soil Ecol.* 18, 193–204. doi: 10.1016/S0929-1393(01)00165-2

Bonkowski, M., and Schaefer, M. (1997). Interactions between earthworms and soil Protozoa: a trophic component in the soil food-web. *Soil Biol. Biochem.* 29, 499–502. doi: 10.1016/S0038-0717(96)00107-1

Bonkowski, M., Scheu, S., and Schaefer, M. (1998). Interactions of earthworms (*Octolasion lacteum*), millipedes (*Glomeris marginata*) and plants (*Hordelymus europeaus*) in a beechwood on a basalt hill: implications for litter decomposition and soil formation. *Appl. Soil Ecol.* 9, 161–166. doi: 10.1016/S0929-1393(98)00070-5

Bouché, M. B. (1977). "Stratégies lombriciennes," in *Soil Organisms as Components of Ecosystems*, eds U. Lohm and T. Persson (Stockholm: Ecological Bulletin vol. 25), 122–132.

Boxshall, G. A., and Self, D. (2011). *UK Taxonomy and Systematics Review—2010. Natural Environment Research Council.* Available online at: http://www.nerc.ac.uk/research/programmes/taxonomy/documents/uk-review.pdf

Bradford, M. A., Tordoff, G. M., Black, H. I. J., Cook, R., Eggers, T., Garnett, M. H., et al. (2007). Carbon dynamics in a model grassland with functionally different soil communities. *Funct. Ecol.* 21, 690–697. doi: 10.1111/j.1365-2435.2007.01268.x

Briones, M. J. I., and Ineson, P. (2002). Use of ^{14}C carbon dating to determine feeding behaviour of enchytraeids. *Soil Biol. Biochem.* 34, 881–884. doi: 10.1016/S0038-0717(02)00010-X

Briones, M. J. I., Ineson, P., and Piearce, T. G. (1997). Effects of climate change on soil fauna; responses of enchytraeids, Diptera larvae and tardigrades in a transplant experiment. *Appl. Soil Ecol.* 6, 117–134. doi: 10.1016/S0929-1393(97)00004-8

Briones, M. J. I., Ineson, P., and Sleep, D. (1999). Use of δ^{13}C values to determine food selection in collembolan species. *Soil Biol. Biochem.* 31, 937–940. doi: 10.1016/S0038-0717(98)00179-5

Briones, M. J. I., Moran, P., and Posada, D. (2009). Are the sexual, somatic and genetic characters enough to solve nomenclatural problems in lumbricid taxonomy? *Soil Biol. Biochem.* 41, 2257–2271. doi: 10.1016/j.soilbio.2009.07.008

Briones, M. J. I., and Schmidt, O. (2004). "Stable isotope techniques in studies of the ecological diversity and functions of earthworm communities in agricultural soils," in *Recent Research Developments in Crop Science*, Vol. 1, ed S. G. Pandalai (Trivandrum: Research Signpost).

Brown, V. K., and Gange, A. C. (1990). Insect herbivory below ground. *Adv. Ecol. Res.* 20, 1–58. doi: 10.1016/S0065-2504(08)60052-5

Brown, W. L. Jr. (2000). "Diversity of Ants. Chapter 5," in *Standard Methods for Measuring and Monitoring Biodiversity*, eds D. Agosti, J. D. Majer, L. Alonso, and R. Schultz (Washington, DC: Smithsonian Institution Press), 45–79.

Brussaard, L. (1998). Soil fauna, guilds, functional groups and ecosystem processes. *Appl. Soil Ecol.* 9, 123–135. doi: 10.1016/S0929-1393(98)00066-3

Brussaard, L., Behan-Pelletier, V., Bignell, D., Brown, V., Didden, W., Folgarait, P., et al. (1997). Biodiversity and Ecosystem functioning in Soil. *Ambio* 26, 563–570.

Burtelow, A., Bohlen, P. J., and Groffmann, P. M. (1998). Influence of exotic earthworm invasion on soil organic matter, microbial biomass and denitrification potential in forest soils of the northeastern US. *Appl. Soil Ecol.* 9, 197–202. doi: 10.1016/S0929-1393(98)00075-4

Chamberlain, P. M., Bull, I. D., Black, H. I. J., Ineson, P., and Evershed, R. P. (2005). Fatty acid composition and change in Collembola fed differing diets: identification of trophic biomarkers. *Soil Biol. Biochem.* 37, 1608–1624. doi: 10.1016/j.soilbio.2005.01.022

Chamberlain, P. M., Bull, I. D., Black, H. I. J., Ineson, P., and Evershed, R. P. (2006). Collembolan trophic preferences determined using fatty acid distributions and compound-specific stable carbon isotope values. *Soil Biol. Biochem.* 38, 1275–1281. doi: 10.1016/j.soilbio.2005.09.022

Chang, C. H., Rougerie, R., and Chen, J. H. (2009). Identifying earthworms through DNA barcodes: pitfalls and promise. *Pedobiologia* 52, 171–180. doi: 10.1016/j.pedobi.2008.08.002

Chauvel, A., Grimaldi, M., Barros, E., Blanchart, E., Desjardins, T., Sarrazin, M., et al. (1999). Pasture damage by an Amazonnian earthworm. *Nature* 398, 32–33. doi: 10.1038/17946

Chesson, P., Pacala, S., and Neuhauser, C. (2002). "Environmental niches and ecosystem functioning," in *Functional Consequences of Biodiversity: Experimental Progress and Theoretical Extensions*, eds A. Kinzig, D. Tilman, and S. Pacala (Priceton, NJ: Princeton University Press), 213–245.

Clapp, J. P., Young, J. P. W., Merryweather, J. W., and Fitter, A. H. (1995). Diversity of fungal symbionts in arbuscular mycorrhizas from a natural community. *New Phytol.* 130, 259–265. doi: 10.1111/j.1469-8137.1995.tb03047.x

Cole, L., Dromph, K. M., Boaglio, V., and Bardgett, R. D. (2004). Effect of density and species richness of soil mesofauna on nutrient mineralization and plant growth. *Biol. Fert. Soils* 39, 337–343. doi: 10.1007/s00374-003-0702-6

Coomans, A. (2002). Present status and future of nematode systematics. *Nematology* 4, 573–582. doi: 10.1163/15685410260438836

Copley, J. (2000). Ecology goes underground. *Nature* 406, 452–454. doi: 10.1038/35020131

Cortet, J., Joffre, R., Elmholt, S., and Krogh, P. H. (2002). Increasing species and throphic diversity of mesofauna affects fungal biomass, mesofauna community structure and organic matter decomposition processes. *Biol. Fert. Soils* 37, 302–317. doi: 10.1007/s00374-003-0597-2

Costello, M. J., May, R. M., and Stork, N. E. (2013). Can we name Earth's species before they go extinct? *Science* 339, 413–416. doi: 10.1126/science.1230318

Coûteaux, M. M., and Darbyshire, J. F. (1998). Functional diversity amongst soil protozoa. *Appl. Soil Ecol.* 10, 229–237. doi: 10.1016/S0929-1393(98)00122-X

Criado, R., García, E., Pedroche, F., and Romance, M. (2013). A new method for comparing rankings through complex networks: model and analysis of competitiveness of major European soccer leagues. *Chaos* 23, 043114. doi: 10.1063/1.4826446

Cuddington, K., Wilson, W. G., and Hastings, A. (2009). Ecosystem engineers: feedback and population dynamics. *Am. Nat.* 173, 488–498. doi: 10.1086/597216

Cummins, K. W. (1974). Structure and function of stream ecosystems. *Bioscience* 24, 631–641. doi: 10.2307/1296676

Davidson, D. A., Bruneau, P. M. C., Grieve, I. C., and Young, I. M. (2002). Impacts of fauna on an upland grassland soil as determined by micromorphological analysis. *Appl. Soil Ecol.* 20, 133–143. doi: 10.1016/S0929-1393(02)00017-3

Dawkins, R. (1982). "The genetical evolution of animal artefacts," in *The Extended Phenotype: The Long Reach of a Gene* (New York, NY: Oxford University Press), 195–208.

De Deyn, G. B., and Van der Putten, W. H. (2005). Linking aboveground and belowground diversity. *Trends Ecol. Evol.* 20, 625–633. doi: 10.1016/j.tree.2005.08.009

De Ruiter, P. C., Moore, J. C., Zwart, K. B., Bouwman, L. A., Hassink, J., Bloem, J., et al. (1993). Stimulation of nitrogen mineralisation in the below-ground food webs of two winter wheat fields. *J. Appl. Ecol.* 30, 95–106. doi: 10.2307/2404274

De Ruiter, P. C., Neutel, A.-M., and Moore, J. C. (1995). Energetics, patterns of interaction strengths and stability in real ecosystems. *Science* 269, 1257–1260. doi: 10.1126/science.269.5228.1257

Decaëns, T. (2010). Macroecological patterns in soil communities. *Global Ecol. Biogeogr.* 19, 287–302. doi: 10.1111/j.1466-8238.2009.00517.x

Decaëns, T., Jiménez, J. J., Gioia, C., Measey, G. J., and Lavelle, P. (2006). The values of soil animals for conservation biology. *Eur. J. Soil Biol.* 42, S23–S38. doi: 10.1016/j.ejsobi.2006.07.001

Decaëns, T., Lavelle, P., and Jiménez, J. J. (2008). Priorities for conservation of soil animals. *CAB Rev. Perspect. Agric. Veterin. Sci. Nutr. Nat. Resour.* 3:14. doi: 10.1079/PAVSNNR20083014

Detsis, V. (2000). Vertical distribution of Collembola in deciduous forests under Mediterranean climatic conditions. *Belg. J. Zool.* 130(Suppl.), 55–59. Available online at: http://www.naturalsciences.be/institute/associations/rbzs_website/pdf/abstracts_130_s1/130_s1_9.pdf

Donovan, S. E., Eggleton, P., and Bignell, D. E. (2001). Gut content analysis and a new feeding group classification of termites. *Ecol. Entomol.* 26, 356–366. doi: 10.1046/j.1365-2311.2001.00342.x

Eggleton, P., and Tayasu, I. (2001). Feeding groups, lifetypes and the global ecology of termites. *Ecol. Res.* 16, 941–960. doi: 10.1046/j.1440-1703.2001.00444.x

Emerson, B. C., Cicconardi, F., Fanciulli, P. P., and Shaw, P. J. (2011). Phylogeny, phylogeography, phylobetadiversity and the molecular analysis of biological communities. *Philos. Trans. R. Soc. Lond. B Biol. Sci.* 366, 2391–2402. doi: 10.1098/rstb.2011.0057

Faber, J. H. (1991). Functional classification of soil fauna: a new approach. *Oikos* 62, 110–117. doi: 10.2307/3545458

Faber, J. H., and Verhoef, H. A. (1991). Functional differences between closely-related soil arthropods with respect to decomposition processes in the presence or absence of pine tree roots. *Soil Biol. Biochem.* 23, 15–23. doi: 10.1016/0038-0717(91)90157-F

Ferris, H. (1982). "The role of nematodes as primary consumers," in *Nematodes in Soil Ecosystems*, ed D. W. Freckman (Austin, TX: University of Texas Press), 3–13.

Fierer, N., Leff, J. W., Adams, B. J., Nielsen, U. N., Bates, S. T., Lauber, C. L., et al. (2012). Cross-biome metagenomic analyses of soil microbial communities and their functional attributes. *Proc. Natl. Acad. Sci. U.S.A.* 109, 21390–21395. doi: 10.1073/pnas.1215210110

Filser, J. (2003). The role of Collembola in carbon and nitrogen cycling in soil. *Pedobiologia* 46, 234–245. doi: 10.1078/0031-4056-00130

Fitter, A. H., Gilligan, C. A., Hollingworth, K., Kleczkowski, A., Twyman, R. M., Pitchford, J. W., et al. (2005). Biodiversity and ecosystem function in soil. *Funct. Ecol.* 19, 369–377. doi: 10.1111/j.0269-8463.2005.00969.x

Folgarait, P. (1998). Ant biodiversity and its relationship to ecosystem functioning: a review. *Biodivers. Conserv.* 7, 1221–1244. doi: 10.1023/A:1008891901953

Folke, C., Holling, C. S., and Perrings, C. (1996). Biological biodiversity, ecosystems and the human scale. *Ecol. Appl.* 6, 1018–1024. doi: 10.2307/2269584

Freckleton, R. P., and Watkinson, A. R. (2000). On detecting and measuring competition in spatially structured plant communities. *Ecol. Lett.* 3, 423–432. doi: 10.1046/j.1461-0248.2000.00167.x

Freckman, D. W., Blackburn, T. H., Brussaard, L., Hutchings, P., Palmer, M. A., and Snelgrove, P. V. R. (1997). Linking biodiversity and ecosystem functioning of soil and sediments. *Ambio* 26, 556–562.

Gibb, H., and Cunningham, S. A. (2011). Habitat contrasts reveal a shift in the trophic position of ant assemblages. *J. Anim. Ecol.* 80, 119–127. doi: 10.1111/j.1365-2656.2010.01747.x

Gisin, H. (1943). Ökologie und Lebensgemeinschaften der Collembolen im Schweizerischen Exkursionsgebiet Basels. *Rev. Suisse Zool.* 50, 183–189.

Gómez, C., Casellas, D., Oliveras, J., and Bas, J. M. (2003). Structure of ground-foraging ant assemblages in relation to land-use change in the northwestern Mediterranean region. *Biodivers. Conserv.* 12, 2135–2146. doi: 10.1023/A:1024142415454

Gorny, M. (1984). "Studies on the relationship between enchytraeids and earthworms," in *Soil Biology and Conservation of the Biosphere*, ed J. Szegi (Budapest: Academiai Kiado), 769–776.

Graefe, U., and Beylich, A. (2003). Critical values of soil acidification for annelid species and the decomposer community. *Newslett. Enchytraeidae* 8, 51–55. Available online at: http://www.ifab-hamburg.de/documents/GraefeBeylich2003.pdf

Grassé, P. P. (1984). *Termitologia*. Paris: Masson.

Gray, N. D., Hastings, R. C., Sheppard, S. K., Loughnane, P., Lloyd, D., McCarthy, A. J., et al. (2003). Effects of soil improvement treatments on bacterial community structure and soil processes in an upland grassland soil. *FEMS Microbiol. Ecol.* 46, 11–22. doi: 10.1016/S0168-6496(03)00160-0

Greenslade, P. J. M., and Halliday, R. B. (1983). Colony dispersion and relationships of meat ants *Iridomyrmex purpureus* and allies in an arid locality in South Australia. *Insect Soc.* 30, 82–99. doi: 10.1007/BF02225659

Hågvar, S. (1998). The relevance of the Rio Convention on biodiversity to conserving the biodiversity of soils. *Appl. Soil Ecol.* 9, 1–7. doi: 10.1016/S0929-1393(98)00115-2

Hartmann, A., Rothballer, M., and Schmid, M. (2008). Lorenz Hiltner, a pioneer in rhizosphere microbial ecology and soil bacteriology research. *Plant Soil* 312, 7–14. doi: 10.1007/s11104-007-9514-z

Hawkins, C. P., and MacMahon, J. A. (1989). Guilds: the multiple meanings of a concept. *Ann. Rev. Ent.* 34, 423–451. doi: 10.1146/annurev.en.34.010189.002231

Hawksworth, D. L., and Mound, L. A. (1991). "Biodiversity databases: the crucial significance of collections," in *The Biodiversity of Microorganisms and Invertebrates: Its Role in Sustainable Agriculture*, ed D. L. Hawksworth (Wallingford, WA: CAB International), 17–31.

Hedde, M., Lavelle, P., Joffre, R., Jimenez, J. J., and Decaëns, T. (2005). Specific functional signature in soil-macroinvertebrate biostructures. *Funct. Ecol.* 19, 785–793. doi: 10.1111/j.1365-2435.2005.01026.x

Heemsbergen, D. A., Berg, M. P., Loreau, M., van Hal, J. R., Faber, J. H., and Verhoef, H. A. (2004). Biodiversity effects on soil processes explained by interspecific functional dissimilarity. *Science* 306, 1019–1020. doi: 10.1126/science.1101865

Heisler, C. L., Wickenbrock, L., and Lübben, H. (1996), Oberflächenstruktur, Aggregatstabilität sowie Durchwurzelbarkeit des Bodens unter dem Einfluss ausgewählter Bodentiergruppen. *Z. Ökol. Natursch.* 5, 97–105.

Hiltner, L. (1904). Ueber neuere Erfahrungen und Probleme auf dem Gebiete der Bodenbakteriologie und unter besonderer BerUcksichtigung der Grundungung und Brache. *Arbeiten der Deutsche Landwirtschafts-Gesellschaft* 98, 59–78.

Hodgson, J. G., Wilson, P. J., Hunt, R., Grime, J. P., and Thompson, K. (1999) Allocating C-S-R plant functional types: a soft approach to a hard problem. *Oikos* 85, 282–294. doi: 10.2307/3546494

Hodkinson, I. D., and Wookey, P. A. (1999). Functional ecology of soil organisms in tundra ecosystems: towards the future. *Appl. Soil Ecol.* 11, 111–126. doi: 10.1016/S0929-1393(98)00142-5

Hooper, D. U., Chapin, F. S. III., Ewel, J. J, Hector, A., Inchausti, P., Lavorel, S., et al. (2005). Effects of biodiversity on ecosystem functioning: a consensus of current knowledge. *Ecol. Monogr.* 75, 3–35. doi: 10.1890/04-0922

Hooper, D. U., and Dukes, J. S. (2004). Overyielding among plant functional groups in a long-term experiment. *Ecol. Lett.* 7, 95–105. doi: 10.1046/j.1461-0248.2003.00555.x

Hooper, D., Hawksworth, D., and Dhillion, S. (1995). "Microbial diversity and ecosystem processes," in *Global Biodiversity Assessment*, ed United Nations Environment Programme (Cambridge: Cambridge University Press), 433–443.

Huang, J., Xu, Q., Sun, Z. J., Tang, G. L., and Su, Z. Y. (2007). Identifying earthworms through DNA barcodes. *Pedobiologia* 51, 301–309. doi: 10.1016/j.pedobi.2007.05.003

Huhta, V. (2007). The role of soil fauna in ecosystems: a historical review. *Pedobiologia* 50, 489–495. doi: 10.1016/j.pedobi.2006.08.006

Huhta, V., Hyvönen, R., Kaasalainen, P., Koskenniemi, A., Muona, J., Mäkelä, I., et al. (1986). Soil fauna of Finnish coniferous forests. *Ann. Zool. Fenn.* 23, 345–360.

Huhta, V., Sulkava, P., and Viberg, K. (1998). Interactions between enchytraeid (*Cognettia sphagnetorum*), microarthropod and nematode populations in forest soil at different moistures. *App. Soil Ecol.* 9, 53–58. doi: 10.1016/S0929-1393(98)00053-5

Hunt, H. W., Coleman, D. C., Ingham, E. R., Ingham, R. E., Elliott, E. T., Moore, J. C., et al. (1987). The detrital foodweb in a shortgrass prairie. *Biol. Fertil. Soils* 3, 57–68.

Jones, C. G., Lawton, J. H., and Shachak, M. (1994). Organisms as ecosystem engineers. *Oikos* 69, 373–386. doi: 10.2307/3545850

Jones, C. G., Lawton, J. H., and Shachak, M. (1997). Positive and negative effects of organisms as physical ecosystem engineers. *Ecology* 78, 1946–1957. doi: 10.1890/0012-9658(1997)078[1946:PANEOO]2.0.CO;2

Jouquet, P., Dauber, J., Lagerlöf, J., Lavelle, P., and Lepage, M. (2006). Soil invertebrates as ecosystem engineers: intended accidental effects on soil and feedback loops. *Appl. Soil Ecol.* 32, 153–164. doi: 10.1016/j.apsoil.2005.07.004

Keudel, M., and Schrader, S. (1999). Axial and radial pressure exerted by earthworms of different ecological groups. *Biol. Fert. Soils* 29, 262–269. doi: 10.1007/s003740050551

Kibblewhite, M. G., Ritz, K., and Swift, M. J. (2008). Soil health in agricultural systems. *Philos. Trans. R. Soc. Lond. B Biol. Sci.* 363, 685–701. doi: 10.1098/rstb.2007.2178

King, R. A., Tibble, A. L., and Symondson, W. O. C. (2008). Opening a can of worms: unprecedented sympatric cryptic diversity within British lumbricid earthworms. *Mol. Ecol.* 17, 4684–4698. doi: 10.1111/j.1365-294X.2008.03931.x

Krištůfek, V., Nováková, A., and Pižl, V. (2001). Coprophilous streptomycetes and fungi—food sources for enchytraeid worms (Enchytraeidae). *Folia Microbiol.* 46, 555–558. doi: 10.1007/BF02818002

Laakso, J., and Setälä, H. (1999). Sensitivity of primary production to changes in the architecture of below-ground food webs. *Oikos* 87, 57–64. doi: 10.2307/3546996

Lavelle, P. (1979) Relations entre types ecologiques et profiles demographiques chez les vers de terre de la savanne de Lamto (Cote d'Ivore). *Rev. Ecol. Biol. Sol.* 16, 85–101.

Lavelle, P. (1983). "The structure of earthworm communities," in *Earthworm Ecology: from Darwin to Vermiculture*, ed J. E. Satchell (London: Chapman and Hall), 449–466. doi: 10.1007/978-94-009-5965-1_39

Lavelle, P. (1988). Earthworms and the soil system. *Biol. Fert. Soils* 6, 237–251. doi: 10.1007/BF00260820

Lavelle, P. (1996). Diversity of soil fauna and ecosystem function. *Biol. Int.* 33, 3–16.

Lavelle, P. (2002). Functional domains in soils. *Ecol. Res.* 17, 551–450. doi: 10.1046/j.1440-1703.2002.00509.x

Lavelle, P., Bignell, D., Lepage, M., Wolters, V., Roger, P., Ineson, P., et al. (1997). Soil function in a changing world: the role of invertebrate ecosystem engineers. *Eur. J. Soil Biol.* 33, 159–193.

Lavelle, P., Blanchart, E., Martin, A., Martin, S., Barois, I., Toutain, F., et al. (1993). A hierarchical model for decomposition in terrestrial ecosystems. Application to soils in the humid tropics. *Biotropica* 25, 130–150. doi: 10.2307/2389178

Liiri, M., Setälä, H., Jaimi, J., Pennanen, H., and Fritze, H. (2002). Soil processes are not influenced by the functional complexity of soil decomposer foodwebs under disturbance. *Soil Biol. Biochem.* 34, 1009–1020. doi: 10.1016/S0038-0717(02)00034-2

Loreau, M., Naeem, S., Inchausti, P., Bengtsson, J., Grime, J. P., Hector, A., et al. (2001). Biodiversity and ecosystem functioning: current knowledge and future challenges. *Science* 294, 804–808. doi: 10.1126/science.1064088

Luxton, M. (1972). Studies on the oribatid mites of a Danish beech-wood soil. I. Nutritional biology. *Pedobiologia* 12, 434–463.

MacFadyen, A. (1963). "The contribution of the fauna to the total soil metabolism," in *Soil Organisms*, eds J. Doeksen and J. van der Drift (Amsterdam: North Holland Publishing Company), 3–17.

MacFadyen, A. (1964). Relations between mites and microorganisms and their significance in soil biology. *Acarologia* 6, 147–149.

Marichal, R., Martinez, A. F., Praxedes, C., Ruiz, D., Carvajal, A. F., Oszwald, J., et al. (2010). Invasion of *Pontoscolex corethrurus* (Glossoscolecidae, Oligochaeta) in landscapes of the Amazonian deforestation arc. *Appl. Soil Ecol.* 46, 443–449. doi: 10.1016/j.apsoil.2010.09.001

Marilley, L., and Aragno, M. (1999). Phylogenetic diversity of bacterial communities from different proximity to *Lolium perenne* and *Trifolium repens* roots. *Appl. Soil Ecol.* 13, 127–136. doi: 10.1016/S0929-1393(99)00028-1

Martinsson, S., and Erséus, C. (2014). Cryptic diversity in the well-studied terrestrial worm *Cognettia sphagnetorum* (Clitellata: Enchytraeidae). *Pedobiologia* 57, 27–35. doi: 10.1016/j.pedobi.2013.09.006

McCann, K. S. (2000). The diversity–stability debate. *Nature* 405, 228–233. doi: 10.1038/35012234

Moody, S. A., Briones, M. J. I., Piearce, T. G., and Dighton, J. (1995). Selective consumption of decomposing wheat straw by earthworms. *Soil Biol. Biochem.* 27, 1209–1213. doi: 10.1016/0038-0717(95)00024-9

Moore, J. C., and De Ruiter, P. C. (1991). Temporal and spatial heterogeneity of trophic interactions within below-ground food webs. *Agr. Ecosyst. Environ.* 34, 371–397. doi: 10.1016/0167-8809(91)90122-E

Moore, J. C., Walter, D. E., and Hunt, H. W. (1988). Arthropod regulation of micro-mesobiota in belowground detrital food webs. *Ann. Rev. Entomol.* 33, 419–439. doi: 10.1146/annurev.en.33.010188.002223

Mulder, C. (2006). Driving forces from soil invertebrates to ecosystem functioning: the allometric perspective. *Naturwissenschaften* 93, 467–479. doi: 10.1007/s00114-006-0130-1

Ostle, N., Briones, M. J. I., Ineson, P., Cole, L., Staddon, P., and Sleep, D. (2007). Isotopic detection of recent photosynthate carbon flow into grassland rhizosphere fauna. *Soil Biol. Biochem.* 39, 768–777. doi: 10.1016/j.soilbio.2006.09.025

Paton, T. R., Humphreys, G. S., and Mitchell, P. B. (1995). *Soils: A New Global View.* New Haven, CT: Yale University Press.

Pawluk, S. (1987). Faunal micromorphological features in moder humus of some western Canadian soils. *Geoderma* 40, 3–16. doi: 10.1016/0016-7061(87)90010-3

Persson, T. (1989). Role of soil animals in C and N mineralisation. *Plant Soil* 115, 241–245. doi: 10.1007/BF02202592

Persson, T., Bååth, E., Clarholm, M., Lundkvist, H., Söderström, B. E., and Sohlenius, B. (1980). "Trophic structure, biomass dynamics and carbon metabolism of soil organisms in a Scots pine forest," in *Structure and Function of Northern Coniferous Forest: An Ecosystem Study*, ed T. Persson (Stockholm: Ecological Bulletins, vol. 32), 419–459.

Petersen, H. (1975). Estimation of dry weight, fresh weight, and calorific content of various Collembolan species. *Pedobiologia* 15, 222–243.

Petersen, H., and Luxton, M. (1982). A comparative analysis of soil fauna populations and their role in decomposition processes. *Oikos* 39, 287–388. doi: 10.2307/3544689

Pfisterer, A. B., Joshi, J., Schmid, B., and Fischer, M. (2004). Rapid decay of diversity-productivity relationships after invasion in experimental plant communities. *Basic Appl. Ecol.* 5, 5–14. doi: 10.1078/1439-1791-00215

Pfisterer, A. B., and Schmid, B. (2002). Diversity-dependent production can decrease the stability of ecosystem functioning. *Nature* 416, 84–86. doi: 10.1038/416084a

Polis, G. A. (1991). Complex trophic interactions in deserts: an empirical critique of food-web theory. *Am. Nat.* 138, 123–155. doi: 10.1086/285208

Ponsard, S., and Arditi, R. (2000). What can stable isotopes (δ^{15}N and δ^{13}C) tell about the food web of soil macro-invertebrates? *Ecology* 81, 852–864. doi: 10.1890/0012-9658(2000)081[0852:WCSINA]2.0.CO;2

Porco, D., Bedos, A., Greenslade, P., Janion, C., Skarzynsky, D., Stevens, M. I., et al. (2012). Challenging species delimitation in Çollembola: cryptic diversity among common springtails unveiled by DNA barcoding. *Invertebr. Syst.* 26, 470–477. doi: 10.1071/IS12026

Porco, D., Decaëns, T., Deharveng, L., James, S. W., Skarzynski, D., Erseus, C., et al. (2013). Biological invasions in soil: DNA barcoding as a monitoring tool in a multiple taxa survey targeting European earthworms and springtails in North America. *Biol. Invasions* 15, 899–910. doi: 10.1007/s10530-012-0338-2

Power, M. E., and Mills, L. S. (1995). The keystone cops meet in Hilo. *Trends Ecol. Evol.* 10, 182–184. doi: 10.1016/S0169-5347(00)89047-3

Pratt, J. R., and Cairns, J. (1985). Functional groups in the Protozoa: roles in differing ecosystems. *J. Protozool.* 32, 415–423. doi: 10.1111/j.1550-7408.1985.tb04037.x

Richard, B. (2008). *Evaluation de la Diversité Specifique des vers de Terre de Haute Normandie par L'utilisation des Codes Barres ADN.* Master's thesis, Lyon: Université de Lyon.

Rigamonte, T. A., Pylro, V. S., and Duarte, G. F. (2010). The role of mycorrhization helper bacteria in the establishment and action of ectomycorrhizae associations. *Braz. J. Microbiol.* 41, 832–840. doi: 10.1590/S1517-83822010000400002

Roesner, J. (1986). Untersuchungen zur Reproduktion von Nematoden in Boden durch Regenwürmer. *Mededelingen van de Faculteit Landbouwwetenschappen Rijkuniversiteit Gent* 51, 1311–1318.

Root, R. B. (1967). The niche exploitation pattern of the blue-gray gnatcatcher. *Ecol. Monogr.* 37, 317–350. doi: 10.2307/1942327

Root, R. B. (1973). Organization of a plant-arthropod association in simple and diverse habitats: the fauna of collards (*Brassica oleracea*). *Ecol. Monogr.* 43, 95–124. doi: 10.2307/1942161

Rougerie, R., Decaëns, T., Deharveng, L., Porco, D., James, S. W., Chang, C.-H., et al. (2009). DNA bar-codes for soil animal taxonomy: transcending the final frontier. *Pesqui. Agropecu. Bras.* 44, 789–801. doi: 10.1590/S0100-204X2009000800002

Rückamp, D., Amelung, W., Theisz, N., Bandeira, A. G., and Martius, C. (2010). Phosphorous forms in Brazilian termite nests and soils: relevance of feeding guild and ecosystems. *Geoderma* 155, 269–279. doi: 10.1016/j.geoderma.2009.12.010

Ruess, L., Haggblom, M. M., Langel, R., and Scheu, S. (2004). Nitrogen isotope ratios and fatty acid composition as indicators of animal diets in belowground systems. *Oecologia* 139, 336–346. doi: 10.1007/s00442-004-1514-6

Rusek, J. (1998). Biodiversity of Collembola and their functional role in the ecosystem. *Biodivers. Conserv.* 7, 1207–1219. doi: 10.1023/A:1008887817883

Scheu, S., and Falca, M (2000). The soil food web of two beech forests (*Fagus sylvatica*) of contrasting humus type: stable isotope analysis of a macro- and a mesofauna-dominated community. *Oecologia* 123, 285–296. doi: 10.1007/s004420051015

Scheu, S., and Setälä, H. (2002). "Multitrophic interactions in decomposer food-webs," in *Multitrophic Level Interactions*, eds B. Tscharntke and B. A. Hawkins (Cambridge: Cambridge University Press), 223–264. doi: 10.1017/CBO9780511542190.010

Schneider, K., Migge, S., Norton, R. A., Scheu, S., Langel, R., Reineking, A., et al. (2004). Trophic niche differentiation in soil microarthropods (Oribatida: Acari): evidence from stable isotope ratios (^{15}N/^{14}N). *Soil Biol. Biochem.* 36, 1769–1774. doi: 10.1016/j.soilbio.2004.04.033

Schulz, E., and Scheu, S. (1994). Oribatid mite mediated changes in litter decomposition: model experiments with ^{14}C-labelled holocellulose. *Pedobiologia* 38, 344–352.

Setälä, H. (1995). Growth of birch and pine seedlings in relation to grazing by soil fauna on ectomycorrhizal fungi. *Ecology* 76, 1844–1851. doi: 10.2307/1940716

Setälä, H. (2000). Reciprocal interactions between Scots pine and soil web structure in the presence and absence of ectomycorrhiza. *Oecologia* 125, 109–118. doi: 10.1007/PL00008881

Setälä, H., and Huhta, V. (1991). Soil fauna increase *Betula pendula* growth: laboratory experiments with coniferous forest floor. *Ecology* 72, 665–671. doi: 10.2307/2937206

Setälä, H., Marshall, V. G., and Trofymow, J. A (1996). Influence of body size of soil fauna on litter decomposition and ^{15}N uptake by poplar in a pot trial. *Soil Biol. Biochem.* 28, 1661–1675. doi: 10.1016/S0038-0717(96)00252-0

Siepel, H. (1994). Life-history tactics of soil microarthropods. *Biol. Fertil. Soils* 18, 263–278. doi: 10.1007/BF00570628

Siepel, H., and De Ruiter-Dijkman, E. M. (1993). Feeding guilds of oribatid mites based on their carbohydrase activities. *Soil Biol. Biochem.* 25, 1491–1497. doi: 10.1016/0038-0717(93)90004-U

Standen, V., and Latter, P. M. (1977). Distribution of a population of *Cognettia sphagnetorum* (Enchytraeidae) in relation to microhabitats in a blanket bog. *J. Anim. Ecol.* 46, 213–229. doi: 10.2307/3957

Stevens, M. H. H., and Carson, W. P. (2001). Phenological complementarity, species diversity and ecosystem functioning. *Oikos* 92, 291–296. doi: 10.1034/j.1600-0706.2001.920211.x

Stoll, P., and Prati, D. (2001). Intraspecific aggregation alters competitive interactions in experimental plant communities. *Ecology* 82, 319–327. doi: 10.1890/0012-9658(2001)082[0319:IAACII]2.0.CO;2

Swift, M. J., Heal, O. W., and Anderson, J. M. (1979). *Decomposition in Terrestrial Ecosystems.* Oxford: Blackwell Science.

Takeda, H. (1995). Changes in the collembolan community during the decomposition of needle litter in a coniferous forest. *Pedobiologia* 39, 304–317.

Tayasu, I., Hyodo, F., Abe, T., Inoue, T., and Spain, A. V. (2002). Nitrogen and carbon stable isotope ratios in the sympatric Australian termites, *Amitermes laurensis* and *Drepanotermes rubriceps* (Isoptera: Termitidae) in relation to their feeding habits and the quality of their food materials. *Soil Biol. Biochem.* 34, 297–301. doi: 10.1016/S0038-0717(01)00181-X

Torsvik, V., Goksoyr, J., Daae, F. L., Sorheim, R., Michalsen, J., and Salte, K. (1994). "Use of DNA analysis to determine the diversity of microbial communities," in *Beyond the Biomass*, eds K. Ritz, J. Dighton, and K. E. Giller (Chichester: John Wiley and Sons), 39–48.

Torsvik, V., and Øvreås, L. (2002). Microbial diversity and function in soil: from genes to ecosystems. *Curr. Opin. Microbiol.* 5, 240–245. doi: 10.1016/S1369-5274(02)00324-7

Uroz, S., Courty, P. E., Pierrat, J. C., Peter, M., Buee, M., Turpault, M. P., et al. (2013). Functional profiling and distribution of the forest soil bacterial communities along the soil mycorrhizosphere continuum. *Microbiol. Ecol.* 66, 404–441. doi: 10.1007/s00248-013-0199-y

Van der Drift, J. (1965). "The effects of animal activity in the litter layer," in *Experimental Pedology*, ed E. G. Hallsworth and D. V. Crawford (London: Butterworths), 227–235.

Verhoef, H. A., and Witteveen, J. (1980). Water balance in Collembola and its relation to habitat selection: cuticular water loss and water uptake. *J. Insect Physiol.* 26, 201–208. doi: 10.1016/0022-1910(80)90081-5

Vineesh, P. J., Sabu, T. K., and Karmaly, K. A. (2007). Community structure and functional group classification of litter ants in the montane evergreen and deciduous forests of Wayanad region of western Ghats, southern India. *Orient. Insects* 41, 427–442. doi: 10.1080/00305316.2007.10417526

Waid, J. (1999). Does soil biodiversity depend upon metabiotic activity and influences? *Appl. Soil Ecol.* 13, 151–158. doi: 10.1016/S0929-1393(99)00030-X

Wall, D. H., Bradford, M. A., John, M. G. S. T., Trofymow, J. A., Behan-Pelletier, V., Bignell, D. E., et al. (2008). Global decomposition experiment shows soil animal impacts on decomposition are climate-dependent. *Glob. Change Biol.* 14, 2661–2677.

Wallwork, J. A. (1970). *Ecology of Soil Animals.* London: McGraw-Hill.

Walter, D. E., Hunt, H. W., and Elliot, E. T. (1988). Guilds or functional groups? An analysis of predatory arthropods from a shortgrass steppe soil. *Pedobiologia* 31, 247–260.

Walter, D. E., and Proctor, H. C. (1999). *Mites: Ecology, Evolution and Behaviour.* Sydney; New York, NY: University of New South Wales Press; CAB International.

Wardle, D. A. (2002). *Communities and Ecosystems.* Princeton, NJ: Princeton University Press.

Wardle, D. A., and Lavelle, P. (1996). "Linkages between soil biota, plant litter quality and decomposition," in *Driven by Nature: Plant Litter Quality and Decomposition*, eds G. Cadisch and K. E. Giller (Wallingford, WA: CAB International), 107–123.

Wolters, V. (2001). Biodiversity of soil animals and its function. *Eur. J. Soil Biol.* 37, 221–227. doi: 10.1016/S1164-5563(01)01088-3

Wood, T. G. (1988) Termites and the soil environment. *Biol. Fert. Soils* 6, 228–236. doi: 10.1007/BF00260819

Yachi, S., and Loreau, M. (1999). Biodiversity and ecosystem productivity in a fluctuating environment: the insurance hypothesis. *Proc. Natl. Acad. Sci. U.S.A.* 96, 1463–1468. doi: 10.1073/pnas.96.4.1463

Yeates, G. W., Bongers, T., de Goede, R. G. M., Freckman, D. W., and Georgieva, S. S. (1993). Feeding habits in soil nematode families and genera—an outline for soil ecologists. *J. Nematol.* 25, 315–331.

Conflict of Interest Statement: The author declares that the research was conducted in the absence of any commercial or financial relationships that could be construed as a potential conflict of interest.

4

Physics of the soil medium organization part 1: thermodynamic formulation of the pedostructure water retention and shrinkage curves

Erik Braudeau[1,2]*, Amjad T. Assi[1,3], Hassan Boukcim[4] and Rabi H. Mohtar[5]

[1] *Qatar Foundation, Qatar Environment and Energy Research Institute, Doha, Qatar*
[2] *Institut de Recherche pour le Développement, Bondy, France*
[3] *Department of Agricultural and Biological Engineering, Purdue University, West Lafayette, IN, USA*
[4] *Valorhiz SAS, Parc Scientifique Agropolis II Bat 6, Montferrier-sur-Lez, France*
[5] *Biological and Agricultural Engineering Department, and Zachry Department of Civil Engineering, Texas A&M University, College Station, TX, USA*

Edited by:
Christophe Darnault, Clemson University, USA

Reviewed by:
Wieslaw Fialkiewicz, Wroclaw University of Environmental and Life Sciences, Poland
Federico Maggi, The University of Sydney, Australia

*Correspondence:
Erik Braudeau, Institut de Recherche pour le Développement, Pédologie Hydrostructurale, 32 Avenue Henri Varagnat, 93140 Bondy, France
e-mail: erik.braudeau@ird.fr

The equations used in soil physics to characterize the hydro-physical properties of the soil medium cannot be other than empirical since they do not take into account the multi-scale functional organization of the soil medium that is described in Pedology. To allow researching the correct formulation of the physical equations describing the soil medium organization and properties, a new paradigm of hydrostructural pedology is being developed. This paradigm is to establish the conceptual link between the classical Pedology and the soil-water physics (hydrostructural characterization and modeling of the soil medium). The paradigm requires the exclusive use of the concept of Structural Representative Elementary Volume (SREV) instead of the classical Representative Elementary Volume (REV) in any physical modeling of the hydrostructural behavior of the soil medium and of the links with the biotic or abiotic processes evolving within it. This article presents the development of the physical equations of the shrinkage curve and the soil water retention curve from the thermodynamic point of view according to the new paradigm. The new equations were tested and the theory validated using data of simultaneous measurement of both curves on a cylindrical soil sample (pedostructure). Implications of these results on the physical modeling in agro environmental sciences are discussed.

Keywords: pedostructure, structural representative elementary volume (SREV) concept, soil shrinkage curve, soil water retention curve, thermodynamic internal energy, Gibbs free energy

INTRODUCTION

The soil system, as it is actually considered in agro-environmental sciences, is represented as a vertical succession of horizontal layers defined as Representative Elementary Volumes (REVs) of the soil medium. This representation resulted in having similar and averaged hydrostructural properties of the soil medium. In this approach, all descriptive variables of the soil medium are average variables referenced to a volume (the REV) of which the internal organization, as well as its own delimitation, is ignored. It is obvious that no physical equation of these hydrostructural properties of the soil medium can be found if one ignores its internal organization; therefore, it is still a difficult task to define and describe any physical interaction between the water molecules and the solid particles making the structure (infrastructure) of this organization.

The use of the REV concept produced several empirical equations that are used in soil physics to characterize the hydro-physical properties of the soil medium, which imposes limitations upon the coupling of the soil physics with other agro-environmental disciplines (Ahuja et al., 2006). For example, the soil water retention curve [h(θ)] which links the volumetric water content (θ) [m³m⁻³ of soil] to the soil matrix pressure head (h)

is still not physically determined and still represented by different parametric equations (El Kadi, 1985; Leij et al., 1997), showing that it is not coming from a unique theory of the soil medium organization. The case is the same with the other curves such as the hydraulic conductivity curve (Leij et al., 1997) and the soil shrinkage curve (Cornelis et al., 2006; Chertkov, 2012).

Braudeau and Mohtar (2009) proposed a new approach for characterizing and modeling the structured soil medium organization and its physical interaction with water. In this new approach, the basic concept of REV was replaced by the concept of Structural Representative Elementary Volume (SREV). This replacement was compulsory for defining a closed hydro-thermo-dynamic systems on the soil structure. Thus, in contrary to the REV, the SREV concept takes into account the soil structure and its hierarchy (**Figure 1**). Another complementary concept was used in the new approach, the pedostructure concept, which was proposed by Braudeau et al. (2004). Pedostructure concept defined and described the soil structure of a soil medium at its first level of organization as an assembly of primary peds and eventually inert mineral grains (sand). All the components present in a certain volume of the pedostructure are referred to the mass of the solid phase

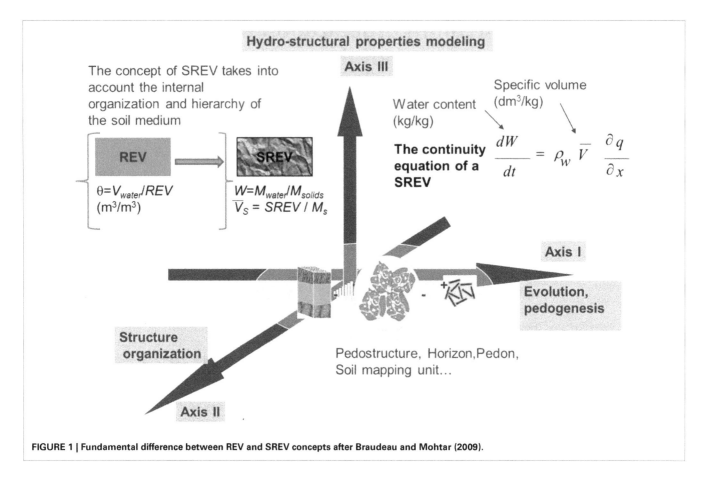

FIGURE 1 | Fundamental difference between REV and SREV concepts after Braudeau and Mohtar (2009).

contained in this volume instead of being referred to the volume itself like in the REV concept (**Figure 1**). The change of REV to SREV leads to a completely different system of descriptive variables for the pedon and its internal organization like the horizons and pedostructures as mentioned in **Table 1**. Using this new system of variables the description of the pedostructure as an assembly of primary peds put in light the two complementary pore systems, said micro (intra primary peds) and macro (inter primary peds), of which the different hydrostructural behaviors are at the basis of all the soil water properties.

These two concepts, pedostructure and SREV, are at the basis of the new paradigm of hydrostructural pedology defining variables and physical units (**Table 1**) related to the pedon and its internal organization represented in **Figure 2**.

The computer model, Kamel® (Braudeau et al., 2009) can be considered as representative of the generated new discipline, highlighting the following two key points:

- The SREV concept and its system of specific variables, extensive and intensive, replaced the REV concept and its system of normalized volumetric variables (densities, water contents in m^3/m^3). Then, it was used for the description and simulation of all dynamic processes inside the pedon.
- The distinction of the pedostructure water content (W) into W_{mi} and W_{ma}, as micro and macro water contents of the

pedostructure, located respectively inside and outside of the plasmic porosity of primary peds.

In this article, and continuing to build on the two key points above, the authors wanted to apply the SREV concept to the thermodynamic theory of soil water written by Sposito (1981) based on the REV concept (hypothesis of homogeneous mixture of a tri-phasic "solids, water, air" soil medium). Therefore, this paper presents a new formulation of the thermodynamic functions and state equations of the soil medium, represented by its pedostructure. In particular, the thermodynamic formulation of the micro and macro pedostructure water contents at equilibrium [$W_{ma}^{eq}(W)$ and $W_{mi}^{eq}(W)$], i.e., at equality of water potentials inside and outside of the primary peds at each equilibrium step of water removal from the pedostructure. The theory is presented hereafter establishing the generalized equations for the soil water retention curve $h(W)$ and the shrinkage curve $\overline{V}(W)$. These equations will be validated in laboratory and compared with the previous equations proposed by Braudeau and Mohtar (2004, 2009), according to the pedostructure concept but without the notions of thermodynamic equilibrium of the soil medium.

THEORY

HIGHLIGHTING THE TWO TYPES OF WATER, INTRA AND INTER PRIMARY PEDS, IN THE SOIL MEDIUM

The distinction between internal and external water of the micro particles of a soil sample (W_{mi} and W_{ma}) was first quantitatively

Table 1 | Pedostructure state variables.

Volume of concern	Specific volume [dm³/kg]	Specific pore volume [dm³/kg]	Specific water content [kg$_{water}$/kg$_{soil}$]	Non saturating water [kg$_{water}$/kg$_{soil}$]	Saturating water [kg$_{water}$/kg$_{soil}$]	Suction [kPa]
Pedostructure	$\overline{V}$	$\overline{Vp}$	W			h
Interpedal porosity		$\overline{Vp}_{ma}$	W_{ma}	w_{st}	w_{ip}	h_{ma}, h_{ip}
Primary peds	$\overline{V}_{mi}$	$\overline{Vp}_{mi}$	W_{mi}	w_{re}	w_{bs}	h_{mi}
Primary particles	$\overline{V}_s$					

Subscripts mi, ma, and s; refer to micro,macro, and solids; ip, st, bs, and re, refer to the name of the corresponding shrinkage phase of the shrinkage curve: interpedal, structural, basic, and residual after Braudeau et al. (2009).

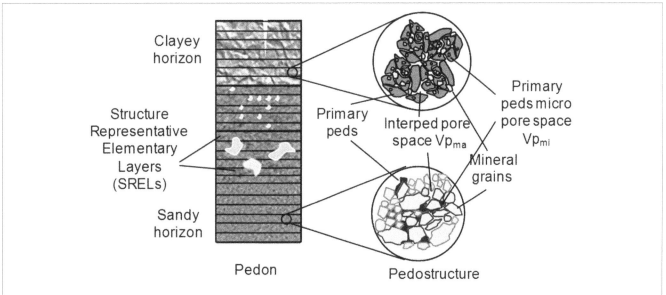

FIGURE 2 | Representation of the internal organization of the pedon, hierarchized into its hydro-functional levels of organization: horizons, pedostructure, primary peds after Braudeau et al. (2009).

established by Braudeau (1988) from the interpretation of the soil shrinkage curve. In such testing, the shrinkage curve was continuously measured on soil samples of approximately 100 cm³ using very sensitive displacement sensors. This distinction of the water content (W) into both deformable micro- and macro-pore systems was quantitatively formulated assuming that the air entry point in the micropore system (primary peds) of the soil, separating *de facto* both nested micro and macro systems, can be accurately read from the measured shrinkage curve at the end of the basic shrinkage phase corresponding to the point B on the curve (**Figure 3**). This was confirmed by mercury porosimetry measurements (Braudeau and Bruand, 1993). In fact, this air entry point delineates conceptually and quantitatively the two nested pore systems: the micro pore system, that of the plasma composing primary peds, and the interpedal macro pore system at the surface outside of the primary peds and complementary to them in the pedostructure system volume (**Figure 3, Table 1**).

Two formulations have been given to these types of water at equilibrium in terms of water content *(W)*: an empirical formulation that resulted in an excellent fit with observed shrinkage curves [$\overline{V}(W)$] (Braudeau, 1988; Braudeau and Bruand, 1993;

Braudeau et al., 1999; Boivin et al., 2004; Boivin, 2007). Then, a physical formulation that based on a probabilistic reasoning by simultaneously removing those two types of water from both systems of the pedostructure during evaporation (Braudeau et al., 2004), such that:

$$W_{ma} = -(1/k_M)\,\mathrm{Ln}\big(1 + \exp(-k_M(W - W_M))\big)$$
$$\text{with } k_M < 0; \qquad (1a)$$

and

$$W_{mi} = W - W_{ma}$$
$$= W + (1/k_M)\,\mathrm{Ln}\big(1 + \exp(-k_M(W - W_M))\big) \quad (1b)$$

In these equations: W_{mi} and W_{ma} are considered complementary in the pedostructure specific pore volume $\overline{Vp}$. While, the pedostructure parameters: k_M and W_M are characteristics of the hydro-structural behavior of the soil medium within the range of the shrinkage curve between the structural and the basic shrinkage phases, of slope K_{st} and K_{ip}, respectively (**Figure 3**).

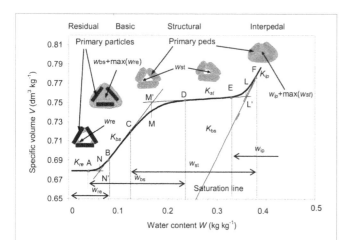

FIGURE 3 | Various configurations of air and water partitioning into the two pore systems, inter and intra primary peds, related to the shrinkage phases of a standard Shrinkage Curve [water content vs. specific volume]. The various water pools [wre, wbs, wst, and wip] are represented with their domain of variations. The linear and curvilinear shrinkage phases are delimited by the transition points (A–F). Points N', M', and L' are the intersection points of the tangents at those linear phases of the Shrinkage Curve. Adapted from Braudeau et al. (2004).

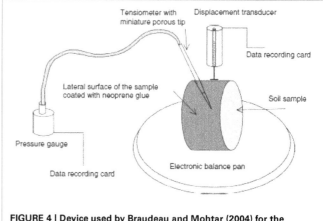

FIGURE 4 | Device used by Braudeau and Mohtar (2004) for the simultaneous measurement of the soil water retention and soil shrinkage curves of a cylindrical soil sample.

Hence, according to (1a) and (1b), W_M is equal in value to the maximum value of W_{mi}, i.e., W_{miSat}, and k_M is such that $k_M = -Ln(2)/W_{maM}$; they are parameters related to the sense and extension of the curvature near point M of the shrinkage curve on **Figure 3**. These parameters can be determined by fitting Equation (1) on the corresponding shrinkage range of the shrinkage curve which, on the other hand, should be continuously measured on soil samples according the methodology of Braudeau et al. (1999).

Then, in the study the soil water retention curve, $h(W)$, associated to the shrinkage curve, $\overline{V}(W)$, of a soil sample using the device shown on **Figure 4**; Braudeau and Mohtar (2004) assumed that $h(W)$ was directly linked to the macropore water content and is, in fact, governed by W_{ma} instead of W. They rewrote different equations of $h(W)$ proposed in the literature, but in terms of W_{ma} rather than W according to the pedostructure concept. The equations tested were physically established by their authors (Berezin et al., 1983; Low, 1987; Rieu and Sposito, 1991) on the basis of different hypotheses made about the hydro-structural properties of the soil medium (fractal and thermodynamic) compatible to the pedostructure terminology (variables and parameters). Among them, the thermodynamic equation of Berezin et al. (1983), was modified into:

$$h_{ma} = \rho_w \overline{E}_{ma} \left(1/(W_{ma} + \sigma) - 1/(W_{maSat} + \sigma)\right) \quad (2a)$$

$$h_{mi} = \rho_w \overline{E}_{mi} \left(1/(W_{mi} - w_{re}) - 1/(W_{miSat} - W_N)\right) \quad (2b)$$

where h_{mi} and h_{ma} are positive and expressed in (kPa); $\overline{E}_{mi}$ and $\overline{E}_{ma}$ were defined as the potential energies of the solid phase resulting from the external surface charge of clay particles, inside and outside the primary peds, in (Jkg^{-1} of solids); σ was defined as a part of the micropore water at interface with interpedal water in (kg water kg^{-1} soil); w_{re} is the residual water content in (kg

water kg^{-1} soil), complementary of the basic water content w_{bs} in W_{mi} (**Table 1**) and defined by the interpretation of the shrinkage curve as shown in Braudeau et al. (2004). The terms: σ, w_{re} and $W_N = w_{reSat}$ in Equation (2b) are pools of water added by Braudeau and Mohtar (2004) to take into account the air entry in the plasmic medium of primary peds.

The fitting results of observed shrinkage and water retention curves as measured in Braudeau and Mohtar (2004) using the device shown on **Figure 4** and the calculated ones based on Equations (1) and (2) were good enough to confirm that W_{ma} is a control variable for these curves, including the shrinkage curve.

However, it was impossible at that time to give any information about $h_{mi}(W_{mi})$, the equilibrium relationship between h_{mi} and h_{ma}, as well as the validity of Equation (2) out of the tensiometer validity range, and other related questions like: the nature of the micro air entry point, the impact on the formulation of the curve, the need or not of parameters σ, and W_N in Equations (2a) and (2b). All these questions will be clarified in the present article. In fact, the challenge is now to introduce the thermodynamic theory, through the SREV concept, into the soil water physics already placed in the frame of the hydrostructural pedology.

THERMODYNAMIC FORMULATION OF W_{ma} AND W_{mi} AT EQUILIBRIUM STATE OF THE PEDOSTRUCTURE

Definition of the hydro-thermodynamic system of the soil medium

Applying the thermodynamic principles to the soil medium such that it can be considered as a physical and organized thermodynamic system exchanging material, liquid, gas, heat, and space with all other systems in contact with it, requires first the acknowledgment of the pedostructure as a thermodynamic system. The SREV concept defined by Braudeau and Mohtar (2009) transforms the virtual delimitation of any representative volume (REV) of the soil medium organization into a hydrostructural functional delimitation of this REV. Then, a SREV can be considered as a thermodynamic system *closed on the mass of the solid phase* of the structure (structural mass) contained in the delimited volume and cut by the delimitation bearing on the external parts of the structure enclosed. Accordingly, the structural mass of a SREV stays constant and can be taken as a reference, instead of the volume, for defining all the extensive variables describing its internal organization.

Thermodynamics variables of the soil medium organization

The definitive change brought by the SREV approach is that the descriptive variables of the internal organization of the soil can all be defined according to the hydro-structural and thermodynamics point of view. **Table 1** lists the thermodynamic variables that describe the hydrostructural state of the soil organization at the pedostructure level of organization while **Figure 2** shows the different hydro-functional levels of organization of the pedon, namely, the horizon, pedostructure SREV, primary peds and primary particles. There are two kinds of variables: intensive variables (temperature, pressure and water potential) and *specific extensive variables* (specific entropy, specific volumes, water and air gravimetric contents) which are extensive variables of the SREV reported to the same mass of reference, M_s: the structural mass of the considered SREV. Thus, all extensive variables of the pedostructure system are referenced to the mass of the solids (M_s) that compose the structure (the oven-dry mass of the pedostructure sample), such that $\overline{V} = V/M_s$, $\overline{A} = A/M_s$, etc. (**Table 1**). We will use the term of pedostructural instead of specific to qualify these variables reported to the mass of the pedostructure SREV. These definitions lead to the following relationships between the extensive variables:

$$\overline{V} = \overline{Vp} + \overline{V_S} = W/\rho_w + \overline{A}/\rho_a + \overline{V}_s \qquad (3a)$$

$$W = W_{mi} + W_{ma}, \overline{A} = \overline{A}_{mi} + \overline{A}_{ma}, \text{ and}$$

$$\overline{Vp} = \overline{Vp}_{mi} + \overline{Vp}_{ma} \qquad (3b)$$

$$\overline{Vp}_{mi} = W_{mi} + \overline{A}_{mi})/\rho_w \text{ and } \overline{Vp}_{ma} = \left(W_{ma} + \overline{A}_{ma}\right)/\rho_w \quad (3c)$$

where $\overline{V}$, $\overline{Vp}$, W, and $\overline{A}$, are, respectively, the apparent pedostructural volume (dm^{3}kg^{-1}) of soil, its total poral volume (dm^{3}kg^{-1}), water mass content (kg$_{water}$ kg$^{-1}_{solids}$), and air mass content (kg$_{air}$ kg$^{-1}_{solids}$); $\overline{V}_s$ is the pedostructural volume of solid particles of soil ($\overline{V}_s = 1/\rho_s = V_s/M_s$); ρ_s, ρ_w, and ρ_a being the densities of the structural solid particles, the water and air phases of the pedostructure in (kg dm^{-3}), respectively. Subscripts mi, ma, w, and s refer to as micro, macro, water, and solids making the soil structure.

Practically, the classic undisturbed soil sample brought to the laboratory for physical analysis (in general a cylinder of about 5 cm diameter, 5 cm high sampled in a soil horizon) can be considered as a SREV of the soil medium of the soil horizon where it was sampled. The hydrostructural characteristics that will be measured on this sample in the laboratory are theoretically those of the soil medium *in situ*, in the corresponding horizon in the field. This makes it possible now to apply the thermodynamic principles to the physical description of the internal organization of the pedon *in situ*.

Thermodynamics of the pedostructure, basic concepts, and terminologies

To apply the pedostructure and SREV concepts to the "thermodynamic theory of water in soil" presented by Sposito (1981)

according to the REV concept, we rewrote equations of the thermodynamic potentials U and G, the internal energy and the Gibbs energy, and their differential formulation, in such manner that: (i) the extensive variables (including the thermodynamic potentials themselves) are all reported to (divided by) the structural mass, M_s, of the pedostructure considered (a SREV of the soil medium organization); (ii) the two types of water W_{mi} and W_{ma} have been considered *as two different components of the liquid water phase* ($W = W_{ma} + W_{mi}$); (iii) the pedostructure at water saturation is the moisture state of reference for the soil water retention h (or soil water pressure) which is written $h = -\rho_w (\mu_w - \mu_{wSat})$, in kPa, with the water density ρ_w in kg/dm^3; and iv) the chemical potentials: μ_w and μ_{air} of the water and air components, in Jkg^{-1}, are accounted negatively by convention. Then, the total differential of the specific internal energy $\overline{U}$ of a pedostructure SREV can be written such as:

$$d\overline{U} = Td\overline{S} - Pd\overline{V} + \mu_{wma}dW_{ma} + \mu_{wmi}dW_{mi} + \mu_{air}d\overline{A}_{air} \quad (4)$$

where $\overline{U}$ is the specific internal energy (Jkg$^{-1}_{soil}$), $\overline{S}$ the specific entropy (JK^{-1} kg$^{-1}_{soil}$), which are, like the other extensive variables of the SREV (**Table 1**), reported to the same structural mass M_s (the mass of solids making the structure of the considered SREV). As for the intensive variables, T is the absolute temperature (K), P is the external pressure (kPa), μ_{wma} and μ_{wmi} are the chemical potentials of the water inside and outside of primary peds (Jkg$^{-1}_{water}$) in the pedostructure.

In the usual terminology, the internal energy of a soil is a thermodynamic potential function U of which the independent variables are S, V, and $\{m_{i\alpha}\}$ which is a set of component masses, i indexing a component and α a phase in the soil medium (Sposito, 1981). As the other thermodynamic potentials obtained by performing the partial Legendre transformation on the internal energy $U = U(S, V, \{m_{i\alpha}\})$, one can consider G as an extensive quantity that is equivalent in all respects to the internal energy U but controlled by the independent variables of state: T, P, and $\{m_{i\alpha}\}$. "A knowledge of a thermodynamic potential is the same as a complete thermodynamic description of a soil" thus we can use U or G according to the set of controlled variables. Note that according to Sposito (1981): $\overline{U} = U/M_s = U\left(\overline{S}, \overline{V}, \{m_{i\alpha}/M_s\}\right)$ or $\overline{G} = G/M_s = G(T, P, \{m_{i\alpha}/M_s\})$ are equivalent to U or G in terms of thermodynamic potential; however, the advantage of using the specific thermodynamic potentials $\overline{U}$ and $\overline{G}$, expressed in units of joule by kg of soil structure, instead of U and G is to allow taking M_s, the mass of solids composing the structure of the SREV, as a reference for all organizational variables of this volume, including the pedostructural volume $\overline{V}$. Thus, considering G as a positive extensive variable, the total differential of the specific free energy $\overline{G}$ of the pedostructure corresponding to $d\overline{U}$ should be written such as:

$$d\overline{G} = \overline{S}dT - \overline{V}dP - \mu_{wma}dW_{ma} - \mu_{wmi}dW_{mi} - \mu_{air}d\overline{A}_{air} \quad (5)$$

and we can say that the hydrostructural and thermodynamic equilibrium of the soil medium is determined at $\overline{U}$ or $\overline{G}$ minimum, i.e., $d\overline{U}$ or $d\overline{G} = 0$.

On the other hand, the Euler equation for $\overline{G}$ as an extensive variable is, according to Sposito (1981) (except for the sign):

$$\overline{G} = -\left[\sum_\alpha \sum_i \left(\mu_{i\alpha} - \mu_{i\alpha}^o \right) m_{i\alpha} \right] / M_S \qquad (6)$$

where α represents the phase (solid, liquid, gaz), i the components in the considered phase, and $\mu_{i\alpha}^o$ chemical potential of the chemical element under Standard State conditions. Subscripts str, wl, air in the following equations refer to as the solid, liquid and gaz phases in the pedostructure.

In the case of the pedostructure system (SREV), Equation (6) becomes:

$$\overline{G} = \overline{G}_{str} + \overline{G}_{wl} + \overline{G}_{air} \qquad (7)$$

where

$$\overline{G}_{str} = -\sum_i \left(\mu_{i\,st} - \mu_{i\,st}^o \right) m_{i\,st} / \sum_i m_{i\,st} \qquad (7a)$$

$$\overline{G}_{wl} = -\left(\mu_{wmi} - \mu_w^o \right) W_{mi} - \left(\mu_{wma} - \mu_w^o \right) W_{ma}$$

$$= \overline{G}_{wmi} + \overline{G}_{wma} \qquad (7b)$$

$$\overline{G}_{air} = \left(\mu_{air} - \mu_{air}^o \right) \overline{A} \qquad (7c)$$

The second law of the thermodynamics says that at equilibrium $\overline{G}$ is minimum and $d\overline{G}_\alpha = 0$ for all phases. Hence, for liquid phase in particular, one can write the following relations:

$$d\overline{G}_{wl} = d\overline{G}_{wmi} + d\overline{G}_{wma} = 0 \qquad (8)$$

$$d\overline{G}_{wl} = -d\left[W_{mi}\left(\mu_{wmi} - \mu_w^o \right) \right]$$
$$- d\left[W_{ma}\left(\mu_{wma} - \mu_w^o \right) \right] = 0 \qquad (9)$$

$$d\overline{G}_{wl} = -W_{mi}d\mu_{mi} - W_{ma}d\mu_{ma} - \left(\mu_{wmi} - \mu_w^o \right)dW_{mi}$$
$$- \left(\mu_{wma} - \mu_w^o \right) dW_{ma} = 0 \qquad (10)$$

The immediate solution of these Equations (8)–(10) is to consider $\overline{G}_{wmi}$ and $\overline{G}_{wma}$ as constant in Equation (8) during the slow decrease in water content (W) by evaporation, which leads to:

$$\left(\mu_{wmi} - \mu_{wmi}^o \right) = -\overline{G}_{wmi}/W_{mi} \text{and} \left(\mu_{wma} - \mu_{wma}^o \right)$$
$$= -\overline{G}_{wma}/W_{ma} \qquad (11)$$

Let us now write the water retention (or suction pressure) in each compartment micro and macro of the pedostructure, h_{mi} and h_{ma}, respectively. They are written by definition such as:

$$h_{mi} = -\rho_w(\mu_{wmi} - \mu_{wmiSat}) \text{ and } h_{ma} = -\rho_w(\mu_{wma} - \mu_{wmaSat}) \qquad (12)$$

where ρ_w is the water bulk density and μ_{wmiSat} and μ_{wmaSat} are the micro and macro water potential at saturation of the pedostructure sample. At every water content W, there is no transfer of water between both compartments if $h_{mi} = h_{ma}$. Replacing the potentials in Equation (12) by their expression in

terms of the water contents W_{mi} and W_{ma} according to Equation (11) gives:

$$h_{mi} = \rho_w\left(\overline{G}_{wmi}/W_{mi} - \overline{G}_{wmi}/W_{miSat} \right) = h_{ma}$$
$$= \rho_w\left(\overline{G}_{wma}/W_{ma} - \overline{G}_{wma}/W_{maSat} \right) \qquad (13)$$

At each infinitesimal departure of water (dW) under constant T and P, there is first a non-equilibrium between h_{mi} and h_{ma} followed by an internal transfer of water between the two pore systems (of poral volume $\overline{V}p_{mi}$and $\overline{V}p_{ma}$) to restore the equilibrium of potentials. This return to equilibrium is in fact controlled by the rate of reorganization (due to swelling or shrinking) of clay particles inside the primary peds, in such a manner that the equality of both water pressures, inside and outside of primary peds, is respected according to Equation (13).

Thus, the important hypothesis that we will have to validate in this article is that a change in water content of the soil sample by evaporation under constant P and T induces (i) a suit of hydrostructural equilibrium states defined by the equality of water retentions inside and outside of primary peds ($h_{mi} = h_{ma}$) for each water content (W) and; and (ii) a particular distribution of W_{mi} and W_{ma} responding to Equations (8)–(13) where $\overline{G}_{wmi}$ and $\overline{G}_{wma}$ are supposed to be constant during this change in water content.

STATE EQUATIONS OF THE PEDOSTRUCTURE IN TERMS OF W
Equilibrium distribution of W_{ma} and W_{mi} according to W

Let us suppose that $\overline{G}_{wmi}$ and $\overline{G}_{wma}$ are energies developed in the water phase by the surface charges of the particles constituting the structure of the considered SREV of the pedostructure. Its organization into an assembly of primary peds imposes a breakdown of surface charges of the structure into the two pore systems: those that are positioned inside of the primary peds, in contact with the microporal water; and the others that are positioned on the outer surface of primary peds, in contact with macro poral water. Since the pedostructure organization remains unchanged with the change of water content, these specific energies $\overline{G}_{wmi}$ and $\overline{G}_{wma}$ (reported to M_s) can be identified to $\overline{E}_{mi}$ and $\overline{E}_{ma}$ (Joule by kg of solids [Jkg^{-1}]), the specific potential energies of the solid phase resulting from the surface charge of clay particles inside and outside of primary peds, as defined by Braudeau and Mohtar (2004). According to Voronin (1980), $E = Q_sRT$, where Q_s is the effective electric charge of the surface or effective exchange capacity in moles/kg of solids; R is the molar gas constant in JmolP^{-1}KP^{-1} and T the absolute temperature. Equation (13) can now be written and used under the following forms:

$$h_{mi} = \rho_w\left(\overline{E}_{mi}/W_{mi} - \overline{E}_{mi}/W_{miSat} \right) \qquad (14a)$$

$$h_{ma} = \rho_w\left(\overline{E}_{ma}/W_{ma} - \overline{E}_{ma}/W_{maSat} \right) \qquad (14b)$$

$$\overline{E}_{ma}/W_{ma} - \overline{E}_{ma}/W_{maSat} = \overline{E}_{mi}/W_{mi} - \overline{E}_{mi}/W_{miSat} \qquad (15)$$

The question is now: what are the values of W_{ma} and W_{mi} that ensure the equality of h_{mi} and h_{ma} reflected by Equation (15) for any value of pedostructure water content? To answer such a question, let's define:

$$\overline{E} = \overline{E}_{mi} + \overline{E}_{ma} \qquad (16a)$$

and the constant

$$A = \overline{E}_{ma}/W_{maSat} - \overline{E}_{mi}/W_{miSat} \qquad (16b)$$

Then, Equation (15) can be rewritten such as:

$$\overline{E}_{mi}/W_{mi} = \left(\overline{E}_{ma} - AW_{ma}\right)/W_{ma} = \left(\overline{E} - AW_{ma}\right)/W \qquad (17)$$

where each member of the three fraction is the sum of the corresponding members of the two first fractions of the equality. Rearranging Equation (17) leads to a quadratic equation as a function of W_{ma} (or W_{mi}):

$$W_{ma}^2 - \left(W + \overline{E}/A\right)W_{ma} + \left(\overline{E}_{ma}/A\right)W = 0 \qquad (18)$$

of which $\left(W_{ma} - \overline{E}/A\right)$ and $\left(W_{mi} + \overline{E}/A\right)$ are the two solutions; giving:

$$W_{ma}^{eq} = (1/2)\left(W + E/A\right) + (1/2)\sqrt{\left(W + \overline{E}/A\right)^2 - 4\left(\overline{E}_{ma}/A\right)W} \qquad (19)$$

and,

$$W_{mi}^{eq} = W - W_{ma}^{eq} = (1/2)\left(W - E/A\right)$$
$$- (1/2)\sqrt{\left(W + \overline{E}/A\right)^2 - 4\left(\overline{E}_{ma}/A\right)W} \qquad (20)$$

The equilibrium conditions imposed by Equation (15) for any value of W, at constant T and P, are satisfied for the values of W_{ma} and W_{mi} given by Equations (19) and (20) leading to a strict equality of micro and macro soil water pressures within the pedostructure organization, such that:

$$h^{eq}(W) = h_{mi}\left(W_{mi}^{eq}\right) = h_{ma}\left(W_{ma}^{eq}\right) \qquad (21)$$

The pedostructure shrinkage curve

The equations of soil shrinkage curve based only on the pedostructure concept.

The original equations of W_{ma} and W_{mi}, (1a) and (1b) used by Braudeau et al. (2004) for modeling the shrinkage curve, were established based only on the pedostructure concept and hence were not thermodynamically based. However, they give the same shapes of curve than the thermodynamically based equations derived in the previous section, namely: Equations (19) and (20), as this will be shown in the results section. They have the same number of parameters: (k_M and W_M) and ($\overline{E}_{ma}/A$ and $\overline{E}/A$), and hence, the parameters of one curve can be calculated from the others by fitting the corresponding equations. Therefore, the measured soil shrinkage curves, that can be considered also as successions of equilibrium states between the micro and macro poral waters during the slow drying by evaporation, could be written in terms of W_{ma}^{eq} and W_{mi}^{eq} using the soil shrinkage curve equation of Braudeau et al. (2004):

$$\overline{V} = \overline{V}_0 + K_{re}w_{re} + K_{bs}w_{bs} + K_{st}w_{st} + K_{ip}w_{ip} \qquad (22)$$

where K_{re}, K_{bs}, K_{st}, and K_{ip} are the slopes at inflection points of the measured soil shrinkage curve when they exist, namely, residual, basic, structural and interpedal, respectively (**Figure 3**). The water contents (w_{re}, w_{bs} and w_{st}, w_{ip}) are pairs of water contents of both micro and macro pore systems (of poral volumes $\overline{Vp}_{mi}$ and $\overline{Vp}_{ma}$). By definition, w_{re} and w_{st} are the water contents responsible for the linear shrinkage phase of slope K_{re} and K_{st} respectively, that withdraw from their respective pore systems, while being partially replaced by the air; whereas their complementary water pools w_{bs} and w_{ip}, responsible for shrinkage phases of slopes K_{bs} and K_{ip}, are removed without air entry at their place. They contribute at varying rates to the water loss during the evaporation of water, starting from the saturated state to the air dry state, depending on their retention in the soil. In their work, Braudeau et al. (2004) established their physical equations based on a probabilistic reasoning such as the Equations (1a) and (1b) for $W_{ma} = w_{st} + w_{ip}$ and $W_{mi} = w_{re} + w_{bs}$; and the following equations for w_{ip} and w_{re}:

$$w_{ip} = (1/k_L)\ln\left[1 + \exp(k_L(W - W_L))\right] \qquad (23)$$
$$w_{re} = W - (1/k_N)\ln\left[1 + \exp(k_N(W - W_N))\right] \qquad (24)$$

Parameters of these equations: k_N, W_N, k_M, W_M, k_L, and W_L, are characteristics of the soil structure and can be obtained from the shrinkage curve at the particular points N, M, and L (**Figure 3**).

Introducing W_{mi}^{eq} and W_{ma}^{eq} in the shrinkage curve equation.

To be consistent with the definition of W_{mi}^{eq} and W_{ma}^{eq} as a couple of pedostructural waters regulated by the equilibrium Equations (19) and (20), we must distinguish this couple (W_{mi}/W_{ma}) from the interpedal water w_{ip} which, when it exists ($K_{ip} \neq 0$), is a water in excess of W_{ma} in the interpedal pore space in the sense that it occupies a new interpedal space acquired by spacing of aggregates.

So we have to consider now the three pedostructural water contents: W_{mi}, W_{ma} and W_{ip}, such that:

$$W = W_{mi} + W_{ma} + W_{ip} \text{ and } W' = W_{mi} + W_{ma} = W - W_{ip} \qquad (25)$$

W' being the pedostructural water content excluding the saturating interpedal water W_{ip} ($\equiv w_{ip}$). This implies that Equations (9) and (10) of the micro and macro water contents of the pedostructure will be now calculated according to W' instead of W, such as:

$$W_{ma}^{eq} = \frac{1}{2}\left(W' + \overline{E}/A\right)$$
$$+ \frac{1}{2}\sqrt[2]{\left(W' + \overline{E}/A\right)^2 - 4W'\overline{E}_{ma}/A} \qquad (26a)$$

$$W_{mi}^{eq} = \frac{1}{2}\left(W' - \overline{E}/A\right)$$
$$- \frac{1}{2}\sqrt[2]{\left(W' + \overline{E}/A\right)^2 - 4W'\overline{E}_{ma}/A} \qquad (26b)$$

Finally, the new equation of the shrinkage curve (16) is written as, neglecting K_{re},

$$\overline{V} = \overline{V}_o + K_{bs}w_{bs}^{eq} + K_{st}w_{st}^{eq} + K_{ip}w_{ip} \qquad (27)$$

where $w_{ip} \equiv W_{ip}$ calculated by Equation (23), $w_{st}^{eq} \equiv W_{ma}^{eq} = W' - W_{mi}^{eq}$, and

$$w_{bs}^{eq} = W_{mi}^{eq} - w_{re} = (1/k_N)\ln\left[1 + \exp\left(k_N\left(W_{mi}^{eq} - W_{miN}^{eq}\right)\right)\right] \qquad (28)$$

where w_{re} has been replaced by its Equation (24) applied to the microporal water W_{mi}^{eq} instead of W.

Air entry point and residual shrinkage phase, interpretation. In Equation (27), K_{bs} appears like a structural coefficient linking the two scales of organization: the primary peds whose volume is $\overline{V}_{mi} \left(= \overline{Vp}_{mi} + \overline{V}_s\right)$ (**Table 1**) and their assembly constituting the pedostructure of volume $\overline{V}$, such that $\rho_w K_{bs} = d\overline{V}/d\overline{V}_{mi} = \rho_w d\overline{V}/dw_{bs}'^{eq}$ at least until the air entry point B. We assume in Equation (27) that, if the pedostructure is stable, K_{bs} stays constant all along the water evaporation and that $\overline{Vp}_{mi}$ decreases while remaining water-saturated from W_{sat} to the supposed air entry point W_B at the end of the basic shrinkage phase of the shrinkage curve (see **Figure 3**), such that: $\overline{Vp}_{mi} = W_{mi}^{eq} = w_{bs} + \max(w_{re})$ for $W \geq W_B$. The point B on the soil shrinkage curve marking the deviation from the basic phase of slope K_{bs}, has been usually interpreted as an air entry into the micropore system (Sposito and Giraldez, 1976; Braudeau et al., 2004). However, instead of assuming an external air entry in the micropore system, one can hypotheses that some water vapor bubbles are created in a saturated site occupied by w_{bs} which cannot shrink anymore because of the steric hindrance of clay particles that necessarily intervenes at a certain time of their re-organization during the shrinkage process. So, for $W < W_B$ in the residual phase, at the same time as a w_{bs} site disappears, a new site of w_{re} is created. Thus, w_{bs} is disappearing progressively because of such steric hindrance between particles and the bubbles of water vapor which are taking place in $\overline{Vp}_{mi}$, at equilibrium with W_{mi} residual. Since the primary peds medium is homogeneous, one can assume that any shrinkage of $\overline{Vp}_{mi}$ is equal in volume to the loss in w_{bs}, regardless the apparition of bubbles in the microporal medium and until it is totally occupied by w_{re} and the water vapor. This process, disappearance of w_{bs} sites in favor of new sites of w_{re} can be described and modeled according to Equation (28), noticing that the relation $\max(w_{re}) = W_{miN}^{eq} \cong W_N$ remains valid since W_{ma}^{eq} is near to 0 in this part of the curve (residual shrinkage).

The pedostructure water retention curve

Case of pedostructural water free of W_{ip}. The important result of the demonstration above is that the thermodynamically based equation of what is traditionally called "the soil moisture characteristic curve" or "the soil water retention curve," $h(W)$, corresponds in fact to the *pedostructure* water retention curve at equilibrium of the water tensions (or pressure) inside and outside of the primary peds in the pedostructure Equation (21).

Hence, the water retention curve of the pedostructure Equation (14) which is a suit of equilibrium states

between the micro and macropore systems, can be written such as:

$$h^{eq}(W) = h_{ma}\left(W_{ma}^{eq}\right) = \rho_w \overline{E}_{ma}\left(1/W_{ma}^{eq} - 1/W_{maSat}\right)(29a)$$

or

$$h^{eq}(W) = h_{mi}\left(W_{mi}^{eq}\right) = \rho_w \overline{E}_{mi}\left(1/W_{mi}^{eq} - 1/W_{miSat}\right) \quad (29b)$$

where W_{ma}^{eq} and W_{mi}^{eq} are given by Equations (19) and (20). We have to note that these latter remain valid as long as parameters E/A and E_{ma}/A do not change with the air entry in the micropore volume. Equations (29a) and (29b) give strictly the same values of matric water pressure $h_{mi/ma}^{eq}$ for a given value of the pedostructure water content. Thus, the characteristic parameters of the pedostructure water retention curve are only: (i) the specific potentials $\overline{E}_{mi}$ and $\overline{E}_{ma}$ in Joules by kg of the structural mass M_s, and (ii) the water contents at saturation W_{miSat} and W_{maSat}.

A singular case occurs when $A = 0$ Equation (16b). It is corresponding to a pedostructure with the characteristics such that: $\overline{E}_{ma}/W_{maSat} = \overline{E}_{mi}/W_{miSat}$. This value of A is not allowed in Equations (19) and (20) of W_{ma}^{eq} and W_{mi}^{eq} but, according to Equation (16b), the equality of potentials $h_{mi} = h_{ma}$ is obtained for values of W_{mi} and W_{ma} proportional to W, such that:

$$W_{ma}^{eq}/\overline{E}_{ma} = W_{mi}^{eq}/\overline{E}_{mi} = W/\overline{E} \qquad (30)$$

However, Equation (20) manifests the only case where the soil water potential can be written directly as:

$$h^{eq} = \rho_w \overline{E}\left(1/W - 1/W_{sat}\right) \qquad (31)$$

It seems, in this case, like there are no aggregates in the considered medium which then would be totally homogeneous without any organization of the porous medium distinguishing rigid unsaturated zones (indexed ma) surrounding deformable saturated zones (indexed mi).

Case of pedostructural water containing W_{ip}. This shrinkage phase is the shrinkage of $\overline{Vp}_{ma}$ due to the W_{ip} removal without any air entry in the system. W_{ip} belongs to only the homogeneous interpedal space under influence of $\overline{E}_{ma}$, the same surface charges environment than W_{ma}, the macroporal water content of the couple (W_{ma}/W_m); thus developing a water pressure h_{ip} that should be written according to Equation (31), such that:

$$h_{ip} = \rho_w \overline{E}_{ma}\left(1/\left(W_{ip}^0 + W_{ip}\right) - 1/\left(W_{ip}^0 + W_{ipSat}\right)\right) \qquad (32)$$

where W_{ip}^0 is a constant corresponding to the air entry pressure head h° in the pedostructure, such as:

$$h^0 = \rho_w \overline{E}_{ma}/\left(W_{ip}^0/W_{ipSat} + 1\right) \qquad (33)$$

Therefore, according to our hypotheses, one can recognize that the pedostructural water $[W]$ is composed of three types of water:

W_{mi}^{eq}; W_{ma}^{eq} and W_{ip}; among which the first two types are associated to the equilibrium water pressure $h_{mi/ma}^{eq}$ and the third one is associated to the swelling interpedal water pressure: h_{ip}. Accordingly, the water retention in the pedostructure is:

$$h^{eq} = h_{mi/ma}^{eq} + h_{ip} \qquad (34)$$

where $h_{mi/ma}^{eq}$ is calculated using Equations (29a) or (29b), h_{ip} using Equation (32), and W_{ma}^{eq} and W_{mi}^{eq} are calculated using Equations (26a) and (26b) in terms of $W' = (W - W_{ip})$.

Table 2 hereafter recapitulates the equations and parameters of the two characteristic curves in the all range of water content.

MATERIALS AND METHODS
SOIL SAMPLES USED FOR VALIDATION OF THE THEORY
We want to validate the equations established in the theory presented above. For that, we need examples of shrinkage curves and water retention curves that were continuously and simultaneously measured on disturbed or undisturbed soil samples representing the pedostructure of soil medium. Three kinds of shrinkage curves found in the literature have been distinguished according to their shape as shown in **Figure 5**:

- The shape of the shrinkage curve is clearly sigmoidal (**Figure 5A**) and there is no saturating interpedal phase due to W_{ip} at the beginning of the curve near saturation. The three

parameters K_{bs}, K_{st}, and K_{ip} of the shrinkage curve Equation (27) are such that: $K_{bs} > 0.3$; $K_{st} < 0.1$ and K_{ip} is equal to zero. This case is the most known and studied (Boivin et al., 2004; Braudeau et al., 2004, 2005; Cornelis et al., 2006; Chertkov, 2012).

- The shape is also sigmoidal but the shrinkage curve begins with a clear linear shrinkage phase parallel to the saturation line, due to the departure of the saturating interpedal water W_{ip} (**Figure 5B**). This case is typically that of reconstituted pedostructure samples from 2 mm sieved aggregates (Braudeau et al., 1999; Betsogo Atoua, 2010).

- There is no sigmoidal part visible on the shrinkage curve and the different shrinkage phases cannot be distinguished (**Figure 5C**). In this case, K_{st} is greater than K_{bs} and can reach values near 1 while K_{ip} is taken equal to zero in the shrinkage curve Equation (27). This kind of shrinkage curve is generally observed on silty loam soils of which the aggregated structure is not well developed (Taboada et al., 2008; Salahat et al., 2012).

Among these three groups, only the first one has examples of simultaneously measured shrinkage and retention curves that had been published (Braudeau and Mohtar, 2004). The device used for these measurements is shown on **Figure 4**. Two of these examples will be used in the present study. The two other groups will be represented in this study by selected soil samples of which the two characteristic curves were measured simultaneously and with a great accuracy using a new device named TypoSoil™ whose

Table 2 | The state variables, equations, and parameters of the Pedostructure soil moisture characteristic curves.

State variables	Equation	Parameters
PEDOSTRUCTURAL WATER CONTENTS (PWC)		
W (PWC)	$W = W_{ma}^{eq} + W_{mi}^{eq} + W_{ip} = W' + W_{ip}$	W_{maSat}^{eq}; W_{miSat}^{eq}; W_{ipSat}
W' (PWC without W_{ip})	$W' = W_{ma}^{eq} + W_{mi}^{eq}$	
W_{Sat} (PWC at saturation)	$W_{Sat} = W_{maSat}^{eq} + W_{miSat}^{eq} + W_{ipSat}$	
PEDOSTRUCTURE WATER CONTENTS AT EQUILIBRIUM (PWCE)		
W_{mi}^{eq} (intra-primary peds PWCE)	$W_{mi}^{eq} = (1/2)\left(W' - \overline{E}/A\right) - (1/2)\sqrt{\left(W' + \overline{E}/A\right)^2 - 4\left(\overline{E}_{ma}/A\right)W'}$	$\overline{E}/A$; $\overline{E}_{ma}/A$
W_{ma}^{eq} (inter primary peds PWCE)	$W_{ma}^{eq} = (1/2)\left(W' + \overline{E}/A\right) + (1/2)\sqrt{\left(W' + \overline{E}/A\right)^2 - 4\left(\overline{E}_{ma}/A\right)W'}$	
W_{ip} (saturating interpedal PWCE)	$W_{ip} = (1/k_L)\ln\left[1 + \exp\left(k_L\left(W - W_L\right)\right)\right]$	W_L; k_L
PEDOSTRUCTURE WATER RETENTION AT EQUILIBRIUM (PWRE)		
h^{eq} (PWRE)	$h^{eq}(W) = h_{ma/ma}^{eq}\left(W_{mi/ma}^{eq}\right) + h_{ip}\left(W_{ip}\right)$	W_{maSat}^{eq}; W_{miSat}^{eq} $\overline{E}_{ma}$; $\overline{E}_{mi}$ W_{ip}^0; W_{ipSat}
	$h^{eq}(W) = \rho_w \overline{E}_{mi/ma}\left(1/W_{mi/ma}^{eq} - 1/W_{mi/maSat}^{eq}\right) + \rho_w \overline{E}_{ma}\left(1/\left(W_{ip}^0 + W_{ip}\right) - 1/\left(W_{ip}^0 + W_{ipSat}\right)\right)$	
PEDOSTRUCTURE SHRINKAGE CURVE VARIABLES AT EQUILIBRIUM		
$\overline{V}$ (pedostructure specific volume at equilibrium)	$\overline{V} = \overline{V}_o + K_{bs}w_{bs}^{eq} + K_{st}w_{st}^{eq} + K_{ip}W_{ip}$	$\overline{V}_o$; K_{bs}; K_{st}; K_{ip}
w_{bs}^{eq} (Basic PWCE)	$w_{bs}^{eq} = W_{mi}^{eq} - w_{re}^{eq}$	k_N; W_N
w_{re}^{eq} (Residual PWCE)	$w_{re}^{eq} = W_{mi}^{eq} - (1/k_N)\ln\left[1 + \exp\left(k_N\left(W_{mi}^{eq} - W_{miN}\right)\right)\right] W_{miN} = W_{mi}^{eq}(W_N)$	
w_{st}^{eq} (Structural PWCE)	$w_{st}^{eq} = W_{ma}^{eq}$	

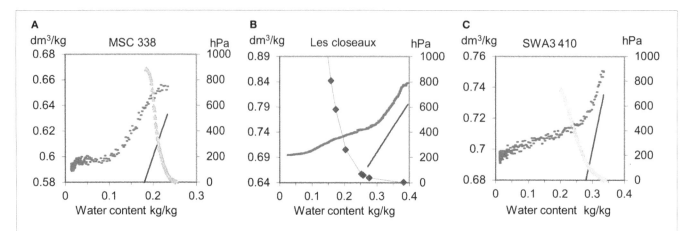

FIGURE 5 | Three types of shrinkage curves (A) sigmoidal; (B) sigmoidal with a saturating interpedal shrinkage phase; and (C) curvilinear structural shrinkage from Betsogo Atoua (2010), Salahat et al. (2012).

usage methodology is described in the second part of this article. It has been recently designed and built as the successor of the retractometer (Braudeau et al., 1999) to which has been added a device for measuring at the same time the shrinkage and the suction pressure of each of the measured samples (**Figure 6**). Thus, the soil samples that have been chosen in this study come from two sources differentiated by the device used for the simultaneous measurement of the two curves. A general description of these soil samples is given below and summarized in **Table 3**.

Source I: Examples of data set measured using the device **Figure 4**. They are available from the studies of Braudeau and Mohtar (2004) and Braudeau et al. (2005). In these studies the authors made continuous and simultaneous measurement of the soil shrinkage curve and the soil water potential curve on undisturbed cylindrical soil samples (100 cm³). The device (**Figure 4**) measured at the same time: the weight of the soil sample, its vertical diameter (as shrinkage of the sample was determined from the vertical change in the soil diameter due to the orientation of the soil core in the apparatus), and its water potential. In general, about 400 sets of measurements of the soil sample weight, diameter, and water potential were recorded for each sample.

Source II: The two characteristic curves were measured simultaneously on soil samples selected in this study using the device represented in **Figure 6**. Two different soil samples have been chosen to represent the two last types of shrinkage curves mentioned above: a soil sample of "Versailles soil" from France, and a soil sample of "Rodah" soil from Qatar.

Two pedostructure samples (prepared in cylinders of 5 cm diameter and height) from a same soil material sieved at 2 mm, known as "Sols du Closeaux à l'INRA de Versailles" (Rousseau, 2003) were reconstituted into the cylinders with a mixture of two different sizes of aggregates: [2–0.2 mm] and [<0.2 mm]. For preparing the soil samples, one kilogram of a disturbed soil was gathered, air dried, and 2 mm-sieved according to the protocol generally used in soil labs to obtain *fine ground* samples (aggregated soil material <2 mm) ready to analysis. Then, two soil samples labeled V4 and V5 were reconstituted by mixing two different sizes of aggregates according to [54% (coarse) – 42% (fine)] and [72 – 28%], respectively (Betsogo Atoua, 2010).

Two undisturbed soil samples were taken from a "Rodah" soil. Rodah soil is potentially the most suitable soil for cultivation in the State of Qatar (Scheibert et al., 2005). It will be more specifically studied in the second part "application" of this article. The soil samples used in this study have a silty clay loam texture and were given the codes RR3 and RR5 as shown in **Table 2**.

FITTING OPTIMIZATION AND ESTIMATION OF CHARACTERISTIC PARAMETERS

In order to test and validate the theory and the resulting equations of state of the pedostructure: $\overline{V}(W)$, $h(W)$, $W_{ma}^{eq}(W)$ and $W_{mi}^{eq}(W)$, we will use the measured data of the shrinkage and water retention curves of soil samples presented above to optimize the adjustment of the modeled curves which are represented by their characteristic parameters which actually are parameters of their physical equation.

The procedure used for the fittings and for determining the characteristic parameters of the curves was similar to the one used in Braudeau et al. (2004) for the shrinkage curve and in Salahat et al. (2012) for the water potential curve using the Microsoft Excel solver, but with considering the new equations established here for both curves. The sum of square of deviation between the measured and calculated curves was the target to minimalize using the Excel solver within a selected range of water content and according to the selected characteristic parameters put as variable parameters in the solver.

RESULTS AND DISCUSSION
DIFFERENT KINDS OF SOIL SHRINKAGE AND WATER POTENTIAL CURVES

Figure 7 presents the measured pedostructural shrinkage and water retention curves of each of the soil samples selected for the study. The fitting of these curves using their theoretical equations established above will be presented and discussed in the following section.

The following comments can be made:

(1) The first soil example (the tropical soil, T140, T340) presents the classical sigmoidal shape of shrinkage curve most known

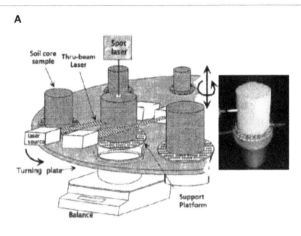

FIGURE 6 | The apparatus TypoSoil™. (A) Schematic drawing for the main components of the apparatus from Braudeau et al. (1999), and **(B)** Picture of the apparatus while working, one can notice that the apparatus can hold up to eight soil samples where the soil sample's height, diameter, weight, and water succion are taken at 10 min intervals and under the same conditions.

Table 3 | General description of the soil samples used in this study.

Sample ID	Soil core	Sample location	Description of the soil sample	Soil texture			ShC and SWCC were measured by
				% Clay	% Silt	% Sand	
T140	Undisturbed	Ivory Coast	The soil sample was taken from 40–60 cm	45.1	5.6	45.3	Device
T740	pedo-structure		depth of B-horizon of ferrallitic soil sequence	22.9	9.0	66.9	**[Figure 4]**
V4	Remade	France	Sampled from the surface horizon	18.2	60.2	21.6	TypoSoil™
V5	pedo-structure		[0–20 cm]				**[Figure 5]**
R3	Undisturbed	Qatar	The soil samples were taken from 0–15 cm	39	52	9	TypoSoil™
R5	pedo-structure		depth of a soil named locally "Rodah"				**[Figure 5]**

in the literature and characterizing well-structured soil mediums (Lauritzen, 1948). The different shrinkage phases: residual, basic, and structural are easily recognizable and the saturating interpedal shrinkage phase due to W_{ip} is absent or weakly represented. This is an indication of the good cohesion between aggregates of the pedostructure for these soils (Braudeau et al., 2005) and the pedostructural water is exclusively constituted of W_{mi} and W_{ma}, such that: $W = W' = W_{mi}^{eq} + W_{ma}^{eq}$ in the Equations (26) and (29) of $h^{eq} = h_{mi/ma}^{eq}$ for this kind of soils (oxisols, kaolinitic, structure stable, and well developed).

(2) The other soil samples do not present this cohesion of structure indicated by the sigmoid shape of the shrinkage curve. However, we must distinguish two cases: (i) the case where a sigmoidal shape of a part of the shrinkage curve is still recognizable such that the points W_N, W_M, and W_L can be roughly positioned on the curve (i.e., Versailles soil "V4 and V5"), and (ii) the case where these points do not appear clearly on the curve as in the Rodah soil "RR3 and RR5." These two cases can be explained such that:

• In the first case [Versailles soil "V4 and V5"]: the soil samples present a shrinkage phase corresponding to the

departure of W_{ip} which is clearly identifiable on the shrinkage curve: parallel to the saturated line and in contrast with the structural shrinkage phase due to water change in w_{st} through evaporation. In this case, W_{ip} cannot be unheeded such that the total pedostructural water content must be considered as $W = W_{mi}^{eq} + W_{ma}^{eq} + W_{ip}$ (Equation 33) where W_{ma}^{eq} and W_{mi}^{eq} are calculated as a function of $W' = W - W_{ip}$ by Equations (26a) and (26b). Also the water retention is in this case: $h^{eq} = h_{mi/ma}^{eq} + h_{ip}$ calculated using Equations (29) and (32).

• In the second case: Rodah soil "RR3 and RR5" on **Figure 7**, the sigmoidal shape does not appear on the shrinkage curve: points N, M, L cannot be positioned. In this case, K_{bs} is very small compared to K_{st} which stays however inferior to 1. Accordingly, W_L is fixed equal to W_{sat} such that $W_{ip} = 0$ and $W = W' = W_{mi} + W_{ma}$ in Equations (26) and (29) for the calculus of the different state variables.

FITTING THE MEASURED CURVES BY THE THEORETICAL EQUATIONS

The pedostructure water retention curve $h(W)$ and shrinkage curve $\overline{V}(W)$ have been established here like state equations corresponding to thermodynamic potential $\overline{G}(T, P, W)$ and

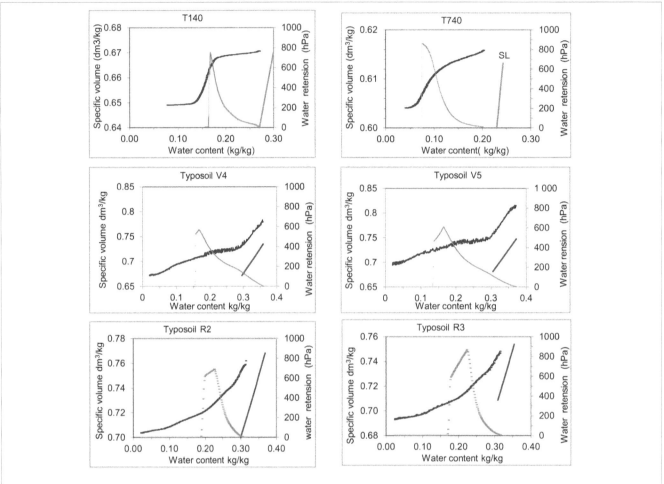

FIGURE 7 | The measured soil shrinkage (blue) and water retention (orange) curves of the soil samples used as examples in this study. The two first samples T140 and T740 were measured in Braudeau and Mohtar (2004) using the device on **Figure 4**, the others were measured in this study using TypoSoil™.

considering T and P constant. These equations are parametric and the adjustment between measured and calculated curves depends thus on the exact value of these parameters, which are also characteristic of the soil structure and its interaction with water. The optimization of the adjustment consists of approaching the closest to the set of the exact values of parameters of the measured curve. If the theoretical equations are correct and the measures accurate, the fit should be almost perfect, validating thus the theory. The procedure of optimization of the adjustment between theoretical and measured curves becomes then a useful mean of determination of the soil characteristic parameters. This procedure of optimization, using Excel and its solver, differs slightly according to the kind of pedostructures quoted above.

Stable (cohesive) and well aggregated pedostructure

This case corresponds to the kaolinitic soils from Ivory Coast [**Figures7A,B**]. There is no interpedal shrinkage phase and $W_{ip} = 0$ ($W_L = W_{sat}$). Results of the fitting of the both characteristic curves of the soil samples T140 and T740 taken as examples are presented on **Figure 8**. The water retention curve is adjusted first. Then, the shrinkage curve is adjusted using its own parameters.

Adjustment of the water retention curve. Parameters of the water retention curve $h(W)$ are: W_{maSat}, W_{miSat}, E_{mi}, and E_{ma}. They are the parameters of Equations (16a) and (16b) for W_{ma}^{eq} and W_{mi}^{eq} and of Equations (29a) and (29b) for h_{ma} and h_{mi}. **Figure 8A** shows the fit of the water retention curve considered alone, using its parameters as variable parameters. Initialization of the parameters before to run the solver, valuable for all types of soil, is generally done as following:

- W_{miSat} is represented on the curve by the point M such that $W_M = W_{miSat}$. This point is placed at h = 400 or 500 hPa on the curve as initial value of the optimization.
- W_{maSat} is replaced by $W_{Sat} = W_{maSat} + W_{miSat}$ as variable parameter; the initial value or fixed value of W_{Sat} is chosen at the beginning or at h = 0 kPa of the water retention curve.
- $\overline{E}_{mi}$ is the quantity of surface charges of the clay-particles inside the primary peds and will be initialized at 40 J/kg.
- $\overline{E}_{ma}$, the charges at the external surface of primary peds, is replaced by $\overline{E}/\overline{E}_{ma}$ as variable parameter which is generally initialized at 100.

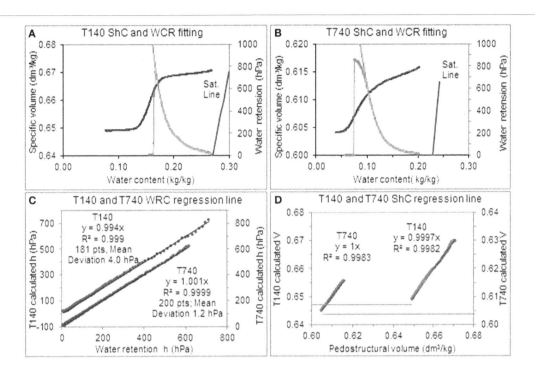

FIGURE 8 | Adjustment of the measured characteristic curves by their theoretical parametric equations for the ferrallitic soil samples T140 and T740. (A,B) Represent for each sample the shrinkage curve (ShC), measured (blue points) and calculated (red line) with its associated water retention curves (WRC), also measured (orange points) and calculated (green line). The quality of the adjustments is illustrated in **(C,D)** which show the regression line between measured and calculated values for both curves.

The adjustment of the water retention curve is always excellent, as shown as example in **Figures 8A,B** for the sample T140. The deviation between the measured and calculated values does not exceed 10 hPa all along the range of validity of the tensiometer curve (here: 8–680 hPa). The values of parameters are given in **Table 4**.

Adjustment of the shrinkage curve. For fitting the shrinkage curve, K_{bs} is fixed to its value read on the curve (slope at point B) and the initialization of parameters concerns k_N, W_N, W_M, $\overline{E}/\overline{E}_{ma}$, that are such that: $k_N = 100$, W_N and W_M take the values read on the shrinkage curve according to the corresponding shrinkage phases (see **Figure 3**) and $\overline{E}/\overline{E}_{ma} = 1.0$. The adjustment of the shrinkage curve considered alone is always excellent as shown on **Figure 8D**; however, as we can see on **Table 4**, the parameters W_M and $\overline{E}/\overline{E}_{ma}$ do not correspond to those found in the water retention adjustment.

We will see further that this discrepancy between the values of W_M and $\overline{E}/\overline{E}_{ma}$ for the two curves does not appear with the data measured using TypoSoil™. It might be caused by the approximate calculus of the volume which used only the vertical measurement of the diameter of the sample (see **Figure 4**) as variable in the old method, compared to the tridimensional measurement of the soil sample in the new method (**Figure 6**).

Weakly consistent and little aggregated pedostructure

The sigmoidal shape is recognizable along with a saturated interpedal shrinkage phase parallel to the saturation line (**Figure 5**). This is the case of [Versailles soil "V4 and V5": **Figures 6B,C**]. For these soil samples, the two types of water: W_{ma}^{eq} ($\equiv w_{st}$) and W_{ip} ($\equiv w_{ip}$), are well distinguished on the shrinkage curve by the point L (**Figure 3**) such that, at saturation, $W_{sat} = W_{miSat}^{eq} + W_{maSat}^{eq} + W_{ipSat}$ with $W_{ipSat} = W_{sat} - W_L$. The pedostructure water retention at equilibrium is $h^{eq} = h_{mi/ma}^{eq} + h_{ip}$ and, in this case, we have to add three new parameters to the four parameters in the previous case: W_L and k_L, parameters of the w_{ip} Equation (23), and W_{ip}^0 which is the constant of adjustment in Equation (32). Initialization of the added parameters is as following: W_L is read approximately on the shrinkage curve, k_L is put at 80 kg/kg and W_{ip}^0 equal to $W_{maSat}/2$.

The adjustment of both curves is made in two times. The shrinkage curve is fitted in first using k_N, W_N, W_M, E/E_{ma}, k_L, and W_L as variable parameters. As for the previous case, K_{bs} is fixed to its value calculated from the curve. The adjustment of the shrinkage curve is made in once without any difficulty. Then, the two curves are fitted jointly minimizing the product of the sum of squares of deviation of the two curves and using the following variable parameters: W_M, $\overline{E}/\overline{E}_{ma}$, E_{mi}, W_{ip}^0, k_L, and W_L; W_{Sat} being fixed at its value at $h = 0$.

Adjustments are excellent as we can see on **Figure 9** for the Versailles soil samples. **Figures 9A,B** show the adjustment of the two curves made conjointly for the two samples V4 and V5. Results are given in **Table 4**. Their characteristic parameters are very similar, indicating the good replicability of the samples manufacturing from the 2 mm sieved soil material. The quality of

Table 4 | Pedostructure thermodynamic equilibrium parameters calculated from the shrinkage curve and the water tension curve of the soil (pedostructure) samples.

Sample ID	k_N [kg/kg]	W_N [kg/kg]	W_M [kg/kg]	E/E_{ma} [-]	E_{mi} [J/kg]	W_{sat} [kg/kg]	K_{bs} [dm³/kg]	K_{st} [dm³/kg]	K_{ip} [dm³/kg]	V_0 [dm³/kg]
T140[1]	175	0.139	0.169[r] 0.171[s]	168[r] 801[s]	159.7	0.292	0.59	0.01	0	0.649
T740[1]	296	0.056	0.124[r] 0.104[s]	78.9[r] 78.8[s]	24.5	0.228	0.178	0.03	0	0.604
V4	41	0.072	0.163	18.2	33	0.363	0.534	0	1	0.674
V5	41	0.072	0.162	17.2	35.4	0.375	0.534	0	1	0.696
R3	58	0.078	0.233	279	249	0.317	0.128	0.41	0	0.694
R5	58	0.078	0.234	270	240	0.309	0.157	0.49	0	0.696

[1] These samples have two values of W_M and E/E_{ma} obtained in the fitting of the water retention curve (r) and the shrinkage curve (s) made separately.

the adjustments can be seen on **Figures 9C,D** that show the regression lines between measured and calculated data for the two curves and for each sample. The fits of the water retention curves are particularly impressive: they are almost perfect within the range of validity of their measure that represents about 280 points: the maximum of absolute deviation being 5 hPa and the average less than 2.2 hPa.

Consistent but non-aggregated pedostructure

No sigmoidal part of the shrinkage curve, it corresponds to the residual, basic and structural shrinkage phases, with $K_{st} > K_{bs}$. This case is that of the Rodah soil samples. There are no mark on curves that might indicate the position of points L and M. The samples will be treated in the fitting of curves like in the first case where the pedostructural water content is composed of only W_{mi} and W_{ma} in equilibrium of potential in the macro and micropore spaces: $W = W_{mi} + W_{ma}$. The difference with the first case, where K_{st} was considered near or equal to zero, is that the removal of the structural macropore water w_{st} provokes a shrinkage of the medium with a strong slope of value intermediary between K_{bs} and 1.

The fit of the water retention curve can be made alone and is made in first, determining the four characteristic parameters: $W_M (= W_{miSat})$, E/E_{ma}, E_{mi} and W_{sat}. The same excellent fits are obtained as above, providing the accurate values of these parameters, which can be seen on the **Figure 10**. Then, the adjustment of the shrinkage curve is done using the parameters already fund and the following variable parameters: k_N, W_N, K_{bs}, and K_{st} of the shrinkage curve only. Initialization of variables is like for the cases above.

Figure 10 shows the good adjustments of the two curves of soils represented by Rodah soils; in particular the measured and modeled water retention curves are almost superimposed on the **Figures 10A,B** for the two samples. The excellent quality of these adjustments can be appreciated on **Figures 10C,D** representing linear regressions between measured and calculated data of the two curves. Moreover, the two Rodah samples that were sampled in the same location have curves of similar shapes; this is reflected by near values of their parameters in **Table 4**. It is an illustration, like in the previous case, of the good reproducibility of the measurement method and of the procedure of adjustment of the curves based on the use of their theoretical parametric equations.

DISCUSSION
COMPARISON WITH THE PREVIOUS EQUATIONS OF THE SOIL WATER POTENTIAL CURVES

For each sample analyzed, the coefficient of regression between measured and calculated water retention curves was very closed to 1. This implies that the parameter σ in Equation (2) of h_{ma} (W_{ma}) proposed before by Braudeau et al. (2004) is not needed. In fact, the role of σ introduced by these authors was to allow the fit of h_{ma} calculated by Equation (2) on the reading of the tensiometer including the zone out of its validity where bubbles of water vapor have taken place in the tensiometer tube. Their hypothesis was that the tensiometer reading corresponded exactly to the water potential of the macropore interpedal water (W_{ma}), which is also our hypothesis, but they considered the relation $h = h_{ma}$ without the condition of $h = h_{ma} = h_{mi}$. Thus, the σ term was introduced in the equation of h_{ma} and was ambiguously defined as transitional micro-poral water content at the interface with the macropore (interpedal) space; it was considered as one of the parameters of the water potential curve in few studies based on the pedostructure concept (Salahat et al., 2012; Singh et al., 2012). However, if we apply the constraint of ($h = h_{mi} = h_{ma}$) in the adjustment of the old equations of W_{ma}, W_{mi}, h_{ma}, and h_{mi} Equations (1a, 1b) and (2a, 2b) on a measured water retention curve $h(W)$, we obtain almost the same results as if the new equations were used. Doing this, the obtained value for σ is near to zero and the calculated curves of w_{st} (old equation) and W_{ma} (new equation) are superimposed as shown on **Figure 11**. The surprising result is this quasi-equivalence of shape between Equations (1) and (19) for W_{ma} and (2) and (20) for W_{mi} (**Figure 11**). The essential difference stays in the fact that Equations (19) and (20) contain by themselves the constraint ($h = h_{mi} = h_{ma}$) so we can use only one of the two equations to express the water retention curve, unlike the former Equations (1) and (2) that must be used together in the fitting of the measured water potential curve. This constraint of equality of potentials was already applied in Kamel® and KamelSoil computer models (Braudeau et al., 2009) and in the other works mentioned

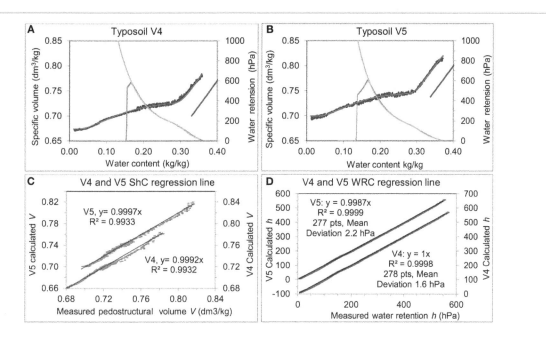

FIGURE 9 | Adjustment of the measured characteristic curves by their theoretical parametric equations for the Versailles soil type, V4 and V5. (A,B) represent for each sample the shrinkage curve (ShC), measured (blue points) and calculated (red line) with its associated water retention curves (WRC), also measured (orange points) and calculated (green line). The quality of the adjustments is illustrated in **(C,D)** which show the regression line between measured and calculated values for both curves.

above referring to the pedostructure concept (Mallory et al., 2011; Salahat et al., 2012; Singh et al., 2012).

VALIDATION OF THE THEORY

The great assumptions that were made in the thermodynamic theory of the soil water (Sposito, 1981) for deducing the two soil moisture characteristic curves, $h(W)$ and $\overline{V}(W)$ (at T and P constant) that have been validated here, were:

(i) The soil medium organization is well described by the pedostructure concept which defines its first levels of organization and the descriptive variables of its hydro-functional components, like W_{mi}, W_{ma}, W_{ip}.

(ii) W_{mi} and W_{ma} are two distinct physical components of the soil water phase whose the thermodynamic potentials $\overline{G}_{mi}(T, P, W_{mi})$ and $\overline{G}_{ma}(T, P, W_{ma})$ are equal to, respectively, $\overline{E}_{mi}$ and $\overline{E}_{ma}$ which are the potential energy of the surface charges of the clay particles, respectively inside and at the surface of primary peds. Therefore, $\overline{G}_{mi}$ and $\overline{G}_{ma}$ stay constant during a change in water content because the pedostructure organization stays stable (and so the clay surface charges distribution) in the wetting-drying cycles.

The excellent adjustments obtained of measured and theoretical curves confirm not only the validity of equations and parameters of these curves but also the systemic and thermodynamic approach that were used for finding their formulation.

HYDRO-FUNCTIONAL TYPOLOGY OF SOILS

We could see that two important characteristics of the clay mineral and of their arrangement in aggregates ($\overline{E} = \overline{E}_{mi} + \overline{E}_{ma}$ and $\overline{E}/\overline{E}_{ma}$) are easily extracted from measured water potential curves, as well as the maximum of water content possible into the both intra and inter primary peds pore space of the pedostructure: W_{miSat} and W_{maSat}. In fact the four parameters [W_{miSat}, W_{maSat}, $\overline{E}_{mi}$, and $\overline{E}_{ma}$] characterize the hydrostructural equilibrium of this arrangement in aggregates, the pedostructure.

An important implication of this hydro-thermodynamic equilibrium modeling of the soil medium is the possibility of establishing a functional typology of soils based on the hydrostructural characterization of its pedostructures. The typology of soils is to restart since pedostructure is now precisely defined as a thermodynamic system which is therefore completely characterized by the knowledge of its equations of equilibrium state [$\overline{V}(W)$ and $h(W)$] represented by their hydrostructural parameters. This thermodynamic typology of pedostructures would allow the linkage between different disciplines of the soil sciences such as soil mapping with remote sensing, biophysical modeling, and especially experiments in the laboratory of such characterized soil sample for analysis of biological processes in the soil medium.

CONCLUSIONS

These initial results demonstrate conclusively that *there is no physical characterization* without the acknowledgement of the hydrostructural and thermodynamic equilibrium of the soil medium organized in primary aggregates made of solid particles, such as clay particles, immersed in water, arranged in layers surrounding these particles, and itself surrounded by the air.

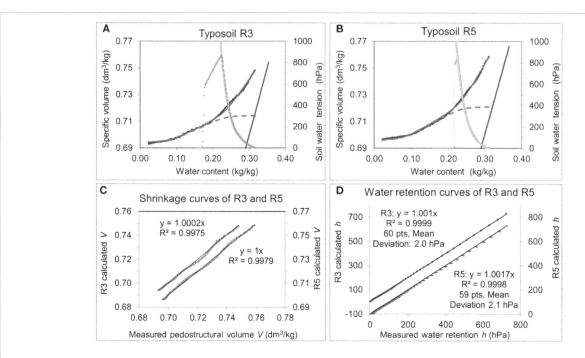

FIGURE 10 | Adjustment of the measured characteristic curves by their theoretical parametric equations for two non-disturbed pedostructure samples of the Rodah soil, R3 and R5. (A,B) represent for each sample the shrinkage curve (ShC), measured (blue points) and calculated (red line) with its associated water retention curves (WRC), also measured (orange points) and calculated (green line). The quality of the adjustments is illustrated in **(C,D)** which show the regression line between measured and calculated values for both curves.

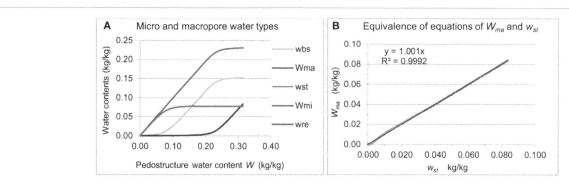

FIGURE 11 | Hydrostructural characterization and modeling of the Rodah soil sample R3. (A) contents according to the total pedostructure water content W, of the different micro and macro-pore water types; and **(B)** comparison of the macropore water content W_{ma} calculated by the new and the old physical equation used for W_{ma} and w_{st} respectively.

We could not find the thermodynamic equation $h(W)$ without resting the fundamentals of the soil water thermodynamics on the soil medium organization and the acknowledge of the two components W_{mi} and W_{ma} of the pedostructural water phase.

In this article, the distinction of the soil water content into two nested fractions, located inside and outside of the clayey plasma of the pedostructure, as previously evidenced by the shrinkage curve, has been physically and mathematically justified using a systemic and thermodynamic approach of the primary soil medium organization: the pedostructure. All the equations presented are based on the concept of pedostructure as a SREV of the soil medium in a soil horizon. The excellent adjustments obtained

of the thermodynamic equations on the measured curves confirm the validity of these equations and their parameters, and also the systemic and thermodynamic approach that were used for finding their formulation. Applying the SREV concept to the current thermodynamic theory of the soil water led us to reformulate the two principal characteristic curves of the soil moisture in the frame of the thermodynamics of organization of the soil medium. These findings pave the way for a thermodynamic formulation of all exchanges (heat, space, solutions, gas) between biological organisms living in the soil and the soil medium itself which offers the physical conditions for their activities and development.

REFERENCES

Ahuja, L. R., Ma, L., and Timlin, D. J. (2006). Trans-disciplinary soil physics research critical to synthesis and modeling of agricultural systems. *Soil Sci. Soc. Am. J.* 70, 311–326. doi: 10.2136/sssaj2005.0207

Berezin, P. N., Voronin, A. D., and Shein, Y. V. (1983). An energetic approach to the quantitative evaluation of soil structure. *Pochvovedeniye* 10, 63–69.

Betsogo Atoua, A. (2010). *Caractérisation hydrostructurale d'un sol limoneux appliquée à l'étude de la croissance d'un champignon filamenteux dans le sol.* Mémoire de Master, Université Paris-Est Créteil/IRD, 66.

Boivin, P. (2007). Anisotropy, cracking, and shrinkage of vertisol samples. Experimental study and shrinkage modeling. *Geoderma* 138, 25–38. doi: 10.1016/j.geoderma.2006.10.009

Boivin, P., Garnier, P., and Tessier, D. (2004). Relationship between clay content, clay type, and shrinkage properties of soil samples. *Soil Sci. Soc. Am. J.* 68, 1145–1153. doi: 10.2136/sssaj2004.1145

Braudeau, E. (1988). Équation généralisée des courbes de retrait d'échantillons de sols structurés. *C. R. Acad. Sci. Paris* 307, 1731–1734.

Braudeau, E., and Bruand, A. (1993). Détermination de la courbe de retrait de la phase argileuse à partir de la courbe de retrait établie sur échantillon de sol non remanié. *C.R. Acad. Sci. Paris* 316, 685–692.

Braudeau, E., Costantini, J. M., Bellier, G., and Colleuille, H. (1999). New device and method for soil SC measurement and characterization. *Soil Sci. Soc. Am. J.* 63, 525–535. doi: 10.2136/sssaj1999.03615995006300030015x

Braudeau, E., Frangi, J. P., and Mothar, R. H. (2004). Characterizing non-rigid dual porosity structured soil medium using its shrinkage curve. *Soil Sci. Soc. Am. J.* 68, 359–370. doi: 10.2136/sssaj2004.3590

Braudeau, E., and Mohtar, R. H. (2004). Water potential in non-rigid unsaturated soil-water medium. *Water Resour. Res.* 40, W05108, 14. doi: 10.1029/2004WR003119

Braudeau, E., and Mohtar, R. H. (2009). Modeling the soil system: bridging the gap between pedology and soil-water physics. *Global Planet. Change J.* 67, 51–61. doi: 10.1016/j.gloplacha.2008.12.002

Braudeau, E., Mohtar, R. H., El Ghezal, N., Crayol, M., Salahat, M., and Martin, P. (2009). A multi-scale "soil water structure" model based on the pedostructure concept. *Hydrol. Earth Syst. Sci. Discuss.* 6, 1111–1163. doi: 10.5194/hessd-6-1111-2009

Braudeau, E., Sene, M., and Mohtar, R. H. (2005). Hydrostructural characteristics of two African tropical soils. *Eur. J. of Soil Sci.* 56, 375–388. doi: 10.1111/j.1365-2389.2004.00679.x

Chertkov, V. Y. (2012). Physical modeling of the soil swelling curve vs. the shrinkage curve. *Adv. Water Res.* 44, 66–84. doi: 10.1016/j.advwatres.2012.05.003

Cornelis, W. M., Corluy, J., Medina, H., Díaz, J., Hartmann, R., Van Meirvenne, M., et al. (2006). Measuring and modelling the soil shrinkage characteristic curve. *Geoderma* 137, 179–191. doi: 10.1016/j.geoderma.2006.08.022

El Kadi, A. I. (1985). On estimating the hydraulic properties of soil, Part 1. Comparison between forms to estimate the soil-water characteristic function. *Adv. Water Resources* 8, 136–147. doi: 10.1016/0309-1708(85)90054-5

Lauritzen, C. W. (1948). Apparent specific volume and shrinkage characteristics of soil materials. *Soil Sci.* 65, 155–179. doi: 10.1097/00010694-194802000-00003

Leij, F., Russell, W., and Lesch, S. M. (1997). Closed-form expressions for water retention and conductivity data. *Ground Water* 35, 848–858. doi: 10.1111/j.1745-6584.1997.tb00153.x

Low, P. F. (1987). Structural component of the swelling pressure of clays. *Langmir* 3, 18–25. doi: 10.1021/la00073a004

Mallory, J., Mohtar, R., Heathman, G., Schulze, D., and Braudeau, E. (2011). Evaluating the effect of tillage on soil structural properties using the pedostructure concept. *Geoderma* 163, 141–149. doi: 10.1016/j.geoderma.2011.01.018

Rieu, M., and Sposito, G. (1991). Relation pression capillaire-teneur en eau dans les milieux poreux fragmente's et identification du caracte 're fractal de la structure des sols. *C. R. Acad. Sci.* 312, 1483–1489.

Rousseau, M. (2003). *Transport préférentiel de particules dans un sol non saturé:de l'expérimentation en colonne lysimetrique à l'élaboration d'un modèle à base physique.* Grenoble: Thèse Institut National Polytechnique de Grenoble.

Salahat, M., Mohtar, R., Braudeau, E., Schulze, D., and Assi, A. (2012). Toward delineating hydro-functional soil mapping units using the pedostructure concept: a case study. *Comput. Electron. Agr.* 86, 15–25. doi: 10.1016/j.compag.2012.04.011

Scheibert, C., Stietiya, M., Sommer, J., Schramm, H., and Memah, M. (2005). "The atlas of soils for the State of Qatar," in *Soil Classification and Land Use Specification Project for the State of Qatar,* eds Ministry of Municipal Affairs and Agriculture, General Directorate of Agricultural Research and Development, Department of Agricultural and Water Research (Doha: Ministry of Municipal Affairs and Agriculture), 19–36.

Singh, J., Mohtar, R., Braudeau, E., Heathmen, G., Jesiek, J., and Singh, D. (2012). Field evaluation of the pedostructure-based model (Kamel). *Comput. Electron. Agr.* 86, 4–14. doi: 10.1016/j.compag.2012.03.001

Sposito, G. (1981). *The Thermodynamics of Soil Solutions.* New York, NY: Oxford University Press, 187–208.

Sposito, G., and Giraldez, J. V. (1976). Thermodynamic stability and the law of corresponding states in swelling soils. *Soil Sci. Soc. Am. J.* 40, 352–358. doi: 10.2136/sssaj1976.03615995004000030016x

Taboada, M. A., Barbosa, O. A., and Cosentino, D. J. (2008). Null creation of air-filled structural pores by soil cracking and shrinkage in silty loamy soils. *Soil Sci.* 173, 130–142. doi: 10.1097/SS.0b013e31815d8e9d

Voronin, A. D. (1980). The structure-energy conception of the hydrophy-sical properties of soils and its practical applications. *Pochvovedeniye* 12, 35–46

Conflict of Interest Statement: (1) The apparatus used in the work presented in the article to validate the theory was recently (2013) patented conjointly by IRD and Valorhiz (of which the CEO is Hassan Boukcim); Erik Braudeau is one of the two inventors. (2) The same for the method of obtaining the pedostructural parameters: patented by IRD and Valorhiz; Erik Braudeau inventor.

5

Physics of the soil medium organization part 2: pedostructure characterization through measurement and modeling of the soil moisture characteristic curves

Amjad T. Assi[1,2], Joshua Accola[1,3], Gaghik Hovhannissian[4], Rabi H. Mohtar[2,5]* and Erik Braudeau[1,4]*

[1] *Qatar Foundation, Qatar Environment and Energy Research Institute, Doha, Qatar*
[2] *Department of Agricultural and Biological Engineering, Purdue University, West Lafayette, IN, USA*
[3] *Department of Biological Systems Engineering, University of Wisconsin-Madison, Madison, WI, USA*
[4] *Institut de Recherche pour le Développement (IRD), Pédologie Hydrostructurale, Bondy, France*
[5] *Biological and Agricultural Engineering Department and Zachry Department of Civil Engineering, Texas A&M University, College Station, TX, USA*

Edited by:
Christophe Darnault, Clemson University, USA

Reviewed by:
Nabeel Khan Niazi, University of Agriculture Faisalabad, Pakistan
Wieslaw Fialkiewicz, Wroclaw University of Environmental and Life Sciences, Poland

*Correspondence:
Rabi H. Mohtar and Erik Braudeau, Biological and Agricultural Engineering Department and Zachry Department of Civil Engineering, Texas A&M University, 302 B Scoates Hall, Mail Stop 2117, College Station, TX 77843-2117, USA
e-mail: erik.braudeau@ird.fr;
mohtar@tamu.edu

Accurate measurement of the two soil moisture characteristic curves, namely, water retention curve (WRC) and soil shrinkage curve (SSC) is fundamental for the physical modeling of hydrostructural processes in vadose zone. This paper is the application part following the theory presented in part I about physics of soil medium organization. Two native Aridisols in the state of Qatar named locally Rodah "räôd'ə" soil and Sabkha "săb'kə" soil were studied. The paper concluded two main results: the first one is about the importance of having continuous and simultaneous measurement of soil water content, water potential and volume change. Such measurement is imperative for accurate and consistent characterization of each of the two moisture characteristic curves, and consequently the hydrostructural properties of the soil medium. The second is about the simplicity, reliability, strength and uniqueness of identifying the characteristic parameters of the two curves. The results also confirmed the validity of the thermodynamic-based equations of the two characteristic curves presented in part I.

Keywords: soils in hyper-arid regions, Rodah soil, Sabkha soil, soil shrinkage curve (SSC), water retention curve (WRC), hydrostructural parameters, pedostructure and SREV concept

INTRODUCTION

The hydrostructural properties of a structured soil medium have been characterized by two fundamental curves: water retention curve (WRC) and the soil shrinkage curve (SSC). These curves have been used to evaluate the soil structure (Haines, 1923; Coughlan et al., 1991; Braudeau et al., 2004, 2005), soil physical quality (Dexter, 2004; Santos et al., 2011), soil deformation (Alaoui et al., 2011) and, in general, to characterize and model the soil-water interaction by linking the soil physical properties with their impact on the water and solute movement through a soil medium. The WRC defines the relationship between the soil-water potential and the water content, while the SSC represents the specific volume changes "or void ratio changes" of a soil due to the changes in its water content. A third fundamental characteristic curve is the unsaturated hydraulic conductivity curve. This curve is difficult to measure (Børgesen et al., 2006); hence, several scientists have used the fitted parameters of the WRC with the available pore size distribution statistical models to predict the unsaturated hydraulic conductivity curve (Burdine, 1953; Brooks and Corey, 1964; Mualem, 1976; van Genuchten, 1980).

The WRC and SSC curves have been measured in laboratory separately and by different apparatus and methods based on their end use. Generally, the measurement range (~0–900 hPa) for the WRC, which is the measurement range of the tensiometer, is suitable for soil water flow and solute transport studies. However, several techniques have been used to measure the SSC. These methods can be characterized into four groups: (i) Archimedes' principle-based approach. The well-known methods of this approach are: the resin-coated method (Brasher et al., 1966), the paraffin-coated method (Lauritzen and Stewart, 1942), and the rubber balloon method (Tariq and Durnford, 1993a). In resin-coated and paraffin-coated methods, the soil samples could be clods (Reeve and Hall, 1978), aggregates (Bronswijk, 1991), or soil cores (Crescimanno and Provenzano, 1999; Cornelis et al., 2006). While, in the rubber balloon method, reconstituted soil cores were used in most studies (Tariq and Durnford, 1993a; Cornelis et al., 2006). In this approach, the soil samples were submerged into water and then the change in the sample volume was determined from the volume of displaced fluid; (ii) physical measurement-based approach: where the soil cores "disturbed or undisturbed" dimensions were measured directly using a vernier caliper (Berndt and Coughlan, 1977; Huang et al., 2011), a linear displacement transducer (Boivin et al., 2004; Braudeau and Mohtar, 2004) or a thin metal stick (Kim et al., 1992); (iii) laser sensors-based approach: where the soil core diameter and height were determined through laser beams such as the retractometer apparatus (Braudeau et al., 1999), (iv) image-based approach: where the volume of the soil sample (either clod or core) was either scanned with a 3-D optical scanner (Sander and Gerke, 2007) or by a simple standard digital camera (Stewart et al., 2012). Several studies have discussed and compared these methods, Cornelis et al. (2006) showed that there were significant differences between the Archimedes' principle-based methods "paraffin-coated and rubber balloon methods" and the physical measurement-based methods "vernier caliper method" where the former produced more accurate and reliable data; however, Sander and Gerke (2007) observed some errors in the resin-coated method that affects the measured volume due to inadequate coating or penetration of the coating materials. Crescimanno and Provenzano (1999) highlighted the problem of anisotropy of vernier caliper method due to the use of confined cores. In general, most of these methods require a continuous measurement follow up for 2–3 weeks (Crescimanno and Provenzano, 1999; Cornelis et al., 2006) and at the end produce 10–20 data pairs.

To fulfill the modeling requirement of the water flow and solute transport through a structured soil medium, the measured discrete data set must be converted into curves by fitting the data with mathematical functions through parameters fittings. Several models were developed to fit the discrete experimental data of the WRC (e.g., El-kadi, 1985; Leij et al., 1997; Groenevelt and Grant, 2004; Fredlund et al., 2011). These WRC models consider the soil as a rigid porous medium whose porosity is represented by equivalent bundle of capillary tubes (Braudeau and Mohtar, 2004; Coppola et al., 2012), their state variables are referenced to a virtual volume, Representative Elementary Volume (REV), which ignores the soil structure (Braudeau and Mohtar, 2009), and their parameters usually have no physical meaning (Chertkov, 2004). However, other researchers (Voronin, 1980; Berezin et al., 1983) followed a thermodynamic-based approach for defining the relationships between the soil water potential and water content by using physiochemical parameters and variables. Several models were developed to define the known shrinkage phases (**Figure 1**) of the SSC termed structural, normal, basic and residual by identifying the inflection point of the assumed S-shape of the

curve (McGarry and Malafant, 1987; Peng and Horn, 2005), transition points between the shrinkage phases which were named as: shrinkage limit point, air entry point, the macropore shrinkage limit point, and the maximum swelling point (e.g., Giráldez et al., 1983; McGarry and Daniells, 1987; McGarry and Malafant, 1987; Kim et al., 1992; Tariq and Durnford, 1993b; Braudeau et al., 1999), the curvature at the transition zones between the shrinkage phases (e.g., Olsen and Hauge, 1998; Peng and Horn, 2005), the slopes of the tangents of the transition points (Peng and Horn, 2005), and the slope of saturation line (Giráldez et al., 1983). Still, other studies (Groenevelt and Grant, 2002; Chertkov, 2003) used empirical coefficients and parameters to model the SSC. Few scientists have tried to integrate these shrinkage characteristic points and/or slopes in modeling the water and solute transport through structured soil medium (Armstrong et al., 2000; Larsbo and Jarvis, 2005; Coppola et al., 2012). However, these models still reference their state variables to virtual volume of soil medium (REV). Moreover, the SSC could be differentiated by the presence or lack of some shrinkage phases. Peng and Horn (2013) identified six types of SSCs based on the number of the existing shrinkage phases in the shrinkage curves by using a large set of experimental data. Finally, few scientists tried to integrate the shrinkage curve.

Compared with the existing studies, this study presents three new issues regarding to: the type of the studied soil; the apparatus used for measuring the WRC and SSC, and the models used for both WRC and SSC. Two native Aridsols, according to US soil taxonomy, in the State of Qatar were investigated in this study. This class of soil has rarely been considered in the shrinkage behavior studies. Then, a new apparatus (Bellier and Braudeau, 2013) was used to continuously and simultaneously measure the data pairs for WRC (gravimetric water content vs. soil suction/potential) and SSC (gravimetric water content vs. specific volume) for eight unconfined soil cores and through a complete drying cycle. Frequent and continuous measurements (approximately every 10 min for 2–3 days under a constant temperature of 40°C) provide essential visual presentations of the two continuous curves (i.e., WRC and SSC) including the shape, the inflection points, and the shrinkage phases which are vital for modeling the curves as discussed before. Such a continuous capture and portrayal of the discrete data can't be obtained by other methods, mainly in the case of the SSC. Also, simultaneous measurements, of the same soil core, ensure having the data pairs for WRC and SSC, which are usually measured separately, under similar conditions (same temperature, humidity, etc.) and same water contents which are considered as the main state variable in all soil water models. None of the existing methods except (Boivin et al., 2004; Braudeau and Mohtar, 2004) give such a measurement. Some studies showed that the different drying rates of the same soil sample affect the WRC behavior (Zhou et al., 2014) and the soil shrinkage and cracking behavior (Tang et al., 2010). These findings strengthen the need for having continuous and simultaneous measurements under consistent conditions which simulates reality. The use of unconfined soil cores decrease the anisotropic shrinkage behavior highlighted by Crescimanno and Provenzano (1999). In addition, the number of the sampled cores (8 samples each time) and the automated measurements

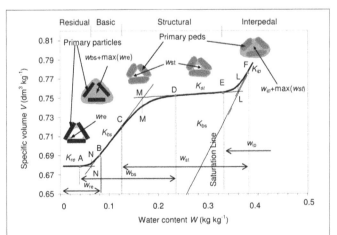

FIGURE 1 | Various configurations of air and water partitioning into the two pore systems, inter and intra primary peds, related to the shrinkage phases of a standard Shrinkage Curve [water content vs. specific volume]. The various water pools [w_{re}, w_{bs}, w_{st}, and w_{ip}] are represented with their domain of variations. The linear and curvilinear shrinkage phases are delimited by the transition points (A–F). Points N', M', and L' are the intersection points of the tangents at those linear phases of the Shrinkage Curve. (Adapted from Braudeau et al., 2004).

reduce the time and effort needed to carry out such measurements. Finally, new models for WRC and SSC were used in this study. These models were developed based on the Pedostructure and SREV concept (Braudeau and Mohtar, 2009) and the Gibbs thermodynamic potential function of the soil medium (Sposito, 1981) and recently presented in Part 1 of this study (Braudeau et al., 2014).

Thus, the objectives of this study were to: (1) introduce a new characterization approach of a soil medium based on continuous measurements of soil water potential and soil shrinkage, (2) establish a methodology for preparing reconstituted and undisturbed soil samples for apparatus' measurements, (3) evaluate the efficiency of this kind of characterization where each parameter has a physical meaning and quantifies a specific hydrostructural property of the Pedostructure. Two native Aridsoils in the State of Qatar were investigated in this study.

MATERIALS AND METHODS
THE WRC AND SSC THERMODYNAMIC EQUATIONS

In Part 1 of this study, Braudeau et al. (2014) were able to derive the following state functions of the pedostructure:

The equation of the pedostructure WRC:

$$
h^{eq}(W) = \begin{cases} h_{mi}\left(W_{mi}^{eq}\right) = \rho_w \bar{E}_{mi}\left(\frac{1}{W_{mi}^{eq}} - \frac{1}{W_{miSat}}\right), \\ \quad \textit{inside the primary peds} \\ h_{ma}\left(W_{ma}^{eq}\right) = \rho_w \bar{E}_{ma}\left(\frac{1}{W_{ma}^{eq}} - \frac{1}{W_{maSat}}\right), \\ \quad \textit{outside the primary peds} \end{cases} \quad (1)
$$

where, W is the pedostructure water content excluding the saturated interpedal water $\left[\text{kg}_{\text{water}} \text{kg}_{\text{soil}}^{-1}\right]$, W_{ma} gravimetric macropore water content "outside the primary peds" $\left[\text{kg}_{\text{water}} \text{kg}_{\text{soil}}^{-1}\right]$, W_{mi} gravimetric micropore water content "inside the primary peds" $\left[\text{kg}_{\text{water}} \text{kg}_{\text{soil}}^{-1}\right]$, $\bar{E}_{ma}$ is potential energy of surface charges positioned on the outer surface of the clay plasma of the primary peds $\left[\text{Jkg}_{\text{solid}}^{-1}\right]$, $\bar{E}_{mi}$ is potential energy of surface charges positioned inside the clay plasma of the primary peds $\left[\text{Jkg}_{\text{solid}}^{-1}\right]$, h_{mi} is the soil suction inside the primary peds [dm ~ kPa], h_{ma} is the soil suction outside the primary peds [dm ~ kPa], ρ_w is the specific density of water $\left[1\text{kg}_{\text{water}} \text{dm}^{-3}\right]$.

The equations of the pedostructure micro and macro pore water contents at equilibrium were derived such that:

$$
W_{ma}^{eq}(W) = \frac{\left(W + \frac{\bar{E}}{A}\right) + \sqrt{\left[\left(W + \frac{\bar{E}}{A}\right)^2 - \left(4\frac{\bar{E}_{ma}}{A}W\right)\right]}}{2} \quad (2a)
$$

and

$$
W_{mi}^{eq}(W) = W - W_{ma}^{eq} = \frac{\left(W - \frac{\bar{E}}{A}\right) - \sqrt{\left[\left(W + \frac{\bar{E}}{A}\right)^2 - \left(4\frac{\bar{E}_{ma}}{A}W\right)\right]}}{2} \quad (2b)
$$

where, A is a constant, such that: $A = \frac{\bar{E}_{ma}}{W_{maSat}} - \frac{\bar{E}_{mi}}{W_{miSat}}$, $\bar{E} = \bar{E}_{mi} + \bar{E}_{ma}$ and W_{miSat} and W_{maSat} are the micro and macro water content at saturation such that $W_{Sat} = W_{miSat} + W_{maSat}$.

Finally, the SSC of the pedostructure was derived such that:

$$
\bar{V} = \bar{V}_0 + K_{bs}w_{bs}^{eq} + K_{st}w_{st}^{eq} + K_{ip}w_{ip} \quad (3)
$$

where, K_{bs}, K_{st}, and K_{ip} are the slopes at inflection points of the measured shrinkage curve at the basic, structural, and interpedal linear shrinkage phases, respectively $\left[\text{dm}^3\text{kg}_{\text{water}}^{-1}\right]$, and w_{bs}, w_{st}, and w_{ip} are the water pools associated to the linear shrinkage phases of the pedostructure in $\left[\text{kg}_{\text{water}} \text{kg}_{\text{soil}}^{-1}\right]$ (Figure 1); $\bar{V}$ is the specific volume of the pedostructure $\left[\text{dm}^3\text{kg}_{\text{soil}}^{-1}\right]$, and $\bar{V}_0$ is the specific volume of the pedostructure at the end of the residual phase $\left[\text{dm}^3\text{kg}_{\text{soil}}^{-1}\right]$.

The values of the water pools associated with the basic shrinkage phase (w_{bs}), the structural shrinkage phase (w_{st}), and the interpedal shrinkage phase (w_{ip}) can be determined as shown in the following relationships:

$$
w_{bs}^{eq} = W_{mi}^{eq} - w_{re} = \frac{1}{k_N}ln\left[1 + \exp\left(k_N\left(W_{mi}^{eq} - W_{miN}^{eq}\right)\right)\right] \quad (4)
$$

$$
w_{st} = W_{ma}^{eq} = W - W_{mi}^{eq} \quad (5)
$$

$$
w_{ip} = \frac{1}{k_L}ln\left[1 + \exp\left(k_L\left(W - W_L\right)\right)\right] \quad (6)
$$

where, k_N and k_L represent the vertical distance between the intersection points N-N′, and L-L′ (Figure 1) on the shrinkage curve $\left[\text{kg}_{\text{soil}} \text{kg}_{\text{water}}^{-1}\right]$, W_{miN}^{eq} is the micro-pore water content calculated by [Equation (2b)] but by using W_N instead of W, W_N is the water content at the intersection point (N′) in Figure 1 and represents the water content of the primary peds at dry state such that $W_N = max(w_{re})$ $\left[\text{kg}_{\text{water}} \text{kg}_{\text{soil}}^{-1}\right]$, w_{re} is the water pool associated with the residual shrinkage phase of the shrinkage curve $\left[\text{kg}_{\text{water}} \text{kg}_{\text{soil}}^{-1}\right]$, W_L is the water content at the intersection point (L′) (Figure 1) such that $W_L = W_M + max(w_{st})$ $\left[\text{kg}_{\text{water}} \text{kg}_{\text{soil}}^{-1}\right]$, and W_M is the water content at the intersection point (M′) (Figure 1) such that $W_M = W_N + max(w_{bs})$ and it represents the saturated water content of the micropore domain $\left[\text{kg}_{\text{water}} \text{kg}_{\text{soil}}^{-1}\right]$.

Anyhow, different soil types have different structures and hence different shapes of the SSC. Peng and Horn (2013) identified six types of shrinkage curves based on the number of shrinkage phases observed in the measured shrinkage curve. The data set used in their study was discrete measurement consisting of 10–30 data pairs. However, as shown in Part 1, Braudeau et al. (2014) identified three main types of the shrinkage curves based on: the shape of the curve (Sigmoidal or not), and the existence of the saturated interpedal water which is responsible for the shrinkage phase parallel to the saturation line (i.e. slope = 1). They used continuous measurements of 200–600 data pairs. The three identified types were: (i) Sigmoidal shrinkage curve without saturated interpedal water, (ii) Sigmoidal shrinkage curve with saturated interpedal water, and (iii) Non-sigmoidal shrinkage curve. However, we are interested in types (i) and (iii) as they represent the shapes of the shrinkage curves of the studied soil.

To summarize, the state variables and the physical parameters describing a structured soil medium can now be characterized in three thermodynamically-based characteristic functions: (1) the micro and macro pedostructure water contents functions $\left[W_{mi}^{eq}(W) \text{ and } W_{ma}^{eq}(W)\right]$, (2) the water retention function $[h(W)]$, and (3) the soil shrinkage function $[\overline{V}(W)]$. However, the question is now how to identify the physical parameters of these functions, this question will be answered in details later in the paper, but let's first recall that in total, there are 12 hydrostructural parameters: $\overline{V}_0$, W_N, k_N, K_{bs}, K_{st}, $\overline{E}_{mi}$, $\overline{E}_{ma}$, W_{miSat}, W_{maSat}, W_L, k_L, K_{ip} where four of them are common parameters for the three characteristic functions ($\overline{E}_{mi}$, $\overline{E}_{ma}$, W_{miSat}, and W_{maSat}) and three of them (W_L, k_L, K_{ip}) are only used in the shrinkage curve of sigmoidal shape with saturated interpedal water. This shape of curves wasn't among the shrinkage curves obtained for the studied soil. Thus, in this study, only 9 hydrostructural parameters were used for characterizing the soil medium organization "pedostructure." These are the parameters of the three characteristic functions, such that:

- The pedostructure micro and macropore water content curves $\left[W_{mi}^{eq}(W) \text{ and } W_{ma}^{eq}(W)\right]$: $\overline{E}/A$, $\overline{E}_{ma}/A$, "i.e., $\overline{E}_{mi}$, $\overline{E}_{ma}$, W_{miSat}, and W_{maSat}."
- The water retention curve $[h(W)]$: $\overline{E}_{mi}$, $\overline{E}_{ma}$, W_{miSat}, and W_{maSat}.
- The soil shrinkage function $[\overline{V}(W)]$: $\overline{V}_0$, W_N, k_N, K_{bs}, K_{st}, $\overline{E}_{mi}/A$, $\overline{E}_{ma}/A$, W_{miSat}, W_{maSat}.

Finally, the continuous and simultaneous measurements provided a strong and reliable visualization of the transition points and slopes of the different shrinkage phases of the shrinkage curves. Such visualization was very imperative for extracting and estimating the hydrostructural parameters as it is discussed later in this paper.

SOILS STUDIED

Two native soils, classified as Aridisols in US Soil Taxonomy, located near Al Khor city in the State of Qatar were used in this study (see **Table 1**). These native soils are: (1) Rodah soil:

"Rodah" is an Arabic word means a garden and it is locally used for the colluvium depressions soils which have been accumulated with recent colluvial materials (mainly calcareous loamy and silty deposits) by the storm water runoff through wadis "ephemeral streams." This soil is potentially the most suitable soil for agricultural uses, and hence most of the farms in the State of Qatar are located over these depressions, (2) Sabkha soil: "Sabkha" is also an Arabic word used for the highly saline depression soils "i.e., salt marshes." The salts accumulation is due to the evaporation of the saline groundwater coming from the sea. In this study, the two native soils were taken from two depressions (Rodah and Sabkha) that are about 1 Km apart from each other (see **Table 1**), both depressions are located over a Haplocalcids great group according to US soil taxonomy (Soil Survey Staff, 1975) where limestone is the dominant outcropping formation (Scheibert et al., 2005). Finally, the Rodah and Sabkha soils used in this study have silty clay loam texture and silty loam texture, respectively.

SAMPLING AND SAMPLES PREPARATION

Disturbed and undisturbed soil cores were considered in this study for each soil type. The soil samples were taken from the top layer (usually 0–10 cm depth) for both types of cores. The procedures for both types of soil core preparation are explained in the following sub-sections.

Disturbed soil samples

In this study, soils were collected from the field then air-dried and then sieved using 200 μm and 2 mm sieves. This range of particle sizes (200 μm–2 mm) was selected in this study to get soil cores with macro-aggregates and hence with good structures, and also to minimize the presence of loose fine sand and silt particles and crystallized salts especially in the case of Sabkha soil. After that, the soil aggregates were filled in thin layers (**Figure 2A**) in a Polyvinyl chloride (PVC) rings ($\Phi = 5$ cm, $h = 5$ cm) whose internal walls were coated with a thin petroleum jelly film to prevent the soil adhering to the wall during construction and to make the removal of the soil cores easier after construction. Note that unconfined soil cores were used in the analyses. During the filling process, the PVC cores were placed in small pans partially filled with water (about 5 mm deep). Whatman filters No. 40 were used to hold the soil aggregates inside the PV cores. The soil aggregates were added in thin layers (about 1 cm) with gentle tapping at the edge of the PVC ring after adding each layer.

Table 1 | General description of the soil samples used in the study.

Soil type	Sampling location		Soil texture				Soil sample IDs	Cores	EC	pH
	Latitude	Longitude	% Clay	% Silt	% Sand	Type			dS/m	–
Rodah	25°26'19"	51°17'51"	39	52	9	Silty clay loam	AM[67–69]	Disturbed [200 μm–2 mm]	1.80 ± 0.010	8.37 ± 0.01
							UDR[1–3]	Undisturbed	0.286 ± 0.016[a]	8.76 ± 0.02
Sabkha	25°45'20"	51°30'14"	15	65	20	Silty loam	AM[1–3]	Disturbed [200 μm–2 mm]	7.61 ± 0.58	8.60 ± 0.08
							UDS[1–3]	Undisturbed	3.62 ± 0.48[a]	8.10 ± 0.23

[a]Note that the EC values for the undisturbed soil samples are very low. This was due to the procedure used in the field by saturating the soil with tap water (EC = 200 μS/cm).

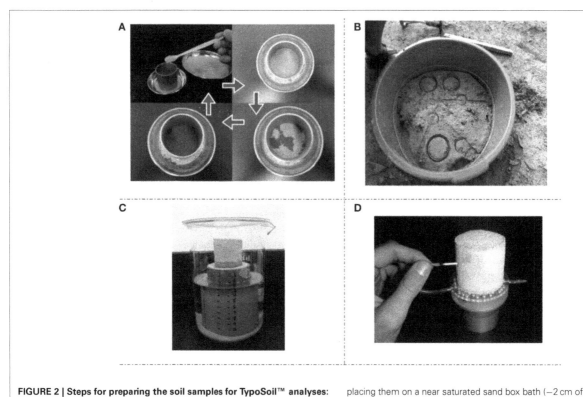

FIGURE 2 | Steps for preparing the soil samples for TypoSoil™ analyses: (A) preparing disturbed soil sample, (B) taking undisturbed soil samples from the field, (C) saturating the unconfined soil samples by capillary through placing them on a near saturated sand box bath (−2 cm of water), (D) placing the soil sample on the perforated supporting platform and inserting the porous ceramic cup mini Tensiometer in the middle of the unconfined core.

The second soil layer had not to be added until the first layer saturated to maintain well-constructed cores without horizontal segmentation. Once the cores construction was finished, a gentle leveling of the soil surface was performed. Then, the soil cores were taken and placed in an oven set at 40°C for approximately 48 h. This process allowed the soil aggregates to shrink until the end of the basic shrinkage phase and enhanced producing a structured soil medium. The time for drying could vary based on the soil type and soil salinity. After that, the soil cores were re-saturated by replacing them in the saturating pans and allowing the cores to saturate. Then, they were dried for a second time by placing them in an oven set at 40°C for about 48 h. This was the procedure for preparing a well-constructed soil aggregates cores.

Undisturbed soil samples

Undisturbed soil samples were also taken from both Rodah and Sabkha soils for better understanding and comparison purposes. After selecting the sampling points, the top soil layer (about 7 cm thick) was saturated by using Infiltrometer. Then, the same PVC rings ($\Phi = 5$ cm, $h = 5$ cm) were used to take the soil samples (see **Figure 2B**). The saturation process was done to eliminate the swelling effect on the soil structure of these confined cores once they were re-saturated in the lab. The soil cores were then removed, labeled, and covered with two caps.

Preparing the soil samples for TypoSoil™ measurements

The soil cores (disturbed and undisturbed) were then saturated by capillary wetting by placing them on a sand box bath (the water level in the bath is 2 cm below the top of the sand box).

As shown in **Figure 2C**, the sand box is simply a ($\Phi = 7.6$ cm, $h = 12$ cm) PVC tube placed in a bath (2000 ml beaker). The saturation process lasted for (1–2 days) and operated under atmospheric pressure (Dickson et al., 1991; Braudeau et al., 1999; Salahat et al., 2012) and with a looking glass on top of the beaker to minimize the evaporation. The PVC rings were removed after a short time. The porous ceramic cup mini Tensiometer and the support platform which contains the pressure gauge were prepared for hosting the soil cores. Both the tensiometer and support platform must be free from any air bubbles. This process was done by flushing the system with degassed, deionized water to minimize the effects of the osmotic potentials in measurement. Then, the tensiometers were inserted (**Figure 2D**).

THE APPARATUS: TypoSoil™ [SOIL TYPOLOGY]

The new apparatus (Bellier and Braudeau, 2013), TypoSoil™, is a device intended to measure continuously and simultaneously the two soil moisture characteristic curves: WRC and SSC for eight (100 cm³) unconfined cylindrical soil cores. The continuous measurement of the two characteristic curves enables the user to identify precisely the characteristic points and/or portions of the curves that are used to predict the soil moisture characteristic functions. The simultaneous measurements guarantee that the soil moisture characteristic curves are measured under identical condition which is not the case in the existing devices.

The WRC is a relationship between soil suction/potential and gravimetric water content, while the SSC is a relationship between the specific volume (volume/mass of solids) and the gravimetric water content. Hence, both curves are related to the gravimetric water content. This point was taken into account while making the measurements in this device. The whole set of measurements (soil weight, dimensions, and suction) are recorded instantaneously when the soil core weight is taken.

TypoSoil™ is a modified device of the retractometer (Braudeau et al., 1999). The later measures only the SSC while the new device measures both SSC and WRC. The new device consists of the following components (**Figure 3**): (i) A biological stove working on a fixed temperature identified by the user (usually 30–40°C). (ii) An electronic analytical balance with MonoBloc weighting cell with a connection point's plate fixed on it. This plate closes the electrical circuit to measure and record the data once contacted with the support platform as explained below (see also **Figure 3**). (iii) Laser sensors: one spot laser sensor that measures the height of the soil sample by triangulation (10 μm resolution) and two thru-beam sensors (5 μm resolution) to measure the diameter

of the soil sample by measuring the portion of the beam not intercepted by the laser, (iv) turning plate which can house 8 cylindrical soil samples (100 cm³ "Φ = 5 cm, h = 5 cm") placed on perforated support platform which contains a pressure gauge inside connected to the soil core by a ceramic-needle tensiometer with an approximate functional range of 0–700 hPa. This rotating plate automatically descends to make a contact between the bottom of the perforated support platform and the connection points fixed on the balance to take and record the measurements (soil weight, dimensions, and suction). The turning plate ascends and rotates to make the measurement for another soil sample. In general, a full cycle of measurements for the 8 cores can be done in 10 min and can be repeated until the sample's weight remains constant (usually it takes 2–3 days at 40°C). Then the sample is oven dried at 105°C to determine its weight M_s (structural mass or mass of solids), (v) control panel with screen for displaying the data during and after the experiment, power and emergency stop

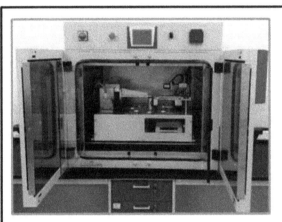

TypoSoil™

Main System Components:

1. **A Biological Stove:** (working at fixed temperature: 30-40 °C).
2. **An Electronic Analytical Balance** with MonoBloc weighting cell.
3. **Turning Plate:** This plate rotates, descends and ascends to take the measurements every time for each soil sample. It can house 8 perforated supporting platform (i.e. 8 Soil Samples). It makes a full cycle in 10 minutes.
4. **Laser Sensors:** One Spot sensor to measure the height and Two Thru-beam sensors to measure the diameter of the soil sample.
5. **Control Panel:** It contains screen, power and emergency stop bottoms, USB, SD Card, and Ethernet connection to download the data.

Turning Plate
can house 8 Perforated Support Platform (Soil Samples)

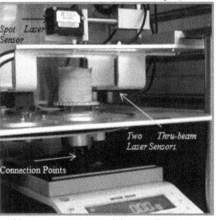

Measurements (core weight, core dimensions and soil suction) are taken when the Turing Plat descends then the connection points fixed on the balance contact with the bottom of the platform

Ceramic-needle Tensiometer

Perforated Support Platform where the ceramic needle Tensiometer is connected to a pressure transducer located inside the stainless steel chamber

FIGURE 3 | TypoSoil™ with its main components and how they function to measure continuously and simultaneously the soil weight (by a balance), soil suction (by a porous ceramic cup mini tensiometer), and the soil dimensions (by one spot laser sensor to measure the height and two thru-beam laser sensors to measure the diameter).

bottoms, USB, SD Card, and Ethernet connections to download the data.

Assuming an isotropic radial shrinkage and uniform distribution of the soil water content throughout the unconfined cylindrical soil cores, the specific volume and the soil water content of the soil core can be identified, respectively, such that:

$$\overline{V} = \frac{\pi D^2 H}{4 M_s} \tag{7}$$

where, $\overline{V}$ is the specific volume of the soil sample $\left[dm^3 Kg_{solid}^{-1}\right]$, D and H are, respectively, the measured diameter and height of the soil sample $[dm]$, M_s is the dry mass of the soil sample at 105°C $[kg_{solid}]$.

$$W = \frac{(m - M_s)}{M_s} \tag{8}$$

where, W is the specific water content of the soil sample $\left[Kg_{water} \, kg_{solid}^{-1}\right]$, m is the measured mass of the soil sample $[kg_{water}]$, M_s is the dry mass of the soil sample at 105°C $[kg_{solid}]$.

The saturation line $[V_{Sat}]$ was calculated by using the following equation (Braudeau et al., 2005):

$$V_{Sat} = \frac{W}{\rho_w} + V_s \tag{9}$$

where, W is the specific water content of the soil sample $\left[Kg_{water} \, kg_{solid}^{-1}\right]$, ρ_w is the specific density of water $\left[1 kg_{water} dm^{-3}\right]$, V_s is the specific volume of the soil phase, estimated from the particle density, $\left[dm^3 Kg_{solid}^{-1}\right]$.

EXTRACTION AND ESTIMATION OF PEDOSTRUCTURE CHARACTERISTIC PARAMETERS [HYDRO-STRUCTURAL PARAMETERS]

The proper modeling of water flow and solute transport through a structured soil medium requires continuous characteristic curves (WRC and SSC) instead of discrete experimental data. In this section, the procedures used for extracting and estimating the hydro-structural parameters of the proposed equations for the pedostructure WRC $[h(W)]$, SSC $[\overline{V}(W)]$, and the micro and macro water contents at equilibrium $\left[W_{mi}^{eq}(W) \text{ and } W_{ma}^{eq}(W)\right]$ from the continuously and simultaneously measured data pairs were discussed. In total, there are twelve parameters: W_{miSat}, W_{maSat}, $\overline{E}_{mi}$, $\overline{E}_{ma}$, $\overline{V}_0$, W_N, k_N, K_{bs}, K_{st}, W_L, k_L, K_{ip}. However, the last three parameters (W_L, k_L, K_{ip}) were not included in this study as they are related to the existence of interpedal water shrinkage phase which was not the case for the types of soil used in this study. **Table 2** provides a comprehensive summary about these parameters, their units and the how they were estimated. The parameters extraction and estimation procedure included the following steps: (i) identify the type of the shrinkage curve, (ii) extract and/or give initial estimates of the values of WRC parameters, $\left(W_{miSat}, W_{maSat}, \overline{E}_{mi}, \overline{E}_{ma}\right)$ (**Figures 4, 5; Table 2**), (iii) minimize the sum of square errors between modeled and measured WRC by using the Microsoft Excel solver, (iv) extract and/or give initial estimates of the values of SSC parameters $\left(W_{miSat}, \overline{E}/\overline{E}_{ma}, \overline{V}_0, W_N, k_N, K_{bs}, K_{st}\right)$ (**Figures 4, 5; Table 2**), and (v) minimize the sum of square errors between modeled and measured SSC by using the Microsoft Excel solver.

The first step in the parameters extraction and estimation process was identifying the shape of the SSC. This step is fundamental because it affects the procedures for the parameters extraction and estimation. In this study, two types were identified: sigmoidal shape without saturated interpedal segment and Non-sigmoidal shapes.

The case of sigmoidal shrinkage curve without saturated interpedal segment

In the case of sigmoidal shrinkage curve without saturated interpedal segment, the different shrinkage phases (residual, basic, and structural) and the transition points among these phases (N and M) could easily be recognized on the measured SSC. This precise distinguishing was only possible due to having continuous measurements of the data pairs (water content and specific volume) of SSC. The following steps were followed for identifying the hydrostructural parameters of the WRC and the SSC:

Extracting and estimating the parameters of WRC. According to Equation (1), the parameters of WRC are: W_{miSat}, W_{maSat}, $\overline{E}_{mi}$, and $\overline{E}_{ma}$. The first two parameters (W_{miSat}, W_{maSat}) represent the water contents of the micropore and macropore volume at saturation, respectively. They were extracted directly from the measured WRC and SSC (**Figure 4**), such that: $W_{miSat} = W_M$, and $W_{maSat} = W_{Sat} - W_{miSat}$. However, the other parameters $\overline{E}_{mi}$, $\overline{E}_{ma}$ represent the potential energy of the surface charges of the clay particles inside and outside the primary peds, respectively. These parameters were given initial values, such that: $\overline{E}_{mi} = 40$ J/kg and $\overline{E}_{ma}$ was replaced by $\overline{E}/\overline{E}_{ma}$ with initial value = 100 J/kg. Finally, the sum of square errors, between the modeled [using the extracted/estimated parameters in Equation (1)] and the measured WRC, was minimized by using the Microsoft Excel solver.

Extracting and estimating the parameters of SSC. According to Equations (3–6), the parameters of SSC are: (W_{miSat}, $\overline{E}/\overline{E}_{ma}$, $\overline{V}_0$, W_N, k_N, K_{bs}, K_{st}). The first two parameters have already been identified from the previous step knowing that $\overline{E} = \overline{E}_{mi} + \overline{E}_{ma}$. Then, $\overline{V}_0$ and K_{bs} which represent respectively the specific volume at the end of the shrinkage curve and the slope of the basic shrinkage phase were extracted precisely from the measured shrinkage curve (**Figure 4**), these two values were assumed fixed and weren't included in the optimization process. However, W_N and K_{st} which represent respectively the water content of the specific pore volume of dry primary peds and the slope of the structural shrinkage phase were estimated as shown in **Figure 4**, then they were included in the optimization process. Finally, k_N was given in initial value of $100 \, kg_s/kg_w$. Thus, only three parameters (W_N, K_{st}, and k_N) were optimized in the process of minimizing the sum of square errors, between the

Table 2 | A summary for all the characteristic parameters, included in this study, for both WRC and SSC.

Parameter	Unit	Description	Extraction/Estimation
W_{miSat}	kg_w/kg_s	It represents the water content of the micropore volume at saturation. Thus, it is a *characteristic transition point*.	Based on the shape of SSC: • S-shape: it equals W_M read directly from SSC. (**Figure 4**) • Non-S-shape: its initial value is read directly from the measured WRC at any point within the range [h∈ (400–500) hPa].
W_{maSat}	kg_w/kg_s	It represents the water content of the macropore volume at saturation. Thus, it is a *characteristic transition point*.	It is estimated such that: $W_{maSat} = W_{Sat} - W_{miSat}$ W_{Sat} corresponds to $h = 0\ hPa$ in the measured WRC. (**Figures 4, 5**).
$\overline{E}_{mi}$	J/kg_s	It represents the potential energy of the surface charges of the clay particles inside the primary peds.	This value is identified by the optimization process. In general, its initial value = 40 J/kg.
$\overline{E}_{ma}$	J/kg_s	It represents the potential energy of the surface charges of the clay particles outside the primary peds.	This value is identified by the optimization process. It is replaced by $\overline{E}/\overline{E}_{ma}$ with an initial value = 100 J/kg.
$\overline{V}_0$	dm^3/kg_s	It represents the specific volume at the end of the shrinkage curve when no further changes in water content can be observed. Thus, it is a *characteristic transition point*.	It is extracted directly from the measured SSC.
W_N	kg_w/kg_s	It represents the water content of the specific pore volume of dry primary ped. Thus, it is a *characteristic transition point*.	An accurate estimate can be extracted directly from the SSC of S-shape (**Figure 4**), while an initial estimate can be extracted from the Non-S-Shape of SSC (**Figure 5**).
k_N	kg_s/kg_w	It represents the vertical distance between N and N′ on **Figure 1**.	This value is identified by the optimization process. In general, one can assume its initial value = 100 kg_s/kg_w.
K_{bs}	dm^3/kg_w	It represents the slope of the basic shrinkage phase of SSC. Thus, it is a *characteristic slope*.	Accurate estimates for these two characteristic slopes can be extracted directly from the SSC of S-shape (**Figure 4**), while initial estimates can be extracted from the Non-S-Shape of SSC (**Figure 5**).
K_{st}	dm^3/kg_w	It represents the slope of the structure shrinkage phase of SSCC. Thus, it is a *characteristic slope*.	

The summary includes: the symbols of the parameters, their units, the physical meaning of each parameter and finally how each of them was extracted/estimated and identified.

modeled [using the extracted/estimated parameters in Equations (3–6)] and the measured SSC, by using the Microsoft Excel solver.

The case of non-sigmoidal shrinkage curve

In the second case, the non-sigmoidal shape of the SSC, one could not identify any mark on the measured shrinkage curve for the positions of the transition points (N, M, and L). However, thanks to the continuous measured data points, the three shrinkage phases of the shrinkage curve (residual, basic, and structural) could be distinguished on the measured curve and hence the slope parameters K_{bs}, K_{st} could be measured directly from the available data (**Figure 5**).

Extracting and estimating the parameters of WRC. Similar procedures for extracting and estimating the hydrostructural parameters of the WRC in the first were followed in this case, but with one exception. The initial value of W_{miSat} was located on the measured WRC between the potential range 400–500 hPa (**Figure 5**). This step was done because it was very difficult to identify the transition point (M) between the basic and the structural phases

on the shrinkage curve in such a case. At the end, the four parameters (W_{miSat}, W_{maSat}, $\overline{E}_{mi}$, $\overline{E}_{ma}$) were included in the optimization process.

Extracting and estimating the parameters of SSC. Similar to the first case, but W_N was roughly estimated on the shrinkage curve (the "X" in **Figure 5**), and K_{bs}, K_{st} were measured from the data as shown in the same figure. Finally, the parameters (W_N, K_{st}, k_N, and K_{bs}) were included in the optimization process. However, K_{bs} was included this time in the optimization process to minimize the effect of the curve's shape.

RESULTS

THE MEASURED WATER RETENTION AND SOIL SHRINKAGE CURVES

The continuous and simultaneous measurements of the WRCs and SSCs for reconstituted and undisturbed Rodah soil samples are shown in **Figure 6**, while **Figure 7** shows the measurements for the reconstituted and undisturbed Sabkha soil samples. Three replicates of disturbed, constructed from aggregates of

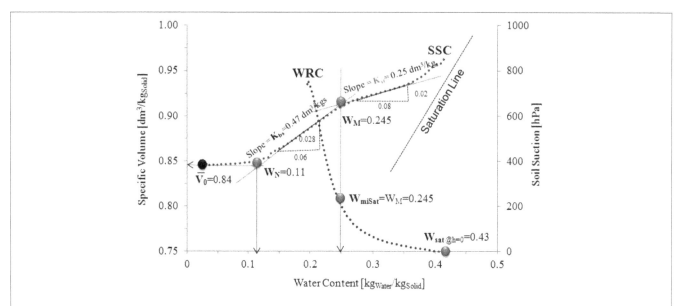

FIGURE 4 | Extracting the characteristic parameters of WRC and SSC in the case of sigmoidal shrinkage curve with no saturated interpedal water [Core#AM69]. The black dot and text indicate that the parameters are fixed to the extracted values, while the gray dots and texts indicate that these parameters could be changed during the optimization.

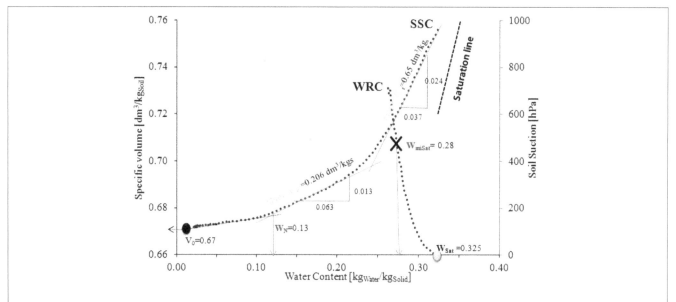

FIGURE 5 | Extracting the characteristic parameters in the case of non-sigmoidal shrinkage curve [Core # UDR2]. The black dot and text indicate that the parameters are fixed to the extracted values, the gray dots and texts indicate that these parameters were optimized by using the MS Excel solver, while the X indicates rough estimate for the parameters.

size range (200 μm–2 mm), and undisturbed soil samples for both soil types: Rodah and Sabkha Soils were analyzed by the TypoSoil™. The measured SSCs for reconstituted soil samples (**Figures 6A, 7A**) showed high departure from the Saturation Line. Such a departure could be a side-effect of such a macro-aggregate soil medium (200 μm–2 mm). During the construction and preparation procedures, the samples were saturated from bottom with (2 cm suction sand box). This could produce some voids filled with air and thus lack complete saturation. However, this was not the case in the undisturbed soil samples

(**Figures 6B, 7B**). Still, the interpedal shrinkage phase parallel to the load line was absent in all samples. Regarding the shrinkage behavior of the soil samples "i.e., the range of the specific volume changes from saturation until dry state," Rodah soil samples showed higher shrinkage amplitude values compared to the Sabkha soil samples. The specific volume changes for reconstituted Rodah soil samples ranged between "0.85 and 0.95" (i.e., the range = 0.1) dm³/kg$_s$, undisturbed Rodah soil samples "0.68–0.75" (i.e., the range = 0.07) dm³/kg$_s$. Reconstituted Sabkha soil samples ranged "0.85–0.88" (i.e., the range = 0.03) dm³/kg$_s$, and

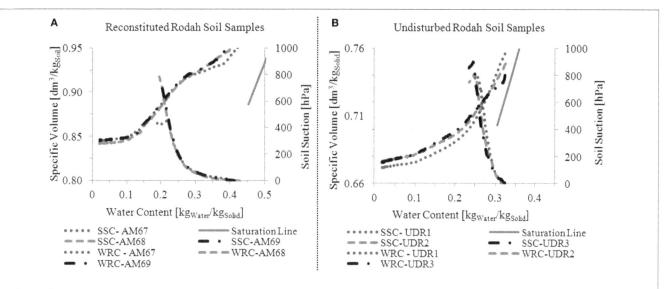

FIGURE 6 | The measured soil moisture characteristic curves for Rodah soil: Water Retention Curve [WRC] and Soil Shrinkage Curve [SSC] for **(A)** the three replicates of reconstituted Rodah soil samples (aggregate size: 200 μm–2 mm); **(B)** the three replicates of undisturbed Rodah soil samples.

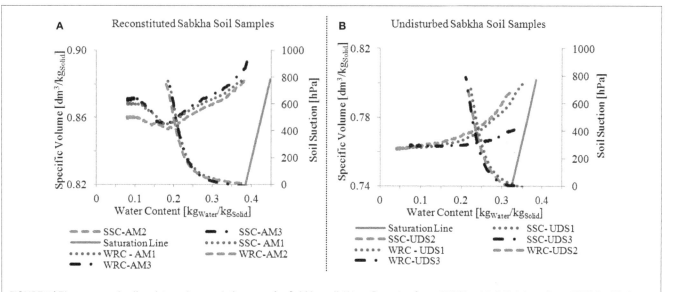

FIGURE 7 | The measured soil moisture characteristic curves for Sabkha soil: Water Retention Curve [WRC] and Soil Shrinkage Curve [SSC] for **(A)** the three replicates of reconstituted Sabkha soil samples (aggregate size: 200 μm–2 mm); **(B)** the three replicates of undisturbed Sabkha soil samples.

the undisturbed Sabkha soil samples "0.76–0.8" (i.e., the range = 0.04) dm^3/kg_s. Two reasons could be considered for explaining the higher shrinkage amplitude of Rodah soil compared with the Sabkha soil; (1) the soil texture for Rodah soil is (silty clay loam) while the Sabkha soil is (silty loam), thus Rodah soil has higher clay content and lower sand content and consequently higher shrinkage amplitude; (2) the higher salinity of Sabkha soil compared with Rodah Soil (**Table 2**) affects the amount of water lost through evaporation during the drying cycle (2–3 days) and reduced the shrinkage of the sample. However, extending the time of drying cycle in the case of Sabkha was meaningless as the samples showed an increase in their specific volumes as shown in **Figure 7**. Moreover, the undisturbed soil samples showed steeper

WRCs for both soil types compared with the reconstituted soil sample; this was a result of the selected aggregate size.

The reconstituted Rodah soil samples [AM67, AM68, and AM69] had almost an S-shape for the SSCs (**Figure 6A**) with steep slopes of the basic shrinkage phases of the shrinkage curves. S-shape usually indicates a good soil structure (Braudeau et al., 1999; Boivin et al., 2004; Peng and Horn, 2005, 2013) and these samples were constructed from macro-aggregates (200 μm–2 mm) excluding all the fine sand and loose silt particles less than 200 μm. **Figure 6A** shows a very good match of the measured WRC and SSC for the three replicates. This good match indicates that samples preparing procedures were reliable. Small variation was observed at the beginning (saturation state) and at the end of the shrinkage curve (dry state), such variations could be due to: (1) different saturated initial states for different samples; (2) different initial volumes of the samples ($\sim 100 \, \text{cm}^3$). Finally, as shown in **Figure 6A**, the measured soil potential for sample (AM67) was only up to 400 hPa.

The reconstituted Sabkha soil samples [AM1, AM2, and AM3] had similar shrinkage and WRCs, but there was a shift in both curves of sample AM2. However, the shrinkage curves of the three samples showed dramatic increase in their specific volumes after a certain point (**Figure 7A**) which had the same water content in the three replicates (this water content $\approx 0.17 \, \text{kg}_{\text{water}}/\text{kg}_{\text{soil}}$). Moreover, it was also noticed that this water content was almost the same water content where the tensiometer readings reached the air entry point of the tensiometers (**Figure 7A**). Actually, such a behavior was reported by Boivin et al. (2006), they interpreted this increase in specific volume as a result of breaking the water meniscus between sand particles due to the drying. However, this particular behavior wasn't observed in reconstituted Rodah soil samples [AM67, AM68, and AM69] which were prepared by similar procedures. Thus, the soil preparation methodology couldn't be blamed for such a behavior. However, the only determinable difference between these samples is the soil salinity. The soil salinity of the reconstituted Rodah soil samples was $1.80 \pm 0.01 \, \text{mS/cm}$, while it was 7.61 ± 0.58 for the reconstituted Rodah soil samples. The soil salinity was measured for the soil samples after being analyzed by TypoSoil™ and by using a multi-parameter meter with soil solution of (1:5 soil/water ratio). Salinity affects the soil flocculation by enhancing the bending of fine particles together (Abu Sharar et al., 1987), side by side with the aggregate size and the drying of the soil samples could lead to disjoining the aggregates and increasing gradually the sample volume. Nevertheless, this part of the shrinkage curve was excluded in the analysis.

The shapes of the SSCs for the undisturbed soil samples for both Rodah (**Figure 6B**) and Sabkha (**Figure 7B**) soils were non-sigmoidal. Moreover, those soil samples had lower saturated water content compared with the reconstituted soil samples as those samples were more compacted compared to the reconstituted ones. Two samples of the undisturbed Rodah [UDR2 and UDR3] showed similar WRC and SSC, while the third one [UDR1] had slightly shifted curves as shown in **Figure 6B**. In the SSC, three shrinkage phases could, somehow, be distinguished. They are: residual, basic, and structural shrinkage phases. This distinction was more difficult in the case of the undisturbed Sabkha soil samples [UDS1, UDS2, and UDS3] as shown in **Figure 7B**. The samples for both undisturbed Rodah and Sabkha soils showed a steep slope for the structural shrinkage phase of the SSC which indicated a high impact of the removal of the structural macropore water on the shrinkage behavior of such types of soil. Soil sample UDS3 was an exception; it didn't show the same behavior nor showed a similar shrinkage curve.

MODELING THE WATER RETENTION AND SOIL SHRINKAGE CURVES

In this section, two processes were evaluated: (1) the extraction and estimation of the hydrostructural parameters from the continuous and simultaneous data pairs of the two characteristic

curves (WRC and SSC); and (2) the efficiency of the thermodynamic and pedostructure based equations to model the two soil moisture characteristic curves (WRC and SSC) using such kind of hydrostructural parameters. In total, there are 12 hydrostructural parameters, while in this study only nine of them were used $(\overline{V}_0, W_N, k_N, K_{bs}, K_{st}, \overline{E}_{mi}, \overline{E}_{ma}, W_{miSat}, W_{maSat})$ due to the obtained shapes of the measured SSC. Nine parameters are still a large number and can be considered as a disadvantage compared to the other existing models. However, the following points should be kept in mind once doing such a comparison: (i) each parameter has a physical meaning and quantifies a specific hydrostructural property of the soil medium; (ii) six parameters $(\overline{V}_0, W_N, K_{bs}, K_{st}, W_{miSat}, W_{maSat})$ out of nine can be extracted from well identified locations (points and slopes) on the measured SSC and WRC (see **Figures 4, 5**) and the accuracy of such a process depends on having continuous and simultaneous measurements, such measurements can be provided by TypoSoil™, and it also depends on the shape of curve as shown in section Extraction and Estimation of Pedostructure Characteristic Parameters [Hydro-structural Parameters]; thus (iii) it ends up having only three parameters $(k_N, \overline{E}_{mi}, \overline{E}_{ma})$ to be optimized and used for modeling both soil moisture characteristic curves (SSC and WRC).

As discussed before in section Extraction and Estimation of Pedostructure Characteristic Parameters [Hydro-structural Parameters], two types of SSCs were observed in this study: sigmoidal without interpedal shrinkage phase and non-sigmoidal shrinkage curves. The procedures for extracting and estimation the hydrostructural parameters were clearly discussed in that section. Comparing the initially extracted parameters from the measured curves with the ones obtained after the optimization process, the Sigmoidal curves were much better than the other type. As shown in **Figure 4**, the extracted parameters of a sigmoidal shrinkage curve sample [AM69] were such that: $W_{miSat} = 0.245$, $W_{maSat} = 0.185$, $W_N = 0.11$, $K_{bs} = 0.47$, $K_{st} = 0.25$, and $\overline{V}_0 = 0.84$. The obtained results from the optimization process were (**Table 3**): $W_{miSat} = 0.24$, $W_{maSat} = 0.19$, $W_N = 0.10$, $K_{bs} = 0.46$, $K_{st} = 0.28$, and $\overline{V}_0 = 0.84$. Such a perfect matching ensured the accuracy of the extraction procedures for those physical parameters and the importance of having continuous and simultaneous measurements of the two characteristic curves. On the contrary, the characteristic points and segments that identify the hydrostructural parameters were not easily identified in the non-sigmoidal shrinkage curve. Still some good pairing was observed between the extracted parameters from the measured curves and the optimized ones **Figure 5** presents an example of this type of curves for sample [UDR2], the extracted parameters from the measured curves were: $W_{miSat} = 0.28$, $W_{maSat} = 0.045$, $W_N = 0.13$, $K_{bs} = 0.206$, $K_{st} = 0.26$, and $\overline{V}_0 = 0.673$; while, the optimized ones were: $W_{miSat} = 0.28$, $W_{maSat} = 0.05$, $W_N = 0.17$, $K_{bs} = 0.37$, $K_{st} = 0.62$, and $\overline{V}_0 = 0.67$. Here, the only parameters that showed variation between the two sets of parameters were W_N and K_{bs} which represent the inflection point between the basic and residual shrinkage phases and the slope of the basic shrinkage curves, respectively. Such a variation was expected due to the difficulty of identifying these characteristic points and segments on such a type of shrinkage

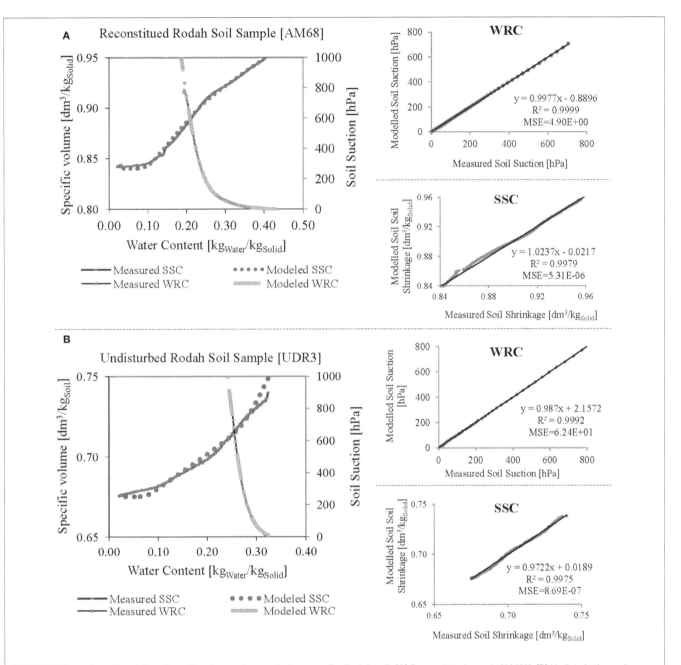

FIGURE 8 | Examples of modeling the soil moisture characteristic curves for Rodah soil. (A) Reconstituted sample [AM68]; **(B)** Undisturbed sample [UDR3]. WRC, Water Retention Curve; SSC, Soil Shrinkage Curve; R^2, coefficient of determination; MSE, Mean Square Errors.

curves. These results highlighted three imperative conclusions. The first is that the extraction methodology of the hydrostructural parameters is adequate. The second is the imperative nature of having continuous and simultaneous measurements of the two characteristic curves. The third is the simplicity of the method in identifying multiple parameters, six parameters out of nine, $(\overline{V}_0, W_N, K_{bs}, K_{st}, W_{miSat}, W_{maSat})$ for two fundamental characteristic curves of the soil medium in one step.

Examples for the modeled WRCs and SSCs by using the extracted/optimized parameters are shown in **Figures 8, 9**. The results were very promising as shown in the curves of the

reconstituted Rodah soil sample [AM68] (**Figure 8A**), undisturbed Rodah soil sample [UDR3] (**Figure 8B**), reconstituted Sabkha soil sample [AM1] (**Figure 9A**), and the undisturbed Sabkha soil sample [UDS2] (**Figure 9B**). A statistical summary of the excellent matching between the measured characteristic curves and modeled ones for all soil samples is shown in **Table 5**. In the case of WRC, 99.80–99.99% of the variation in the measured data could be explained by the modeled ones with root mean square errors (RMSEs) ranging between (2.2 and 4.5) hPa. While, for the SSC, 98.4–99.98% of the variation in the measured data could be explained by the modeled ones with RMSEs ranging

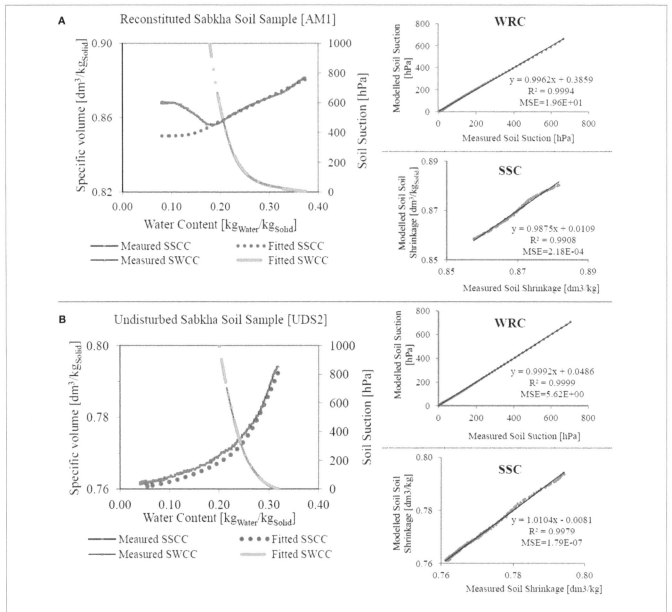

FIGURE 9 | Examples of modeling the soil moisture characteristic curves for Sabkha soil. (A) Reconstituted sample [AM1]; **(B)** Undisturbed sample [UDS2]. WRC, Water Retention Curve; SSC, Soil Shrinkage Curve; R^2, coefficient of determination; MSE, Mean Square Errors.

between $(9.3 \times 10^{-4} - 1.5 \times 10^{-2})$ dm^3/kg$_{solid}$. It should be kept in mind that the measurement and modeled WRC curves were for the range of the readings of the tensiometers. This excellent matching between the measured characteristic curves and modeled ones supports the validity, adequacy and reliability of the extraction methodology of the hydrostructural parameters and the thermodynamic-based equations of the two characteristic curves.

Finally, the issue of uniqueness of these hydrostructural parameters for a specific soil type was evaluated. "Uniqueness" is whether the three replicates produced the same hydrostructural parameters and whether these parameters are different from other soil types. A statistical summary of the hydrostructural

parameters for Rodah soil samples and Sabkha soil samples are shown in **Tables 3**, **4**, respectively. This summary provided sufficient evidence to support the uniqueness of the hydrostructural parameters for each soil type, these evidences were: (i) the parameters $(\overline{E}_{mi}, W_{miSat}, \overline{V}_0, k_N,$ and $W_N)$ were almost the same for three replicates in each group, see the values of standard deviations (SD) and coefficient of variation (CV), (ii) the observed variations in the parameters $(W_{maSat}, K_{bs}, K_{st})$ are justifiable. It was obvious from the measured curves (**Figures 6, 7**) that there was one sample of each group that had different soil shrinkage or retention curves, these samples were: (a) sample [AM67] in the reconstituted Rodah soil group showed a different shape of the structural shrinkage phase, which was characterized by K_{st} and

Table 3 | Hydro-structural parameters for the water retention curves and soil shrinkage curves of the reconstituted and undisturbed Rodah soil samples.

Soil core ID	Hydro-structural characterization for Rodah soil samples								
	Water retention curve[a] [WRC]				Soil shrinkage curve [SSC]				
	E_{mi} [J/kg$_s$]	E_{ma} [J/kg$_s$]	W_{miSat} [kg$_w$/kg$_s$]	W_{maSat} [kg$_w$/kg$_s$]	V_0 [dm^3/kg$_s$]	k_N [kg$_w$/kg$_s$]	W_N [kg$_w$/kg$_s$]	K_{bs} [dm^3/kg$_s$]	K_{st} [dm^3/kg$_s$]
DISTURBED RODAH SOIL SAMPLES: AGGREGATES SIZE [200 μm–2 mm]									
AM67	82	0.58	0.24	**0.23**	0.85	1.23	0.10	0.47	**0.19**
AM68	78	0.57	0.24	0.18	0.84	1.24	0.10	0.47	0.27
AM69	80	0.66	0.24	0.19	0.84	1.26	0.10	0.46	0.28
Average	*80*	*0.60*	*0.24*	*0.20*	*0.84*	*1.24*	*0.10*	*0.47*	*0.25*
SD	*2.20*	*0.05*	*0.00*	*0.03*	*0.00*	*0.01*	*0.00*	*0.01*	*0.05*
CV	*0.03*	*0.08*	*0.00*	*0.15*	*0.00*	*0.01*	*0.00*	*0.01*	*0.19*
UNDISTURBED RODAH SOIL SAMPLES									
UDR1	181	0.19	0.29	**0.03**	0.67	0.20	0.18	**0.47**	**1.07**
UDR2	186	0.49	0.28	0.05	0.68	0.20	0.17	0.37	0.62
UDR3	183	0.40	0.27	0.06	0.68	0.20	0.15	0.40	0.68
Average	*184*	*0.36*	*0.28*	*0.05*	*0.67*	*0.20*	*0.17*	*0.41*	*0.79*
SD	*2.14*	*0.16*	*0.01*	*0.01*	*0.00*	*0.00*	*0.02*	*0.05*	*0.25*
CV	*0.01*	*0.44*	*0.03*	*0.27*	*0.01*	*0.00*	*0.12*	*0.12*	*0.31*

[a] The hydro-structural parameters of WRC are also parameters for the SSC; SD: Standard Deviation; CV: Coefficient of Variation.

Table 4 | Hydro-structural parameters for the water retention curves and soil shrinkage curves of the reconstituted and undisturbed Sabkha soil samples.

Soil core ID	Hydro-structural characterization for Sabkha soil samples								
	Water retention curve[a] [WRC]				Soil shrinkage curve [SSC]				
	E_{mi} [J/kg$_s$]	E_{ma} [J/kg$_s$]	W_{miSat} [kg$_w$/kg$_s$]	W_{maSat} [kg$_w$/kg$_s$]	V_0 [dm^3/kg$_s$]	k_N [kg$_w$/kg$_s$]	W_N [kg$_w$/kg$_s$]	K_{bs} [dm^3/kg$_s$]	K_{st} [dm^3/kg$_s$]
DISTURBED SABKHA SOIL SAMPLES: AGGREGATES SIZE [200 μm–2 mm]									
AM1	77.7	0.68	0.22	0.17	0.85	1.00	0.14	0.14	0.13
AM2	68.6	0.65	0.22	0.19	0.85	1.00	0.17	0.15	0.13
AM3	71.4	0.39	0.23	0.14	0.85	1.00	0.15	0.16	0.16
Average	*72.6*	*0.57*	*0.22*	*0.17*	*0.85*	*1.00*	*0.16*	*0.15*	*0.14*
SD	*4.65*	*0.16*	*0.01*	*0.02*	*0.00*	*0.00*	*0.02*	*0.01*	*0.02*
CV	*0.06*	*0.22*	*0.03*	*0.12*	*0.00*	*0.00*	*0.10*	*0.07*	*0.12*
UNDISTURBED SABKHA SOIL SAMPLES									
UDS1	106	0.52	0.25	0.09	0.76	0.20	0.10	0.06	0.30
UDS2	90	0.61	0.25	0.07	0.76	0.20	0.08	0.05	0.35
UDS3	102	0.42	0.25	0.08	0.76	0.20	0.08	**0.00**	**0.12**
Average	*99.8*	*0.51*	*0.25*	*0.08*	*0.76*	*0.20*	*0.09*	*0.04*	*0.26*
SD	*7.97*	*0.10*	*0.00*	*0.01*	*0.00*	*0.00*	*0.01*	*0.03*	*0.12*
CV	*0.08*	*0.19*	*0.00*	*0.15*	*0.00*	*0.00*	*0.13*	*0.89*	*0.47*

[a] The hydro-structural parameters of WRC are also parameters for the SSC; SD: Standard Deviation; CV: Coefficient of Variation.

highly affect W_{ma} as shown in Equations (3) and (5). Ignoring these two parameters K_{st} and W_{maSat} of this sample (highlighted bold in **Table 3**), the other two replicates have similar values, such that: $K_{st} = 0.27$ and 0.29, and $W_{maSat} = 0.18$ and 0.19, respectively for the samples [AM68] and [AM69]. Moreover, the values of K_{bs} for all samples were the same, $K_{bs} = 0.47$, 0.47 and 0.46, respectively, for the samples [AM67, 68, 69]; (b) sample [UDR1] in the undisturbed Rodah soil group showed a different shape and a shift of the structural and basic shrinkage phases (**Figure 6B**), these phases were characterized by K_{st} and K_{bs} and highly affected W_{ma}. Ignoring these three parameters K_{st}, K_{bs} and W_{maSat} of this sample (highlighted bold in **Table 3**), the other two replicates provided similar values, such that: $K_{st} = 0.62$ and 0.68, $K_{bs} = 0.37$ and 0.40 and $W_{maSat} = 0.05$ and 0.06, respectively

Table 5 | Statistical summary of the comparison between the measured and modeled water retention curves and soil shrinkage curves of the Rodah and Sabkha soil samples.

Soil sample	Core ID	Water retention curve [WRC]		Soil shrinkage curve [SSC]	
		R^2	MSE	R^2	MSE
Reconstituted Rodah soil samples	AM67	0.9996	4.58E+00	0.9975	2.86E−06
	AM68	0.9999	4.90E+00	0.9979	5.31E−06
	AM69	0.9994	1.22E+01	0.9969	7.61E−06
Undisturbed Rodah soil samples	UDR1	0.9980	1.10E+02	0.9969	2.04E−06
	UDR2	0.9983	9.38E+01	0.9975	1.78E−06
	UDR3	0.9992	6.24E+01	0.9975	8.69E−07
Reconstituted Sabkha soil samples	AM1	0.9994	1.96E+01	0.9908	2.18E−04
	AM2	0.9989	3.71E+01	0.9868	6.64E−05
	AM3	0.9995	1.57E+01	0.9836	2.85E−04
Undisturbed Sabkha soil samples	UDS1	0.9999	5.34E+00	0.9944	1.43E−06
	UDS2	0.9999	5.62E+00	0.9979	1.79E−07
	UDS3	0.9995	1.04E+01	0.9829	2.41E−07

for the samples [UDR2] and [UDR3]; (c) In the case of reconstituted Sabkha soil, sample [AM3] had a different shape of the structural shrinkage phase and thus slightly affected the parameters K_{st} and W_{maSat}; and (d) In the case of undisturbed Sabkha soil, it was obvious from **Figure 7B**, that the shape of the SSC of soil sample [UDS3] was totally different than the others, this was reflected mainly in the two characteristic parameters of the slopes of shrinkage phases K_{st} and K_{bs}. Ignoring this sample, the other two replicates provided similar values, such that: K_{st} = 0.30 and 0.35; K_{bs} = 0.06 and 0.05, respectively, for the samples [UDS1] and [UDS2]. Finally, (e) the parameter $\overline{E}_{ma}$ showed some variations among the three replicates, but it was noticed that the highest variations were for the same samples that had problems as identified before (i.e., UDR1, AM3, and UDS3). Two points could be concluded from these results. There is power in the physical meaning of the used parameters in justifying and explaining any observed variations in the measured characteristic curves (WRC and SSC), and the identified hydrostructural parameters were unique for each soil sample group.

DISCUSSION

This paper presents a comprehensive work for measuring, characterizing, and modeling the soil shrinkage and WRCs based on the thermodynamic and Pedostructure concepts. It includes: (1) a methodology for preparing disturbed and undisturbed soil samples to be ready for the analyses by the new device, TypoSoil™. This device provides simultaneous and continuous measurements for the data pairs required to construct the two curves; (2) presentation of a simple and reliable method for extracting and estimating the hydrostructural parameters for the two curves; and (3) an application of the proposed work of Braudeau et al. (2014), in a hyper-arid region soil including very salty soil, for modeling the WRC and SSC according to the thermodynamic theory of the Pedostructure. To conduct this study, two native Aridisols in the

state of Qatar, named locally "Rodah soil" and "Sabkha soil" were used. Three replicates of reconstituted and undisturbed soil cores were prepared and studied in this paper.

The results of the measured curves by the new device proved that the procedure for preparing the samples is reliable and reproducible. Minor variations were observed among the WRC and SSC of the three replicates of each soil type, except one case (UDS3). Moreover, the paper highlighted the importance of having continuous and simultaneous measurements for this kind of characterization where each parameter not only quantifies a specific physical property of the soil medium but also, in most cases, extracted from specific characteristic points and segments on the measured SSC, such as: $(\overline{V}_0, W_N, k_N, K_{bs}, K_{st})$, or identified at once from the two curves, such as: (W_{miSat}, W_{maSat}).

The paper portrays the simplicity, robustness, uniqueness, and accuracy of the method used for extraction of the 9 hydrostructural parameters. Very accurate values of multiple parameters (6 out of 9 parameters!) were extracted at once and in a simple way from the measured curves. These parameters were: $(\overline{V}_0, W_N, K_{bs}, K_{st}, W_{miSat}, W_{maSat})$. In general, the extracted parameters for the sigmoidal shrinkage curves were more accurate than the case of non-sigmoidal shrinkage curves. This is due to the difficulty of identifying the characteristic points on the shrinkage curve. Moreover, the results evidenced the uniqueness of these parameters for each soil type.

Finally, excellent matching was observed between the measured characteristic curves and modeled ones based the thermodynamic theory of the Pedostructure. There is enough supporting evidence for the validity, adequacy and reliability of the extraction methodology of the hydrostructural parameters and the thermodynamic-based equations of the two characteristic curves. Moreover, the final values of the hydrostructural parameters proved the uniqueness of these parameters for each soil type.

These promising results open the doors for further studies, such as: using the power and uniqueness of these parameters for other studies related to the water flow and solute transports and soil remediation.

ACKNOWLEDGMENTS

Input from the following people from Valorhiz SAS company is highly acknowledged: Hassan Boukcim and Estelle Hedri.

REFERENCES

Abu Sharar, T. M., Bingham, F. T., and Rhoades, J. D. (1987). Stability of soil aggregates as affected by electrolyte concentration and composition. *Soil Sci. Soc. Am. J.* 51, 309–314. doi: 10.2136/sssaj1987.03615995005100020009x

Alaoui, A., Lipiec, J., and Gerke, H. H. (2011). A review of the changes in the soil pore system due to soil deformation: a hydrodynamic perspective. *Soil Till. Res.* 115–116, 1–15. doi: 10.1016/j.still.2011.06.002

Armstrong, A. C., Matthewes, A. M., Portwood, A. M., Leeds-Harrison, P. B., and Jarvis, N. J. (2000). CRACK-NP: a pesticide leaching model for cracking clay soils. *Agric. Water Manag.* 44, 183–199. doi: 10.1016/S0378-3774(99)00091-8

Bellier, G., and Braudeau, E. (2013). *Device for Measurement Coupled with Water Parameters of Soil.* Geneva: WO 2013/004927 A1, World Intellectual Property Organization, European Patent Office.

Berezin, P. N., Voronin, A. D., and Shein, Y. V. (1983). An energetic approach to the quantitative evaluation of soil structure. *Pochvovedeniye* 10, 63–69.

Berndt, R. D., and Coughlan, K. I. (1977). The nature of changes in bulk density with water content in a cracking clay. *Aust. J. Soil Res.* 15, 27–37. doi: 10.1071/SR9770027

Boivin, P., Garnier, P., and Tessier, D. (2004). Relationship between clay content, clay type, and shrinkage properties of soil samples. *Soil Sci. Soc. Am. J.* 68, 1145–1153. doi: 10.2136/sssaj2004.1145

Boivin, P., Garnier, P., and Vauclin, M. (2006). Modeling the shrinkage and water retention curves with the same equations. *Soil Sci. Soc. Am. J.* 70, 1082–1093. doi: 10.2136/sssaj2005.0218

Børgesen, C., Jacobsen, O., Hansen, H., and Schaap, M. (2006). Soil hydraulic properties near saturation, an improved conductivity model. *J. Hydrol.* 324, 40–50. doi: 10.1016/j.jhydrol.2005.09.014

Brasher, B. R., Franzmeier, D. P., Valassis, V., and Davidson, S. E. (1966). Use of saran resin to coat natural soil clods for bulk density and water retention measurements. *Soil Sci.* 101:108. doi: 10.1097/00010694-196602000-00006

Braudeau, E., Assi, A. T., Boukcim, H., and Mohtar, R. H. (2014). Physics of the soil medium organization part 1: thermodynamic formulation of the pedostructure water retention and shrinkage curves. *Front. Environ. Sci.* 2:4. doi: 10.3389/fenvs.2014.00004

Braudeau, E., Costantini, J. M., Bellier, G., and Colleuille, H. (1999). New device and method for soil shrinkage curve measurement and characterization. *Soil Sci. Soc. Am. J.* 63, 525–535. doi: 10.2136/sssaj1999.03615995006300030015x

Braudeau, E., Frangi, J. P., and Mothar, R. H. (2004). Characterizing non-rigid dual porosity structured soil medium using its shrinkage curve. *Soil Sci. Soc. Am. J.* 68, 359–370. doi: 10.2136/sssaj2004.0359

Braudeau, E., and Mohtar, R. H. (2004). Water potential in non-rigid unsaturated soil-water medium. *Water Resour. Res.* 40, 1–14. doi: 10.1029/2004WR003119

Braudeau, E., and Mohtar, R. H. (2009). Modeling the soil system: bridging the gap between pedology and soil-water physics. *Glob. Planet. Change J.* 67, 51–61. doi: 10.1016/j.gloplacha.2008.12.002

Braudeau, E., Sene, M., and Mohtar, R. H. (2005). Hydrostructural characteristics of two African tropical soils. *Eur. J. Soil Sci.* 56, 375–388. doi: 10.1111/j.1365-2389.2004.00679.x

Bronswijk, J. J. B. (1991). Relationship between vertical soil movements and water content changes in cracking clays. *Soil Sci. Soc. Am. J.* 55, 1120–1226. doi: 10.2136/sssaj1991.03615995005500050004x

Brooks, R. H., and Corey, A. T. (1964). *Hydraulic Properties of Porous Media.* Hydrology Paper No. 3. Fort collins, CO: Colorado State University.

Burdine, N. T. (1953). Relative permeability calculations from pore-size distribution data. *J. Petrol. Technol.* 5, 71–78. doi: 10.2118/225-G

Chertkov, V. Y. (2003). Modelling the shrinkage curve of soil clay pastes. *Geoderma* 112, 71–95. doi: 10.1016/S0016-7061(02)00297-5

Chertkov, V. Y. (2004). A physically based model for the water retention curve of clay pastes. *J. Hydrol.* 286, 203–226. doi: 10.1016/j.jhydrol.2003.09.019

Coppola, A., Gerke, H., Comegna, A., Basile, A., and Comegna, V. (2012). Dual-permeability model for flow in shrinking soil with dominant horizontal deformation. *Water Resour. Res.* 48:W08527. doi: 10.1029/2011WR011376

Cornelis, W. M., Corluy, J., Medina, H., Díaz, J., Hartmann, R., Van Meirvenne, M., et al. (2006). Measuring and modelling the soil shrinkage characteristic curve. *Geoderma* 137, 179–191. doi: 10.1016/j.geoderma.2006.08.022

Coughlan, K. J., McGarry, D., Loch, R. J., Bridge, B., and Smith, D. (1991). The measurement of soil structure. *Aust. J. Soil Res.* 29, 869–889. doi: 10.1071/SR9910869

Crescimanno, G., and Provenzano, G. (1999). Soil shrinkage characteristic curve in clay soils: measurement and prediction. *Soil Sci. So. Am. J.* 63, 25–32. doi: 10.2136/sssaj1999.03615995006300010005x

Dexter, A. R. (2004). Soil physical quality: part II. Friability, tillage, tilth and hardsetting. *Geoderma* 120, 215–225. doi: 10.1016/j.geoderma.2003.09.005

Dickson, E. L., Rasiah, V., and Groenvelt, P. H. (1991). Comparison of four prewetting techniques in wet aggregate stability determination. *Can. J. Soil Sci.* 7, 67–72. doi: 10.4141/cjss91-006

El-kadi, A. I. (1985). On estimating the hydraulic properties of soil, part 1. Comparison between forms to estimate the soil- water characteristic function. *Adv. Water Resour.* 8, 136–147. doi: 10.1016/0309-1708(85)90054-5

Fredlund, D. G., Sheng, D., and Zhao, J. (2011). Estimation of soil suction from the soil-water characteristic curve. *Can. Geotech. J.* 48, 186–198. doi: 10.1139/T10-060

Giráldez, J. V., Sposito, G., and Delgado, C. (1983). A general soil volume change equation: I. The two-parameter model. *Soil Sci. Soc. Am. J.* 47, 419–422. doi: 10.2136/sssaj1983.03615995004700030005x

Groenevelt, P. H., and Grant, C. D. (2002). Curvature of shrinkage lines in relation to the consistency and structure of a Norwegian clay soil. *Geoderma* 106, 235–245. doi:10.1016/S0016-7061(01)00126-4

Groenevelt, P. H., and Grant, C. D. (2004). A new model for the soil-water retention curves that solves the problem of residual water content. *Eur. J. Soil Sci.* 55, 479–485. doi: 10.1111/j.1365-2389.2004.00617.x

Haines, W. B. (1923). The volume changes associated with variations of water content in soil. *J. Agric. Sci. Camb.* 13, 296–311. doi: 10.1017/S0021859600003580

Huang, C., Shao, M., and Tan, W. (2011). Soil shrinkage and hydrostructural characteristics of three swelling soils in Shaanxi, China. *J. Soils Sediments* 11, 474–481. doi: 10.1007/s11368-011-0333-8

Kim, D. J., Vereecken, H., Feyen, J., Boels, D., and Bronswijk, J. J. B. (1992). On the characterization of properties of an unripe marine clay soil: I. Shrinkage processes of an unripe marine clay soil in relation to physical ripening. *Soil Sci.* 153, 471–481. doi: 10.1097/00010694-199206000-00006

Larsbo, M., and Jarvis, N. J. (2005). Simulating solute transport in a structured field soil: uncertainty in parameter identification and predictions. *J. Environ. Qual.* 34, 621–634. doi: 10.2134/jeq2005.0621

Lauritzen, C. W., and Stewart, A. J. (1942). Soil-volume changes and accompanying moisture and pore-space relationships. *Soil Sci. Soc. Am. Proc.* 6, 113–116. doi: 10.2136/sssaj1942.036159950006000C0019x

Leij, F., Russell, W., and Lesch, S. M. (1997). Closed-form expressions for water retention and conductivity data. *Ground Water* 35, 848–858. doi: 10.1111/j.1745-6584.1997.tb00153.x

McGarry, D., and Daniells, I. G. (1987). Shrinkage curve indices to quantify cultivation effects on soil structure of a Vertisol. *Soil Sci. Soc. Am. J.* 51, 1575–1580. doi: 10.2136/sssaj1987.03615995005100060031x

McGarry, D., and Malafant, K. W. J. (1987). The analysis of volume change in unconfined units of soil. *Soil Sci. Soc. Am. J.* 51, 290–297. doi: 10.2136/sssaj1987.03615995005100020005x

Mualem, Y. (1976). A new model for predicting the hydraulic conductivity of unsaturated porous media. *Water Resour. Res.* 12, 513–522. doi: 10.1029/WR012i003p00513

Olsen, P. A., and Hauge, L. E. (1998). A new model of the shrinkage characteristic applied to some Norwegian soils. *Geoderma* 83, 67–81. doi: 10.1016/ S0016-7061(97)00145-6

Peng, X., and Horn, R. (2005). Modeling soil shrinkage curve across a wide range of soil types. *Soil Sci. Soc. Am. J.* 77, 372–372. doi: 10.2136/sssaj2004.0146

Peng, X., and Horn, R. (2013). Identifying six types of soil shrinkage curves from a large set of experimental data. *Soil Sci. Soc. Am. J.* 69, 584–592. doi: 10.2136/sssaj2011.0422

Reeve, M. J., and Hall, D. G. M. (1978). Shrinkage in clayey sub-soils of contrasting structure. *J. Soil Sci.* 29, 315–323. doi: 10.1111/j.1365-2389.1978.
tb00779.x

Salahat, M., Mohtar, R. H., Braudeau, E., Schulze, D., and Assi, A. (2012).
Toward delineating hydro-functional soil mapping units using the pedostructure concept: a case study. *Comput. Electron. Agric.* 86, 15–25. doi:
10.1016/j.compag.2012.04.011

Sander, T., and Gerke, H. H. (2007). Noncontact shrinkage curve determination for
soil clods and aggregates by three-dimensional optical scanning. *Soil Sci. Soc.
Am. J.* 71, 1448–1454. doi: 10.2136/sssaj2006.0372

Santos, G. G., Silva, E. M., Marchaõ, R. L., Silveira, P. M., Braund, A., James, F.,
et al. (2011). Analysis of physical quality of soil using the water retention curve:
validity of the S-index. *C. R. Geosci.* 343, 295–301. doi: 10.1016/j.crte.2011.
02.001

Scheibert, C., Stietiya, M., Sommer, J., Schramm, H., and Memah, M. (2005). *The
Atlas of Soils for the State of Qatar, Soil Classification and Land Use Specification
Project for the State of Qatar, Ministry of Municipal Affairs and Agriculture,
General Directorate of Agricultural Research and Development, Department of
Agricultural and Water Research.* Doha: Ministry of Municipal Affairs and
Agriculture.

Soil Survey Staff. (1975). *Soil Taxonomy: A Basic System Soil Classification for
Making and Interpreting Soil Surveys. USDA-SCS Agriculture Handbook 436.*
Washington, DC: U.S. Government Printing Office.

Sposito, G. (1981). *The Thermodynamic of Soil Solution.* New York, NY: Oxford
University Press.

Stewart, R. D., Abou Najm, M. R., Rupp, D. E., and Selker, J. S. (2012).
An image-based method for determining bulk density and the soil shrinkage curve. *Soil Sci. Soc. Am. J.* 76, 1217–1221. doi: 10.2136/sssaj2011.
0276n

Tang, C., Cui, Y., Tang, A., and Shi, B. (2010). Experiment evidence on the temperature dependence of desiccation cracking behavior of clayey soils. *Eng. Geol.*
114, 261–266. doi: 10.1016/j.enggeo.2010.05.003

Tariq, A. R., and Durnford, D. S. (1993a). Soil volumetric shrinkage measurements:
a simple method. *Soil Sci.* 155, 325–330. doi: 10.1097/00010694-199305000-
00003

Tariq, A. R., and Durnford, D. S. (1993b). Analytical volume change model
for swelling clay soils. *Soil Sci. Soc. Am. J.* 57, 1183–1187. doi: 10.2136/
sssaj1993.03615995005700050003x

van Genuchten, M. Th. (1980). A closed-form equation for predicting the hydraulic
conductivity of unsaturated soils. *Soil Sci. Soc. Am. J.* 44, 892–898. doi:
10.2136/sssaj1980.03615995004400050002x

Voronin, A. D. (1980). The structure-energy conception of the hydro-physical
properties of soils and its practical applications. *Pochvovedeniye* 12, 35–46.

Zhou, A., Sheng, D., and Li, J. (2014). Modelling water retention and volume
change behaviors of unsaturated soils in non-isothermal conditions. *Comput.
Geotech.* 55, 1–13. doi: 10.1016/j.compgeo.2013.07.011

Conflict of Interest Statement: The authors declare that the research was conducted in the absence of any commercial or financial relationships that could be
construed as a potential conflict of interest.

6

A framework for soil-water modeling using the pedostructure and Structural Representative Elementary Volume (SREV) concepts

Erik F. Braudeau[1]* and Rabi H. Mohtar[2]

[1] Hydrostructural Pedology, Centre de Recherche Ile de France, IRD France Nord, Institut de Recherche pour le Développement, Bondy, France
[2] Department of Biological and Agricultural Engineering, Zachry Department of Civil Engineering, Texas A&M University, College Station, TX, USA

Edited by:
Juergen Pilz, Alpen-Adria Universität Klagenfurt, Austria

Reviewed by:
Yuichi S. Hayakawa, The University of Tokyo, Japan
Shawn P. Serbin, Brookhaven National Laboratory, USA

*Correspondence:
Erik F. Braudeau, Institut de Recherche pour le Développement, Pédologie Hydrostructurale, 32 Avenue Henri Varagnat, 93140 Bondy, France
e-mail: erik.braudeau@ird.fr

Current soil water models do not take into account the internal organization of the soil medium and consequently ignore the physical interaction between the water film at the surface of solids that form the soil structure and the structure itself. In this sense, current models deal empirically with the physical soil properties, which are all generated from this soil water and soil structure interaction. As a result, the thermodynamic state of the soil water medium, which constitutes the local physical conditions of development for all biological and geochemical processes within the soil medium, is still not well defined and characterized. This situation limits modeling and coupling the different processes in the soil medium since they all thermodynamically linked to the soil water cycle. The objective of this article is to present a complete framework for characterizing and modeling the internal soil organization and its hydrostructural properties resulting from interaction of its structure with the soil water dynamics. The paper builds on the pedostructure concept, which allowed the integration of the soil structure into equations of water equilibrium and movement in soils. The paper completes the earlier framework by introducing notions of soil-water thermodynamics that were developed in application to the concept of the Structural Representative Elementary Volume (SREV). Simulation of drainage after infiltration in the Yolo loam soil profile, as compared to measured moisture profile using the measured soil characteristic parameters, showed a high degree of agreement. This new modeling framework opens up new prospects in coupling agro-environmental models with the soil medium, recognizing that the soil organization, hydro-structural, and thermodynamic properties are the foundation for such coupling.

Keywords: soil water modeling, multi-scale soil water processes, pedostructure, hydrostructural soil properties, soil hydrodynamics, soil water thermodynamics

INTRODUCTION

Representing the structured medium of top soil, with special attention to its hierarchy of scales, is imperative for understanding and modeling the dynamics of water flow and storage in the vadose zone. This is of importance in hydrology, agronomy, and geochemistry, and in studies involving the environmental fate of solutes at field, farm, and watershed scales. Single pore representation of the soil water medium is very common in agronomic models such as GRASIM (Mohtar et al., 1997) or CropSyst (Stöckle et al., 2003). The tendency in modeling water flow and solute transport in structured soil is to distinguish two domains, micro- and macro-pore, in an implicit soil horizon REV (Representative Elementary Volume) of the soil medium (Othmer et al., 1991; Chen et al., 1993; Gerke and van Genuchten, 1993; Katterer et al., 2001; Logsdon, 2002; Simunek et al., 2003, 2008).

However, none of these models consider the soil medium as a structured medium with aggregates, and thus, they are unable to take into account the notion of primary peds, which represent the first level of organization of the primary particles into aggregates (Brewer, 1964), and the notion of pedostructure (Braudeau et al., 2004) which is constituted by their assembly. Consequently, the swelling-shrinkage properties of the primary peds within the soil structure that were quantified by Braudeau and Bruand (1993), and also the resulting hydro-structural properties of the soil medium (Braudeau et al., 2004, 2005; Braudeau and Mohtar, 2006), cannot be considered and modeled in current agro-environmental modeling.

Thus, as long as the hydro-structural behavior of soil organization and structure remain undefined, the modeling of the numerous biophysical and chemical processes that take place within the soil medium cannot be fully physically modeled. It is why Braudeau and Mohtar (2009) introduced a new paradigm for modeling soil water on the basis of two new concepts in soil science: the pedostructure concept and the Structure Representative Elementary Volume (SREV) concept. The latter allows transformation of the pedostructure organization into a thermodynamic system consisting of two nested and complementary pore sub-systems (micro and macro) that refer to the inside and outside of the primary ped media. These micro- and macro-pore systems are in a quasi-equilibrium state of water retention when slow changes in water content (such as by evaporation or drainage) occur, whatever the water content.

A computer model simulating the multi scale hydro-structural functioning of the pedon, named Kamel®, was introduced by Braudeau et al. (2009) based on the pedostructure concept only, i.e., without integrating the recent advances in soil water thermodynamics following application of the SREV concept to the pedostructure organization. The objective of the paper is to present the complete theory behind Kamel® updated with the recent thermodynamic development of its state equations of equilibrium. Specifically, we will introduce (i) the basic principles of the soil-water modeling framework represented by Kamel®, including the pedostructure concept, the SREV concept, and the new thermodynamics equations resulting from it; and (ii) the implementation of equations into the computer model using the Simile® software environment. Application of this framework is presented followed by a discussion on the novelty of this modeling framework and its potential applications to current environmental issues.

THEORY

PEDON REPRESENTATION

Primary peds and pedostructure

Brewer (1964) introduced the concepts of peds, primary ped, and the S-matrix. He defined a ped as "an individual natural soil aggregate consisting of a cluster of primary particles and separated from adjoining peds by surfaces of weakness which are recognizable as natural voids or by occurrence of cutans." Primary peds are the simplest peds occurring in a soil material. They are not divided into smaller peds, but they may be packed together to form compound peds of higher level of organization. The S-matrix of a soil material "is the material within the simplest (primary) peds, or composing apedal soil materials, in which the pedological features occur; it consists of plasma, skeleton grains, and voids that do not occur in pedological features other than plasma separations."

Braudeau et al. (2004) completed this morphological definition of primary peds with a hydro-functional definition based on the determination of an air entry point in the clayey plasma using the shrinkage curve that was continuously measured by the apparatus detailed in Braudeau et al. (1999). They introduced the term "pedostructure" which is the assembly of primary peds that is represented by two nested and complementary pore systems named as *micropore* and *macropore* corresponding respectively to the intra and inter primary ped pore space. Furthermore, Braudeau and Mohtar (2009) proposed the notion of "Structural Representative Elementary Volume" (SREV) to introduce the pedostructure concept in the systemic approach of the natural organization of the "soil-plant-atmosphere" continuum. The new paradigm generated enables characterization and modeling of the hydrostructural properties of soil horizons as hydro-functional elements of the pedon, itself representative of the soil mapping unit as schematized on **Figure 1**.

According to this new paradigm a new system of descriptive variables and characteristic parameters were defined for the two distinct media of the pedostructure: micro and macro pore volumes corresponding to the inside and outside of the primary peds. The pedostructure variables are listed in **Table 1**: $\overline{V}_{mi}$, $\overline{Vp}_{mi}$, W_{mi}, h_{mi}, k_{mi}, $\overline{V}_{ma}$, $\overline{Vp}_{ma}$, W_{ma}, h_{ma}, k_{ma} (nomenclature and definitions are given in the text below).

Structural representative elementary volume (SREV) of the pedostructure

Unlike REV, SREV is virtually delimited by an enclosure permeable to air, water, and solute fluxes, but not to the solid particles that compose the structure and of which the mass (structural mass, M_s) contained in the SREV stays constant. This description defines any SREV as a volume V, virtually sampled in the homogeneous medium that it represents, and such that its structural mass, M_s, and the relative organization of solids particles between them remain unchanged with a change in water content. This allows taking M_s as reference to which all specific extensive variables of the SREV are reported; for example in **Table 1**, the total water content ($W = M_w/M_s$) and the pedostructural specific volume ($\overline{V}_{ps} = V_{ps}/M_s$). We notice that this latter variable cannot exist in the REV-based system of variables since the volume itself of the REV is taken as reference for its extensive variables. Moreover, not only the organizational variables are reported to the structural mass M_s, but also the thermodynamic functions that are specific of the considered SREV. For example, $\overline{U}$ is the total internal energy, and $\overline{G}_{wl}$, the total Gibbs free energy of water in its liquid phase, both contained in the SREV of pedostructure considered and reported to the structural mass M_s of this SREV. These properties are called, respectively: (i) pedostructural (instead of specific) internal energy, and (ii) pedostructural Gibbs free energy of the liquid water phase, counted positively (Braudeau et al., 2014):

$$d\overline{U} = Td\overline{S} - Pd\overline{V} + \mu_{mi}dW_{mi} + \mu_{ma}dW_{ma} + \mu_{air}d\overline{A} \quad (1a)$$

and

$$\overline{G}_{wl} = \overline{G}_{wmi} + \overline{G}_{wma} = -W_{mi}\mu_{mi} - W_{ma}\mu_{ma} \quad (1b)$$

where $[\overline{S} = S/M_s;\ \overline{V} = V/M_s;\ \overline{G}_{wmi} = G_{wmi}/M_s;\ \overline{G}_{wma} = G_{wma}/M_s;\ W_{mi} = M_{wmi}/M_s;\ W_{ma} = M_{wma}/M_s]$ are the extensive variables of the pedostructure divided by its structural mass M_s; related to the water components (micro and macro) of the liquid phase, respectively: the pedostructural entropy, volume, Gibbs free energy of the micro and macro liquid water phases and pedostructural micro and macro water contents. The intensive variables T, P, μ_{wmi} and μ_{wma}, and μ_{air}, are respectively: the absolute temperature, the pressure exerted on the considered SREV, the water chemical potential (<0) inside and outside of primary peds, and the chemical potential of the air.

A decrease in water content from water saturation induces a water suction pressure, or water retention, h_{mi} and h_{ma}, in the both complementary pore systems of the pedostructure SREV. They are defined such as:

$$h_{mi} = -\rho_w^0 (\mu_{wmi} - \mu_{wmiSat}) \text{ and}$$
$$h_{ma} = -\rho_w^0 (\mu_{wma} - \mu_{wmaSat}) \quad (1c)$$

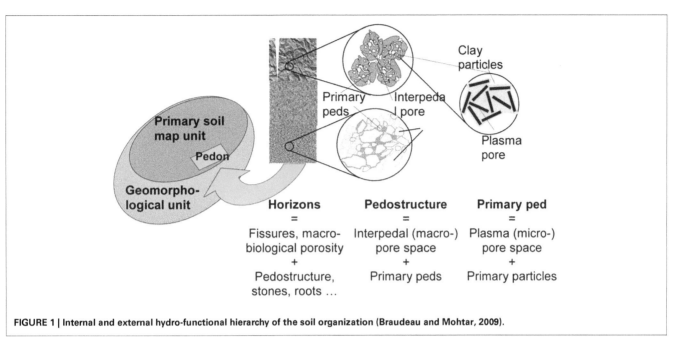

FIGURE 1 | Internal and external hydro-functional hierarchy of the soil organization (Braudeau and Mohtar, 2009).

Table 1 | Pedostructure state variables.

Functional levels linked to the pedostructure	Pedostructural volumes (dm^3/kg)	Pedostructural pore volumes (dm^3/kg)	Pedostructural water contents (kg/kg)	Soil water retention (kPa)	Hydraulic Conductivity (dm/s)	Non-saturating water contents (kg/kg)	Saturating water contents (kg/kg)
Pedostructure SREV	$\overline{V}_{ps}$		W_{ps}	h	k_{ps}		
Interpedal pore space		$\overline{Vp}_{ma}$	W_{ma}	h_{ma}	k_{ma}	w_{st}	w_{ip}
Primary peds		$\overline{Vp}_{mi}$	W_{mi}	h_{mi}	k_{mi}	w_{re}	w_{bs}
Primary particles	$\overline{V}_s$						

Subscripts mi and ma, hor, fiss, and s; refer to micro and macro, horizon, fissures, and solids; ip, st, bs, and re, refer to the name of the corresponding shrinkage phase of the shrinkage curve: interpedal, structural, basic, and residual.

where μ_{wmiSat} and μ_{wmaSat} are the water chemical potential at saturation.

Equations (1) determine the hydrostructural and thermodynamic equilibrium states of any pedostructure SREV defined in a soil medium and under the controlled variables: T, P, W, and μ_{air}. Braudeau et al. (2014) and Assi et al. (2014) could then show the two following points: (i) $\overline{G}_{wl}$, $\overline{G}_{wmi}$, and $\overline{G}_{wma}$ stay constant with any change in water content, and (ii) at each equilibrium state depending of W, there is necessarily equality between the water retention inside and outside of primary peds, such that: $h_{mi} = h_{ma} = h$ at any value of W at equilibrium. They could deduce then the physical equations of the pedostructural shrinkage curve and the water retention curve that will be presented in the next section.

This new approach leads to the following qualifications and properties of the soil medium in each horizon of the pedon, considering that, at the global scale of the soil horizon, the pedostructure is only a component of the soil medium organization but which imposes to the horizon its hydrostructural properties:

(1) The soil medium is physically described by its pedostructural state variables $\overline{S}$, $\overline{V}$, $\overline{A}$, W_{mi}, and W_{ma} as functions of the pedostructural water content W at equilibrium under the intensive variables T, P, h_{mi}, h_{ma}, and μ_{air}. These intensive variables are assumed to have the same value within each of the successive layers that have been discretized as horizontal SREVs in the soil horizon (see next section) and named SRELs (**Table 1**).

(2) The *homogeneity* of the soil medium in a soil horizon can be defined as the stability all over the horizon, of parameters of the pedostructural state equations: $\overline{S}(W)$, $\overline{V}(W)$, $\overline{A}(W)$, and $h(W)$, as parametric functions of the variable W at T, P, and μ_a constant. Practically, knowing that $\overline{A}/\rho^0_{air} = \overline{V} - W/\rho^0_w - \overline{V}_s$ for the pedostrucure ($\overline{V}_s$ being the pedostructural volume of the solid phase), a soil horizon will be considered structurally homogeneous in regard to its physical properties with water if the three functions: the pedostructural entropy curve, $\overline{S}(W)$, shrinkage curve, $\overline{V}(W)$ and water retention curve, $h(W)$, keep their hydrostructural parameters constant everywhere in the medium of the horizon, for given P and T. From the thermodynamic point of view, and in contrary to the current vision of the soil characterization, there is not only one soil moisture characteristic curve [the water retention curve $h(W)$], but at least three soil

moisture characteristic curves that are the equations of state of the free energy equation at equilibrium depending on W: $\overline{S}(W)$, $\overline{V}(W)$ and $h(W) = h_{mi}(W_{mi}) = h_{ma}(W_{ma})$.

(3) Every pedostructural extensive (or organizational) variable of a SREV, is hierarchically nested inside the SREV with respect to the hierarchical organization of the medium represented by the SREV. This allows us to define the descriptive variables of the soil medium at its various functional levels, namely; the pedon, horizon, pedostructure, primary peds, and primary particles (**Figure 1**, Braudeau and Mohtar, 2009).

Discretization of the pedon according to the SREV concept

Simulation of the gravimetric water transfer and related processes within and through the pedon requires discretization of the medium into representative elementary layers. Assuming that a soil horizon has homogenous hydro-structural properties, and that it is a SREV of the corresponding horizon in the soil mapping unit, then each soil horizon can be discretized into thin horizontal layers that have the physical properties (pedostructure parameters) of the horizon considered, and that are at different equilibrium states depending on the water content all over the width of the pedon (**Figure 2**). These layers are called Structural Representative Elementary Layers (SRELs). In general, fine layers having a minimum thickness of 0.2 dm can be considered for the simulation of water transfer in soils. Moreover, the pedon must be wide enough so it can be representative of the soil with regards to pedologic features such as fissures, stones, etc., that may be observed at the pedon scale in the field.

Descriptive variables of an SREL include those of the pedostructure plus other features such as roots, biogenic macropores, stones, etc., whose volumes are related to the total pedostructure mass of the layer (M_{psL}). **Table 2** presents the different hydro-functional subsystems of the pedon, like the SRELs, of which variables and parameters should be defined with respect to the SREV concept to be compatible with the other levels of organization of the pedon. In particular, the mass of pedostructure included in the SREL is conveniently taken as reference rather than the total mass of solids belonging to the layer. This allows us to keep the variables and properties of the pedostructure as part of the new SREL set of variables. For example, suppose that an SREL

in one soil horizon consists of a certain volume of pedostructure and other organizational volumes like volumes of stones, roots, biogenic macro-pores, etc., all these volumes, including also the pedostructure volume, should be estimated as fractions of the total volume of the SREL, while the pedostructure mass contained in the SREL, M_{psL}, should be used as mass of reference for extensive variables of the SREL. Therefore, all SRELs of the horizon should have the same characteristic parameters of the entire horizon.

If that $a = \left(V_{ps} + V_{fiss}\right)/V_{horizon}$ is the volumetric proportion of pedostructure and fissures due to shrinkage in the considered horizon SREV, and $\overline{V}_{ps} = V_{ps}/M_{ps}$ (dm^3/kg), is the pedostructure specific volume measured in the laboratory on a soil sample representing the pedostructure in this horizon's SREV, then, the pedostructure mass in the discretized layer volume V_{lay}, SREL of the horizon, will be:

$$M_{psL} = aV_{layer}/\overline{V}_{ps} \tag{2}$$

where M_{psL} is all the pedostructure mass in the layer, V_{lay} is the volume of the SREL, and $\overline{V}_{ps}$ is the specific volume of the pedostructure. Accordingly, if we set b, the volumetric fraction of stones in the horizon, c, the volumetric fraction of biological voids, d, the proportion of roots, etc., therefore the specific volume of a SREL of the horizon will be:

$$\overline{V}_{layer} = \overline{V}_{ps} + Vp_{fiss}/M_{psL} + V_{stone}/M_{psL} + Vp_{bio}/M_{psL}$$
$$+ V_{roots}/M_{psL} \tag{3a}$$

$$\overline{V}_{layer} = \overline{V}_{ps} + \overline{Vp}_{fiss} + \overline{V}_{stone} + \overline{Vp}_{bio} + \overline{V}_{roots} \tag{3b}$$

Setting:

$$\left(\overline{V}_{ps} + \overline{Vp}_{fiss}\right) = a\overline{V}_{layer}; \overline{V}_{stone} = b\overline{V}_{lay}; \overline{Vp}_{bio} = c\overline{V}_{layer}; \text{ and}$$

$$\overline{V}_{roots} = d\overline{V}_{layer} \tag{3c}$$

defines coefficients a, b, c, d as parameters of a horizon at the pedon scale that have to be determined by observations of soil profiles *in-situ*; they are such that their sum is equal to 1. As well as we have distinguished the poral volumes $\overline{Vp}_{fiss}$ and $\overline{Vp}_{bio}$ of the layer, in addition to the poral volumes $\overline{Vp}_{mi}$ and $\overline{Vp}_{ma}$ of the pedostructural volume $\overline{V}_{ps}$, we distinguish also the water contents in these volumes, such as, respectively, W_{fiss}, W_{bio}, and W_{ps}. Among the specific volumes in equations (3) only $\left(\overline{V}_{ps} + \overline{Vp}_{fiss}\right)$ is directly dependent on the pedostructural water content W_{ps} according to the pedostructural shrinkage curve $\overline{V}_{ps}(W_{ps})$.

HYDRO-THERMODYNAMIC EQUILIBRIUM OF THE PEDOSTRUCTURE
The pedostructural water retention curve

Braudeau et al. (2014) derived the expressions of the pedostructure water retention inside and outside of primary peds (h_{mi} and h_{ma}) from Equations (1) stated above. The pedostructure water retention h at equilibrium, such that $h = h_{mi} = h_{ma}$, is determined by the constant (stable) repartition of the surface charges

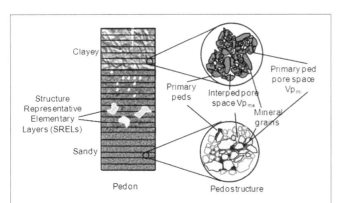

FIGURE 2 | Soil medium organization and discretization of the pedon into SRELs.

Table 2 | List of the hydro-functional subsystems of a pedon SREV, their internal components, and the corresponding parameters.

SREVs of concern	Internal components	Morphological parameter	Functional parameters
Pedon	-Surface layer -Horizons	S_{xy}, L_z	Bottom conditions Surface conditions
Surface layer	-Pedostructure, clodes -Macro inter aggregate space	H_{surf}	$K_{satSurf}$, $V_{surfSat}$, a_{surf}, b_{surf}, c_{surf}, d_{surf}
Horizons	-Succession of SRELS	H_{hor}, Depth	SREL parameters
SRELs	Pedostructure V_{ps}, macropore volumes (V_{pbio}, V_{pfiss}, ...) and solid elements (V_{stones}, V_{roots})	H_{iSat}, S_{xy}	-Field saturated hydraulic conductivity K_{sat} -Volumetric % of components a, b, c, d ... in the horizon -Pedostructure parameters

Parameters are explained in the text.

of the solids particles between inside and outside of primary peds ($\overline{E}_{mi}$ and $\overline{E}_{ma}$) when the water content is changing:

$$h_{mi} = \rho_w^0 \overline{E}_{mi} \left(1/W_{mi} - 1/W_{miSat}\right) \quad (4)$$

$$h_{ma} = \rho_w^0 \overline{E}_{ma} \left(1/W_{ma} - 1/W_{maSat}\right) \quad (5)$$

where ρ_w^0 is the specific density of water; $\overline{E}_{mi}$ and E_{ma} are the total potential energies (positive) relative to a fixed number of the surface charges of the solid particles (clays) inside of primary peds (for $\overline{E}_{mi}$) and outside at their surface in the interpedal macropore space (for $\overline{E}_{ma}$). $\overline{E}_{mi}$ and $\overline{E}_{ma}$ are in joules/kg of solids equal to the corresponding Gibbs free energies $\overline{G}_{mi}$ and $\overline{G}_{ma}$ developed in the water phase surrounding the clay particles of the pedostructure. Both parameters W_{miSat} and W_{maSat} are, respectively, the micro and the macro water content of the pedostructure at saturation, when $W_{ps} = W_{psSat}$. Thus, neglecting for Kamel the contribution h_{ip} of the saturation interpedal water content W_{ip} which appear in certain soils weakly structured (Braudeau et al., 2014), parameters of the soil (pedostructure) water retention curve are $\overline{E}_{mi}$, $\overline{E}_{ma}$, W_{miSat}, and W_{maSat}.

Pedostructural micro and macro water contents at equilibrium

The equality of h_{mi} and h_{ma} (Equations 4, 5) at each value of W_{ps} is the condition for a reversible and slow change in water content that can be considered as a suit of hydro structural equilibrium states function of W_{ps}. Braudeau et al. (2014) derived the unique solution of equality of Equations (4, 5), the couple $\left[W_{ma}^{eq}(W); W_{mi}^{eq}(W)\right]$, the equations of which are given below:

Considering that:

$$\overline{E} = \overline{E}_{ma} + \overline{E}_{mi} \quad (6)$$

and

$$A = \overline{E}_{ma}/W_{maSat} - \overline{E}_{mi}/W_{miSat} = constant \quad (7)$$

W_{ma}^{eq} is the positive solution of the quadratic equation:

$$A W_{ma}^2 - \left(\overline{E} + A W_{ps}\right) W_{ma} + \overline{E}_{ma} W_{ps} = 0 \quad (8)$$

leading to:

$$W_{ma}^{eq} = \frac{1}{2}\left(W_{ps} + \overline{E}/A\right) + \frac{1}{2}\sqrt[2]{\left(W_{ps} + \overline{E}/A\right)^2 - 4W_{ps}\overline{E}_{ma}/A} \quad (9)$$

then

$$W_{mi}^{eq} = W_{ps} - W_{ma}^{eq} = \frac{1}{2}\left(W_{ps} - \overline{E}/A\right)$$
$$- \frac{1}{2}\sqrt[2]{\left(W_{ps} + \overline{E}/A\right)^2 - 4W_{ps}\overline{E}_{ma}/A} \quad (10)$$

Equations (9) and (10) give values of W_{mi}^{eq} and W_{ma}^{eq} at equilibrium for each value of W. That means that, when they are reported in Equations (4, 5) of h_{mi} and h_{ma}, these last have *exactly* the same value h. **Figure 3** shows, as example, a measured soil water retention curve by tensiometer with the fitted curve $h(W_{ps})$ and the corresponding curves of micro and macro pedostructural water contents at equilibrium.

Physical equations and parameters of the pedostructure hydrostructural equilibrium

According to the new developments in thermodynamics of the soil medium presented here above, the pedostructure shrinkage curve already modeled in Kamel using equations developed by Braudeau et al. (2004) were updated with the thermodynamic equations of W_{mi}^{eq} and W_{ma}^{eq}. This update of the shrinkage curve is described in Supplementary Material, Appendix 2. A recapitulation of the whole equations and parameters describing the hydrostructural equilibrium states of the pedostructure in terms of its water content W_{ps} is presented in **Table 3**.

SOIL WATER DYNAMICS AT THE PEDOSTRUCTURE SCALE
The Richards equation for the pedostructure medium

There are two types of water movement that are considered in the pedostructure and correlatively in the SRELs of the soil profile: (1) a local micro-macro transport corresponding to the water exchange between both pore systems, inside and outside of primary peds of the pedostructure, and (2) a transport through the pedostructure and between SRELs that involves only the interpedal water, W_{ma}. The transfer equations are presented by Braudeau and Mohtar (2009); Braudeau et al. (2009):

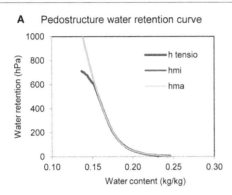

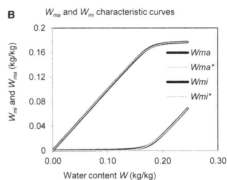

FIGURE 3 | (A) Example of measured soil water retention curve using a tensiometer ($h(W)$) fitted by the theoretical water retention curve at equilibrium ($h = h_{ma} = h_{mi}$); and **(B)**, the corresponding curves $W_{ma}^{eq}(W)$ and $W_{mi}^{eq}(W)$ at equilibrium of suction pressures, calculated from Equations (9, 10), and W_{ma}^* and W_{mi}^*, from Equations (A2b) and the sum (A2c, A2d).

Table 3 | State variables, equations, and parameters of the pedostructure.

State variables	Equation	Equation Number	Parameters
	Pedostructural Water Contents		
W_{ps}	$W_{ps} = W_{ma}^{eq} + W_{mi}^{eq} + W_{ip} = W_{ps}' + W_{ip}$		
W_{ps}'	$W_{ps}' = W_{ma}^{eq} + W_{mi}^{eq}$		
W_{sat}	$W_{sat} = W_{maSat}^{eq} + W_{miSat}^{eq} + W_{ipSat}$		$W_{maSat}^{eq}; W_{miSat}^{eq}; W_{ipSat}$
	Pedostructure Water Contents at Equilibrium		
W_{mi}^{eq}	$W_{mi}^{eq} = (1/2) (W_{ps}' - E/A)$	Equation (10)	$\overline{E}/A; \overline{E}_{ma}/A$
	$-(1/2)\sqrt{(W_{ps}' + \overline{E}/A)^2 - 4(\overline{E}_{ma}/A)W_{ps}'}$		
W_{ma}^{eq}	$W_{ma}^{eq} = (1/2) (W_{ps}' + E/A)$	Equation (9)	
	$+(1/2)\sqrt{(W_{ps}' + \overline{E}/A)^2 - 4(\overline{E}_{ma}/A)W_{ps}'}$		
W_{ip}	$W_{ip} = (1/k_L)\ln\left[1 + \exp\left(k_L(W_{ps} - W_L)\right)\right]$	Equation (A.2)	$W_L; k_L$
	Pedostructure Water Retention		
$h^{eq}(W_{ps})$	$h^{eq}(W) = h_{ma/ma}^{eq}(W_{mi/ma}^{eq}) + h_{ip}(W_{ip})$		$W_{maSat}^{eq}; W_{miSat}^{eq}; \overline{E}_{ma};$
	$h^{eq}(W) = \rho_w \overline{E}_{mi/ma}(1/W_{mi/ma}^{eq} - 1/W_{mi/maSat}^{eq})$	Equations (4) and (5)	$\overline{E}_{mi}; W_{ip}^0; W_{ipSat}$
	$+\rho_w \overline{E}_{ma}\left(1/(W_{ip}^0 + W_{ip}) - 1/(W_{ip}^0 + W_{ipSat})\right)$		
	Pedostructure Shrinkage		
$\overline{V}_{ps}$	$\overline{V}_{ps} = \overline{V}_{ps}^O + K_{bs}w_{bs}^{eq} + K_{st}w_{st}^{eq} + K_{ip}W_{ip}$	Equation (A6)	$\overline{V}_{ps}^O; K_{bs}; K_{st}; K_{ip}$
w_{bs}^{eq}	$w_{bs}^{eq} = W_{mi}^{eq} - w_{re}^{eq}$		
w_{re}^{eq}	$w_{re}^{eq} = W_{mi}^{eq} - (1/k_N)\ln\left[1 + \exp\left(k_N\left(W_{mi}^{eq} - W_{miN}\right)\right)\right]$	Equation (A7)	$k_N; W_N$
	$W_{miN} = W_{mi}^{eq}(W_N)$		
w_{st}^{eq}	$w_{st}^{eq} = W_{ma}^{eq}$	Equation (A4)	

$$dW_{ma}/dt = \rho_w^0 \overline{V}_{ps} \, \partial \left[k_{kma}(-dh_{ma}/dz + 1)\right]/\partial z$$
$$-k_{mi}(h_{mi} - h_{ma}) \quad (11)$$

and

$$dW_{mi}/dt = k_{mi}(h_{mi} - h_{ma}) \quad (12)$$

where z is the depth (upwardly positive); h_{mi} and h_{ma} are the water retention (positive values) inside and outside primary peds

expressed in dm of water height (equivalent to a pressure in kPa), z in dm and k_{ma} in dm/s. The water pressures h_{mi} and h_{ma} are determined by Equations (4, 5) in terms of W_{mi} and W_{ma} out of equilibrium; equation and parameters for $k_{ma}(W_{ma})$ and k_{mi} are given here after.

Pedostructural hydraulic conductivity equation

The hydraulic conductivity k_{ma} in Equation (11) is exclusively related to the macro-pore water content W_{ma} of the pedostructure. The water present outside of the pedostructure, such as the one present in fissures and biogenic pores defined in Equation (3) for the SREL at the horizon scale, does not constitute any part of the pedostructural water content W_{ps}. Thus, we can assume that the movement of water within the pedostructural part of the horizon is always slow enough to consider changes in organization as a suit of equilibrium states in the same manner that the shrinkage and the soil retention curves were considered as W_{ps}-dependent characteristics of equilibrium states of the soil medium.

Summing up Equations (11, 12) leads to a new shape of the Richards equation, not only concerning the use of SREV instead of REV variables, but also the controlling variables themselves: h_{ma} and k_{ma} that should be expressed in terms of W_{ma}. Following Davidson et al. (1969) who showed that the unsaturated hydraulic conductivity, from saturation to field capacity, is well described by an exponential equation of W, instead of h (Gardner, 1958) generally used today (Bruckler et al., 2002), we experimentally observed the following equation for k_{ma} that wil be called pedostructural hydraulic conductivity k_{ps} (Assi et al., Submitted):

$$k_{ma} \equiv k_{ps} = k_{ps}^0 exp \left(\alpha_{ps} W_{ma} \right) \quad (13a)$$

or

$$k_{ps} = k_{psSat} exp \left(\alpha_{ps} (W_{ma} - W_{maSat}) \right) \quad (13b)$$

where $k_{ps}^0 = k_{psSat}/exp \left(\alpha_{ps} W_{maSat} \right)$ and α_0 are the two characteristic parameters of this equation.

Equation (13) can also be expressed in terms of h such as, according to Equation (5):

$$k_{ps} = k_{ps}^0 exp \left(\alpha_{ps} \rho_w \overline{E}_{ma} / \left(h + \rho_w \overline{E}_{ma} / W_{maSat} \right) \right) \quad (14)$$

of form: $k_{ps} = k_{ps}^0 exp \left(\alpha_h / (h + \beta_h) \right)$
where $\alpha_h = \alpha_{ps} \rho_w \overline{E}_{ma}$, $\beta_h = \rho_w \overline{E}_{ma} / W_{maSat}$ and $k_{ps}^0 = k_{psSat}/exp \left(\alpha_h / \beta_h \right)$
Equation (13b) is used in Kamel® for the unsaturated hydraulic conductivity of the pedostructure in terms of W_{ma} limited to W_{st}, that means excluding the contribution of the saturated interpedal water W_{ip}. The latter will be taken into account while measuring the saturated hydraulic conductivity by one of the standard laboratory methods such as falling head or constant head permeability tests.

Swelling rate of the pedostructure

The swelling and shrinkage dynamics of the pedostructure are governed by the same conceptual process that governs the water exchange between the primary peds and the interpedal pore space. Braudeau and Mohtar (2006) validated a particular case (aggregates immersed in water) of Equation (12) expressing the water exchange between the two media as proportional to the difference in their swelling pressure. In this equation, k_{mi} is the transfer rate coefficient (kg_{water} kg_{soil}^{-1} $kPa^{-1}s^{-1}$) for the absorption-desorption of the interped water by the primary peds. This coefficient expresses the velocity of the last layer of water

on the surface of the clay particles entering or leaving the primary peds. We assume that k_{mi} is constant in the entire range of water content, from saturation to the micro air entry point B of the shrinkage curve (**Figure 4**) and that Equation (12) can be generalized to the shrinkage ($dw_{bs}/dt < 0$) as well as the swelling ($dw_{bs}/dt > 0$).

SOIL STRUCTURE AND WATER POOLS DYNAMICS AT THE HORIZON AND PEDON SCALE

Discretization of the pedon into SRELs allows us to model the internal processes that could not be modeled using the classical REV concept, where structural volumes are not defined and determined. Equations (3) define the different specific volumes of pore spaces that can be observed complementarily to the pedostructure at the horizon or pedon scale. Fissures and cracks are directly linked to the hydrostructural state of the pedostructure while pores of biological origin are generally fixed, which induces different kinds of hydraulic conductivity that should be distinguished from the pedostructural hydraulic conductivity.

Vertical fissures

The opening of the vertical porosity ($\overline{Vp}_{fiss}$) when a wetted soil is drying is modeled as follows in the new Kamel® which has to take into account the volume of pore systems other than that of the pedostructure in a pedon SREL. Actually, position ($z = \Sigma H_i$) and thickness (H_i) of the SRELs are governed by $\overline{V}_{layer} = H_i \cdot S$ (where S is a fixed section of the pedon) which is calculated through equations (3) in which $\overline{Vp}_{fiss}$ and $\overline{V}_{ps}$ can be known in terms of W_{ps} according to the pedostructure shrinkage curve Equation A.2 (in Supplementary Material, Appendix 2).

When the water is removed by evaporation or drainage, from the soil medium initially water saturated, vertical fissures appear at the soil surface, then through the SRELs with depth, when the water content decreases under W_{ps} corresponding to the end of the saturated interpedal shrinkage phase, of slope $K_{ip} = 1$, and the beginning of shrinkage of the primary peds in the pedostructure with the decrease of W_{mi}^{eq}.

Because of the difference between the one-dimensional volume change of the SRELs, according to their content in pedostructural volume $\overline{V}_{ps}$, and the three-dimensional volume change of the pedostructure itself, the relationship between the two nested SREVs (soil layer and pedostructure) is, according to Equation (3d) and assuming an isotropic shrinkage of the pedostructure:

$$\overline{V}_{layer}/\overline{V}_{layerL} = a \left(\overline{V}_{ps} + \overline{V}_{fiss} \right)/a\overline{V}_{psL} = H_{ps}S_{ps}/H_{psL}S_{psL} \quad (15)$$

where H_{ps} and S_{ps} are the height and surface of the volume occupied by the pedostructure in the layer. Since the shrinkage of $\overline{V}_{ps} + \overline{Vp}_{fiss}$ is only vertical, the horizontal section of this volume stays constant with the change of H_{ps} such that $S_{ps} = S_{psL}$ in Equation (15).

The shrinkage of the pedostructure is tri-dimensional, so:

$$\overline{V}_{ps}/\overline{V}_{psL} = \left(H_{ps}/H_{psL} \right)^3 = \left(\triangle V_{ps} + V_{psL} \right)/V_{psL} \quad (16)$$

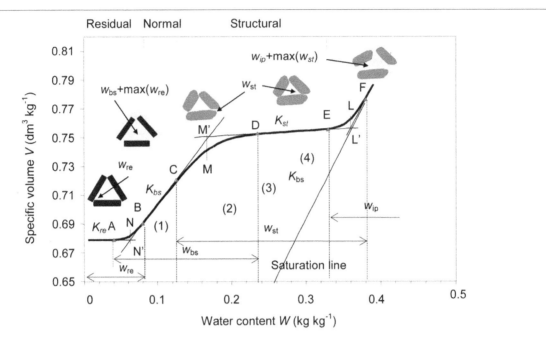

FIGURE 4 | Different configurations of air and water partitioning into the two pore systems, inter and intra primary peds, related to the shrinkage phases of a standard Shrinkage Curve. The various water pools, w_{re}, w_{bs}, w_{st}, w_{ip}, are represented with their domain of variation. The linear and curvilinear shrinkage phases are delimited by the transition points (A, B, C, D, E, and F). Points N', M', and L' are the intersection points of the tangents at those linear phases of the SC (Braudeau et al., 2004).

and

$$H_{ps}/H_{psL} = \left(\Delta V_{ps}/\overline{V}_{psL}+1\right)^{\frac{1}{3}} \cong \left(1 + \Delta \overline{V}_{ps}/3\overline{V}_{psL}\right)$$
$$= \left(2\overline{V}_{psL} + \overline{V}_{ps}\right)/3\overline{V}_{psL} \qquad (17)$$

Combining Equations (15, 17) leads to:

$$\overline{V}_{layer}/\overline{V}_{layerL} = \left(2\overline{V}_{psL} + \overline{V}_{ps}\right)/3\overline{V}_{psL} \qquad (18)$$

and

$$\overline{Vp}_{fiss} = \overline{V}_{psL}H_{ps}/H_{psL}-\overline{V}_{ps} = 2\left(\overline{V}_{psL} - \overline{V}_{ps}\right)/3 \qquad (19)$$

Thus, the pedostructural volume of the fissures present in a layer representative of a soil horizon is directly related to the pedostructural volume $\overline{V}_{ps}$.

Hydraulic conductivity of a SREL at the pedon scale

When the soil surface is subject to rain or irrigation, the vertical porosity, $\overline{Vp}_{fiss}$ and the biological macropores $(\overline{Vp}_{bio})$ must be taken into account in the water infiltration balance. Both are different and must be treated separately. Actually, in a SREL, the fissural pore volume, $\overline{Vp}_{fiss}$ is directly dependent on $\overline{V}_{ps}$. (Equation 19) and thus is function of the pedostructural water content W_{ps} while the biological macropores volume $\overline{Vp}_{bio}$ can be considered constant. The latter can be morphologically highlighted and measured in the field (Abou Najm et al., 2010; Sanders et al., 2012).

Water movement in biological macropores, (W_{pbio}). This water is considered moving out of the pedostructure but in contact with it in the SREL. Out of the pedostructure the water is free and submitted to gravity. The hydraulic conductivity k_{pbio} can be assumed, according to the works of Chen and Wagenet (1992) and Chen et al. (1993), proportional to the saturation level of the biological porosity and equal to:

$$k_{pbio} = \left(K_{Sat} - k_{psSat}\right)\left(W_{pbio}/W_{pbioSat}\right) \qquad (20)$$

where K_{Sat} is the traditional hydraulic conductivity of the soil at saturation which is measured on saturated soil samples large enough to include the macrobiological porosity.

Thus, the movement of the free water in this poral system is the sum of two fluxes: a vertical Darcian flux in the gravitational field and a lateral flux due to the absorption of this free water by the pedostructural macroporosity:

$$dW_{pbio}/dt = \rho_w\overline{Vp}_{bio}\frac{\partial}{\partial x}\left[\left(k_{Sat} - k_{psSat}\right)\left(W_{pbio}/W_{pbioSat}\right)\right]$$
$$-\rho_w\overline{Vp}_{bio}k_{ps}h_{ma}/y_{bio} \qquad (21)$$

where y_{bio} is a length (dm) parameter, representing a mean distance between the biological porosity and the pedostructure medium.

Water movement in fissures and cracks, (W_{fiss}). The presence of water in the fissures created by the pedostructure shrinkage does not play any role in the vertical transfer of the pedostructural water W_{ps} through the pedostructure and the biological macroporosity of the SRELs. This water stored in the pedostructural fissures after a rain or irrigation will be locally absorbed by primary peds of the pedostructure. Let W_{lay} the water content of a SREL as the sum of W_{ps} and W_{pbio}, excluding the water stored in cracks and fissures W_{fiss} which is considered as external to the soil medium organization. Absorption of this water by the soil medium of the $SREL_i$ can be written such as:

$$dW_{lay\_i}/dt = k_{mi} \left(h_{mi} + h_{fiss\_i} \right) \quad (22)$$

where h_{fiss_i} is the pressure head in the fissural water at the depth z_i of the $SREL_i$ considered. Therefore, W_{fiss} must be distinguished from the other SREL water contents and in particular from W_{lay}, the calculation of h_{fiss_i} being a function of the rainfall intensity, the open volume of fissures at the soil surface, $\overline{V}_{fiss_i = 0}$, and the fissural water storage at time t: $\sum_{i0}^{iend} \overline{V}_{fiss\_i} - \sum_{i0}^{iend} W_{fiss\_i}$. We can notice that during this lateral absorption of W_{fiss} by the pedostructure, the height of the water in fissures stays the same, the absorbed water being replaced by a same volume of the pedostructure swelling.

IMPLEMENTATION OF THE THEORY IN SOIL WATER MODELING

UPDATE OF THE SOIL WATER MODEL KAMEL

The earlier version of Kamel® (Braudeau et al., 2009) was based on the pedostructure concept which took into account the hydrofunctionality of primary peds and of their assembly; it followed the system's approach for the definition of the hierarchically nested variables listed in **Table 1**. Thus, its update taking account of the advances in thermodynamics of the soil medium presented here above consists of two improvements:

(1) Replacing the previous equations of the pedostructure hydrostructural functioning by the new equations of thermodynamic origin, in particular for $\overline{V}(W)$; $h(W)$; W_{mi}^{eq}; W_{ma}^{eq} and W_{ip}. This update will provide what we call *Kamel-Core*, for a use in laboratory, modeling the thermodynamic equilibriums of the soil medium (pedoclimate) as the conditions of development of the biological processes studied in laboratory.

(2) Introducing the new organization variables needed for the description of subsystems other than the pedostructure that compose a soil horizon and thus, the SRELs: such as the volumes of stones, roots, biological macropores, etc. This improvement will lead to a soil water model for field applications, named *Kamel-field*. It can simulate a pedon, organized in horizons containing subsystems surrounded by the pedostructure, like they are usually described in a soil profile on the field, in particular for the estimation of the field parameters like coefficients a, b, c, etc. and the dimensions of the horizons according to **Table 2**. Thus, the modeling of a pedon by Kamel-field can be considered as a physical modeling of the soil mapping unit of which the pedon is representative. This means that a soil map should contain all the parameters used by *Kamel-field*, at both pedostructure and

horizon scales, to be totally characterized and modeled in the soil-plant-atmosphere system.

In both cases, the pedostructure parameters are needed; **Table 3** recapitulates the new equations along with their respective parameters. These parameters could be obtained by fitting the new equations on simulated characteristic curves obtained using pedotransfer functions. However, the physical approach is of course to measure the 4 characteristic curves of the pedostructure: shrinkage curve, water retention curve, soil swelling rate, and hydraulic conductivity, and extract their parameters by adjustment with the theoretical equations of these curves. Implementation of the theory concerns not only the model in itself but also the methodology of measurement of the parameters required. The next section presents the new methodology used for getting the Kamel parameters.

SPECIFIC METHODOLOGY FOR THE KAMEL PARAMETERS DETERMINATION

We have to distinguish between the pedostructural parameters that all can be measured in the laboratory on standard soil samples of near 100 cm³, and the pedon parameters that must be estimated in the field.

Pedostructure parameters

A complete hydrostructural characterization of the pedostructure for Kamel® requires the accurate and continuously measurement in laboratory of the four pedostructural characteristic curves mentioned above: shrinkage curve $\overline{V}_{ps}(W_{ps})$, water retention curve $h(W_{ps})$, conductivity curve $k_{ps}(W_{ps})$ and the time dependent swelling curve $\overline{V}_{ps}(t)$. Their equations and parameters are summarized in **Table 3**.

Hydrostructural equilibrium parameters. These are 12 parameters:

$\overline{V}_{ps}^o$, K_{bs}, K_{st}, K_{ip}, k_N, W_N, $\overline{E}_{ma}$, $\overline{E}_{mi}$, W_{miSat}, W_{maSat}, k_L, and W_L

for the two hydro structural soil moisture characteristic curves (pedostructural shrinkage curve and water retention curve).

A new apparatus TypoSoil™ (Bellier and Braudeau, 2013) has been recently built to fulfill standards conditions of measurement of the shrinkage curve and the water retention curve (using micro-tensiometers). Measures of weight, water tension pressure, diameter and high of the sample, are made simultaneously and almost continuously (10 min of intervals) on an unconfined cylindrical soil sample (5 cm diameter, 5 cm in height), from water saturation to the dry state. Braudeau et al. (2014) and Assi et al. (Submitted), described the theory and accordingly the whole procedure using TypoSoil™ for preparing the samples, measuring the characteristic curves and for extracting the 12 parameters representing the two curves, whatever the shape of the shrinkage curve.

Hydrostructural dynamic parameters. These are k_{ps}^o and α_{ps} of the Equation (13a) and k_{mi} of Equation (12). Parameters k_{ps}^o and α_{ps} can be measured accurately using the Hyprop™ device (Assi et al., Submitted) or using the wind method (Wendroth et al.,

1993) in condition of having preliminarily the pedostructural parameters of the sample analyzed by these methods.

As for k_{mi}, Braudeau and Mohtar (2006) showed how the micro-macro water exchange coefficient, k_{mi}, can be determined from the measurement of the swelling rate of a soil sample immersed in water.

Reference mass of solids for the SREV variables

According to the SREV and SREL definitions, the reference mass of solids in a layer is its pedostructural mass, M_{ps}; that is, the mass of solids in the total volume of pedostructure V_{ps}, present in the considered layer of volume V_{layer}. Thus, each layer has a fixed M_{ps}, depending on its delimitation, when the horizon is discretized in layers. Therefore, this discretization of the pedon in layers (**Figure 2**) must be conducted in such a manner that the pedostructural mass of solids in each layer can be calculated. Given that the section, S_{xy}, of the pedon is constant and that the position (depth, z_i) and thickness (H_i) of SRELs are variables controlled by the specific volume of layers ($\overline{V}_{layer}$), the horizons in the pedon must be considered, for discretization, at the total water saturation state (W_{laySat}), where no fissure due to shrinkage is open and where all state variables are assumed homogeneously distributed in the horizons. Accordingly, M_{psl} could be calculated such as:

$$M_{psL} = aV_{laySat}/\overline{V}_{psSat} = aS_{lay}H_{iSat}/\overline{V}_{psSat} \quad (23)$$

where the horizontal section of the pedon (S_{lay}), the volumetric proportion of pedostructure in the horizon (a), and the chosen thickness of the layers at saturation of the horizons (H_{iSat}), are the three organizational parameters of the pedon required to determine the pedostructural mass of solids present in each SREL i.

SREL and pedon parameters

Parameters of the pedon, its horizons and the SRELs coming from the discretization of the horizons, have already been presented in **Table 2**, separately from parameters of the pedostructure (**Table 3**). They describe the internal organization of the pedon into horizons and SRELs according to observations made at the field scale on a soil profile. These parameters characterize the soil surface boundaries, the volumetric fractions a, b, c, etc. of each of the terms of Equation (3a) for each horizon. Aside these morphological or organizational parameters, the hydrodynamic parameters like the saturated hydraulic conductivity at

field scale of the soil horizons must be measured through field measurements (Libardi et al., 1980; Angulo-Jamarillo et al., 2000).

ALGORITHM OF SIMULATION

Updating Kamel®-Core using Kamel® original algorithm where only the equations of the hydrostructural functioning of the pedostructure are updated, does not require changes in the algorithm used in Kamel® (Braudeau et al., 2009) for simulating the water movement in the soil horizons. This is because there are no other systemic components sharing the space inside the SRELS with the pedostructure. On the other hand, in the development of Kamel®-Field using Kamel®, the organizational variables such as W_{ps}; $\overline{V}_{ps}$; W_{fiss} etc. should be specified in the equations used in the simulation of the different water fluxes and their contribution to the pedostructural water cycle.

The discretization of the soil horizons into soil layers and the nomenclature used are shown in **Figure 5**, where the fluxes $F1$ and $F2$ are represented for each layer at every time step along with the thickness H_i, the water pressure $h_i(=h_{ma}>0)$ and the conductivity $k_i(=k_{ps})$ of the pedostructure in the layer.

Each layer is small enough (2 cm thick is recommended) such that the state variables kept the same values everywhere in the layer and the resulting conductivity of the portion delimited by two dashed lines ($k_{i-1/2}$ or $k_{i+1/2}$) can be approximated by the arithmetic average of the two corresponding conductivities:

$$k_{i-1/2} = (k_{i-1} + k_i)/2 \text{ and } k_{i+1/2} = (k_i + k_{i+1})/2. \quad (24)$$

At each time step, for the layer i, fluxes through the upper and lower surfaces of the soil layer i, $F1_i$, and $F2_i$ are calculated:

$$F1_i = \frac{k_{i-1} + k_i}{H_{i-1} + H}\left(h_{i-1} - h_i - \frac{H_{i-1} + H_i}{2}\right) \quad (25)$$

and

$$F2_i = \frac{k_i + k_{i+1}}{H_i + H_{i+1}}\left(h_i - h_{i+1} - \frac{H_i + H_{i+1}}{2}\right) \quad (26)$$

The conditions are such that $F2_{i-1} = F1_i$ and $F2_i = F1_{i+1}$, and that $F2$ and $F1$ cannot be greater than the available space in the receiving layer nor can they extract more water than possible from the providing layer. At the inferior limit of the profile ($i =$ end),

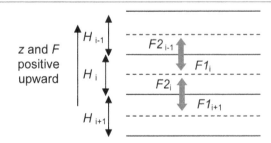

FIGURE 5 | Representation of the Darcian flow of the pedostructural macro pore water (W_{ma}) through the structural representative elementary layers (SRELs) of the discretized pedon.

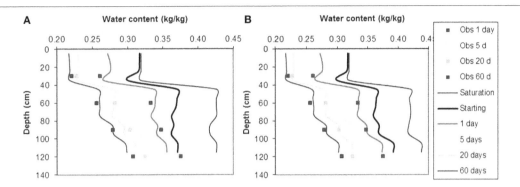

FIGURE 6 | Simulation of drainage after infiltration up to equilibrium (black line "starting") in the Yolo loam soil profile (Davidson et al., 1969). Squares represent the measured observations and lines represent the simulated moisture profile using the measured soil characteristic parameters given in the article: **(A)** without any change of these parameters and **(B)** with a correction of the parameter K_{psSat} for the second and fourth horizon corresponding to the deviation of water content at one day observed on the first graph (−0.006 and +0.02 kg/kg, respectively); (Braudeau et al., 2009).

$F2_{end}$ is calculated as if the layer outside has the same characteristics as the last layer of the profile but with k and h being taken as k_{end} and h_{end} calculated at the previous time step t_{i-1}.

Under these conditions, changes in W_{ps_i} due to $F1$ and $F2$ for each step is then calculated using:

$$\triangle W_{ps\_i} = \rho_w \overline{V}_{ps\_i} (F2_i - F1_i)/H_i \qquad (27)$$

noticing that $\overline{V}_{ps}$ should be used here instead of $\overline{V}_{layer}$ in the earlier version in which the pedostructure occupied all the volume of the layer.

The change in W_{ps} is initially considered as a change in W_{ma}, which provokes temporarily a change in h_{ma} and then an imbalance between the retention of W_{mi} and W_{ma}. There is therefore, during the time step, a transfer of water between the two pore systems which is calculated using the following equation:

$$\triangle W_{mi} = -\triangle W_{ma} = k_{mi} (h_{mi} - h_{ma}) \qquad (28)$$

During the same time step, we may have an eventual contribution to W_{ma} coming from the W_{pbio} cycle and also to W_{mi} coming from W_{fiss}, as it is mentioned above.

MODEL EVALUATION

This section presents Kamel® evaluation (old version) based on field observation of moisture profile as well as comparison with Hydrus 1D model. The intent is to show evidence of the model utility in characterization and modeling soil water medium. More detailed assessment of the model is available in Braudeau et al. (2009) and Singh et al. (2012).

Kamel® was first applied by Braudeau et al. (2009) to a case study of a field experiment of internal drainage published by Davidson et al. (1969). Estimation of parameters for the pedostructural shrinkage curve, water retention curve, and hydraulic conductivity curve was fully detailed in Braudeau et al. (2009). Kamel® was further evaluated at the field scale in comparison to the soil water model Hydrus-1D (Simunek et al., 2008) by Singh et al. (2012). In this new version of Kamel® the formulation of the four hydro-structural characteristic curves of the

pedostructure have been changed but the descriptive variables used and the shape of these curves are all the same. Therefore, the results obtained in these evaluations suffice for the evaluation of the updated version. These evaluations are summarized hereafter. They show high level of agreement between Kamel® simulations and field measurements and Hydrus 1-D simulations. The sections below describe both evaluations:

COMPARISON WITH MOISTURE PROFILE FIELD DATA

Figure 6 shows the computed moisture profiles of Kamel® at 0, 1, 5, 20, and 60 days of drainage, without evaporation from the surface and starting from the saturated state after infiltration. The figure also shows the field measured data of Davidson et al. (1969); the soil water content at 30, 60, 90, and 120 cm depth after 1 day, 5, 20, and 60 days of drainage without evaporation for comparison. The simulation were conducted without any calibration, simulation results show a good agreement with measured values where the first day, the difference between simulated and observed water content is $\triangle W1day = -0.006$ at 60 cm and 0.02 kg/kg at 120 cm. Less than 0.01 kg/kg are observed at 5 and 20 days for horizons 3 and 4 (90 and 120 cm) (Braudeau et al., 2009).

COMPARISON WITH HYDRUS-1D

Measured and predicted soil moisture by Kamel® and Hydrus-1D® at the 5 cm depth for the 4 month time period are presented in **Figure 7** (Singh et al., 2012). Rainfall during this period is plotted on the secondary Y-axis. Since this is the surface layer, measured water content responds quickly to rainfall and evapotranspiration. The figure shows that Hydrus-1D® soil moisture prediction was in good agreement with observed water content.

CONCLUSION

The principal function of Kamel® is to be a computational model that represents a soil pedon and its internal hydrological functioning at the different levels of hydro-functional organization of the soil medium. Integrating these levels of hydro-functional organization is significant in linking the internal functioning of the soil to its behavior and properties with the other elements of eco- or agro systems when subject to external climatic conditions. The update of Kamel® planned here introduces new prospects

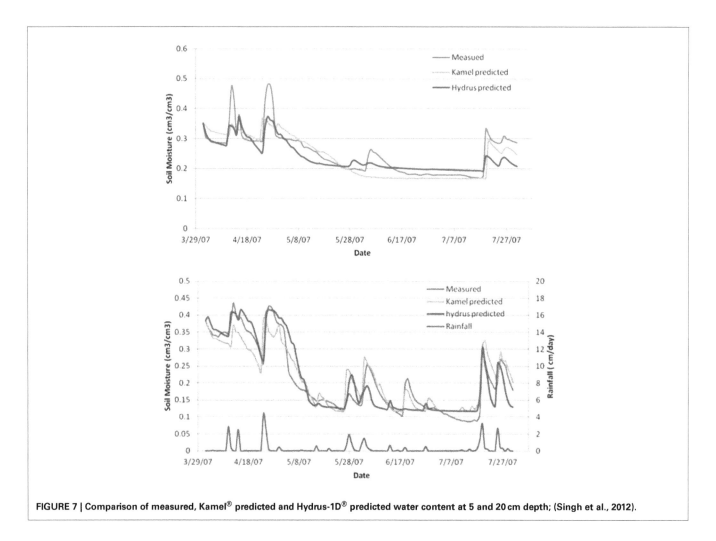

FIGURE 7 | Comparison of measured, Kamel® predicted and Hydrus-1D® predicted water content at 5 and 20 cm depth; (Singh et al., 2012).

in multi-scale modeling of the natural environment discussed hereafter.

(1) As portrayed, Kamel® is a model that simulates two physical phenomena: (i) the hydrostructural and thermodynamic equilibrium states of the pedostructure, depending on the pedostructural water content, and (ii) the joint dynamics of the water and the soil structure at the different functional scales of the pedon (soil hydro-structural dynamics).

(2) Accordingly, Kamel®'s parameters and equations must be physically-based according to a new paradigm that can be called "hydrostructural pedology" where all the hydrostructural properties of the soil medium are thermodynamically defined and quantified. In this way, a soil mapping unit can be characterized by its representative pedon providing with the hydrostructural parameters of its horizons.

(3) Moreover, using this paradigm, the thermodynamic definition of the pedostructure system allowed us to quantitatively distinguish the two kinds of water fluxes inside and outside of the pedostructure within the soil horizons.

(4) Thus, Kamel® differentiates between two kinds of water, a pedostructural water, the water that takes part in the hydro-thermodynamic equilibrium of the soil-plant-water natural continuum, and a free water that is only subject to gravity

out of the thermodynamic conditions laid down by the pedostructure.

(5) This distinction between a *pedostructural water* and a free water in the soil medium that could not be separated and modeled as such before. This distinction allows for the differentiation of what have been called green water and blue water resources.

(6) Finally, the soil as it will be modeled by Kamel®-field, constitutes the lieu of transformation of the blue water, driven down by gravity, to the green water or thermodynamic water of the soil plant atmosphere continuum. Therefore, Kamel® is a computer model that integrates hydrology and pedology in modeling the transfer of both types of water in the soil, the free water coming from rainfall or irrigation and the pedostructural water consumed by plants or evaporating at the soil surface to the atmosphere. This modeling determines the availability of space, water and air in the soil medium that condition the development of biotic and abiotic soil processes.

ACKNOWLEDGMENTS

The authors express their appreciation to the Graduate Research Associates Amjad Assi and Guy Bou Lahdou for their involvement in this project. The authors also wish to acknowledge QScience.com and Dr. Christopher Leonard for their assistance in editing the manuscript.

REFERENCES

Abou Najm, M., Jabro, J., Iverson, W., Mohtar R., and Evans, R. (2010). New method for the characterization of three-dimensional preferential flow paths in the field. *Water Resour. Res.* 46, 1–18. doi: 10.1029/2009WR008594

Angulo-Jamarillo, R., Vandervaere, J. P., Roulier, S., Thony, J. L., Gaudet, J. P., and Vauclin, M. (2000). Field measurement of soil surface hydraulic properties by disc and ring infiltrometers: a review and recent developments. *Soil Till. Res.* 55, 1–29. doi: 10.1016/S0167-1987(00)00098-2

Assi, A., Accola, J., Hovhannissian, G., Mohtar, R. H., and Braudeau, E. (2014). Physics of the soil medium organization part2: pedostructure characterization through measurement and modeling of the soil moisture characteristic curves. *Front. Environ. Sci.* 2:5. doi: 10.3389/fenvs.2014.00005

Bellier, G., and Braudeau, E. (2013). *Device for Measurement Coupled with Water Parameters of Soil. WO 2013/004927 A1.* Geneva: World Intellectual Property Organization, WIPO.

Braudeau, E., Assi, A. T., Boukcim, H., and Mohtar, R. H. (2014). Physics of the soil medium organization part 1: thermodynamic formulation of the pedostructure water retention and shrinkage curves. *Front. Environ. Sci.* 2:4. doi: 10.3389/fenvs.2014.00004

Braudeau, E., and Bruand, A. (1993). Détermination de la courbe de retrait de la phase argileuse à partir de la courbe de retrait établie sur échantillon de sol non remanié. *C. R. Acad. Sci. Paris* 316, 685–692.

Braudeau, E., Costantini, J., Bellier, G., and Colleuille, H. (1999). New device and method for soil shrinkage curve measurement and characterization. *Soil Sci. Soc. Am. J.* 63, 525–535. doi: 10.2136/sssaj1999.0361599500630003015x

Braudeau, E., Frangi, J. P., and Mohtar, R. H. (2004). Characterizing non-rigid dual porosity structured soil medium using its Shrinkage Curve. *Soil Sci. Soc. Am. J.* 68, 359–370. doi: 10.2136/sssaj2004.3590

Braudeau, E., and Mohtar, R. H. (2006). Modeling the swelling curve for packed soil aggregates using the pedostructure concept. *Soil Sci. Soc. Am. J.* 70, 494–502. doi: 10.2136/sssaj2004.0211

Braudeau, E., and Mohtar, R. (2009). Modeling the soil system: bridging the gap between pedology and soil-water physics. *Glob. Planet. Change J.* 67, 51–61. doi: 10.1016/j.gloplacha.2008.12.002

Braudeau, E., Mohtar, R. H., El Ghezal, N., Crayol, M., Salahat, M., and Martin, P. (2009). A multi-scale "soil water structure" model based on the pedostructure concept. *Hydrol. Earth Syst. Sci. Discuss.* 6, 1111–1163. doi: 10.5194/hessd-6-1111-2009

Braudeau, E., Sene, M., and Mohtar, R. H. (2005). Hydrostructural characteristics of two African tropical soils. *Eur. J. Soil Sci.* 56, 375–388. doi: 10.1111/j.1365-2389.2004.00679.x

Brewer, R. (1964). *Fabric and Mineral Analysis of Soils.* New York, NY: John Wiley and Sons, 482.

Bruckler, L., Bertzzi, P., Angulo-Jamarillo, R., and Ruy, S. (2002). Testing an infiltration method for estimating soil hydraulic properties in the laboratory. *Soil Sci. Soc. Am. J.* 66, 384–395. doi: 10.2136/sssaj2002.3840

Chen, C., Thomas, D. M., Green, R. E., and Wagenet, R. J. (1993). Two-domain estimation of hydraulic properties in macropore soils. *Soil Sci. Soc. Am. J.* 57, 680–686. doi: 10.2136/sssaj1993.03615995005700030008x

Chen, C., and Wagenet, R. J. (1992). Simulation of water and chemicals in macropore soils. Part I Representation of the equivalent macropore influence and its effect on soil water flow. *J. Hydrol.* 130, 105–126. doi: 10.1016/0022-1694(92)90106-6

Davidson, J. M., Stone, L. R., Nielson, D. R., and Larue, M. E. (1969). Field measurement and use of soil water properties. *Water Resour. Res.* 5, 1312–1321. doi: 10.1029/WR005i006p01312

Gardner, W. R. (1958). Some steady-state solutions of the unsaturated moisture flow equation with application to evaporation from a water table. *Soil Sci.* 85, 228–232. doi: 10.1097/00010694-195804000-00006

Gerke, H., and van Genuchten, M. (1993). A dual-porosity model for simulating the preferential movement of water and solutes in structured porous media. *Water Resour. Res.* 29, 305–319. doi: 10.1029/92WR02339

Katterer, T., Schmied, B., Abbaspour, K.C., and Schulin, R. (2001). Single- and dual-porosity modelling of multiple tracer transport through soil columns: effects of initial moisture and mode of application. *Eur. J. Soil Sci.* 52, 25–36. doi: 10.1046/j.1365-2389.2001.00355.x

Libardi, P. L., Reichardt, K., Nielsen, D. R., and Biggar, J. W. (1980). Simple field method for estimating soil hydraulic conductivity. *Soil Sci. Soc. Am. J.* 44, 3–7. doi: 10.2136/sssaj1980.03615995004400010001x

Logsdon, S. D. (2002). Determination of preferential flow model parameters. *Soil Sci. Soc. Am. J.* 66, 1095–1103. doi: 10.2136/sssaj2002.1095

Mohtar, R. H., Jabro, J. D., and Buckmaster, D. (1997). A grazing simulation model: GRASIM B: Field Testing. *Trans. ASABE* 40, 1495–1500.

Othmer, H., Diekkrüger, B., and Kutilek, M. (1991). Bimodal porosity and unsaturated hydraulic conductivity. *Soil Sci.* 152, 139–149. doi: 10.1097/00010694-199109000-00001

Sanders, E. C., Abu Najm, M. R., Mohtar, R. H., Kladivco, E., and Shulze, D. (2012). Field method for separating the contribution of surface-connected preferential flow pathways from flow through the soil matrix. *Water Resour. Res.* 48. doi: 10.1029/2011WR011103

Simunek, J., Jarvis, N. J., van Genuchten, M. T., and Gardenas, A. (2003). Review and comparison of models for describing non-equilibrium and preferential flow and transport in the vadose zone. *J. Hydrol.* 272, 14–35. doi: 10.1016/S0022-1694(02)00252-4

Simunek, J., Sejna, M., Saito, H., Sakai, M., and van Genuchten. M. (2008). *The HYDRUS-1D Software Package for Simulating the Movement of Water, Heat, and Multiple Solutes in Variably Saturated Media, Version 4.08, HYDRUS Software Series 3.* Riverside, CA: Department of Environmental Sciences, University of California Riverside.

Singh, J., Mohtar, R., Braudeau, E., Heathmen, G., Jesiek, J., and Singh, D. (2012). Field evaluation of the pedostructure-based model (Kamel®). *Comput. Electron. Agr.* 86, 4–14. doi: 10.1016/j.compag.2012.03.001

Stöckle, C. O., Donatelli, M., and Nelson, R. (2003). CropSyst, a cropping systems simulation model. *Eur. J. Agron.* 18, 289–307. doi: 10.1016/S1161-0301(02)00109-0

Wendroth, O., Ehlers, W., Hopmans, J. W., Kage, H., Halbertsma, J., and Wösten, J. H. M. (1993). Reevaluation of the evaporation method for determining hydraulic functions in unsaturated soils. *Soil Sci. Soc. Am. J.* 57, 1436–1443.

Conflict of Interest Statement: This article is an extension and an update (some fundamental equations) of an old version of the Kamel model which was registered at AP Paris in 2006 by IRD: Braudeau, E. F., 2006. Kamel® Simile version. Agence pour la Protection des Programmes, Paris, IDDN.FR.001.390019.000.S.P.2006.000.31500

7

An integrated, probabilistic model for improved seasonal forecasting of agricultural crop yield under environmental uncertainty

Nathaniel K. Newlands[1,2]*, David S. Zamar[3], Louis A. Kouadio[1], Yinsuo Zhang[4], Aston Chipanshi[5], Andries Potgieter[6], Souleymane Toure[7] and Harvey S. J. Hill[8]

[1] Science and Technology Branch, Agriculture and Agri-Food Canada, Lethbridge Research Centre, Lethbridge, AB, Canada
[2] Department of Statistics, University of British Columbia, Vancouver, BC, Canada
[3] Department of Chemical and Biological Engineering, University of British Columbia, Vancouver, BC, Canada
[4] Science and Technology Branch, National Agroclimate Information Service, Agriculture and Agri-Food Canada, Ottawa, ON, Canada
[5] Science and Technology Branch, National Agroclimate Information Service, Agriculture and Agri-Food Canada, Regina, SK, Canada
[6] Queensland Alliance for Agriculture and Food Innovation, University of Queensland, Toowoomba, QLD, Australia
[7] Habitat Conservation Management Division, Canadian Wildlife Service, Environment Canada, Montreal, QC, Canada
[8] Science and Technology Branch, National Agroclimate Information Service, Agriculture and Agri-Food Canada, Saskatoon, SK, Canada

Edited by:
Mohammad Mofizur Rahman Jahangir, The University of Dublin, Ireland

Reviewed by:
Holger Hoffmann, Leibniz Universität Hannover, Germany
Sangeeta Lenka, Indian Institute of Soil Science, India
Mohamed Samer, Cairo University, Egypt

*Correspondence:
Nathaniel K. Newlands, Science and Technology Branch, Agriculture and Agri-Food Canada, Lethbridge Research Centre, 5403 1st. Ave. S., P.O. Box 3000, Lethbridge, AB, T1J 4B1, Canada
e-mail: nathaniel.newlands@agr.gc.ca

We present a novel forecasting method for generating agricultural crop yield forecasts at the seasonal and regional-scale, integrating agroclimate variables and remotely-sensed indices. The method devises a multivariate statistical model to compute bias and uncertainty in forecasted yield at the Census of Agricultural Region (CAR) scale across the Canadian Prairies. The method uses robust variable-selection to select the best predictors within spatial subregions. Markov-Chain Monte Carlo (MCMC) simulation and random forest-tree machine learning techniques are then integrated to generate sequential forecasts through the growing season. Cross-validation of the model was performed by hindcasting/backcasting and comparing forecasts against available historical data (1987–2011) for spring wheat (*Triticum aestivum* L.). The model was also validated for the 2012 growing season by comparing forecast skill at the CAR, provincial and Canadian Prairie region scales against available statistical survey data. Mean percent departures between wheat yield forecasted were under-estimated by 1–4% in mid-season and over-estimated by 1% at the end of the growing season. This integrated methodology offers a consistent, generalizable approach for sequentially forecasting crop yield at the regional-scale. It provides a statistically robust, yet flexible way to concurrently adjust to data-rich and data-sparse situations, adaptively select different predictors of yield to changing levels of environmental uncertainty, and to update forecasts sequentially so as to incorporate new data as it becomes available. This integrated method also provides additional statistical support for assessing the accuracy and reliability of model-based crop yield forecasts in time and space.

Keywords: agriculture, Bayesian, climate, crop yield, forecasting, regional, uncertainty

1. INTRODUCTION

1.1. MOTIVATION

There is increasing worldwide concern over the social, environmental and economic destructive potential of extreme climatic events and cumulative environmental impacts (e.g., climate variability, invasive pests) (FAO, 2011). Extreme events associated with climate variability disrupt water, energy and food production, supply and availability (Jentsch et al., 2007). Climate extremes are having a major impact on inter-annual wheat production worldwide (USDA, 2013). Unanticipated extreme events can be large enough in some countries to offset a significant portion of the increases in average yields that arose from technology, carbon dioxide fertilization, and other factors (Lobell et al., 2011). Reliable crop forecasts have the potential to greatly aid decision-makers in identifying potential risks and benefits to increase crop production and to gauge uncertainty, particularly during times where production is uncertain, or across regions where production is highly variable (Hansen, 2005; Littell et al., 2011). Across Africa, food aid imports and emergency assistance, strategic food reserves, and the granting of private firm licences for food import and exports all rely on crop forecasts (Jayne and Rashid, 2010). Extreme climatic events (i.e., drought/flood, extreme cold/heat-waves) impact major food-producing regions in Canada, United States, Russia, China, Australia and North/South Korea, thereby raising the prospect of higher commodity prices and localized food shortages in the future (Global Infomation and Early-Warning System (GIEWS), Food and Agriculture Organization of the United Nations/FAO[1]). As more extreme weather events are

_____

[1] www.fao.org/giews/

anticipated to accompany a warming climate, improved methods for forecasting crop production and its response to climate and other agronomic factors, are becoming increasingly important to guide agricultural producers in making more informed in-season crop management and financial decisions (IPCC, 2013).

The benefit and value of crop forecasts varies according to a range of criteria, namely, relevance, reliability, stakeholder engagement, holism and accuracy (Hammer et al., 2001; McIntosh et al., 2007). These aspects consider the type, availability, coverage of expert knowledge and empirical data, including the required lead-time, time-step, and duration for forecasting. Utilizing available satellite remote-sensing data, alongside observer-based field survey data provides a more rapid and less costly approach to repeated sampling and updating of regional or national-scale forecasts over large cropland areas of interest, while also helping to increasing forecast accuracy, reliability and consistency. Furthermore, crop yield and production forecasting is gaining increasing interest in ecological science because of the availability of large volumes of data from observational monitoring networks and satellite remote-sensing platforms, increases in computational power, and advances in multivariate optimization methodologies. There is also increasing attention on addressing societal needs for better (i.e., more robust) strategies for managing and exploiting natural resources sustainably, given significant global economic, social and environmental change (Luo et al., 2011).

1.2. CROP YIELD FORECASTING

A forecasting system has two primary functions: generation and control, for which there are many different types of designs with varying capabilities and levels of data and knowledge integration. Forecast generation includes acquiring data to revise the forecasting model, producing a statistical forecast and presenting results to the user. Forecast control involves monitoring the forecasting process to detect out-of-control conditions and identifying opportunities to improve forecasting performance. Some systems use simple, calibrated reference curves or semi-empirical equations, while others use statistical models or more complex agro-ecosystem process models. Because of the complex interplay of variables affecting crop yield, a general auto-regressive, integrated moving-average time-series (ARIMA) type model for forecasting is often not suitable. Different levels of engagement relate to the operational cost to survey crops in relation to gain in forecast accuracy, available input data streams, and choice, complexity and extent of innovation of a given model-data assimilation approach (Hammer et al., 2001; Stone and Meinke, 2005). To supplement our focus on crop yield forecasting in the Canadian context, we refer interested readers to supplementary information provided on existing operational systems for crop yield forecasting at various spatial and temporal scales (Supplementary Material).

1.3. AGROECOSYSTEM DYNAMICS AND MEASUREMENT UNCERTAINTY

Agroecosystems are systems that comprise dynamic processes that interact and take place over multiple spatial and temporal scales. As a consequence, there are significant changes in

the leading predictors that control the underlying state for an observed response at any point in time and space. As leading predictors change, so to, can the signal to noise ratio (SNR) of the underlying processes, introducing loss and gain of information for statistical estimation and forecasting. The SNR can be estimated as a ratio of total variance output from a model to the total variance of model inputs. Recently, the uncertainty associated with temperature-driven processes in regional-scale crop models has been reported to be, on average, 12% higher than climate model uncertainty (Koehler et al., 2013). Reducing such uncertainty requires focusing not only on the climate inputs, but also on testing structural model assumptions on crop development that include: changing senescence, the influence that crop water status and changing carbon dioxide levels can have on canopy temperature and heat stress on a crop, crop growth duration and length of day (Hoffmann and Rath, 2013).

Integrating data from different measurement platforms can help to hedge the risk of relying on a single source of information, especially under situations of high data sparsity and environmental variability. Spatial-based models aid in identifying particular crops, times and regions that are most affected by environmental impacts and to assist efforts to measure and analyze ongoing efforts to adapt (Challinor et al., 2003; Hansen, 2005; Matthews et al., 2013). Ecological models consider agroecosystems as an inter-connected system of plant, soil, atmospheric and other underlying, interacting processes. Statistical models, on the other hand, typically focus on representing and understanding individual components, specific processes and/or scales of interactions within an agroecosystem. Nonetheless, statistical models are being increasingly developed and applied to understand whole system dynamics and behavior and "big data" problems. Traditionally, ecological forecasting has typically been based on process-oriented models, informed by data in arbitrary ways. Although most ecological models incorporate some representation of mechanistic processes, many models are generally not adequate to quantify real-world dynamics and provide reliable forecasts with accompanying estimates of uncertainty (Littell et al., 2011). This is because such mechanistic models can be easily over-parameterized by requiring tens to hundreds of parameters and variables to explain an observed pattern or response "signal," without substantially increasing the SNR by explaining a greater fraction of unexplained "noise." In contrast, probabilistic approaches seek to optimize the smallest number of parameters and variables required to explain a signal and minimize the unexplained noise concurrently. This ensures that the SNRs of model predictions and forecasts decrease and are more robust to any changes in input parameters and variables. Moreover, it may be the case that there are limits to crop yield predictability in terms of limits in the achievable SNR. Such limits may be not depend on whether deterministic/probabilistic model is used, nor model complexity in terms of the number of parameters or variables in a model.

1.4. KNOWLEDGE GAPS

Crop yield is a complex variable that is influenced by different climate-drivers, and underlying environmental and genetic factors prior to actual harvest. While the observed variability of major crop commodities can be determined using historical data alone, many other considerations are required to calibrate and generate predictions of future production and its uncertainty. The advancement and changes in crop genetics, farming technology and efficiency, improved agronomic management practices, and ecological dynamics (i.e., soil, water and air/climate) requires one to add a sufficient level of detail into the construction of models that are able to more reliably forecast crop yield and production (Challinor, 2009; Matthews et al., 2013). Moreover, statistical methods aid in accounting for and minimizing measurement bias within climate time-series data that is used in agricultural models and yield forecasting (Hoffmann and Rath, 2012). Applying models at different spatial and temporal scales also contributes additional scaling-based uncertainty, so that the accuracy of a forecasting model at one spatial scale may not be the same, when the same model is applied at another scale (Ewert et al., 2011).

Many current model-based forecast methods are very complex, operate only at the point or field scale, and/or utilize crop phenology information or remote-sensing data, but not both. These aspects limit the application of mechanistic, dynamic crop and agroecosystem models when predicting at the regional-scale, and favor a statistical (i.e., probabilistic) approach that generates predictions based on a smaller subset of leading predictors. In summary, current challenges in using models for forecasting, include: (1) high environmental variability, volatility/stochasticity and spatial heterogeneity, (2) the need for forecasts to be assessed in a dynamic framework, (3) high levels of bias and mismatch of spatial and temporal scales between coupled ocean-atmosphere climate- and crop growth- model forecasts, and, (4) whether required institutional and policy arrangements exist to provide a broader environmental-economic agricultural risk framework to ensure crop forecasts offer a beneficial, viable option for producers to improve their economic livelihood and adopt in the longer term. Such arrangements must provide sufficient flexibility in agricultural management to be able to respond to different levels of perceived, real needs and risks. Many crop forecasting systems only consider a single source of variability. This fails to account for crop yield variability that may be explained by the combined effect of both agroclimate and remote sensing indices. Furthermore, models based on remote-sensing indices alone are only able to provide accurate forecasts late in the growing season when crop growth becomes visible. Likewise, forecasts that do incorporate agroclimate indices, such as precipitation and temperature, fail to include information from remote-sensing data and/or do not make use of the spatial dependence of crop yields.

1.5. RESEARCH OBJECTIVE

Our research objective was to devise a probabilistic model to forecast crop yield and to increase forecast skill in time and space by integrating both agroclimate and satellite remote-sensing data. This involved tracking uncertainty across large regions and between subregions, and updating of forecasts sequentially through growing season. This approach addresses several knowledge gaps, namely: the need to utilize satellite remote-sensing data and to incorporate spatially-dependent environmental variation influencing crop development and growth, and the need to provide a probabilistic, adaptable approach that can more

objective tune, adjust and sequentially improve its forecasts under changes in the amount of data available and level of environmental variability. The model integrates state-of-the-art Bayesian statistical forecasting techniques employing Markov-Chain Monte Carlo (MCMC)-based simulation, robust regression, variable-selection and random forest-tree machine learning. Specifically, (1) we conducted a sensitivity analysis of the model, comparing its performance under simulated changes in the data that is available for forecasting, (2) compared model performance between different subregions as the model selects and ranks different explanatory or predictor variables to reduce uncertainty in its forecasts, and (3) cross-validated the model was performed by hindcasting/backcasting it and comparing its forecasts against available historical data (1987–2011) for spring wheat (*Triticum aestivum* L.). It was also validated for the 2012 growing season, comparing its forecast skill at the CAR, provincial and Canadian Prairie region scales against available statistical survey data. Finally, we identify ways the model could be further enhanced, improved and made more reliable when being applied to different crops and large crop production areas.

2. MATERIALS AND METHODS

2.1. STUDY REGION

Our study region was the Canadian Prairies encompassing the provinces of Alberta, Saskatchewan and Manitoba (**Figure 1**). This region accounts for roughly 91% of Canada's total wheat production (Statistics Canada, 2012a). The boundaries of the 40 Census Agricultural Regions (CARs) spanning the study region originated from the 2011 census of agriculture (Statistics Canada, 2012c). The Prairies are the northernmost branch of the Great Plains of North America with a flat or rolling plains landform and is the major region where wheat is grown in Canada and the most altered of Canada's ecozones. The mountains to the west block precipitation that would otherwise fall within the region, making the western portion very dry. Strong Chinook winds also reduce humidity and deliver brief episodes of spring-like conditions in the western regions during winter. Precipitation generally increases eastward. Temperatures are extreme due to negligible ocean buffering. Winter temperatures average $-10\,°C$ and summers average $15\,°C$ and are very cold. The growing season is defined as the period of each year when crops can be grown. In Canada the growing season is usually determined by the days between last and first frost (i.e., May to October). Growing season precipitation varies between 300 and 400 mm. The dominant soils within this region are the Chernozemic soils. The Chernozemic soils are strongly determined by the accumulation, decomposition and transformation of soil organic matter within the topsoil (or A horizon), whereby native grassland vegetation and climate both influence the amount and nature of organic matter retained within the soil. Deposition of plant material belowground in the grassland system has been the primary factor whereby soil organic matter accumulates (Fuller, 2010). Regional variations in climate and vegetation distribution have formed the distinct soil zones, namely the: Brown, Dark Brown, Black, and Dark Gray zones that have different depths and soil organic matter content within their surface layers (Pennock et al., 2011).

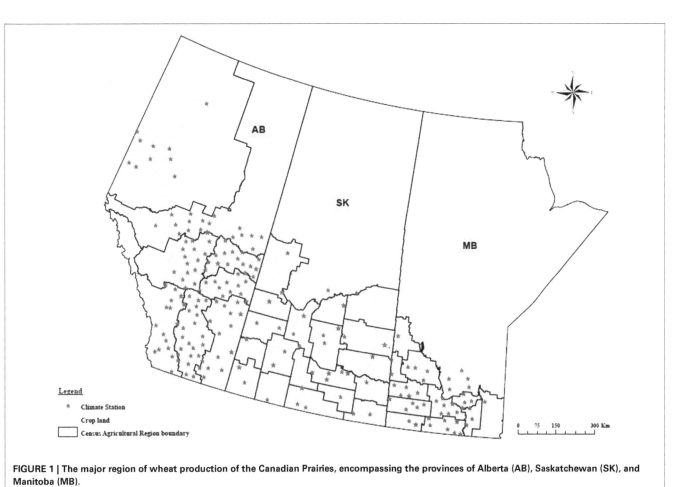

FIGURE 1 | The major region of wheat production of the Canadian Prairies, encompassing the provinces of Alberta (AB), Saskatchewan (SK), and Manitoba (MB).

2.2. DATA SOURCES

2.2.1. Agro-climate

Historical crop yield data from 1987 to 2011 for each of the CARs was obtained from the Field Crop Reporting Series of the Canadian Crop Condition Assessment Program (CCAP) of the Agriculture Division, Statistics Canada (Statistics Canada, 2007, 2012a,b). Station-based daily minimum/maximum temperatures and precipitation data were provided by Environment Canada and other partner institutions through the Drought Watch program, operated by the National Agro-climate Information Service (NAIS) of Agriculture and Agri-Food Canada (AAFC). This data was quality controlled and temporally-interpolated to provide a continuous, historical daily time-series from 1987 to 2012. A total of 259 climate stations were involved in the study (**Figure 1**). Agro-climatic variables considered in the model were growing-degree days (GDD) above an ambient air temperature of 5 °C, soil water availability (SWA), precipitation (P), and crop water deficit index (WDI) defined as $WDI = (1 - AET/PET)$, where AET and PET are actual and potential evapotranspiration (Moran et al., 1994). AET and PET were computed using the Baier and Robertson algorithm (Baier and Robertson, 1996). Soil water availability (SWA) was defined, at any given time, as the percentage of available soil water holding capacity (AWHC). AWHC at the location of each climate station was determined

from soil data obtained from the Soil Landscapes of Canada database from AAFC's Canadian System of Soil Classification (CANSIS, SLC Version 3.2, http://sis.agr.gc.ca/cansis/nsdb/slc/v3.2/intro.html). Crop-specific AWHC depends on soil type and is defined as the difference between soil water field capacity and wilting point, estimated as the cumulative amount of soil water to a maximum of 1 m soil depth or root-restricting soil layer under a difference in volumetric water at hydrostatic pressure of -33 and -500 kPa (Bootsma et al., 2005). Thermal-time (i.e., GDD-based) indices are widely referenced and used in crop development and growth response to temperature as well as crop heat stress and as an indicator of critical temperature thresholds for vernalization (e.g., 3°C of cold time) and for tracking crop developmental and growth. For example, in the case of cereal crops like barley and wheat indices are used to track the phenological stages of emergence, tillering, stem elongation, heading/flowering/anthesis, ripening, or for oilseeds like canola: sowing/germination, emergence, leaf initiation, stem elongation, bud formation, flowering, pod development, maturity (Ferris et al., 1998; Bootsma et al., 2005; Carew et al., 2009; Robertson et al., 2013). Nonetheless, other indices or measures of thermal-time requirements for crops such as biometeorological time (BMT) may be more accurate when crop end-use quality aspects are of interest, which for wheat includes: flour protein, farinograph dough development time and farinograph stability (Saiyed et al., 2009). Station-based minimum/maximum temperatures, precipitation and AWHC were input into a crop-specific soil water balance equation called the Versatile Soil Moisture Budget (VSMB) to generate the corresponding agro-climatic indices for each climate station at a daily time step (i.e., GDD, SWA, AET, PET) (DeJong, 1988; Baier and Robertson, 1996). More than one climate station was referenced within most CARs, but the number of stations within a given CAR did vary. In generating the agro-climate indices using VSMB, the start and end of growing season is determined by the accumulation of sufficient heat units and soil moisture. Daily agro-climatic values were temporally-averaged by month. We provide a summary of our data sources used in simulating, training and validating the forecasting model in **Table 1**.

2.2.2. Satellite, remotely-sensed indices

Normalized-difference vegetation index (NDVI) was also included as a predictor of yield. It is defined as NDVI= (ρNIR − ρRED)/(ρNIR + ρRED), where ρNIR and ρRED are the near-infrared and infrared portions of the electromagnetic spectrum of (0.75–1.5 µm or 841–876 nm) and (0.6–0.7 µm or 620–670 nm), respectively. This index is based on the contrast between the maximum absorption in the red due to chlorophyll pigments (0.4–0.7 µm) and the maximum reflection in the infrared by leaf cellular structure (0.7–1.1 µm) (Habourdane et al., 2004). The NDVI historical time-series were then combined to generate weekly composites south of 60° latitude (i.e., across the agricultural land area of Canada). The values of ρVIS and ρNIR are bounded between 0 and 1. NDVI itself varies between −1 (indicating no vegetation) and 1 (indicating dense vegetation). Weekly NDVI imagery data was compiled from historical data from the Advanced Very High Resolution Radiometer (AVHRR, 1 km resolution), U.S. National Oceanographic and Atmospheric Administration (NOAA) for years 1987–2012. Further quality

control and processing in generating NDVI weekly composites was conducted (Reichert and Caissy, 2002). NDVI values were not crop-specific. Weekly NDVI indices from Julian week 18 to 36 (i.e., May to September) of each year were aggregated into 3-week moving means.

2.3. FORECASTING METHODOLOGY

The integrated methodology includes possible crop-specific input data, and additional auxiliary indices that can be integrated and involved various modules and different computational steps (**Figure 2**). The forecast model was coded using the R Statistical Language (Version 2.15.3, R Development Core Team) (R Foundation, 2013) and made use of the following R library packages: MASS, robustbase, impute, WGCNA, np, mnormt, mcmc, tmvtnorm, Caret, and randomforest. The model's leading, tuning parameters and their library association, description and default values are summarized in **Table 2**.

2.3.1. Model calibration and training

The forecasting equations are based on Bayesian theory, and are an adaptation of a Bayesian auxiliary variable approach of Tanner and Wong (1987). These equations are used to jointly estimate and predict the underlying "environmental" state (i.e., prescribed by a set of agro-climate variables), the set of model parameters, including any additional auxiliary variables (e.g., remotely-sensed indices). Our statistical problem involves the prediction of future state and response, so is hereafter termed forecasting. Some ambiguity or discrepancy in terminology exists in the ecological and statistical sciences in regard to forecasting to better distinguish relevant terminology. Here, we adopt the terminology of Luo et al. (2011) in defining forecasting, prediction, projection, and prognosis, where forecasting involves a probabilistic statement on future states of an ecological system after data are assimilated into a model, prediction generates future states of an ecological system based on logical consequences of model structure, projection

---

Table 1 | Data integrated for simulating, training and validating the forecasting model across 1987–2012.

Type	Source	Original scale	Re-scaling (temporal/spatial averaging)
CROP YIELD DATA			
Historical field-reporting survey (kg/ha)	Statistics Canada	Monthly, CAR[a]	None
CLIMATE DATA			
Precipitation (mm/day) Min/max air temperature (°C)	Environment Canada	Daily, station/point-based	Daily-averaged, spatially-averaged across all stations within a given CAR with equal weighting
AGRO-CLIMATE INDICES			
Crop water deficit index (WDI) (no units) Growing degree days (GDD) (days) Soil water availability (SWA) (no units)	AAFC, NAIS[b] VSMB[c] model	Daily, station/point-based	Monthly-averaged, Spatially-averaged across all stations within a given CAR with equal weighting
REMOTE SENSING INDEX			
Normalized-difference vegetation index (NDVI) (no units)	NOAA- AVHRR[d]	Weekly, 1 km	Three-week averages from week 18 to 36 Spatially-averaged to CAR scale

[a] Census agricultural region (CAR).

[b] Agriculture and agri-food canada (AAFC), National agro-climate information service (NAIS).

[c] Versatile soil moisture budget (VSMB).

[d] National oceanographic and atmospheric administration (NOAA)—advanced very high resolution radiometer (AVHRR).

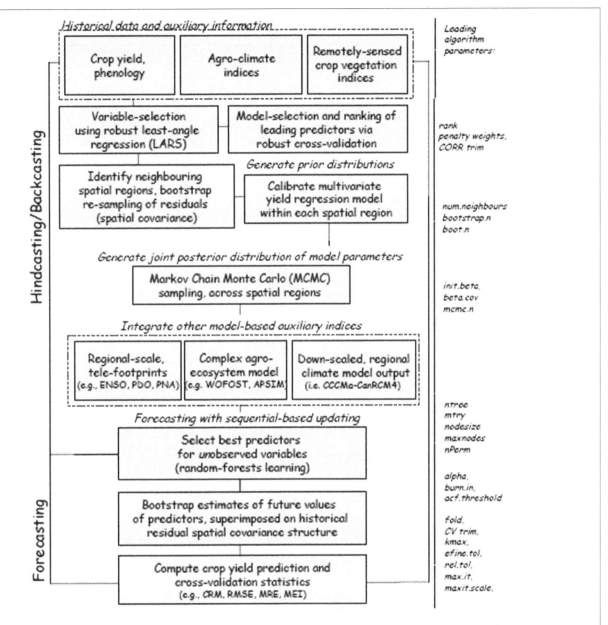

FIGURE 2 | Major components of the sequential, spatial-based Bayesian forecasting model of regional-scale crop yield. Model parameters that are tuned and fixed within each model component are identified to the right. The procedure consists of calibration and a prediction steps. Hindcasting/ backcasting via leave-one-out cross-validation (LOOCV) is performed using the input data as training data. Sequential-based forecasting is accomplished via the random-forests learning algorithm. This methodology is able to integrate other possible auxiliary historical indices, such as regionally-downscaled indices derived from El-Nino Southern Oscillation (ENSO) index, Pacific Decadal (PDO), and the Pacific North American (PNA) climate teleconnection index. Forecast indices generated from complex agroecosystem models, such as WOFOST (WOrld FOod STudies, Wageningen UR) agricultural production model and APSIM, the Agricultural Production Systems sIMulator. Forecast indices derived from downscaled future CCCma-CanRCM4 denotes CanRCM4 regional climate model (RCM) scenario output produced by the Canadian Centre for Climate Modeling and Analysis (CCCma) for the North American region with a horizontal grid resolution of approx. 25 km and available at: http://www.cccma.ec.gc.ca/data/canrcm/CanRCM4/ could, in the future, also be integrated for generating regional yield forecasts. Adapted from Kouadio and Newlands (2014), with permission.

generates future states of an ecological system conditioned upon scenarios, and prognosis is a subjective judgment of future states of an ecological system. There are numerous methodological and practical advantages conferred by introducing auxiliary variables in an estimation and forecasting problem: (1) when computing the likelihood density of the observed data, $f(y|\theta)$ is analytically intractable, (2) when the inclusion of the auxiliary variables simplifies or improves the computation using Markov-Chain Monte-Carlo (hereafter, MCMC), whereby, $f(y|x, \theta)$ is a more complete data likelihood, and can be computed faster than the observed data likelihood, $f(y|\theta)$, (3) when a statistical problem involves latent or hidden variables, (4) when underlying states

Table 2 | Leading parameters (name, R library association, description, setting) of the integrated forecasting model.

Parameter	Description	Setting (default)
rank	Maximum num. of predictors (Equation 1)	5 (5)
penalty.weights	Penalty weights (assumed equal) for predictors (Equation 1)	Real vector
CORR trim (calibration)	Threshold to remove highly correlated variables	0.95 (0.95)
num.neighbors	Num. neighbors, building empirical prior distributions	3
bootstrap.n	Num. bootstrap re-sampling of neighboring CAR residuals	100
boot.n	Num. of bootstrap samples, building empirical prior distributions	100 (500)
init.beta	Starting vector, MCMC chain	real vector
beta.cov	Covariance vector, MCMC proposal distribution (MVN)	real vector
mcmc.n, mc.size	Length (size) of MCMC chain	1000 (5000)
ntree	No. of regression trees to grow	500 (500)
mtry	No. of variables randomly sampled as candidates at each split	3 (3)
nodesize	Minimum size of terminal nodes (increases growth of smaller trees)	5 (5)
maxnodes	Maximum no. of terminal nodes	4 (max possible)
nPerm	No. of permutations of data per tree	1 (1)
α	$(1 - \alpha)$ % confidence level (i.e., 0.10 is 90 %)	0.10 (0.10)
burn.in	burn-in size for MCMC	0 (0)
acf.threshold	autocorrelation threshold, batch means size (MCMC SE errors)	0.05 (0.10)
fold	No. cross-validation folds	5 (5)
CV trim (forecasting)	Threshold upper percentile for outlier detection	0 (0.04)
kmax	Maximal number of refinement steps, S-estimation	5000 (5000)
refine.tol	Relative convergence tolerance (cross-validation), S-estimation	0.00001 (0.00001)
max.it	Maximum no. IRWLS iterations, M-estimation,	5000 (5000)
rel.tol	Relative convergence tolerance (cross-validation), M-estimation	0.00001 (0.00001)
maxit.scale	Maximum number of adaptive scale iterations.	500 (500)

No. denotes number.

MVN denotes multivariate normal distribution.

IRWLS denotes iteratively re-weighted least-squares.

MCMC denotes Markov-Chain Monte-Carlo.

are controlled by dynamic processes occurring at multiple spatial and temporal scales (i.e., hierarchical problem), (5) when spatial interpolation and/or upscaling of point- or site-based data is impractical due to data sparsity and/or high spatial heterogeneity, and fine- and coarser-grained information must instead be integrated, (6) a response has a strong spatial dependence and estimation and forecasting is needed across a large spatial region, (7) when auxiliary variables track a response variable faster (i.e., near-real time), reducing delay, and (8) to improve a sequential forecast, by conditioning it at a given fixed or variable time-step.

2.3.2. Bayesian inference

Bayesian inference uses Bayes' rule to update the probability estimate for a hypothesis as additional evidence is learned, given by;

$$p(H|E) \propto p(E|H)p(H), \tag{1}$$

where H is any hypothesis whose probability is influenced by data or observational evidence E, $p(H)$ is the *prior* probability (probablity of H before E is observed), $p(H|E)$ is the *posterior* probability of H after E is observed. $p(E|H)$ is the probablity of observing E given H, also called the *likelihood*, where $p(E)$ is the marginal likelihood or evidence. Bayes' rule therefore states that the "posterior probability is proportional to the prior times likelihood," or similarly, "posterior is prior times likelihood over evidence." The posterior is the result of updating our prior information with data (i.e., learning). Since the denominator does not depend on any H, it acts as a proportionality or normalization constant, and is obtained by averaging the "prior times likelihood" (i.e., numerator) over all possible hypotheses, H, such that,

$$p(H|E) = \frac{p(E|H)p(H)}{p(E)}; \quad p(E) = \int_H p(E|H)p(H)dH, \tag{2}$$

If we observe data, y, from a sampling density (i.e., distribution of the data), $p(y|\theta)$, where θ is a vector of parameters, and assign θ a prior $p(\theta)$, then, following from Equations (1, 2), the posterior density of θ is then;

$$p(\theta|y) = \frac{p(y|\theta)p(\theta)}{p(y)} \propto p(y|\theta)p(\theta);$$

$$p(y) = \int_\theta p(y|\theta)p(\theta)d\theta. \tag{3}$$

We introduce underlying state variables, q, (referred to as a Bayesian data augmentation, auxiliary or latent variable approach). The resultant joint posterior probability, taking into account both the θ parameters and the state variables, is given by;

$$p(\theta, q|y) \propto p(y|q, \theta)p(q|\theta)p(\theta) \tag{4}$$

The posterior of model parameters is then;

$$p(\theta|y) = \int_q p(\theta, q|y)dq \tag{5}$$

Because the necessary integration to obtain the posterior distribution is analytically intractable, MCMC using the Metropolis-Hastings algorithm is typically used to obtain a sample from the joint posterior distribution of q and θ (Albert, 2009).

We assume that the parameters, θ, have a multivariate normal distribution (MVN) as the conjugate prior density. The Monte-Carlo standard errors and confidence intervals are computed using the method of batch-means (i.e., that generally requires posterior samples of at least 1000) (Jones et al., 2006).

For sequential-based forecasting, we adopt a state-space modeling framework, and express the probability densities above, as functions of a set of observed, $p(y_t|x_t, \theta)$, and underlying process, $p(x_t|x_{t-1}, \theta)$ discrete states of a dynamic system, for a sequence of times $t = (1, \ldots, T)$, that evolve from an initial state, $p(x_1, \theta)$.

If we denote and discretize any auxiliary explanatory variables, denoted as z_t, then, Equation (4), can be re-expressed (Tanner and Wong, 1987; King, 2012), as;

$$p(\theta, z|y, x) \propto p(y|x, \theta, z)p(x|\theta, z)p(z|\theta)p(\theta) \qquad (6)$$

where,

$$p(y|x, \theta, z) = \prod_{t=1}^{T} p(y_t|x_t, \theta, z_t); \quad p(x|\theta, z) = \prod_{t=1}^{T} p(x_t|z_t);$$

$$p(z|\theta) = p(z_1) \prod_{t=2}^{T} p(z_t|z_{t-1}). \qquad (7)$$

The x_t denote environmental states (i.e., a vector $\mathbf{X}$ of agro-climate variables or indices), and z_t denote states of additional auxiliary explanatory variables (i.e., a vector $\mathbf{Z}$ of remotely-sensed variables or indices). At each iteration of an MCMC algorithm, the model parameters and auxiliary variables are updated. One can further identify different spatial regions, s, such that $x_t \leftarrow x_{s,t}$ and $z_t \leftarrow z_{s,t}$.

The posterior predictive or forecast density is either the replication of y given the model (denoted y^{rep}) or the prediction of a new and unobserved y, (denoted y^{new}), given the model. This is the likelihood of the replicated or predicted data, averaged over the posterior distribution $p(\theta|y)$, given by;

$$p\left(y^{rep}|y\right) = \int_{\theta} p\left(y^{rep}|\theta\right) p(\theta|y)d\theta; \text{ or}$$

$$p\left(y^{new}|y\right) = \int_{\theta} p\left(y^{new}|\theta\right) p(\theta|y)d\theta \qquad (8)$$

We assume a linear model that, in matrix notation, for a general response vector, $\mathbf{Y}$, is given by;

$$\mathbf{Y} = \mathbf{D}\beta + \epsilon, \qquad (9)$$

under the distributional assumptions,

$$Y|\beta, \sigma^2, \mathbf{D} \sim N_n(\mu, \Sigma) \sim N_n(\mathbf{D}\beta, \sigma^2 \mathbf{I}) \qquad (10)$$

where $\mathbf{D}$ is the *design matrix*, β is a vector of unknown model parameters, and ϵ is a vector of independently and identically-distributed normal random errors with mean zero and variance σ^2. Our design matrix, $\mathbf{D}$, involves two sets of explanatory variables, such that $\mathbf{D}=(\mathbf{X}|\mathbf{Z})$, having columns $x_1, \ldots, x_{n_p}$ augmented with columns $z_1, \ldots, z_{n_q}$. N_n denotes the MVN of dimension, n, with mean μ, variance-covariance matrix Σ and $\mathbf{I}$ the identity matrix.

Crop yield within each CAR region was modeled as a multivariate regression equation with spatially-varying coefficients (Banarjee et al., 2004),

$$E(\mathbf{Y}|\beta, \mathbf{X}, \mathbf{Z}) = \hat{y}_{i,j} = (\gamma_{i,0} + \gamma_{i,1} \times j) + \alpha_i y_{i,j-1} + \sum_{l=1}^{n_p} \beta_{i,j}^{(l)} x_{i,j}^{(l)}$$

$$+ \sum_{l=n_p+1}^{n} \beta_{i,j}^{(l)} z_{i,j}^{(l)} \qquad (11)$$

where $\hat{y}_{i,j}$ denotes the estimated or expected value of $y_{i,j}$, the crop yield for year j (i.e., to distinguish calibration yearly time-step from the forecast monthly time-step, denoted by t), where $j = (2, \ldots, T)$, within a given CAR, i, where $i = (1, \ldots, C)$. $x_{i,j}^{(l)}$ and $z_{i,j}^{(l)}$ denote the l predictor variables for i at time j. The total number of predictor (i.e., n_p agro-climate and n_q auxiliary remotely-sensed) is $n = n_p + n_q$. The coefficients, $\beta_{i,j}^{(l)}$ are spatially and temporally-varying. Uncertainty, $\varepsilon_{s,i}$ is independent and normally distributed (random error) with mean zero and variance σ_i^2. The regression coefficients, $\gamma_{i,0}$ (yield intercept), and $\gamma_{i,1}$ (technology trend coefficient) are used to de-trend the yield data and α is a lag-1 autoregressive term. The technology trend accounts for historical increases in yield from genetics, management practices to improve soil fertility, water conservation and minimize soil erosion and nutrient leaching. The technology trend in yield was assumed to be linear. The inter-annual autocorrelation in yield was assumed to vary across CARs.

For the ith region, the design matrix, for fixed i, is given by $\mathbf{D}_i = \left(X_i^1, X_i^2, \ldots, X_i^{n_p}|Z_i^{n_p+1}, Z_i^{n_p+2}, \ldots, Z_i^n\right)$, which is associated with the full/complete model parameter vector, $\Theta_i = \left(\gamma_0, \gamma_1, \alpha, \beta_i, \sigma_i^2\right)$. In matrix notation,

$$\mathbf{D}_i = \begin{pmatrix} x_{i,2}^{(1)} & x_{i,2}^{(2)} & \cdots & x_{i,2}^{(n_p)} & z_{i,2}^{(n_p+1)} & z_{i,2}^{(n_p+2)} & \cdots & z_{i,2}^{(n)} \\ x_{i,3}^{(1)} & x_{i,3}^{(2)} & \cdots & x_{i,3}^{(n_p)} & z_{i,3}^{(n_p+1)} & z_{i,3}^{(n_p+2)} & \cdots & z_{i,3}^{(n)} \\ \vdots & & & \vdots & \vdots & \vdots & & \vdots \\ x_{i,T}^{(1)} & x_{i,T}^{(2)} & \cdots & x_{i,T}^{(n_p)} & z_{i,T}^{(n_p+1)} & z_{i,T}^{(n_p+2)} & \cdots & z_{i,T}^{(n)} \end{pmatrix}$$

$$(12)$$

2.3.3. Robust regression

Robust regression is a specialized technique used in the model to ensure it is less sensitive to outliers and may be applied to situations of unequal variance (Khan et al., 2007, 2010). This technique was applied to account for heteroscedasticity and outliers in the historical data during model training and calibration. Heteroscedasticity occurs when the variance of an explanatory or predictor variable is dependent on its value. Robust regression also provides a flexible and general technique for modeling based on residuals, because it is less influenced by the presence of outliers. Variable-selection is then employed for prediction in the case where there are no significant outliers. Robust regression is a compromise between excluding outliers entirely from the analysis and treating all the data points equally, as it is done in ordinary least squares (OLS) regression. The idea of robust regression is to weigh the observations differently based on how well-behaved the observations are. Several approaches to robust estimation have been proposed, including R-estimators and L-estimators. However, M-estimators are more widely used because of their generality, high breakdown point, and efficiency. M-estimators are a generalization of maximum likelihood estimators (MLEs). Standard MM-type regression estimates use a bi-square re-descending score function and returns a highly robust and efficient estimator (with 50% breakdown point and 95% asymptotic efficiency for normal errors). The tuning parameters of lmrob, the R package we used to implement robust regression, comprises an MM-type robust linear regression estimator that consists of an initial S-estimate, followed by an M-estimate via regression that enables one to specify a breakdown point and asymptotic efficiency consistent with normal distributional assumptions (Koller and Stahel, 2011).

2.3.4. Bootstrap least-angle regression

Bootstrap robust least angle regression (B-RLARS) was applied in selecting the leading $m \leq n_q$ potential predictors from $z^1, z^2, \ldots, z^{n_q}$ after adjusting for the effects of $x^1, x^2 \ldots, x^{n_p}$. Let $z^{1,*}, z^{2,*}, \ldots, z^{m,*}$ denote the leading n_q ranked variables. A final, reduced model, with $m^* \leq n_q$ predictors, is obtained with robust leave-one-out cross validation (LOOCV), by selecting auxiliary predictors from $z^{1,*}, z^{2,*}, \ldots, z^{n_q,*}$, after adjusting for the influence of $x^1, x^2 \ldots, x^{n_p}$ (Khan et al., 2007, 2010).

Let the complete model design matrix, as $\mathbf{D}_{C,T,(n_p+m^*)}$ following the selection of best-fit predictors. Let Ψ be the vector of model parameters corresponding to the joint distribution of $\mathbf{D}_{C,T,(n_p+m^*)}$ (recall the model parameter vector is $\Theta_i = (\gamma_0, \gamma_1, \alpha, \beta_i, \sigma_i^2)$). The conditional likelihood function of $(y_2, y_3, \ldots, y_n)$ given y_1 is then;

$$f\left(y_2, \ldots, y_n | y_1, \Theta, \mathbf{D}_{C,T,(n_p+m^*)}\right)$$

$$= f\left(y_2|y_1\right) f\left(y_3|y_1, y_2\right) \cdots f\left(y_n|y_1, y_2, \ldots, y_{n-1}\right) \quad (13)$$

$$= N\left(\gamma_0 + 2\gamma_1 + \sum_{l=1}^{n_p}\beta_2^{(l)}x_2^{(l)} + \sum_{l=n_p+1}^{n}\beta_2^{(l)}z_2^{(l)} + \alpha y_1, \sigma^2\right)$$

$$\times N\left(\gamma_0 + 3\gamma_1 + \sum_{l=1}^{n_p}\beta_3^{(l)}x_3^{(l)} + \sum_{l=n_p+1}^{n}\beta_3^{(l)}z_3^{(l)} + \alpha y_2, \sigma^2\right)$$

$$\times \cdots \times N\left(\gamma_0 + n\gamma_1 + \sum_{l=1}^{n_p}\beta_n^{(l)}x_n^{(l)}\right.$$

$$\left. + \sum_{l=n_p+1}^{n}\beta_n^{(l)}z_n^{(l)} + \alpha y_{n-1}, \sigma^2\right),$$

where the CAR subscript, i, was omitted. Given that $(y_2, \ldots, y_n)$ is independent of Ψ given y_1, Θ, and $\mathbf{D}_{C,T,(n_p+m^*)}$, it follows that;

$$p\left(y_2, \ldots, y_n, \mathbf{D}_{C,T,(n_p+m^*)} | y_1, \Theta, \Psi\right)$$

$$= p\left(y_2, \ldots, y_n | y_1, \Theta, \mathbf{D}_{C,T,(n_p+m^*)}\right)$$

$$p\left(\mathbf{D}_{C,T,(n_p+m^*)} | \Psi\right).$$

Under a fully Bayesian approach, prior distributions for both Θ and Ψ must be specified. We assumed a separation of variables, such that, $p(\Theta, \Psi) = p(\Theta) p(\Psi)$, and constructed an empirical prior distribution for Θ by residual bootstrapping the data from neighboring CARs. Information from neighboring CARs was also considered, employing an approach previously outlined and cross-validated by Bornn and Zidek (2012). This boostrapping procedure was as follows: the neighboring CARs of each given CAR were identified and the calibrated yield regression equation was fit to the given CAR and other surrounding, neighboring CARs simultaneously. The fitted models from the set of CAR neighbors were then cross-validated using the data for the given CAR. The top k ranked CARs, which could comprise a single or multiple neighbors to a given CAR, were then finally selected based on obtaining the minimal cross-validation error. This method was found to generate a more meaningful prior distribution of the model parameters by providing additional spatial covariance support (i.e., considers the residual spatial covariance between CARs). The joint posterior density function can be expressed as;

$$p(y^{new}|y) = \int_{\theta} p(y^{new}|\beta, \sigma^2) \, p(\beta, \sigma^2|y) \, d\beta d\sigma^2 \quad (14)$$

The model predicted future values of the model variables (i.e., forecasted) within a given CAR using only the information from selected model predictors. This assumed a Gaussian prior distribution for each predictor variable, as a conjugate prior for the joint multivariate Gaussian posterior likelihood distribution. It is well-known that choosing a conjugate prior ensures that the resulting posterior distribution is of the same distribution family as the prior (with a closed-form solution), and helps to avoid over-fitting on small training samples. However, the selected predictor variables were determined not to be good predictors of one another and were uncorrelated (results not shown here). For this reason, the forecast method instead uses the entire set of available variables when selecting the best subset of predictors that jointly estimates the unobserved values of future variables. This was first accomplished for each CAR using the multi-variate adaptive regression splines (MARS) non-parametric algorithm, but was later substituted to use the Random Forests algorithm (Chipanshi et al., 2012; Newlands and Zamar, 2012; Kouadio and Newlands, 2014). The random forests algorithm proved to be more computationally efficient and accurate because it created multiple bootstrapped regression trees without pruning and averages the outputs, very effective in reducing variance and error in high dimensional data sets (Breiman, 2001; Chen et al., 2012).

Also, by incorporating non-parametric Bayesian priors, we realized that our methodology would be more flexible and generally applicable in modeling with a wide set of variables and indices. It would also not have to assume a conjugate prior, which in many realistic or real-world cases is inappropriate. Model complexity is automatically determined by the model selection method we used because it specifies a maximum number of predictors (R-LARS followed by robust cross-validation).

2.4. SENSITIVITY ANALYSIS AND VALIDATION METHODOLOGY

Dynamic sensitivity analysis was conducted by simulating the model for different input settings of the leading parameters. We also experimented by including and withholding precipitation (P) as an input variable (i.e., a variable with high stochasticity) to evaluate the change this had on forecast error. Ranges for these simulations were set by considering both default values provided in statistical literature linked with the R library code and values prescribed from the application of various algorithms to real-world data, as in the case of the random forest algorithm parameters.

The goal of cross-validation was to estimate the expected level of fit of a model to a data set that was independent of the data that were used to train the model. We cross-validated our model forecasts against historical data (1987–2011) by back-casting/hindcasting involving statistical bootstrapping and then leave-one-out cross-validation (LOOCV). The LOOCV selects a single (i.e., year) observation from the original full record as a sample of validation data, and the remaining observations as the training data. This was repeated such that every observation that exists in the sample (i.e., year) was used once as validation data. We performed an additional test of this cross-validation procedure by removing more than just 1 year of data at a time. This considered samples (K) of 1–5 years, following a K-fold cross-validation procedure. This additional test evaluated how robust the forecasts were to removal of training data/loss of input information termed forecast degeneracy. The model was also forecasted for the 2012 growing season and its output was compared to available statistical survey data across a range of spatial scales. In 2012, a total of 64% of land across the Great Plains/Midwestern United States is under abnormally dry to exceptional drought conditions (United States Department of Agriculture (USDA) Drought Monitor). NOAA has reported that the January–June period in 2012 was the warmest first half of any year on record for the contiguous United States, and the warmest the area has ever experienced since the dawn of record-keeping in 1895 (United States National Climatic Data Center, NCDC). In 2012, nearly 50% of corn and 37% of the soybeans grown in the United States were rated poor to very poor, with three-quarters of US cattle acreage in drought-affected areas. As this year exhibited such extreme conditions, it provided data with high environmental variability and a strong test of the accuracy and robustness of the model's forecasting methodology.

3. RESULTS

3.1. OBSERVED VARIABILITY

The observed variability of the agroclimate and remotely-sensed vegetation variables was considerable when pooled across CARs and years within the study region (**Figure 3**). In these box-plots (and throughout this paper), the solid line indicates the median error, with ends of the box indicating the upper and lower quartiles as a measure of data spread spanning 50% of the dataset and eliminating the influence of outliers. The whiskers are the two lines outside of the box that extend to the highest and lowest data values. An increasing trend in growing season mean-monthly precipitation (mm) (see inset a) was evident, explained by frequent and larger convective-driven storm events deliver rainfall across the region, accompanying increased summer temperature. Over-winter precipitation and accompanying spring thaw in May raised crop available water and reduces the soil capacity to store available water (inset b) until later in the season, when rainfall decreases (inset a) and significant soil water evaporative losses typically occur. This pattern of variability was consistent with the observed increase of crop water deficit index (WDI) (inset c). Lower/upper quartiles for GDD ranged between 180 and 450 GDD over the growing season, with 180 on average accumulated by end of May. Over the years in the climate station record, many of the CARs experienced extreme rainfall (as outliers in the box-plot of distribution quartiles). CARs also experience extreme temperatures in the middle of the growing season, signified by numerous outlier values in June from the plot of the GDD index (degree days above 5°C) (inset d). The variability of AVHRR-based NDVI (**Figure 3**, inset E) identified many outlier CARs that are situated far outside of the main distribution quartiles, indicating a general tendency for over-estimation of NDVI early in the growing season, and under-estimation mid-season, likely due to surface water backscatter.

3.2. MODEL SENSITIVITY

The sensitivity of forecasts was tested by simulating the model under changes in the: (1) number of predictors, (2) threshold for removing correlation variables (CORR trim), (3) threshold for the upper percentile for identifying outliers in cross-validation (CV trim), and (4) chain-size specified in the MCMC-simulation sampling from the joint posterior distribution (Equation 14) (**Figure 4**). These results tracks the change in CAR-based root-mean-square error (RMSE) uncertainty in forecasted yield versus each of the leading design parameters, whereby CORRtrim must be greater than 0.93 and a CV trim must be a value between 0.01 and 0.04 to minimize RMSE. These parameters were regionally-calibrated or tuned and set to 0.95 and 0.04, respectively. MCMC chain-size did not have a large effect on the median, or 10% and 90% percentiles or forecast RMSE uncertainty. The sensitivity to MCMC chain-size did not appear to change through the growing season, as the curve slopes were mostly flat for July, August and September. Therefore, we set the chain-size to a reasonable, mid-point value of 1000 to reduce run-time without any substantial loss of precision. A minimum of four predictors minimized RMSE. The inclusion of additional predictors, such as standard deviation of each of the predictor variables, did not make a significant difference (i.e., neither raised nor lowered forecast uncertainty). An additional sensitivity test of model on the effect of including precipitation (P) on forecast uncertainty was conducted (**Figure 5**). Selected regression results are summarized for two CARs that generated the best and worst regression fits

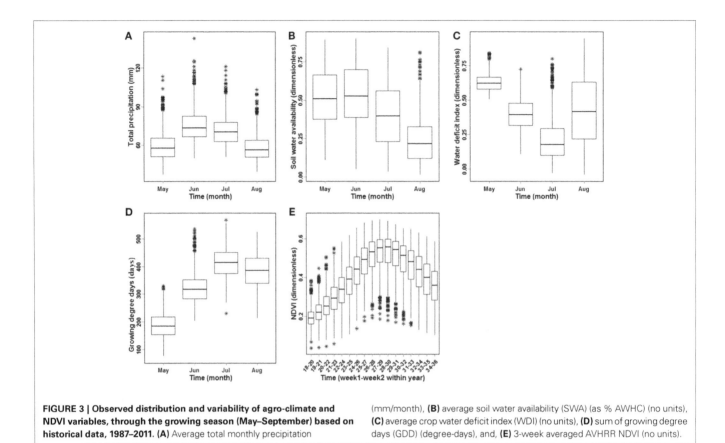

FIGURE 3 | Observed distribution and variability of agro-climate and NDVI variables, through the growing season (May–September) based on historical data, 1987–2011. (A) Average total monthly precipitation (mm/month), **(B)** average soil water availability (SWA) (as % AWHC) (no units), **(C)** average crop water deficit index (WDI) (no units), **(D)** sum of growing degree days (GDD) (degree-days), and, **(E)** 3-week averaged AVHRR NDVI (no units).

(**Figure 6**). Scatterplots of observed versus predicted yield (**A,B**), residual versus predicted yield (**C,D**), and the relative importance or proportion of observed variance explained by each of the best-fit selected predictors (**E,F**) were generated. Model error reduces when considering from one to up to three predictors. Model accuracy increased when NDVI was included with the agroclimate variables (i.e., WDI, GDD, SWA) and year. Based on these findings, the default maximum number of model predictors was set to be five to regionally-calibrating yield across all the CARs in the study region. This sensitivity analysis identified whether all or a subset of these variables are necessary to obtain a best-fit in each CAR. For instance, with both agroclimate and NDVI indices as predictors, WDI, year and NDVI were selected in CAR 4741. When we constrained model input to NDVI only, RMSE for this region became very large, and is why CAR 4741 appears as an outlier for the "NDVI," and is no longer an outlier in the "agroclimate and NDVI indices" cross-validation run (see **Figure 5**, right inset). This was also the case for CAR 4607 and 4609, where NDVI and year were the only predictors selected by the variable-selection algorithm of the forecast model. Further sensitivity-checks were made by re-running the forecast model and examining the change of RMSE of the output for different values of other design parameters listed in **Table 2**. In many instances, no appreciable effect was measured to support or justify changing the default settings obtained from the scientific literature, as was the case for the random forest tree algorithm. Although planting date did vary across our region (mostly by 15 days min to max), it was not selected as a leading covariate of yield

and when forced into the model also did not have a significant effect on RMSE uncertainty, so we fixed this variable.

3.3. MODEL VALIDATION

The spatial pattern of model forecast uncertainty (RMSE) across all the CARs was generated for different combinations of the input predictors (**Figure 7**). Change in RMSE for many of the CARs was dependent on how the model was constrained in selecting only the prescribed maximum number and type of predictor variable. Combining both NDVI and agroclimate predictors enabled minimal uncertainty in most CARs (i.e., within 150–250 kg/ha). However, there were CARs in northern Alberta (4860), central Saskatchewan (4730), and south-eastern Manitoba (4609 and 4610) for which uncertainty could not be appreciably reduced further.

Mean percent departures between historically observed and 2012 forecasted yield varied between under-estimation of 1–4% at the mid-point (i.e., July) of the May-October crop growing season, to over-estimation of less than 1% at end of season (i.e., September) (**Table 3**). This compares to a slightly higher mean percent departure based on independent, field crop-reporting yield data in 2012, of under-estimation of 3.6% mid-season, to less than 1% of over-estimation end-season (**Table 4**). Yield was forecasted for 2012 to be 3386, 2471, 2559 kg/ha in Alberta, Saskatchewan and Manitoba, respectively. Across the Canadian Prairies, forecasted yield was 2803 kg/ha, varying between 2513 and 3092 at the 90% confidence level (**Figure 8**). Further inspection of yield variability within the two CARs with the highest

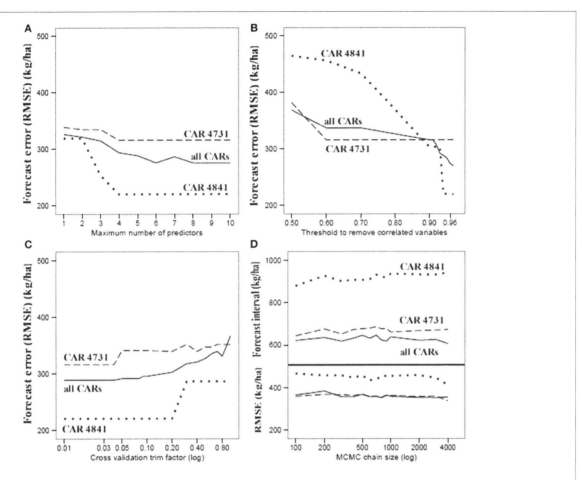

FIGURE 4 | Sensitivity of forecasted error in yield (RMSE, kg/ha) (y-axes) to leading model design parameters (x-axes). The solid line, dotted, and dashed lines track the median yield across: all CARs, CAR 4841 (low sensitivity case), and CAR 4731 (high sensitivity case), respectively, for: **(A)** maximum number of predictors (trim factor to 0.04 and number of predictors varies from 1 to 10), **(B)** CORR trim factor (i.e., threshold for removing correlated variables), with the maximum number of predictors set to 5, **(C)** CV trim factor (i.e., a parameter of the lmrob() R library algorithm of the robustbase package that is a percentage for the upper percentile of observations to drop based on their cross-validated errors). The correlation trim factor was set to 0.04 on this sensitivity run, and number of predictors was set to 5 and the x-axis is in log scale, and **(D)** shows two panels divided by the solid horizontal line—the upper panel shows the range defined by the difference between the 10% and 90% percentiles of forecast yield in relation to changes in the MCMC chain size (mcmc.n). This considered all CARs and years of historical input data (i.e., 1987–2011). The three lines are the results for July (dotted), August (dashed), and September (solid). The lower panel tracks RMSE in forecasted yield corresponding to the same monthly forecasts in the upper panel.

forecast error (i.e., CV > 18%) for CARs (4610 and 4860) revealed that they exhibit unique patterns that departed from our model assumptions and variability observed within other CARs (**Figure 9**). Sample autocorrelation functions (ACFs) of historical crop yield that track the strength of the inter-annual dependence in yield between successive years, revealed an extended (i.e., 3-year duration) correlation with a decaying or tapering pattern for CAR 4610, indicating an autoregressive random process of order 3, AR(3), with a positive coefficient. For CAR 4860, an alternating and tampering pattern was identified, indicating an autoregressive process of order 1, AR(1), with a negative slope coefficient.

The RMSE uncertainty associated with the median forecast within CARs was 330 kg/ha, varying between 280 and 360 kg/ha at the 25% and 75% percentile, respectively, from LOOCV analysis of forecasts across all historical years. For the 2012 growing season, the median was less at 270 kg/ha, ranging between 230

and 300 kg/ha. When cross-validating the model forecasts, the number of years that were left out was varied from 1 to up to 5 years at a time (**Figure 10**). While we utilized a LOOCV procedure with 1 year left out at a time. The additional cross-validation results from the K-fold cross-validation led support for this specification, indicating that specifying any higher number of years left out at a time, only reduced the model's forecast accuracy (RMSE error) by less than 5%.

4. DISCUSSION
4.1. MODEL PERFORMANCE
Model forecasts were sensitive to both the choice of design parameters and the set of input explanatory (predictor) variables. Agroclimate variables contributed significantly to improving overall multivariate regression fits across most CARs. The CARs, where the model performed the worst, only had NDVI and year as explanatory variables. However, NDVI did explain

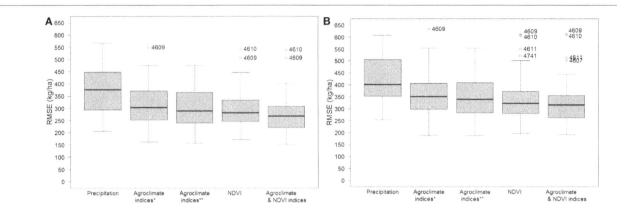

FIGURE 5 | Sensitivity of forecast error (i.e., Root-mean-squared error variance, RMSE) for different combinations of selected agroclimate variables and NDVI. (A) Forecasted yield in 2012, and **(B)** cross-validated (LOOCV) (i.e., backcasted) forecast yield. The predictors sets considered in each run were: Precipitation (P); agroclimate variables of crop water deficit index (WDI), growing degree-days (GDD), soil water availability (SWA), where * denotes *P is not* included, and ** denotes *P is* included; NDVI; WDI, GDD, SWA, and NDVI. Note: seeding date was assumed fixed, and year, *t*, was included as an additional input variable in all the sensitivity runs. The ID's of outlier CARs are indicated.

a significant portion of the observed yield variance in time and space (**Figure 6**). Because the sensitivity of forecasts to changes in the leading model design parameters was successfully minimized. This suggests that selecting the best predictors of yield within each CAR region, thresholding the detection of cross-correlation between predictors (so as to avoid multi-collinearity), combining information between neighboring CAR regions, and using robust regression to detect outliers, significantly improved robustness of the model's forecasts across the Canadian Prairies (i.e., a large study region).

A simple linear regression model to predict Canadian spring wheat crop yield and applied at the provincial scale achieved 53–77% accuracy relying solely on agro-climate indices (i.e., growing degree day (GDD), precipitation (P), actual/potential evapotranspiration (AET, PET) and crop water deficit index WDI) (Qian et al., 2009). At the provincial scale, percent departure values for 2012, and LOOCV results across the entire region (all CARs) supports that our integrated model forecasts attains a higher overall or forecast skill (CV < 10% or accuracy > 90%), even accuracy does vary between CARs. With further adjustment to our model to allow for alternating autocorrelation with negative slope coefficients, we infer the two CARs with the highest uncertainty would also be greatly improved. Seeded area, irrigation and/or pests could also explain why these two CARs depart from all others, warranting further investigation. Also, without using agroclimate data, and using only satellite remote-sensing data (i.e., NDVI) was shown to predict hstorical wheat yield ranging between 47 and 80% accuracy at the CAR scale and across the Canadian Prairies (Mkhabela et al., 2011). Our findings show that while NDVI is a strong predictor of yield, combining agroclimate and NDVI indices substantially increases the accuracy and confidence in regional forecasts. One can also better discriminate between key ecological variables (i.e., temperature and available soil water) driving yield variability in time and space, which would otherwise, not be possible when only considering a single index.

Globally, wheat accounts for 20% of calories consumed. It is one of the most important crops, alongside rice and maize (corn). Our integrated model forecasted spring wheat yield for 2012 of 2803 kg/ha (Canadian Prairies) and associated uncertainty of 2513–3092 at the 90% confidence level agrees to close to 0.1% of end-of-season field survey yield, and 4% at the start of the growing season. This forecasted yield for 2012 compares reasonably well to independent crop insurance data for 1965–2007 on spring wheat yield across the Prairies accounting for a progressive increase of 0.8–1%/year technology-trend in yield during the period of 2007–2012. Crop insurance data indicated highest and lowest average yields of 4589 and 80.9 kg/ha in 2005 and 1988, respectively, and an overall historical average of 1991.2 kg/ha (Robertson et al., 2013). Total factor productivity (TFP) growth rate in crop production (1980–2004) for the Canadian Prairies was estimated to be 1.77%, where technology change contributes close to 83%, in addition to the smaller contributions from scale effects and changes in the degree of technical efficiency (e.g., improved machinery or crop genetics). TFP has, historically, increased the most in Manitoba, explained in part by reduced use of summer fallow, but historically has lagged within Alberta (Stewart et al., 2009). Also, Canadian Plant Breeder Rights (PBR) qualifying wheat cultivars and improved soil quality have increased wheat yield by 37.2 (1%) to 54.5 kg/ha (2%) within the Canadian Prairies (Carew et al., 2009). Our forecast model could also better account for technology-change, rather than assuming a fixed, linear rate of productivity growth within each CAR based on historical data and trends. Instead, a more detailed equation involving percent of wheat seeded area, spatial cultivar diversity index, average cultivar age and annual cropland planted could be introduced (Carew et al., 2009) in addition to the year time trend already considered (that captures advances in non-genetic technology). However, this could make our regional-scale model too complex. The model generated robust forecasts of spring wheat yield for this extreme year 2012 based on historical data. It was also able to pinpoint two CARs in southern Manitoba

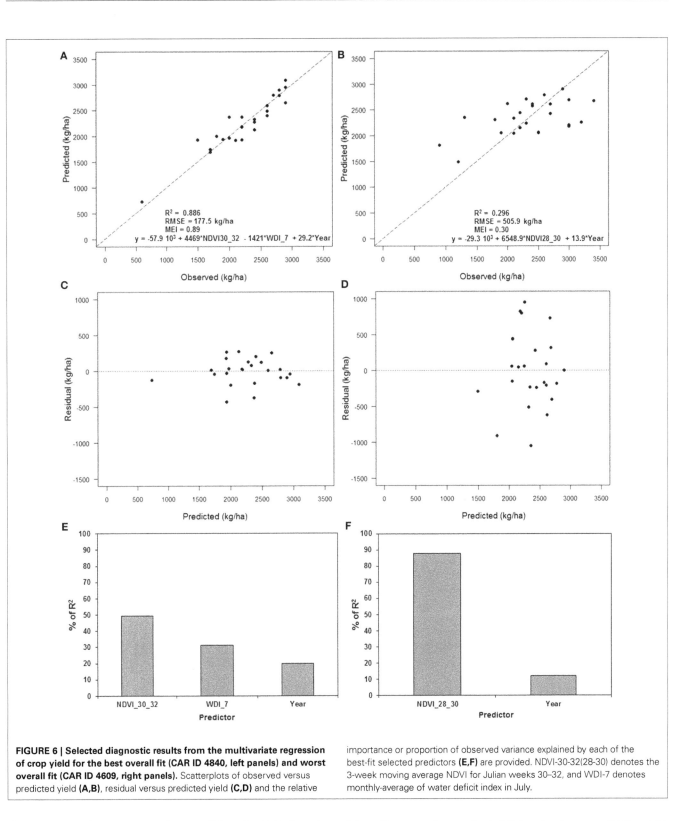

FIGURE 6 | Selected diagnostic results from the multivariate regression of crop yield for the best overall fit (CAR ID 4840, left panels) and worst overall fit (CAR ID 4609, right panels). Scatterplots of observed versus predicted yield **(A,B)**, residual versus predicted yield **(C,D)** and the relative importance or proportion of observed variance explained by each of the best-fit selected predictors **(E,F)** are provided. NDVI-30-32(28-30) denotes the 3-week moving average NDVI for Julian weeks 30–32, and WDI-7 denotes monthly-average of water deficit index in July.

that experienced extreme growing conditions in 2012, but also identify that yield within this area exhibits extended correlation with a duration of up to 3 years—very likely attributed to the impact of "persistent" drought and flooding conditions. While TFP has increased most in Manitoba, monthly rainfall variability is high within these two CARs, varying between 17 and 20%.

More climate stations and rainfall data for these two CARs might therefore help to reduce forecast error in this region.

4.2. MODEL LIMITATIONS

Our model currently does not consider crop genetics and does not input cultivar-specific data on yield or their developmental

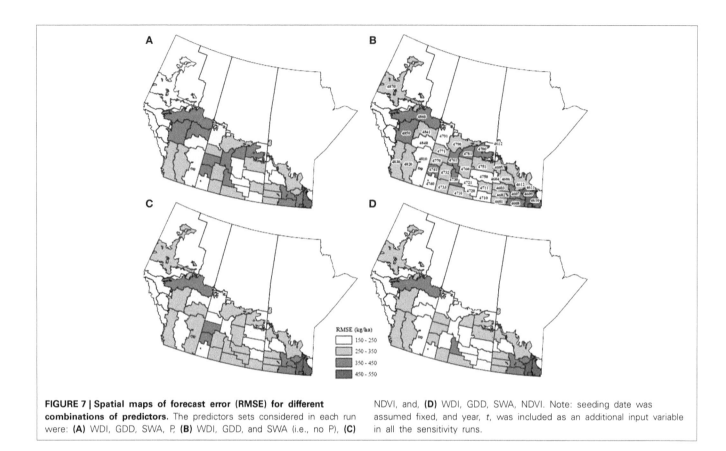

FIGURE 7 | Spatial maps of forecast error (RMSE) for different combinations of predictors. The predictors sets considered in each run were: **(A)** WDI, GDD, SWA, P, **(B)** WDI, GDD, and SWA (i.e., no P), **(C)** NDVI, and, **(D)** WDI, GDD, SWA, NDVI. Note: seeding date was assumed fixed, and year, t, was included as an additional input variable in all the sensitivity runs.

Table 3 | Forecasting of 2012 spring wheat crop yield (kg/ha) across the Canadian Prairie provinces (AB—Alberta, SK—Saskatoon, MB—Manitoba, All—Prairies).

		Obs.	A		B		C	
			Pred.	(L90, U90),%	Pred.	(L90, U90),%	Pred.	(L90, U90),%
JUL	AB	3200	3232	(2753, 3711), +1.00	3225	(2722, 3727), +0.780	3235	(2774, 3696), +1.09
	SK	2403	2410	(1941, 2879), +0.291	2555	(2220, 2891), +6.33	2347	(1917, 2778), −2.33
	MB	3031	2736	(2258, 3213), −9.73	2327	(1863, 2791), −23.2	2633	(2226, 3039), −13.1
	All	2772	2744	(2270, 3218), −1.01	2673	(2214, 3132), −3.57	2698	(2260, 3136), −2.67
AUG	AB	3200	3399	(3065, 3733), +6.22	3301	(2934, 3669), +3.16	3371	(3015, 3728), +5.34
	SK	2403	2403	(2049, 2757), 0.00	2417	(2057, 2778), +0.583	2450	(2162, 2738), +1.96
	MB	3031	2578	(2214, 2943), −15.0	2550	(2236, 2865), −16.0	2549	(2212, 2886), −16.0
	ALL	2772	2776	(2427, 3124), +0.144	2745	(2389, 3101), −0.974	2785	(2466, 3104), +0.469
SEP	AB	3200	3397	(3060, 3733), +6.16	3312	(2991, 3633), +3.50	3386	(3064, 3707), +5.81
	SK	2403	2394	(2056, 2732), −0.375	2492	(2209, 2775), +3.70	2471	(2216, 2726), +2.83
	MB	3031	2579	(2211, 2947), −14.9	2555	(2232, 2877), −15.7	2559	(2224, 2894), −15.6
	ALL	2772	2771	(2429, 3112), −0.036	2787	(2485, 3089), +0.541	2803	(2513, 3092), +1.12

The relative change in forecast error is benchmarked through the growing season (JUL—July, AUG—August, SEP—September), by varying the quality and quantity of the input data: (A) agro-climate only, (B) NDVI only, and (C) agro-climate and NDVI. Percent departure (%) between model (Pred.) and observed (Obs.) crop yield, and lower (L90) and upper (U90) 90% confidence intervals are provided. Departure values are reported to three significant figures.

and growth requirements, and does not select agroclimate or remotely-sensed predictors according to such cultivar-specific requirements. So, there is a further need to better understand how cultivar differences and genetics × management × environment interactions (i.e., G × M × E) could be used to temporally segment or partition input remote-sensing data for early, mid and late-season (or quick, medium and slow maturing) varieties. Mapping crop suitability spatially for different crop varieties

Table 4 | Comparison of model forecasts (using both agro-climate and NDVI indices) to Statistics Canada field crop-reporting survey data (CCAP), upscaled to the provincial scale.

	Survey data	Forecast model		
	SEP	JUL	AUG	SEP
AB	(3200; 7,429,900)	(3235; 7,596,497), +1.10	(3371; 7,824,064), +5.34	(3386; 7,852,666), +5.81
SK	(2400; 8,068,000)	(2347; 7,791,587), −2.21	(2450; 8,266,783), +2.08	(2471; 8,386,307), +2.96
MB	(3000; 2,946,000)	(2633; 2,572,017), −12.2	(2549; 2,685,150), −15.03	(2559; 2,682,153), −14.7
All	(2800; 18,539,900)	(2698; 18,047,675), −3.64	(2785; 18,855,267), −0.536	(2803; 19,000,476), +0.107

Model forecasts of spring-wheat crop yield (Y) (kg/ha) and production (P) (tonnes or metric tons, here denoted MT) in 2012 are provided for months within the growing season. Percentage departure (D %) between observed and forecast crop yield is computed. Entries are provided as (Y,P), (D %). Note, for wheat: 1 bu/ac = 67.25 kg/ha = 0.0673 t/ha (metric t). Departure values are reported to three significant figures.

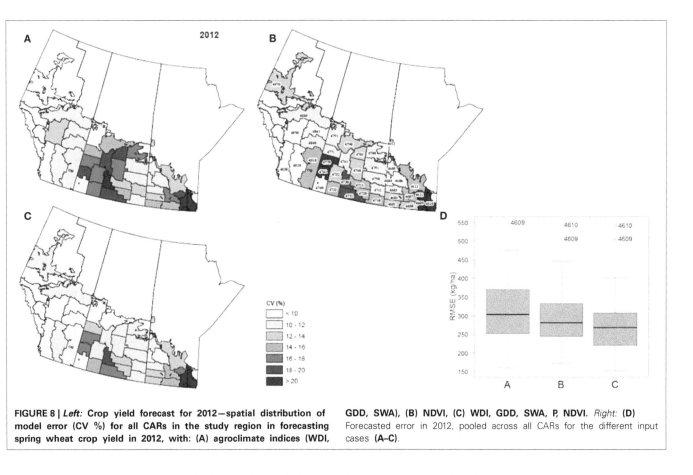

FIGURE 8 | Left: Crop yield forecast for 2012—spatial distribution of model error (CV %) for all CARs in the study region in forecasting spring wheat crop yield in 2012, with: **(A)** agroclimate indices (WDI, GDD, SWA), **(B)** NDVI, **(C)** WDI, GDD, SWA, P, NDVI. *Right:* **(D)** Forecasted error in 2012, pooled across all CARs for the different input cases **(A–C)**.

could also enable a determination of the best spatial resolution to integrate agroclimate, remote-sensing vegetation and phenology indices in generating regional-scale forecasts.

Recent advances in multivariate time series segmentation methods, such as that described in Graves and Pedrycz (2009) could then be applied to historical remote sensing data to help automate the identification of transition points between stages directly from spatially-referenced, longitudinal time-series of yield obtained from culitvar crop-breeding field trials. We aim to further explore how the performance of our forecast model could be improved by considering crop genetics by including an auxiliary index based on APSIM agroecosystem model

forecasts across a series of site-specific long-term cropping sites across the Canadian Prairies. The use of auxiliary indices derived from more complex models of agroecosystem soil, water, air and agronomic management interactions could offer further potential reductions in forecast uncertainty, and an ability to extend our seasonal forecasts and link them to longer-term yield forecasts.

Given that our model does not yet integrate crop phenology, cultivar differences, pest infestations, the interpretation and broader application of our current model to other regions and crops is limited. However, representing such differences at a regional-scale is still a big challenge due to lack of such data

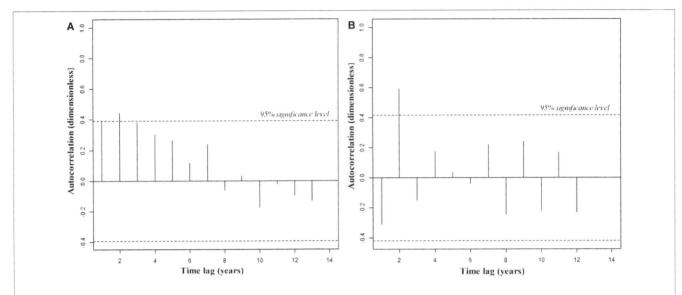

FIGURE 9 | Identified patterns of autocorrelation functions (ACFs) of historical crop yield for CARs with high RMSE forecast error. (A) CAR 4860 with AR(3) process with positive slope, and **(B)** CAR 4610 with AR(1) process with negative slope.

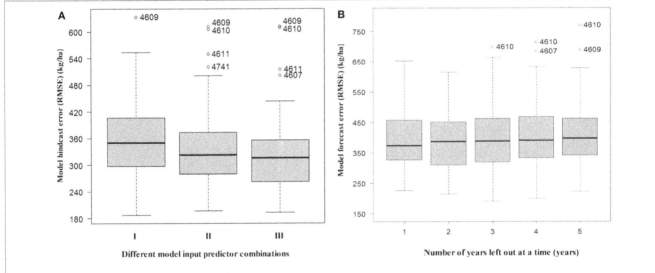

FIGURE 10 | (A) Variation in model hindcast/backcast error from leave-one-out cross-validation (LOOCV) (RMSE in kg/ha) for three different combinations of input predictors: (I) Agro-climate indices (WDI, GDD, SWA), (II) Remote-sensing index (NDVI), and (III) Agroclimate and NDVI index (WDI, GDD, SWA, NDVI). A year effect is included in all cases. For case (III), the CAR 4610 is an outlier (as per the 2012 model forecast output results shown in **Figure 5**). Unlike the single year model forecast results for 2012, the LOOCV procedure that considers all possible combinations of missing data (i.e., years) in the historical input data, also identifies CAR 4609 as an outlier. **(B)** Variation in model forecast error (i.e., skill) in terms of "degeneracy" in the input data, simulated by increasing the number of years at a time that were left out in the LOOCV cross-validation procedure. Results correspond to a total of 1000 MCMC runs. CAR 4606 was excluded as it had only 7 years of missing data.

across large regions, so applying our model to other areas could generate useful insights for decision-makers, where a lack of data exists. Building a component to spatial track pest infestations from pest monitoring data, like that of wheat midge (*Sitodiplosis mosellana*), a common agricultural pest within most areas of the world where non-resistant wheat varieties are still grown, could also improve the reliability of the forecast model. With further improvement and extension of input data, and refinement of its design, we anticipate our forecast model could

provide reliable and robust regional-scale crop yield forecasts. Alongside using our model to forecast crop yield, participatory workshops and stakeholder knowledge-sharing and engagement is critical for enabling open discussion of any real and perceived differences in model and farmer forecast skill. Such approaches have proved invaluable for improving forecast models and their relevance, impact and broader applicability (Pease et al., 1993; Potgieter et al., 2003; Roncoli, 2006; Crane et al., 2010).

4.3. IMPROVING FORECAST SKILL

Several key improvements for our forecasting methodology and use of statistical algorithms for variable- and model- selection could be further explored. A more general autocorrelation function, which can account for extended patterns, could be tested. Also, heavier penalties for selecting predictor variables whose values are available late in the growing season could also be tested. This would make the selection of the best set of predictors through the growing season to be less constrained and dependent on historical patterns and thus more responsive to future environmental changes affecting yield. Also, tuning of the random forest algorithm parameters could be explored using the tuneRF R library set of algorithms. Even though future changes may be uncertain and depart significantly from historical trends, the necessity of an adequate length and quality of historical data does impose a requirement to partially- or fully-calibrated model-based forecasts using our integrated methodology. Recently, the accuracy in forecasting crop yield (i.e., wheat, barley and canola) also using NDVI remote-sensing and agro-climate data and employing multivariate linear regression and a variety of machine learning techniques other than the random forest tree algorithm, such as model-based recursive partitioning (MOB) and Bayesian neural networks (BNN) has been investigated (Johnson, 2013). Hierarchical clustering across the CAR regions was used for variable-selection. Based on a Mean Absolute Error (MAE)-based forecast skill score, reportedly multiple linear regression (MLR) achieves the highest skill for spring wheat and canola, while BNN and MOB are able to further increasing skill for barley. This may indicate that if new auxiliary agroclimate or remote-sensing indices are utilized which are non-linearly related or introduce a non-normal (e.g., highly skewed or multi-modal) probability prior distribution, or regional-scale climate and yields become more non-linearly related in the future, forecasting algorithms that are able to better deal with non-linearity may need to be substituted to ensure quality in crop forecasts. Here, substituting our MCMC algorithm with an integrated nested Laplace approximation (INLA) that could offer a computationally cheaper alternative. The INLA algorithm would enable representing agro-climate, remote-sensing or other predictors as functions that can be each fit to a different model with a different form of spatial dependence and degrees of non-linearity.

There are also several key improvements to the input data that could be undertaken. Assimilating additional climate station data within certain CARs that exhibit strong spatial heterogeneity (compared to other CARs) in temperature and precipitation could provide further reductions in forecast uncertainty. This could help reduce uncertainty, in addition to accounting for autocorrelation in these variables, within the two CAR's in southern Manitoba that showed highest RMSE that we speculate is due to localized drought and flooding conditions that have occurred there. Validating the agroclimate index for soil water availability (SWA) that was derived from the VSMB soil water model against other independent data would be informative before testing other available remotely-sensed indices (i.e., other than NDVI), such as EVI2 (Enhanced Vegetation Index) for this region could improve forecasting there. Integration of EVI2 remote-sensing index would incorporate the blue portion of the light spectrum and help to better track crop yield by reducing sensitivity of the NDVI to this backscatter. Also, summer storms can deliver large amounts of rainfall in the Canadian Prairies and even in semi-arid environments, sufficient long periods of sustained high rainfall can raise the water table (upper level of saturated soil-water zone) causing groundwater flooding and prolonged conditions of excess surface water, ponding and runoff. Integrating other remotely-sensed indices could be used to track waterlogging (excess water) between the development stages of tillering and anthesis. This would increase the reliability of the forecast model to recognize, respond and forecast yield under extreme conditions of increase rainfall intensity expected to occur in the future due to increased climate variability. Currently, precipitation was removed as a direct, leading predictor given it did not improve forecast accuracy because of its high variance. Instead, precipitation was indirectly factored into the WSI index. Nonetheless, precipitation, early in the growing season, can have a leading influence on attainable yield (He et al., 2013). For example, Bolton and Friedl (2013) have recently reported that significant precision in predicting corn yield in the United States can be achieved using the Normalized Difference Water Index (NDWI) ($R^2 = 0.69$ in semi-arid areas) as well as the two-band EVI2 of NASA's MODerate-resolution Imaging Spectroradiometer (MODIS) onboard NOAA's Terra satellite, predicting corn and soybean yield better than NDVI. From 2000 onwards, NDVI data, available the MODIS could also be utilized (MODIS, 250 m resolution, Level-2 Gridded (L-2G) surface reflectance data (collection V005)) (NASA, 2012). A statistical regression-based comparison of MODIS with AVHRR 16-day (i.e., bi-weekly) NDVI composite data indicates that for the row crop and small grains land cover classes, over 90% of the variation observed in MODIS NDVI is associated with variation in AVHRR NDVI values (Gallo et al., 2004). Findings from this comparison also indicate that there remains some residual cloud contamination in both data sets that contributes to outliers. Therefore, replacing AVHRR NDVI with MODIS NDVI post-2000 may not reduce the occurrence of outliers, unless cloud contamination is dealt with first. For this reason, AVHRR was deemed adequate for the purposes of our modeling, pending further quality control (i.e., explanation and subsequent reduction) of NDVI residual variance. Cloud removal, cropland area and boundary masking and other quality control measures for MODIS NDVI are described in Davidson et al. (2009). Leaf area index (LAI), soil-adjusted vegetation index (SAVI) could also be integrated and help track yield across different developmental stages based on existing evidence they correlate well with the different crop phenological stages (Kumar et al., 1999; Graves and Pedrycz, 2009). Other combined and improved spectral indices have been developed, validated to better incorporate non-linearity between NDVI and crop biophysical parameters and under water-limited conditions (Habourdane et al., 2004; Eitel et al., 2008; Smith et al., 2008). Further work could also optimize the model to the time period within the growing season that is the most sensitive and most significant for establishing final yield. So, instead of referencing WSI across the entire season, WSI for the most sensitive time period of crop growth could be used. It is important to highlight that inclusion of remote-sensing indices is not to complete replace field crop survey data in training and validating seasonal crop forecasts, but to supplement, improve and add value to such forecasts in both a spatial and temporal context.

5. CONCLUDING REMARKS

Our forecast modeling findings demonstrate that forecast skill at the regional-scale can be improved with an integrated approach involving the integration of different auxiliary indices (e.g., agroclimate and remotely-sensed) within a probabilistic, Bayesian framework that incorporates both data and model structural uncertainty. The need for improved operational forecasting methods is particularly urgent because forecasting models are so crucial for guiding and informing agricultural crop production, management and policy decisions. Models also provide forecasts with extended "early-warning" or lead-time, enabling stake-holders to better respond to potential impacts and emerging risks. Recent results from the world's largest standardized inter-comparison of ensemble-based model projections of climate change impacts on wheat crop yield calibrated against >300 field data-sets, reveals net-uncertainties of 20–30% CV when partially calibrated, and 2–7% CV for fully-calibrated models (Asseng et al., 2013). Such uncertainty is explained, in part, due to observed variability in climate, soil water holding capacity, sowing date and agronomic management (i.e., fertilizer application) (Asseng et al., 2013; Carter, 2013). Our findings indicate that by coupling satellite, remote-sensing with agroclimate data, forecast uncertainty in model-based crop yield forecasts can be further reduced to within the range of 1–4% (i.e., for spring wheat in the Canadian Prairies). This integrated methodology offers a consistent, generalizable approach for sequentially forecasting crop yield at the regional-scale. It provides a statistically robust, flexible way to concurrently adjust to data-rich and data-sparse situations, adaptively select different predictors of yield to changing levels of environmental uncertainty, and to update forecasts sequentially so as to incorporate new data as it becomes available. It also provides additional statistical information (i.e., bias, variance, sensitivity and cross-validation statistics) for better assessing the reliability of generated crop yield forecasts in time and space.

We aim to further apply our integrated forecast methodology to generate crop forecasts across Canada by embedding it within an operational forecasting system called the Integrated Canadian Crop Yield Forecaster (ICCYF). This operational decision-support tool with provide forecasts for all major Canadian crops with national coverage across all major agricultural areas. This will require the validation of our model for other major Canadian crops such as barley, canola and soybean. Operational crop outlooks likely will delivered by AAFC and publically-released in partnership with Statistic Canada's Crop Condition Assessment Program. With further extension and validation testing of this forecasting methodology to other major Canadian grain and oilseed crops (e.g., barley and canola), this method will provide a reliable approach for generating more rapid and lower cost crop forecasts each year, within the growing season, and across Canada's agricultural extent. Alongside our continued research work, further international collaborative efforts will seek to improve the design of a consistent and reliable international framework for crop yield and production forecasting and outlook reporting that integrates remote-sensing

and agroclimate predictors (Nikolova et al., 2012). Our findings reported here demonstrate that our integrated methodology can provide robust and accurate regional-scale forecasts of spring wheat yield. It offers a flexible, scalable approach for integrating additional auxiliary indices, independent of spatial or temporal scale. In this way, it can integrate remotely-sensed vegetation or finer-scale indices based on field site data. Oursensitivity and cross-validation model findings also provide rigorous analytical testing of the relative benefits of using and combining difference and diverse sources of information in terms of better explaining and reducing uncertainty in model-based forecasts. This integrated method also provides additional statistical support for assessing the accuracy and reliability of model-based crop yield forecasts in time and space. We aim, in the future, to collaborate internationally in applying and validating the integrated forecasting methodology to other countries and continents that experience different regional climate and cropping conditions.

ACKNOWLEDGMENTS

This research was funded under the Growing Forward One, Sustainable AGriculture Environmental Systems (SAGES) Program of Agriculture and Agri-Food Canada (AAFC), with additional support from AAFC's National Agro-climate Information Service (NAIS) and the Growing Forward Two, Canadian federal funding program. We thank Charles Serele (Federal Department of National Defence), Budong Qian and Richard Warren (AAFC) for their assistance in acquiring and manipulating agro-climate input data. We thank Andrew Davidson for help in processing and preparing the NDVI remote-sensing/earth observational data. We also acknowledge feedback from a broader set of government collaborators involved in the operational development and delivery of the crop outlook prototype tool: C. Champagne, P. Cherneski, B. Daneshfar, A. Davidson, X. Geng, E. Gorelov, D. Qi, R. Rieger, and D. Waldner from AAFC and F. Bedard and G. Reichert from Statistics Canada for their support and contributions. We thank R. Armstrong (AAFC) for analyzing the autocorrelation in the wheat data, and R. Carew, L. Townley-Smith, and B. MacGregor for feedback and discussions on this research linked with ecological and economic aspects. We also thank three anonymous reviewers for their feedback that helped to improve our manuscript. A prototype version of this forecast model for generating operational crop outlooks is called the Integrated Canadian Crop Yield Forecaster (ICCYF) has been coded and tested using the open source software and statistical libraries provided by the R Statistical Software (R Development Core Team, 2013). ArcGIS[TM], (ESRI[TM], Version 10, 2010) was used to visualize model output, processing spatial data and generating crop outlook maps. A user-guide is available with the R code for non-commercial and academic research. With further extension and testing of the forecasting method to other major crops and across Canada's agricultural extent, it is anticipated that a web-based portal will, in the future, deliver seasonal, operational crop outlook reports.

REFERENCES

Albert, J. (2009). *Bayesian Computation with R*. New York, NY: Springer Science-Business Media. doi: 10.1007/978-0-387-92298-0

Asseng, S., Ewert, F., Rosenweig, C., Jones, J., Hatfield, J., Ruane, A., et al. (2013). Uncertainty in simulating wheat yields under climate change. *Nat. Clim. Change* 3, 827–832. doi: 10.1038/nclimate1916

Asseng, S., Keating, B., Fillery, I., Gregory, P., Bowden, J., Turner, N., et al. (1998). Performance of the APSIM-wheat yield in Western Australia. *Field Crops Res.* 57, 163–179. doi: 10.1016/S0378-4290(97)00117-2

Baier, W., and Robertson, G. (1996). Soil moisture modelling - conception and evolution of the VSMB. *Can. J. Soil Sci.* 76, 251–261. doi: 10.4141/cjss96-032

Banarjee, S., Carlin, B., and Gelfand, A. (2004). *Hierarchical Modeling and Analysis for Spatial Data*. Monographs on statistics and applied probability, Vol. 101. New York, NY: CRC/Chapman and Hall (Taylor and Francis LCC).

Bastiaanssen, W., and Ali, S. (2003). A new crop yield forecasting model based on satellite measurements applied across the Indus Basin, Pakistan. *Agric. Ecosyst. Environ.* 94, 321–340. doi: 10.1016/S0167-8809(02)00034-8

Bolton, D., and Friedl, M. (2013). Forecasting crop yield using remotely sensed vegetation indices and crop phenology metrics. *Agric. For. Meteorol.* 173, 74–84. doi: 10.1016/j.agrformet.2013.01.007

Bootsma, A., Gameda, S., and McKenney, D. (2005). Potential impacts of climate change on corn, soybeans and barley yields in Atlantic Canada. *Can. J. Soil Sci.* 85, 345–357. doi: 10.4141/S04-019

Bornn, L., and Zidek, J. V. (2012). Efficient stabilization of crop yield prediction in the Canadian Prairies. *Agric. For. Meteorol.* 153, 223–232. doi: 10.1016/j.agrformet.2011.09.013

Breiman, L. (2001). Random forests. *Mach. Learn.* 45, 5–32. doi: 10.1023/A:1017934522171

Carew, R., Smith, E., and Grant, C. (2009). Factors influencing wheat yield and variability: evidence from Manitoba, Canada. *J. Agric. Appl. Econ.* 41, 625–639. Available online at: http://ageconsearch.umn.edu/bitstream/56649/2/jaae159.pdf

Carter, T. (2013). Multi-model yield projections. *Nat. Clim. Change* 3, 784–786. doi: 10.1038/nclimate1995

Challinor, A. (2009). Towards development of adaptation options using climate and crop yield forecasting at seasonal to multi-decadal timescales. *Environ. Sci. Pol.* 12, 453–465. doi: 10.1016/j.envsci.2008.09.008

Challinor, A., Slingo, J., Wheeler, T., Craufurd, P., and Grimes, D. F. (2003). Toward a combined seasonal weather and crop productivity forecasting system: determination of the working spatial scale. *J. Appl. Meteorol.* 42, 175–192. doi: 10.1175/1520-0450(2003)042<0175:TACSWA>2.0.CO;2

Chen, J., Li, M., and Wang, W. (2012). Statistical uncertainty estimation using random forests and its applications to drought forecast. *Math. Probl. Eng.* 2012, 1–12. doi: 10.1155/2012/786281

Chipanshi, A., Zhang, Y., Newlands, N., H. Hill, H., and Zamar, D. (2012). "Introducing the integrated canadian crop yield forecaster (ICCYF)," in *Proceedings of the Workshop on the Application of Remote Sensing and GIS Technology on Crops Productivity Among APEC Economies*, 1–14.

Crane, T., Roncoli, C., Paz, K., Breuer, N., Broad, K., Ingram, K., et al. (2010). Forecast skill and farmer's skill: seasonal climate forecasts and agricultural risk management in the southeastern united states. *Weather Clim. Soc.* 2, 44–59. doi: 10.1175/2009WCAS1006.1

Davidson, A., Howard, A., Sun, K., Pregitzer, M., Rollin, P., and Aly, Z. (2009). *A National Crop Monitoring System Prototype (NCMS-P) Using MODIS Data: Near-Real-Time Agricultural Assessment from Space*. Technical report, Agriculture and Agri-Food Canada (AAFC).

DeJong, R. (1988). Comparison of two soil-water models under semi-arid growing conditions. *Can. J. Soil Sci.* 68, 17–27. doi: 10.4141/cjss88-002

Eitel, J., Gessler, P., Long, D., and Hunt, E. (2008). Combined spectral index to improve ground-based estimates of nitrogen status in dryland wheat. *Agron. J.* 100, 694–1702. doi: 10.2134/agronj2007.0362

Ewert, F., Ittersum, M. V., Heckelei, T., Therond, O., Bezlepkina, I., and Andersen, E. (2011). Scale changes and model linking methods for integrated assessment of agri-environmental systems. *Agric. Ecosyst. Environ.* 142, 6–17. doi: 10.1016/j.agee.2011.05.016

FAO. (2011). *The State of the World's Land and Water Resources for Food and Agriculture (SOLAW) – Managing Systems at Risk – Summary Report*. Rome, Italy: Food and Agriculture Organization of the United Nations.

Ferris, R., Ellis, R., Wheeler, T., and Hadley, P. (1998). Effect of high temperature stress at anthesis on grain yield and biomass of field-grown crops of wheat. *Ann. Bot.* 82, 631–639. doi: 10.1006/anbo.1998.0740

Fuller, L. (2010). Chernozemic soils of the prairie region of Western Canada. *Prairie Soils Crops* 3, 37–45. Available online at: http://www.prairiesoilsandcrops.ca/articles/volume-3-6-screen.pdf

Gallo, K., Ji, L., Reed, B., Dyer, J., and Eidenshink, J. (2004). Comparison of MODIS and AVHRR 16-day normalized difference vegetation index composite data. *Geophys. Res. Lett.* 31, 1–4. doi: 10.1029/2003GL019385

Graves, D., and Pedrycz, W. (2009). "Multivariate segmentation of time series with differential evolution," in *Proceedings of the European Society for Fuzzy Logic and Technology (EUSFLAT)* (Lisbon), 1108–1113.

Habourdane, D., Miller, J., Pattey, E., Zarco-Tejada, P., and Strachan, I. (2004). Hyperspectral vegetation indices and novel algorithms for predicting green LAI of crop canopies: modeling and validation in the context of precision agriculture. *Remote Sens. Environ.* 90, 337–352. doi: 10.1016/j.rse.2003.12.013

Hammer, G., Hansen, J., Phillips, J., Mjelde, J., Hill, H., Love, A., et al. (2001). Advances in application of climate prediction in agriculture. *Agric. Syst.* 70, 515–553. doi: 10.1016/S0308-521X(01)00058-0

Hammer, G., Meinke, H., and Potgieter, A. (2000). "The use of seasonal forecasts, climate data, and models to improve the profitability and sustainability of cropping systems," in *Emerging Technologies in Agriculture: From Ideas to Adoption*, 1–12.

Hansen, J. (2005). Integrating seasonal climate prediction and agricultural models for insights into agricultural practice. *Philos. Trans. R. Soc. B* 360, 2035–2047. doi: 10.1098/rstb.2005.1747

He, Y., Wei, Y., DePauw, R., Qian, B., Lemke, R., Singh, A., et al. (2013). Spring wheat yield in the semiarid Canadian Prairies: effects of precipitation timing and soil texture over recent 30 years. *Field Crops Res.* 149, 329–337. doi: 10.1016/j.fcr.2013.05.013

Hoffmann, H., and Rath, T. (2012). Meteorologically consistent bias correction of climate time series for agricultural models. *Theor. Appl. Climatol.* 110, 129–141. doi: 10.1007/s00704-012-0618-x

Hoffmann, H., and Rath, T. (2013). Future bloom and blossom frost risk for *Malus domestica* considering climate model and impact model uncertainties. *PLoS ONE* 8:e75033. doi: 10.1371/journal.pone.0075033

IPCC. (2013). *Climate Change 2013: The Physical Science Basis*. Report, Intergovernmental Panel on Climate Change (IPCC).

Jayne, T., and Rashid, S. (2010). "The value of accurate crop production forecasts," in *Forth African Agricultural Markets Program (AAMP) Policy Symposium - Agricultural Risks Management in Africa: Taking Stock of What Has and Hasn't Worked* (Lilongwe), 1–14.

Jentsch, A., Kreyling, J., and Beierkuhnlein, C. (2007). A new generation of climate-change experiments: events, not trends. *Front. Ecol. Environ.* 5:365–374. doi: 10.1890/1540-9295(2007)5[365:ANGOCE]2.0.CO;2

Johnson, M. D. (2013). *Crop Yield Forecasting on the Canadian Prairies by Satellite Data and Machine Learning Methods*. Master's thesis, University of British Columbia, Atmospheric Science.

Jones, G., Haran, M., Caffo, B., and Neath, R. (2006). Fixed-width output analysis for Markov-chain Monte Carlo. *J. Am. Stat. Assoc.* 101, 1537–1547. doi: 10.1198/016214506000000492

JRC. (2012). *MARS Crop Yield Forecasting System (MCYFS)*. Technical report, Joint Research Centre, European Commission.

Keating, B., Carberry, P., Hammer, G., Probert, M., Robertson, M., Holzworth, D., et al. (2003). An overview of APSIM, a model designed for farming systems simulation. *Eur. J. Agron.* 18, 267–288. doi: 10.1016/S1161-0301(02)00108-9

Khan, J., Aelst, S. V., and Zamar, R. (2007). Robust linear model selection based on least angle regression. *J. Am. Stat. Assoc.* 102, 1289–1299. doi: 10.1198/016214507000000950

Khan, J., Aelst, S. V., and Zamar, R. (2010). Fast robust estimation of prediction error based on resampling. *Comput. Stat. Data Anal.* 54, 3121–3130. doi: 10.1016/j.csda.2010.01.031

King, R. (2012). A review of bayesian state-space modelling of capture-recapture recovery data. *Interface Focus* 2, 190–204. doi: 10.1098/rsfs.2011.0078

Koehler, A.-K., Challinor, A. J., Hawkins, E., and Asseng, S. (2013). Influences of increasing temperature on indian wheat: quantifying limits to predictability. *Environ. Res. Lett.* 8, 034016. doi: 10.1088/1748-9326/8/3/034016

Koller, M., and Stahel, W. (2011). Sharpening wald-type inference in robust regression for small samples. *Comput. Stat. Data Anal.* 55, 2504–2515. doi: 10.1016/j.csda.2011.02.014

Kouadio, L., and Newlands, N. (2014). *Data Hungry Models in a Food Hungry World – An Interdisciplinary Challenge Bridged by Statistics*, chapter 21. New York, NY: CRC/Chapman and Hall (Taylor and Francis, LCC), 371–385.

Kumar, V., Shaykewich, C., and Haque, C. (1999). "Phenological stages-based NDVI in spring wheat yield estimation for the Canadian Prairies," in *Geoscience and Remote Sensing Symposium (IGARSS) 2011 IEEE International* (Hamburg), 2330–2332.

Lansigan, F., Salvacion, A. R., Paningbatan, E. P. Jr., Solivas, E., and Matienzo, E. (2007). "Developing a knowledge -based crop forecasting system in the Philippines," in *Proceedings of the 10th National Convention on Statistics (NCS)*, 1–14.

Littell, J., McKenzie, D., Kerns, B. K., Cushman, S., and Shaw, C. G. (2011). Managing uncertainty in climate-driven ecological models to inform adaptation to climate change. *Ecosphere* 360, 1–19.

Lobell, D., Schlenker, W., and Costa-Roberts, J. (2011). Climate trends and global crop production since 1980. *Science* 333, 616–620. doi: 10.1126/science.1204531

Luo, Y., Ogle, K., Tucker, C., Fei, S., Gao, C., LaDeau, S., et al. (2011). Ecological forecasting and data assimilation in a data-rich era. *Ecol. Appl.* 21, 1429–1442. doi: 10.1890/09-1275.1

Matthews, R. B., Rivington, M., Muhammed, S., Newton, A. C., and Hallett, P. D. (2013). Adapting crops and cropping systems to future climates to ensure food security: the role of crop modelling. *Glob. Food Secur.* 2, 24–28. doi: 10.1016/j.gfs.2012.11.009

McIntosh, P., Poo, M., Risbey, J., Lisson, S., and Rebbeck, M. (2007). Seasonal climate forecasts for agriculture: towards better understanding and value. *Field Crops Res.* 104, 130–138. doi: 10.1016/j.fcr.2007.03.019

Mkhabela, M., Bullock, P., Raj, S., Wang, S., and Wang, Y. (2011). Crop yield forecasting on the Canadian Prairies using MODIS NDVI data. *Agric. For. Meteorol.* 151, 385–393. doi: 10.1016/j.agrformet.2010.11.012

Moran, M., Clarke, T., Inoue, Y., and Vidal, A. (1994). Estimating crop water deficit between surface-air temperature and spectral vegetation index. *Remote Sens. Environ.* 46, 246–263. doi: 10.1016/0034-4257(94)90020-5

NASA. (2012). *MOD 13 - Gridded Vegetation Indices (NDVI & EVI) [Data]*. Technical report, U.S. National Aeronautics and Space Administration.

Newlands, N., and Zamar, D. (2012). "In-season probabilistic crop yield forecasting - integrating agro-climate, remote-sensing and crop phenology data," in *Proceedings of the Joint Statistical Meetings (JSM), "Statistics—Growing to Serve a Data Dependent Society," Section on Statistics and the Environment, 2012 (CD-ROM Digital/Online Library)* (San Diego, CA), 1–15.

Nikolova, S., Bruce, S., Randall, L., Barrett, G., Ritman, K., and Nicholson, M. (2012). *Using Remote Sensing Data and Crop Modelling to Improve Production Forecasting: A Scoping Study*. Technical Report 12.3, Department of Agriculture, Fisheries and Forestry, Australian Bureau of Agricultural and Resource Economics and Sciences (ABARES), Australian Government.

Pease, W., Wade, E., Skees, J., and Shrestha, C. (1993). Comparisons between subjective and statistical forecasts of crop yields. *Rev. Agric. Econ.* 15, 339–350. doi: 10.2307/1349453

Pennock, D., Bedard-Haughn, A., and Viaud, V. (2011). Chernozemic soils of Canada: genesis, distribution, and classification. *Can. J. Soil Sci.* 91, 719–747. doi: 10.4141/cjss10022

Potgieter, A., Everingham, Y., and Hammer, G. (2003). On measuring quality of a probablistic commodity forecast for a system that incorporates seasonal climate forecasts. *Int. J. Climatol.* 23, 1195–1210. doi: 10.1002/joc.932

Potgieter, A., Hammer, G., and Doherty, A. (2006). *Oz-Wheat: A Regional-Scale Crop Yield Simulation Model for Australian Wheat*. Technical report, Queensland Department of Primary Industries and Fisheries.

Potgieter, A., Hammer, G., Doherty, A., and de Voil, P. (2005). A simple regional-scale model for forecasting sorghum yield across North-eastern Australia. *Agric. For. Meteorol.* 1, 143–153. doi: 10.1016/j.agrformet.2005.07.009

Qian, B., Jong, R. D., Warren, R., Chipanshi, A., and Hill, H. (2009). Statistical spring wheat yield forecasting for the Canadian Prairie provinces. *Agric. For. Meteorol.* 149, 1022–1031. doi: 10.1016/j.agrformet.2008.12.006

Raes, D., Steduto, P., Hsiao, T., and Fereres, E. (2009). AquaCrop – The FAO crop model to simulate yield response to water: Ii. Main algorithms and software description. *Agron. J.* 101, 438–447. doi: 10.2134/agronj2008.0140s

R Foundation. (2013). *R: A Language and Environment for Statistical Computing*. Technical report, R Foundation for Statistical Computing, Vienna, Austria.

Reichert, G., and Caissy, D. (2002). *A Reliable Crop Condition Assessment Program (CCAP) Incorporating NOAA AVHRR Data, a Geographical Information System, and the Internet*. Technical report, Statistics Canada, Ottawa.

Robertson, S., Jeffrey, S., Unterschultz, J., and Boxall, P. (2013). Estimating yield response to temperature and identifying critical temperatures for annual crops in the Canadian Prairie region. *Can. J. Plant Sci.* 93, 1237–1247. doi: 10.4141/cjps2013-125

Roncoli, C. (2006). Ethnographic and participatory approaches to research on farmers' responses to climate predictions. *Clim. Res.* 33, 81–99. doi: 10.3354/cr033081

Saiyed, I., Bullock, P., Sapirstein, H., Finlay, G., and Jarvis, C. (2009). Thermal time models for estimating wheat phenological development and weather-based relationships to wheat quality. *Can. J. Plant Sci.* 89, 429–439. doi: 10.4141/CJPS07114

Smith, A., Bourgeois, G., Teillet, P., Freemantle, J., and Nadeau, C. (2008). A comparison of NDVI and MTVI2 for estimating LAI using CHRIS imagery: a case study in wheat. *Can. J. Remote Sens.* 34, 539–548. doi: 10.5589/m08-071

Statistics Canada. (2007). *Census Agricultural Regions Boundary Files for the 2006 Census of Agriculture - Reference Guide*. Technical report, Statistics Canada.

Statistics Canada. (2012a). *1976–2011 Crops Small Area Data*. Technical report, Field Crop Reporting Series, Agriculture Division, Statistics Canada.

Statistics Canada. (2012b). *Definitions, Data Sources and Methods of Field Crop Reporting Series*. Technical report, Agriculture Division, Statistics Canada.

Statistics Canada. (2012c). *Census Agricultural Regions Boundary Files of the 2011 Census of Agriculture*. Agriculture Statistics Division, Statistics Canada.

Steduto, P., Hsiao, T., Raes, D., and Fereres, E. (2009). AquaCrop – the FAO crop model to simulate yield response to water: I. Concepts and underlying principles. *Agron. J.* 101, 426–437. doi: 10.2134/agronj2008.0139s

Stephens, D., Butler, D., and Hammer, G. (2000). *Using Seasonal Climate Forecasts in Forecasting the Australian Wheat crop*, Vol. 21. Atmospheric and Oceanographic Sciences Library.

Stephens, D., Lyons, T., and Lamond, M. (1989). A simple model to forecast wheat yield in Western Australia. *J. R. Soc. Western Austr.* 37, 77–81.

Stewart, B., Veeman, T., and Unterschultz, J. (2009). Crops and livestock productivity growth in the Prairies: the impacts of technical change and scale. *Can. J. Agric. Econ.* 57, 379–394. doi: 10.1111/j.1744-7976.2009.01157.x

Stone, R., and Meinke, H. (2005). Operational seasonal forecasting of crop performance. *Philos. Trans. R. Soc. B* 360, 2109–2124. doi: 10.1098/rstb.2005.1753

Tanner, M., and Wong, W. (1987). The calculation of posterior distribution by data augmentation. *J. Am. Stat. Soc.* 82, 528–540. doi: 10.1080/01621459.1987.10478458

USDA. (2013). *Production, Supply and Distribution: Electronic Database*. Technical report, United States Department of Agriculture (USDA).

Wit, A. D., and van Diepen, C. (2007). Crop model data assimilation with the Ensemble Kalman filter for improving regional crop yield forecasts. *Agric. For. Meteorol.* 146, 38–56. doi: 10.1016/j.agrformet.2007.05.004

Wu, B., Meng, J., Li, Q., Yan, N., Du, X., and Zhang, M. (2013). Remote sensing-based global crop monitoring: experiences with China's CropWatch system. *Int. J. Digit. Earth* 7, 113–137. doi: 10.1080/17538947.2013.783131

Conflict of Interest Statement: The authors declare that the research was conducted in the absence of any commercial or financial relationships that could be construed as a potential conflict of interest.

8

Network topology reveals high connectance levels and few key microbial genera within soils

Manoeli Lupatini¹,², Afnan K. A. Suleiman¹, Rodrigo J. S. Jacques¹, Zaida I. Antoniolli¹,
Adão de Siqueira Ferreira³, Eiko E. Kuramae² and Luiz F. W. Roesch⁴*

¹ Departamento de Solos, Universidade Federal de Santa Maria, Santa Maria, Brazil
² Department of Microbial Ecology, Netherlands Institute of Ecology (NIOO-KNAW), Wageningen, Netherlands
³ Instituto de Ciências Agrárias, Universidade Federal de Uberlândia, Uberlândia, Brazil
⁴ Universidade Federal do Pampa, São Gabriel, Brazil

Edited by:
Christophe Darnault, Clemson University, USA

Reviewed by:
Tancredi Caruso, Queen's University of Belfast, UK
Ulisses Nunes Da Rocha, Vrije Universiteit Amsterdam, Netherlands
Zhili He, University of Oklahoma, USA
Aidan M. Keith, Centre for Ecology and Hydrology, UK

*Correspondence:
Luiz F. W. Roesch, Centro Interdisciplinar de Pesquisas em Biotecnologia (CIP-Biotec), Universidade Federal do Pampa, Avenida Antonio Trilha, 1847, São Gabriel, RS 97300-000, Brazil
e-mail: luizroesch@unipampa.edu.br

Microbes have a central role in soil global biogeochemical process, yet specific microbe–microbe relationships are largely unknown. Analytical approaches as network analysis may shed new lights in understanding of microbial ecology and environmental microbiology. We investigated the soil bacterial community interactions through cultivation-independent methods in several land uses common in two Brazilian biomes. Using correlation network analysis we identified bacterial genera that presented important microbial associations within the soil community. The associations revealed non-randomly structured microbial communities and clusters of operational taxonomic units (OTUs) that reflected relevant bacterial relationships. Possible keystone genera were found in each soil. Irrespective of the biome or land use studied only a small portion of OTUs showed positive or negative interaction with other members of the soil bacterial community. The more interactive genera were also more abundant however, within those genera, the abundance was not related to taxon importance as measured by the Betweenness Centrality (BC). Most of the soil bacterial genera were important to the overall connectance of the network, whereas only few genera play a key role as connectors, mainly belonged to phyla *Proteobacteria* and *Actinobacteria*. Finally it was observed that each land use presented a different set of keystone genera and that no keystone genus presented a generalized distribution. Taking into account that species interactions could be more important to soil processes than species richness and abundance, especially in complex ecosystems, this approach might represent a step forward in microbial ecology beyond the conventional studies of microbial richness and abundance.

Keywords: network analysis, community ecology, keystone species, soil microbial interactions, high-throughput sequencing

INTRODUCTION

Understanding the interaction among different taxa within a soil microbial community and their responses to environmental changes is a central goal in microbial ecology and very important to better explore the complexity of soil processes. Soil microbial ecologists have borrowed several complex ecological theories from macroecology, including competitive strategies (Prosser et al., 2007) and biogeography (Griffiths et al., 2011). Most of the statistical techniques adapted to microbial systems have been used to test these theories however; they are only focused on single properties of the microbial communities. The studies have been focused on microbial alpha and/or beta diversity to answer fundamental ecological questions (e.g., to understand how different soil management types affect the bacterial community diversity and composition). On the other hand, interactions among associated taxon could contribute more to ecosystem processes and functions than species diversity in soil environmental processes (Zhou et al., 2011).

Within a microbial community, interactions can be visualized as ecological networks, in which interactive taxa are linked together, either directly or indirectly through intermediate species. The study of networked systems has received great attention in the last years, especially in the mathematical and social sciences, mainly as result of the increasing availability to obtain and analyse large datasets. These methods have been applied to the study of various biological contexts including healthy microbiota in human microbiome (Duran-Pinedo et al., 2011; Faust and Raes, 2012), cancer (Choi et al., 2005), food webs (Estrada, 2007), marine microbial community (Steele et al., 2011), and recently this technique have been used to better understand soil microbial processes by examining complex interactions among microbes (Prasad et al., 2011; Roesch et al., 2012). The use of network analysis in microbial ecology has the potential for exploring inter-taxa correlations allowing an integrated understanding of soil microbial community structure and the ecological rules. This approach can truly be applied to large soil microbial datasets offering new insights into the microbial community structure and the ecological rules guiding community assembly (Barberán et al., 2012).

The networks analysis could be essential to explain several fundamental questions still unclear about microbial ecological theories. A good example is related to presence or not of keystone species. The concept of keystone species was introduced in microbial ecology and to date the identification of keystone taxa or populations is a critical issue in soil microbial ecology given the extreme complexity, high diversity, and uncultivated status of the large portion of community (Zhou et al., 2011). Keystones are important to maintain the function of the microbial community and their extinction might lead to community fragmentation (Martín González et al., 2010). Another important issue that network analysis could explain is the importance of the abundance of taxa for supporting the structure and function of the soil microbial community. So far, most of literature studies have focused the attention on dominant species is soil ecosystems (Campbell and Kirchman, 2013). However, low abundant taxa should participate significantly in ecosystem functioning despite their low abundance and therefore some of them may be considered as keystones (Rafrafi et al., 2013).

In order to gain understanding on the organization of a complex microbial communities, here we used correlation network analysis to study soil microbial organization. Specifically we addressed the following questions: (i) Is it possible to detect keystone bacterial taxa in soils? (ii) If yes, are the keystone taxa exclusive to each land use or they are the same in most land uses? (iii) Are the most abundant taxa more important to connect distincts operational taxonomic units (OTUs) and maintain the structure of microbial interactions in soil? To answer those questions we performed a large-scale pyrosequencing-based analysis of the 16S rRNA gene on replicate samples from two biomes in Brazil and implemented microbial ecological network analysis to examine how the microbial community members interact with each other and which members are important to support the microbial community structure in the land uses studied. Our central objective was to characterize and to understand ecological networks pattern in soil microbial communities based on high-throughput sequencing data.

MATERIALS AND METHODS
SAMPLING SITES AND SAMPLE COLLECTION
To analyse the soil bacterial community interactions, soil samples were collected within two biomes in Brazil: one site was located within the Pampa biome which covers an area shared by Brazil, Argentina, and Uruguay in the southern of South America and is characterized by typical vegetation of native grassland, with sparse shrub and tree formations (Overbeck et al., 2007). The soils from this biome came from two sites. At site A, soil samples were collected in areas with four different land uses: natural pasture (30° 00′ 38.2″ S and 54° 50′ 17.4″ W)—currently used for grazing of cattle; native forest (30° 00′ 39.7″ S and 54° 50′ 05.6″ W)—used only for preservation of wildlife; soybean field (30° 00′ 40.3″ S and 54° 50′ 13.2″ W)—cultivated under no-tillage system on oat straw; 9-years-old Acacia tree plantation (*Acacia mearnsii* Willd.) (30° 00′ 27.5″ S and 54° 50′ 10.2″ W) (for more details about areas and sampling see Lupatini et al.,

2013—Raw sequences were submitted to the NCBI Sequence Read Archive under the study number SRP013204, experiment number SRX255448). At site B, soil samples were collected from a natural forest (30° 24′ 09.3″ S and 53° 52′ 59.1″ W) and 8-years-old pasture (30° 24′ 08.9″ S and 50° 53′ 05.9″ W) used for grazing of cattle (for more details about areas and sampling see Suleiman et al., 2013—Raw sequences were submitted to the NCBI Sequence Read Archive under the study number SRP013204, experiment number SRX148308). Composite samples (four sub-samples per sampling point) were collected during the spring of 2010 by taking 5 cm diameter, 0–5 cm depth cores. Equal masses of sub-samples removed from cores were pooled and mixed. Four biological repetitions were taken per each land use. DNA was isolated from at least 1 g of soil using the PowerSoil® DNA Isolation Kit (MO BIO Laboratories Inc., Carlsbad, CA, USA), according to the manufacturer's instructions.

The second sampling site was located within the Brazilian Savanna biome, also known as Cerrado. The Cerrado is a representative biome in central Brazil and the second largest biome in species diversity of South America. It is characterized by high diversity of plants with over 10,000 species (nearly half are endemic) and different vegetation types including forest formations, savannas, and grasslands (Oliveira and Marquis, 2002). The soil sampling at Cerrado biome was carried out in a natural forest (19° 20′ 41″ S and 48° 00′ 58″ W); 20-years-old pasture used for grassing (19° 20′ 42″ S and 48° 05′ 22″ W); 15-years-old sugarcane field (19° 20′ 43″ S and 48° 05′ 49″ W); and Pinus plantation (19° 04′ 39″ S and 48° 10′ 19″ W) (for more details about areas and sampling see Rampelotto et al., 2013—Raw sequences were submitted to the NCBI Sequence Read Archive under the study number SRP017965, experiment number SRX217724). Each soil sample was taken as a cut out measuring 30 × 20 × 5 cm (L × W × D). Four subsamples were collected randomly within this cut out and were passed through a 3.35-mm sieve. Genomic DNA was extracted from 250 mg of soil sample using Soil DNA Isolation Kit (Norgen, Canada) as described by the manufacturer.

16S rRNA GENE AMPLIFICATION AND PYROSEQUENCING
The 16S rRNA gene fragments were sequenced using 454 GS FLX Titanium (Lib-L) chemistry for unidirectional sequencing of the amplicon libraries. Barcoded primers allow for combining amplicons of multiple samples into one amplicon library and, furthermore, enable the computational separation of the samples after the sequencing run. Independent PCR reactions were performed for each soil sample to amplify the V1-V2 region (311 nucleotides) with the primers 27F and 338R. The primers were attached to the GS FLX Titanium Adaptor A-Key (5′-CCATCTCATCCCTGCGTGTCTCCGACTCAG-3′) and Adaptor B-Key (5′-CCTATCCCCTGTGTGCCTTGGCAGTCTCAG-3′) sequences, modified for use with GS FLX Titanium Em PCR Kits (Lib-L) and a two-base linker sequence was inserted between the 454 adapter and the 16S rRNA primers to reduce any effect the composite primer might have on PCR efficiency. PCR reactions were carried out in triplicate with the GoTaq PCR core system (Promega, Madison, WI, USA). The mixtures contained 5 μl of

10× PCR buffer, 200 mM dNTPs, 100 mM of each primer, 2.5 U of Taq polymerase, and approximately 100 ng of DNA template in a final volume of 50 μl. The PCR conditions were 94°C for 2 min, 30 cycles of 94°C for 45 s; 55°C for 45 s; and 72°C for 1 min extension; followed by 72°C for 6 min. The PCR products were purified and combined in equimolar ratios with the quantitative DNA binding method (SequalPrep Kit, Invitrogen, Carlsbad, CA, USA) to create a DNA pool for pyrosequencing on a Roche GS-FLX 454 automated pyrosequencer (Roche Applied Science, Branford, CT, USA).

SEQUENCE PROCESSING AND NETWORK ANALYSIS

The raw sequences were processed using Mothur v.1.30.2 (Schloss et al., 2009). Briefly, the multiplexed reads were filtered for quality and assigned to corresponding soil samples. The filtering criteria removed any sequence, which the longest homopolymer was greater than 8 nucleotides, contained ambiguous base call, had more than one mismatch to the barcode sequence, had more than two mismatches to the primer sequence and were smaller than 200 bases in length. In addition the sequences were trimmed by using a moving window that was 50 bases long and average quality score higher than 30. The dataset was simplified by obtaining a non-redundant set of sequences that were further aligned against the SILVA reference alignment (http://www.arb-silva.de/). To maximize the number of sequences that overlap over the longest span, the sequence that started after the position that 85% of the sequences did, or ended before the position that 85% of the sequences did, were removed from the alignment. The alignment was then trimmed since we need they overlap in the same alignment space. Finally, to reduce sequencing noise a pre-clustering step was applied (Huse et al., 2010) and the chimeric sequences were checked by chimera.slayer script in Mothur v.1.30.2. The command lines with the parameters used here are available in the Supplementary Material.

For network analysis, the OTUs were grouped at genus and only those genera with more than five sequences were considered in the following analysis. The choice for genus aimed to generate consistent OTUs with high abundances for subsequent analyses based on correlations. This approach also circumvents the potential taxonomic misclassifications due to sequencing bias. Since the networks comprised a set of share taxa within a soil, the bacterial genera represented by zero sequences in a sample were excluded from data analysis. The pipeline used for developing this study is presented in the Supplementary Material.

Associations between the microbial communities were examined by calculating all possible Pearson rank correlations between bacterial genera using the Otu.association script from Mothur v.1.30.2. A valid interaction event was considered to be a robust correlation if the Pearson correlation coefficient (p) was either equal or greater than 0.9 or −0.9 and statistically significant (p-value equal or smaller than 0.05—calculated as the proportion of the r-values generated from randomized data that are larger than the Pearson correlation coefficient that was calculated from the original data). The cutoff correlation of 0.9 or −0.9 was chosen to increase the confidence for strong bacterial interactions. To describe the topology of the resulting networks, a set of measures (average clustering coefficient, average path length,

and modularity) were calculated (Newman, 2006). The network structure was explored and visualized with the interactive platform gephi (Bastian et al., 2009) using directed network (where edges have direction) and the Fruchterman–Reingold layout.

To determine whether our webs were not random networks and really represented the actual bacterial interactions in soil, we compared random networks of equal size (same number of nodes and edges) to the networks obtained by this study. One thousand random networks were calculated by the Erdös–Rényi

Table 1 | Total number of high-quality sequences and sequencing coverage for taxonomic genus level in land uses in Pampa and Cerrado biomes.

Land use	Total sequences	Coverage genus level
PAMPA BIOME		
Site A		
Acacia plantation 1	8380	0.98
Acacia plantation 2	9083	0.98
Acacia plantation 3	7802	0.98
Acacia pantation 4	7327	0.98
Natural forest 1	9262	0.98
Natural forest 2	12,684	0.99
Natural forest 3	9083	0.98
Natural forest 4	14,654	0.99
Natural pasture 1	8971	0.98
Natural pasture 2	6665	0.98
Natural pasture 3	10,798	0.98
Natural pasture 4	10,384	0.99
Soybean field 1	6412	0.98
Soybean field 2	3686	0.96
Soybean field 3	2407	0.95
Soybean field 4	2276	0.94
Site B		
Naturalforest 1	14,516	0.99
Natural forest 2	14,884	0.99
Natural forest 3	34,724	0.99
Natural forest 4	16,223	0.99
Natural pasture 1	11,543	0.99
Natural pasture 2	13,143	0.99
Natural pasture 3	23,388	0.99
Natural pasture 4	27,167	0.99
CERRADO BIOME		
Sugarcane field 1	13,213	0.99
Sugarcane field 2	13,923	0.99
Sugarcane field 3	14,347	0.99
Natural forest 1	13,216	0.99
Natural forest 2	13,921	0.99
Natural forest 3	14,347	0.99
Pinus plantation 1	13,209	0.99
Pinus plantation 2	13,918	0.99
Pinus plantation 3	14,347	0.99
Pasture 1	5291	0.99
Pasture 2	14,800	0.99
Pasture 3	7935	0.99

Table 2 | Global network statistics for microbial association networks from land uses in Pampa and Cerrado biome.

Biome	Pampa						Cerrado			
	Site A				Site B					
Land use	Acacia plantation	Soybean field	Natural forest	Natural pasture	Natural Forest	Natural pasture	Sugarcane field	Pinus plantation	Natural forest	Pasture
Total number of OTUs[a]	724	611	807	780	912	900	755	629	714	748
Number of nodes or OTUs[b]	107 (14.8*)	56 (9.16)	154 (19.1)	146 (17.7)	197 (21.6)	197 (21.8)	107 (14.2)	70 (11.1)	121 (16.9)	84 (11.2)
Total number of significant correlations	718	148	1499	1048	2715	2389	1730	703	2257	1079
Number of significant positive correlations	353 (49.2**)	33 (22.3)	586 (39.1)	535 (51.1)	1030 (37.9)	1117 (46.7)	946 (54.7)	296 (42.1)	817 (36.2)	426 (39.5)
Number of significant negative correlations	365 (50.8)	115 (77.7)	913 (60.9)	513 (48.9)	1685 (62.1)	1272 (53.3)	784 (45.3)	407 (57.9)	1440 (63.8)	653 (60.5)
Avg. Clustering Coefficient[c]	0.59	0.53	0.62	0.58	0.62	0.61	0.76	0.79	0.76	0.77
Avg. Path Length[c]	3.11	3.80	2.90	3.14	2.85	2.90	2.30	2.41	2.30	2.29
Modularity[c]	0.45	0.61	0.50	0.58	0.44	0.47	0.40	0.53	0.40	0.50

[a]All OTUs presented in soil samples in Pampa and Cerrado biomes and grouped according to genus.

[b]Only the OTUs selected by $p \geq 0.9$ or $p \leq -0.9$, P-value ≤ 0.05.

[c]Calculated p-values in each land use equals to 0.001.

*The numbers in parentheses indicate the percentage of OTUs that show interactions related to the total number of OTUs found in land uses or **related to total number of significant correlations between OTUs in a network.

model [G(n,m)] using an script wrote in R (available on the Supplementary Material). From each random network, values of average clustering coefficient, average path length and modularity were calculated. The proportion of those values that were larger than the values calculated based on the original data were computed to get a *p*-value for the null hypothesis that the networks were obtained at random. This approach is based on using a fixed number of links to connect randomly chosen nodes and serves as point of reference against which our real biological networks might be compared (Vick-Majors et al., 2014). To measure the relative importance (how influential a taxon is within a network) of each taxon within the network we calculated two measures of centrality: Betweenness Centrality (BC) (Martín González et al., 2010; Vick-Majors et al., 2014) and Closeness Centrality (CC) (Freeman, 1979). BC counts the fraction of shortest paths going through a given bacterial taxon to another. The BC of a taxon in a network reflects the importance of control that the taxon exerts over the interactions of other taxons in the network (Martín González et al., 2010; Vick-Majors et al., 2014). CC denotes the proximity of a node to all other nodes in the network quantifying how many steps away genus *i* is from all others in the web (Freeman, 1979). Taxa with high CC are likely to have a pronounced effect on microbial community because it can rapidly affect other species in a community (Martín González et al.,

2010). Finally, to identify possible patterns between taxon abundance *vs.* CC or BC we use dispersion graphs to describe the relationship between these pair of variables.

RESULTS

The number of high-quality sequences obtained after sequence processing in each sample and the sequence coverage are presented in **Table 1**. An average of 12,164 sequences ($\geq$200 bases and $\geq$30 quality score) were obtained per sample. The smallest sequence coverage at the genus level was 94% however most of the samples presented a sequence coverage of 99%. The coverage indicated that the number of sequences obtained from each soil sample was sufficient to reveal most of the taxonomic units indicating that the samples were well represented by the number of sequences obtained and that we could perform the following OTU-based analysis.

The second step in data analysis was to verify whether the networks obtained were non-random networks. In order to test it, we compared our networks with 1000 randomly generated networks (Erdös–Rényi model; Erdös and Rényi, 1959) using the values of observed average clustering coefficient, average path length and modularity from each of our networks (**Table 2**). The *p*-values for average clustering coefficient, modularity, and path length were 0.001. This indicated that our webs were more organized

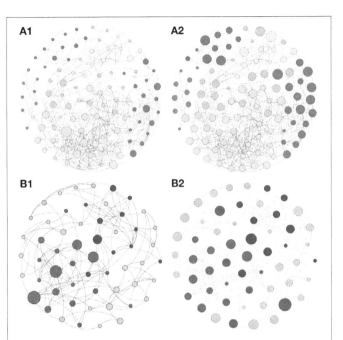

FIGURE 1 | Network interactions of soil bacterial genus found in Acacia plantation (A1, A2) and in the Soybean plantation (B1, B2) from site A on Pampa biome. A connection stands for a strong Pearson's correlation ($p \geq 0.9$ and P-value ≤ 0.05). Each circle (usually called node) represents a bacterial genus and the sizes of the circles are proportional to the value of betweenness centrality in **(A1)** and **(B1)**. In **(A2)** and **(B2)** the sizes of the circles are proportional to value of closeness centrality. Lines connecting two bacterial genera represent the interactions between them. Blue lines represent the positive significant correlations and red lines represent a negative significant correlation. The colors of the circles represent the bacterial modules. For clarity, the OTU's identity was omitted. Detailed networks containing the identity of each node can be observed in the Supplementary Figures S1–S5 and Supplementary Table S1.

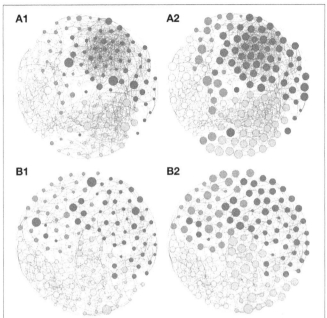

FIGURE 2 | Network interactions of soil bacterial genus found in the Natural forest (A1, A2) and in the Natural pasture (B1, B2) from site A on Pampa biome. A connection stands for a strong Pearson's correlation ($p \geq 0.9$ and P-value ≤ 0.05). Each circle (usually called node) represents a bacterial genus and the sizes of the circles are proportional to the value of betweenness centrality in **(A1)** and **(B1)**. In **(A2)** and **(B2)** the sizes of the circles are proportional to value of closeness centrality. Lines connecting two bacterial genera represent the interactions between them. Blue lines represent the positive significant correlations and red lines represent a negative significant correlation. The colors of the circles represent the bacterial modules. For clarity, the OTU's identity was omitted. Detailed networks containing the identity of each node can be observed in the Supplementary Figures S1–S5 and Supplementary Table S1.

than would be expected by a random network with identical size of nodes and edges and showed that our networks were non-random. Once established that we obtained adequate sequencing coverage and non-random networks we further explored the positive and negative interactions between co-occurrent bacterial taxons.

Based on the global network statistics presented in **Table 2** and irrespective of the biome or land use studied only a small portion of OTUs (9.16 to 21.8%) showed positive or negative interaction with other members of the soil bacterial community. Those interactive OTUs were the most abundant ones making up about 68 to 92% of the total number of taxonomic units found in the soils tested. The proportion of positive correlations was variable according to the land use and ranged from 22.3% (soybean field from Pampa biome) to 54.7% (sugarcane field from Cerrado). In average, the number of negative correlations was higher than the number of positive correlations in most land uses tested (**Table 2**).

Based on the high BC score few possible keystone taxa were detected (**Figures 1A–5A** and Supplementary Table S1). The OTUs considered keystone species (depicted as nodes with larger sizes in the network) mainly belonged to different genus of the phylum *Proteobacteria* and *Actinobacteria*, the main bacterial

phyla found in soils. Taxonomic units belonging to *Chloroflexi*, *Bacteroidetes*, and *Firmicutes* were also characterized as keystone taxa. These keystone taxa were not the same between or within biomes and appeared to be unique to each sampling location. The five keystone genus selected by the greatest value of BC from each of the soil sites are presented on **Table 3**. Based on the CC ranking, a larger number of OTUs were identified as highly important (high CC) for connectance of the microbial network since the values of CC did not present a high variation among the OTUs. No keystone genera were detected by this measurement denoting similar proximity of all genera within the network (**Figures 1B–5B** and Supplementary Material).

Studies in soil microbial ecology suggest that abundant microorganisms might have high impact on microbial structure and function. To understand how taxon abundance and the centrality measures are related, a dispersion graph with the relative abundance of all OTUs vs. the values of betweenness and closeness was constructed (**Figure 6**). Despite the abundance of genera seems to be an important parameter that define the interactions between taxonomic members of the soil bacterial community, the diagrams indicates that there is no strength relation between taxon abundance and centrality measures. However, it's possible

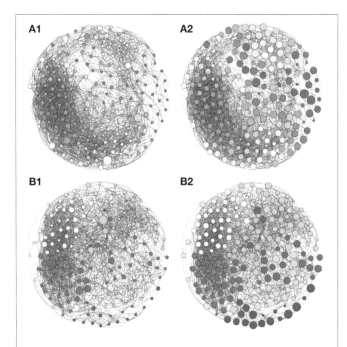

FIGURE 3 | Network interactions of soil bacterial genus found in Natural forest (A1, A2) and in the Natural pasture (B1, B2) from site B, on Pampa biome. A connection stands for a strong Pearson's correlation ($p \geq 0.9$ and P-value ≤ 0.05). Each circle (usually called node) represents a bacterial genus and the sizes of the circles are proportional to the value of betweenness centrality in **(A1)** and **(B1)**. In **(A2)** and **(B2)** the sizes of the circles are proportional to value of closeness centrality. Lines connecting two bacterial genera represent the interactions between them. Blue lines represent the positive significant correlations and red lines represent a negative significant correlation. The colors of the circles represent the bacterial modules. For clarity, the OTU's identity was omitted. Detailed networks containing the identity of each node can be observed in the Supplementary Figures S1–S5 and Supplementary Table S1.

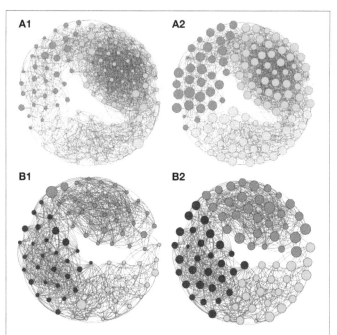

FIGURE 4 | Network interactions of soil bacterial genus found in Natural forest (A1, A2) and in the Pasture (B1, B2) on Cerrado biome. A connection stands for a strong Pearson's correlation ($p \geq 0.9$ and P-value $\leq$ 0.05). Each circle (usually called node) represents a bacterial genus and the sizes of the circles are proportional to the value of betweenness centrality in **(A1)** and **(B1)**. In **(A2)** and **(B2)** the sizes of the circles are proportional to value of closeness centrality. Lines connecting two bacterial genera represent the interactions between them. Blue lines represent the positive significant correlations and red lines represent a negative significant correlation. The colors of the circles represent the bacterial modules. For clarity, the OTU's identity was omitted. Detailed networks containing the identity of each node can be observed in the Supplementary Figures S1–S5 and Supplementary Table S1.

to note that few abundant taxa presented a slightly tendency to have high values of CC (**Figure 6B**).

DISCUSSION

In this study, we focused on microbial community associations within two ecologically important biomes in Brazil. We collected soil samples from a set of biological replicates, allowing us to detect patterns on ecological interaction using network analyses, which describe who is present and who affects whom positively or negatively. Positive correlations between microbial populations suggest the occurrence of a mutualistic interaction while negative correlations might suggest the presence of competition for hosts or predation relationship between microorganisms (Steele et al., 2011). Those interactions are strongly attached to important to soil process. For instance, a mutualistic relationship between ammonia-oxidizing bacteria (AOB) and nitrite-oxidizing bacteria (NOB) is essential to the stability of soil nitrification process, a key reaction of the global nitrogen cycle (Graham et al., 2007). On the other hand, species of *Myxobacteria* are a group of micropredator bacteria metabolically active in the soil ecosystems that play a key role in the turnover of carbon (Lueders et al., 2006). Neutral interactions can not be interpreted with this network-based approach.

Every approach presents positive and negative aspects. Before following the discussion, it is appropriated to consider some limitations of this work in order to better interpret the results: (i) unlike other studies, only correlations with $r \geq \pm 0.9$ ($p \leq 0.05$) were used to generate the networks. According to Taylor (1990) the correlation coefficient (a linear association between two variables) is an abstract measure and not given to a direct precise interpretation. Low values of r does not explain or account for significant variation in the value of the dependent variable (y). Conservative cutoffs increase the confidence for detecting only strong interactions. Less stringent cutoffs decrease the reliability of the results; (ii) PCR-based and massive sequencing techniques introduce biases related to primer mismatches, insertion/deletion (indels) sequencing errors, and chimeric PCR artifacts which can affect the interpretations of microbial community structure and diversity (Pinto and Raskin, 2012); (iii) the copy number of the 16S rRNA gene varies greatly per bacterial genome (from one in many species up to 15 in some bacteria) and these differences induce to errors in relative abundance measurements (Klappenbach et al., 2001); (iv) the proportion of inactive bacterial cells from soils ranged from 61 to 96% (Lennon and Jones, 2011). Inactive or dormant members of the microbial community

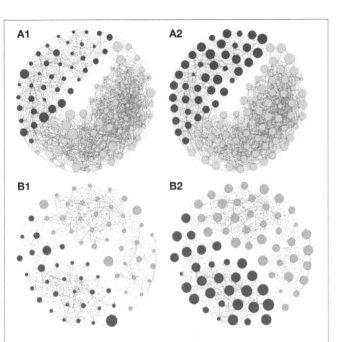

FIGURE 5 | Network interactions of soil bacterial genus found in Sugarcane (A1, A2) and in the Pinus plantation (B1, B2) on Cerrado biome. A connection stands for a strong Pearson's correlation ($p \geq 0.9$ and P-value ≤ 0.05). Each circle (usually called node) represents a bacterial genus and the sizes of the circles are proportional to the value of betweenness centrality in **(A1)** and **(B1)**. In **(A2)** and **(B2)** the sizes of the circles are proportional to value of closeness centrality. Lines connecting two bacterial genera represent the interactions between them. Blue lines represent the positive significant correlations and red lines represent a negative significant correlation. The colors of the circles represent the bacterial modules. For clarity, the OTU's identity was omitted. Detailed networks containing the identity of each node can be observed in the Supplementary Figures S1–S5 and Supplementary Table S1.

might persist in DNA samples potentially masking the active constituents of the community. This could explain why a large amount of taxons found in different land uses did not present interactions with other member of the community; (v) the network analysis is considered an OTU-based approach since it relies on detection of correlation between taxonomic unities. According to Lemos et al. (2011), in order to apply such an approach, a large sampling intensity (coverage $\geq 90\%$) is needed to get reliable results. Datasets with low number of sequences are likely to present a low sequence coverage that in turn will make it more unlikely to found OTUs correlation; (vi) finally, another drawback related to microbial network construction is the faulty prediction of a relationship between two taxa since interspecies interactions might be affected by third-party organisms in prokaryotic ecosystems (Haruta et al., 2009). Within this study, we attempted to overcome these biases as much as possible. Although those biases may not be neglected, considering the high levels of robustness and resolution of our methodology, the low variation among replicates from each land use and the quality of the results, we believe these biases were minimized and our findings are consistent.

Linking the structure of microbial communities to soil ecosystem has been a challenge in ecology. The extent, specificity,

and stability of microbial associations are difficult to assess systematically in the environment (Chaffron et al., 2010) however, co-occurrence network analysis (primarily based on statistically significant tests of correlation) were successfully applied to at least partially solve this problem (Barberán et al., 2012; Faust et al., 2012; Friedman and Alm, 2012; Gilbert et al., 2012; Rodriguez-Lanetty et al., 2013). According to Faust and Raes (2012), after abundance data have been obtained, it is possible to predict microbial relationships under the premise that strongly non-random distribution patterns are mostly due to ecological reasons. Studies on ecosystem function are traditionally limited to measurements of changes in species diversity and composition limiting our ability to link the structure of communities to the function of natural ecosystems (Philippot et al., 2013; Rudolf and Rasmussen, 2013). An important benefit of networks to study microbial ecology is the ability to understand which organisms are most important in maintaining the structure and interactions of microbial communities in soils. Due to the choice of a linear model (Pearson correlation) to describe how the taxa of a soil microbial community interact with each other, the network analysis allows only the detection of positive and/or negative interactions. While we acknowledge that not all correlations between bacterial genera found in this study might be valid, empirical evidence that correlated microbial species might actually been interactive were already demonstrated. Duran-Pinedo et al. (2011) provided an important evidence of accuracy and usefulness of this kind of analysis by isolating a not-yet-cultivated organism based on the network analysis results. The authors showed that network analysis could facilitated the cultivation of a previously uncultivated organism (*Tannerella* sp. OT286) and proved that certain species that did not grow in artificial media alone could form colonies in the presence of other microorganisms. Due to the limitations of this approach, here we adopted the term "theoretical" network association to express the positive and/or negative interaction between soil microbial genera (for an extensive revision about the difficulties and pitfalls about the use of network inference to assess microbial interactions see Faust and Raes, 2012). The application of theoretical network modeling to real microbial ecological network provide insight into the complex organization levels of microbes and identify key microbial populations or key functional genes in soil ecosystem. Using theoretical network model, based on random matrix theory (RMT) approach to delineate the network interactions, it was identified that the structure of the networks under typical and elevated CO_2 levels was substantially different in terms of network topology, node overlap, module preservation, and network hubs, suggesting that the network interactions among different phylogenetic groups/populations were markedly changed (Zhou et al., 2011).

In this study we attempted to answer three fundamental questions: (i) Is it possible to detect keystone bacterial taxa in soils? (ii) If yes, are the keystone taxa exclusive to each land use or they are the same in most land uses? (iii) Are the most abundant taxa more important to connect distincts OTUs and maintain the structure of microbial interactions in soil? Many approaches attempted to detect different aspects of network topology and thus provide different information for better understanding how the microbial communities are arranged in the soil. The effective center (or

Table 3 | The five genera selected by the greatest values of Betweenness Centrality (BC) found in each of the sampling sites.

Id	BC	CC	Abundance (%)	Taxonomy
PAMPA BIOME—SITE A				
Acacia plantation				
Otu048	167.86	1.74	0.41	"Firmicutes"; "Clostridia"; "Thermoanaerobacterales"; "Thermoanaerobacteraceae"; Desulfovirgula
Otu075	124.91	2.07	0.28	"Proteobacteria"; AlphaproteoRhizobiales; Bradyrhizobiaceae; Agromonas
Otu074	113.81	1.93	0.28	"Proteobacteria"; GammaproteoXanthomonadales; Xanthomonadaceae; Luteimonas
Otu052	113.40	1.70	0.39	"Bacteroidetes"; "Sphingobacteria"; "Sphingobacteriales"; "Chitinophagaceae"; Segetibacter
Otu090	94.85	3.11	0.22	"Actinobacteria"; ActinoActinomycetales; Streptosporangiaceae; Thermopolyspora
Soybean field				
Otu023	31.83	1.25	0.87	"Acidobacteria"; Acidobacteria_Gp7; unclassified; unclassified; unclassified
Otu036	28.67	1.67	0.46	"Proteobacteria"; AlphaproteoRhizobiales; Beijerinckiaceae; Methylocapsa
Otu053	23.83	2.33	0.34	"Proteobacteria"; BetaproteoMethylophilales; Methylophilaceae; Methylotenera
Otu039	21.00	2.08	0.42	"Proteobacteria"; GammaproteoXanthomonadales; Xanthomonadaceae; Luteimonas
Otu090	18.50	2.57	0.20	"Proteobacteria"; AlphaproteoRhodospirillales; Acetobacteraceae; Acidicaldus
Natural forest				
Otu044	373.99	2.14	0.49	"Proteobacteria"; AlphaproteoRhodospirillales; Rhodospirillaceae; Caenispirillum
Otu091	360.95	2.64	0.22	"Actinobacteria"; ActinoActinomycetales; Nocardioidaceae; Pimelobacter
Otu102	360.62	2.39	0.18	"Firmicutes"; "Clostridia"; Clostridiales; "Lachnospiraceae"; Catonella
Otu052	305.71	2.44	0.41	"Bacteroidetes"; "Sphingobacteria"; "Sphingobacteriales"; "Chitinophagaceae"; Terrimonas
Otu086	291.67	2.39	0.23	"Actinobacteria"; ActinoActinomycetales; Geodermatophilaceae; Blastococcus
Natural pasture				
Otu065	253.60	2.79	0.31	"Proteobacteria"; GammaproteoThiotrichales; Piscirickettsiaceae; Sulfurivirga
Otu074	246.91	2.28	0.26	"Bacteroidetes"; "Sphingobacteria"; "Sphingobacteriales"; "Chitinophagaceae"; Lacibacter
Otu024	232.90	2.77	0.89	"Firmicutes"; "Clostridia"; Clostridiales; "Ruminococcaceae"; Ethanoligenens
Otu022	218.44	2.00	0.92	"Thermodesulfobacteria"; ThermodesulfoThermodesulfobacteriales; Thermodesulfobacteriaceae; Caldimicrobium
Otu023	210.93	2.56	0.90	"Bacteroidetes"; "Sphingobacteria"; "Sphingobacteriales"; "Chitinophagaceae"; Terrimonas
PAMPA BIOME—SITE B				
Natural forest				
Otu114	756.52	2.34	0.17	"Actinobacteria"; ActinoAcidimicrobiales; Acidimicrobiaceae; Ferrimicrobium
Otu116	630.83	2.82	0.16	"Proteobacteria"; BetaproteoBurkholderiales; Burkholderiales_incertae_sedis; Thiomonas
Otu110	479.81	1.91	0.17	"Actinobacteria"; ActinoActinomycetales; Micromonosporaceae; Asanoa
Otu054	446.16	2.26	0.34	"Acidobacteria"; Acidobacteria_Gp22; unclassified; unclassified; unclassified
Otu087	444.01	2.44	0.21	"Acidobacteria"; Acidobacteria_Gp11; unclassified; unclassified; unclassified
Natural pasture				
Otu106	580.47	2.24	0.34	"Bacteroidetes"; "Sphingobacteria"; "Sphingobacteriales"; "Chitinophagaceae"; Flavisolibacter
Otu063	437.57	2.66	0.30	"Actinobacteria"; ActinoActinomycetales; Microbacteriaceae; Microterricola
Otu103	403.97	2.27	0.19	"Proteobacteria"; GammaproteoChromatiales; Chromatiaceae; Nitrosococcus
Otu124	332.86	2.05	0.14	"Bacteroidetes"; "Sphingobacteria"; "Sphingobacteriales"; "Chitinophagaceae"; Lacibacter
Otu130	330.74	2.00	0.13	"Bacteroidetes"; "Sphingobacteria"; "Sphingobacteriales"; "Chitinophagaceae"; Segetibacter
CERRADO BIOME				
Sugarcane field				
Otu085	182.27	1.72	0.23	"Proteobacteria"; BetaproteoBurkholderiales; Oxalobacteraceae; Undibacterium
Otu058	179.22	1.74	0.35	"Proteobacteria"; BetaproteoBurkholderiales; Comamonadaceae; Rhodoferax
Otu104	172.96	2.65	0.18	"Actinobacteria"; ActinoActinomycetales; Sporichthyaceae; Sporichthya
Otu097	157.86	1.96	0.18	"Proteobacteria"; AlphaproteoCaulobacterales; Caulobacteraceae; Caulobacter
Otu030	155.65	1.86	0.59	"Actinobacteria"; ActinoActinomycetales; Kineosporiaceae; Kineosporia
Pinus plantation				
Otu047	180.34	2.52	0.39	"Actinobacteria"; ActinoActinomycetales; Sporichthyaceae; Sporichthya
Otu042	142.37	2.24	0.47	"Proteobacteria"; BetaproteoBurkholderiales; Oxalobacteraceae; Duganella
Otu028	124.23	1.40	0.71	"Bacteroidetes"; "Sphingobacteria"; "Sphingobacteriales"; "Chitinophagaceae"; Terrimonas
Otu060	120.65	2.09	0.28	"Proteobacteria"; Alphaproteobacteria_order_incertae_sedis; Alphaproteobacteria_incertae_sedis; Elioraea
Otu057	119.63	1.87	0.30	"Proteobacteria"; BetaproteoBurkholderiales; Oxalobacteraceae; Janthinobacterium

(Continued)

Table 3 | Continued

Id	BC	CC	Abundance (%)	Taxonomy
Natural forest				
Otu052	276.09	2.17	0.24	"Actinobacteria"; ActinoAcidimicrobiales; Acidimicrobiaceae; Acidimicrobium
Otu080	180.84	2.03	0.13	"Actinobacteria"; ActinoActinomycetales; Acidothermaceae; Acidothermus
Otu050	172.54	2.26	0.25	"Proteobacteria"; AlphaproteoRhizobiales; Beijerinckiaceae; Methylovirgula
Otu067	159.45	1.85	0.18	"Proteobacteria"; Alphaproteobacteria_order_incertae_sedis; Alphaproteobacteria_incertae_sedis; Elioraea
Otu048	156.83	2.36	0.27	"Acidobacteria"; Acidobacteria_Gp4; unclassified; unclassified; unclassified
Natural pasture				
Otu071	259.25	2.20	0.21	"Planctomycetes"; "Planctomycetacia"; Planctomycetales; Planctomycetaceae; Zavarzinella
Otu032	208.29	1.47	0.50	"Bacteroidetes"; "Sphingobacteria"; "Sphingobacteriales"; Sphingobacteriaceae; Mucilaginibacter
Otu022	139.44	1.70	1.02	"Proteobacteria"; BetaproteoBurkholderiales; Burkholderiales_incertae_sedis; Methylibium
Otu038	121.18	2.07	0.37	"Proteobacteria"; GammaproteoOceanospirillales; Litoricolaceae; Litoricola
Otu061	120.52	2.68	0.25	"Proteobacteria"; AlphaproteoRhodospirillales; Acetobacteraceae; Rhodovarius

ID, OTU identification; BC, betweenness centrality; CC, closeness centrality.

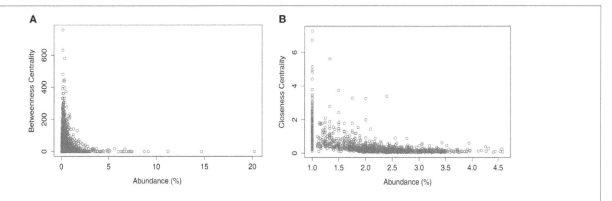

FIGURE 6 | Relationship between taxon relative abundance data from the total number of OTUs at genus level found in different land uses in Pampa and Cerrado biome and betweenes (A) and closeness (B) centrality.

centers) of a network, also called "hubs" might represent keystones species as predicted from network theory (Montoya et al., 2006) however, the network structure is very complex and there is no unifying approach for identifying such hubs. A number of studies have been performed using the degree centrality to identify hubs in networks but we decided to use BC and CC because the degree is a local quantity which does not inform about the importance of a node in the network (Barthélemy, 2004). Our analysis of centrality illustrates that most of soil bacterial taxons are important to the overall connectance of the network (presented high CC), whereas only few taxons play a key role as connectors (presented high BC). Eiler et al. (2012) also detected numerous phylogenetic groups with high number of associations, which may represent groups with particular strong interdependencies. They suggested that in a highly complex environment, like soil, there may be hundreds of such keystones species. The keystone species in soil environment play an exceptionally important role in determining the structure and function of ecosystems. Rudolf and Rasmussen (2013) showed that differences in food network structure were significantly correlated with changes in all ecosystem processes.

The most widely used definition for keystone species is one "whose impact on its community or ecosystem is large, and disproportionately large relative to its abundance" (Power et al., 1996). According to our network analysis, only a fraction of the total number of OTUs presented either positive or negative interactions (**Table 2**) however, the more interactive taxa were also found in more abundance within the soil samples. On the other hand, the interactive taxa did not presented any relationship with the two measures of centrality applied in this study (see **Figure 6**). Recently, Campbell and Kirchman (2013) and Zhang et al. (2013) suggested that abundant and easily detectable organisms might have a high impact on microbial structure, function, and nutrient cycling. Our network analysis corroborated such findings however, the role of less-abundant organisms is not easily understood and might not be neglected. Less abundant members from soil microbial community contributed to biogeochemical process as important sulfate reducers in a long-term experimental peatland field site (Pester et al., 2010). In addition, these rare or only less abundant microorganisms might act as keystone species in complex soil bacterial communities and could serve as a reservoir of genetic and functional diversity and/or buffer ecosystems against species loss or environmental change (Brown et al., 2009). Finally, it was observed that each land use presented a different set of keystone genera and that no keystone genera presented a generalized distribution.

In this study, we investigated the inter-taxa associations in complex microbial soil ecosystems applying systems biology principles. Such approach is essential to explain the persistence of microbial species in a constantly changing ecosystems, and the tolerance of current ecosystems to natural gains and losses of species as well as their vulnerability to unnaturally inflated extinction rates (Montoya et al., 2006). Species interactions could be more important to soil processes than species richness and abundance, especially in complex ecosystems. The visualization of microbial networks allowed us to detect microbial hubs, which are key microbes or microbial behaviors that let us comprehend the complex microbial systems in which they are found. Ultimately, such network models will be able to predict the outcome of community alterations and the effects of perturbations. Although exploring such ecological networks is essential to our better understanding of microbial ecology, more investigations are needed to circumvent important methodological limitations such as prediction of a relationship between two genera through inference of correlations. The technique will benefit from the incorporation of a less simplistic model that take into account not only the relationship between two microbial genera but also the effect of third-party microorganisms in the system and random processes. In addition, the network approach could be used to text the microbial assemblage theories, neutral and niche theories. This approach proves to be valuable to practical community-level conservation biology and represents a step forward in microbial ecology beyond the conventional studies of microbial richness and abundance.

ACKNOWLEDGMENTS

The authors acknowledge the National Council for Scientific and Technological Development (CNPq—Brazil) and the Coordination for the Improvement of Higher Education Personnel (CAPES—Brazil) for their financial support. This work was supported by the Fundação de Amparo à Pesquisa do Estado do Rio Grande do Sul (FAPERGS process no. 1012030) and by the National Council for Scientific and Technological Development (CNPq—Brazil process no. 476762/2010-30 and 479133/2012-3). Publication number 5593 of Netherlands Institute of Ecology (NIOO-KNAW).

REFERENCES

Barberán, A., Bates, S. T., Casamayor, E. O., and Fierer, N. (2012). Using network analysis to explore co-occurrence patterns in soil microbial communities. *ISME J.* 6, 343–351. doi: 10.1038/ismej.2011.119

Barthélemy, M. (2004). Betweenness centrality in large complex networks. *Eur. Phys. J. B* 38, 163–168. doi: 10.1140/epjb/e2004-00111-4

Bastian, M., Heymann, S., and Jacomy, M. (2009). "Gephi: an open source software for exploring and manipulating networks," in *International AAAI Conference on Weblogs and Social Media* (San Jose, CA).

Brown, M. V., Philip, G. K., Bunge, J. A., Smith, M. C., Bissett, A., Lauro, F. M., et al. (2009). Microbial community structure in the North Pacific ocean. *ISME J.* 3, 1374–1386. doi: 10.1038/ismej.2009.86

Campbell, B. J., and Kirchman, D. L. (2013). Bacterial diversity, community structure and potential growth rates along an estuarine salinity gradient. *ISME J.* 7, 210–220. doi: 10.1038/ismej.2012.93

Chaffron, S., Rehrauer, H., Pernthaler, J., and von Mering, C. (2010). A global network of coexisting microbes from environmental and whole-genome sequence data. *Genome Res.* 20, 947–959. doi: 10.1101/gr.104521.109

Choi, J. K., Yu, U. S., Yoo, O. J., and Kim, S. (2005). Differential coexpression analysis using microarray data and its application to human cancer. *Bioinformatics* 21, 4348–4355. doi: 10.1093/bioinformatics/bti722

Duran-Pinedo, A. E., Paster, B., Teles, R., and Frias-Lopez, J. (2011). Correlation network analysis applied to complex biofilm communities. *PLoS ONE* 6:e28438. doi: 10.1371/journal.pone.0028438

Eiler, A., Heinrich, F., and Bertilsson, S. (2012). Coherent dynamics and association networks among lake bacterioplankton taxa. *ISME J.* 6, 330–342. doi: 10.1038/ismej.2011.113

Erdös, P., and Rényi, A. (1959). On random graphs. *Publ. Math. Debrecen.* 5, 290–297.

Estrada, E. (2007). Characterization of topological keystone species local, global and "meso-scale" centralities in food webs. *Ecol. Complex.* 4, 48–57. doi: 10.1016/j.ecocom.2007.02.018

Faust, K., and Raes, J. (2012). Microbial interactions: from networks to models. *Nat. Rev. Microbiol.* 10, 538–550. doi: 10.1038/nrmicro2832

Faust, K., Sathirapongsasuti, J. F., Izard, J., Segata, N., Gevers, D., Raes, J., et al. (2012). Microbial co-occurrence relationships in the human microbiome. *PLoS Comput. Biol.* 8:e1002606. doi: 10.1371/journal.pcbi.1002606

Freeman, L. C. (1979). Centrality in social networks, conceptual clarification. *Soc. Netw.* 1, 215–239. doi: 10.1016/0378-8733(78)90021-7

Friedman, J., and Alm, E. J. (2012). Inferring correlation networks from genomic survey data. *PLoS Comput. Biol.* 8:e1002687. doi: 10.1371/journal.pcbi.1002687

Gilbert, J. A., Steele, J. A., Caporaso, J. G., Steinbrück, L., Reeder, J., Temperton, B., et al. (2012). Defining seasonal marine microbial community dynamics. *ISME J.* 6, 298–308. doi: 10.1038/ismej.2011.107

Graham, D. W., Knapp, C. W., Van Vleck, E. S., Bloor, K., Lane, T. B., and Graham, C. E. (2007). Experimental demonstration of chaotic instability in biological nitrification. *ISME J.* 1, 385–393. doi: 10.1038/ismej.2007.45

Griffiths, R. I., Thomson, B. C., James, P., Bell, T., Bailey, M., and Whiteley, A. S. (2011). The bacterial biogeography of British soils. *Environ. Microbiol.* 13, 1642–1654. doi: 10.1111/j.1462-2920.2011.02480.x

Haruta, S., Kato, S., Yamamoto, K., and Igarashi, Y. (2009). Intertwined interspecies relationships: approaches to untangle the microbial network. *Environ. Microbiol.* 11, 2963–2969. doi: 10.1111/j.1462-2920.2009.01956.x

Huse, S. M., Welch, D. M., Morrison, H. G., and Sogin, M. L. (2010). Ironing out the wrinkles in the rare biosphere through improved OTU clustering. *Environ. Microbiol.* 12, 1889–1898. doi: 10.1111/j.1462-2920.2010.02193.x

Klappenbach, J. A., Saxman, P. R., Cole, J. R., and Schmidt, T. M. (2001). rrndb: the Ribosomal RNA Operon Copy Number Database. *Nucl. Acids Res.* 29, 181–184. doi: 10.1093/nar/29.1.181

Lemos, L. N., Fulthorpe, R. R., Triplett, E. W., and Roesch, L. F. W. (2011). Rethinking microbial diversity analysis in the high throughput sequencing era. *J. Microbiol. Methods* 86, 42–51. doi: 10.1016/j.mimet.2011.03.014

Lennon, J. T., and Jones, S. E. (2011). Microbial seed banks: the ecological and evolutionary implications of dormancy. *Nat. Rev. Microbiol.* 9, 119–130. doi: 10.1038/nrmicro2504

Lueders, T., Kindler, R., Miltner, A., Friedrich, M. W., and Kaestner, M. (2006). Identification of bacterial micropredators distinctively active in a soil microbial food web. *Appl. Environ. Microbiol.* 72, 5342–5348. doi: 10.1128/AEM.00400-06

Lupatini, M., Suleiman, A. K. A., Jacques, R. J. S., Antoniolli, Z. I., Kuramae, E. E., Camargo, F. A. D., et al. (2013). Soil-Borne bacterial structure and diversity does not reflect community activity in Pampa biome. *PLoS ONE* 8:e76465. doi: 10.1371/journal.pone.0076465

Martín González, A. M., Dalsgaard, B., and Olesen, J. M. (2010). Centrality measures and the importance of generalist species in pollination networks. *Ecol. Complex.* 7, 36–43. doi: 10.1016/j.ecocom.2009.03.008

Montoya, J. M., Pimm, S. L., and Sole, R. V. (2006). Ecological networks and their fragility. *Nature* 442, 259–264. doi: 10.1038/nature04927

Newman, M. E. J. (2006). Modularity and community structure in networks. *Proc. Natl. Acad. Sci. U.S.A.* 103, 8577–8582. doi: 10.1073/pnas.0601602103

Oliveira, P. S., and Marquis, R. J. (2002).*Cerrados of Brazil: Ecology and Natural History of a Neotropical Savanna*. New York, NY: Columbia Press.

Overbeck, G. E., Mueller, S. C., Fidelis, A., Pfadenhauer, J., Pillar, V. D., Blanco, C. C., et al. (2007). Brazil's neglected biome: the South Brazilian Campos. *Perspect. Plant Ecol. Evol. Syst.* 9, 101–116. doi: 10.1016/j.ppees.2007.07.005

Pester, M., Bittner, N., Deevong, P., Wagner, M., and Loy, A. (2010). A "rare biosphere" microorganism contributes to sulfate reduction in a peatland. *ISME J.* 4, 1591–1602. doi: 10.1038/ismej.2010.75

Philippot, L., Spor, A., Henault, C., Bru, D., Bizouard, F., Jones, C. M., et al. (2013). Loss in microbial diversity affects nitrogen cycling in soil. *ISME J.* 7, 1609–1619. doi: 10.1038/ismej.2013.34

Pinto, A. J., and Raskin, L. (2012). PCR biases distort bacterial and archaeal community structure in pyrosequencing datasets. *PLoS ONE* 7:e43093. doi: 10.1371/journal.pone.0043093

Power, M. E., Tilman, D., Estes, J. A., Menge, B. A., Bond, W. J., Mills, L. S., et al. (1996). Challenges in the quest for keystones. *Bioscience* 46, 609–620. doi: 10.2307/1312990

Prasad, S., Manasa, P., Buddhi, S., Singh, S. M., and Shivaji, S. (2011). Antagonistic interaction networks among bacteria from a cold soil environment. *FEMS Microbiol. Ecol.* 78, 376–385. doi: 10.1111/j.1574-6941.2011.01171.x

Prosser, J. I., Bohannan, B. J. M., Curtis, T. P., Ellis, R. J., Firestone, M. K., Freckleton, R. P., et al. (2007). The role of ecological theory in microbial ecology. *Nat. Rev. Microbiol.* 5, 384–392. doi: 10.1038/nrmicro1643

Rafrafi, Y., Trably, E., Hamelin, J., Latrille, E., Meynial-Salles, I., Benomar, S., et al. (2013). Sub-dominant bacteria as keystone species in microbial communities producing biohydrogen. *Int. J. Hydrogen Energ.* 38, 4975–4985. doi: 10.1016/j.ijhydene.2013.02.008

Rampelotto, P. H., Ferreira, A. D. S., Muller Barboza, A. D., and Wurdig Roesch, L. F. (2013). Changes in diversity, abundance, and structure of soil bacterial communities in Brazilian savanna under different land use systems. *Microb. Ecol.* 66, 593–607. doi: 10.1007/s00248-013-0235-y

Rodriguez-Lanetty, M., Granados-Cifuentes, C., Barberan, A., Bellantuono, A. J., and Bastidas, C. (2013). Ecological inferences from a deep screening of the Complex Bacterial Consortia associated with the coral, *Porites astreoides. Mol. Ecol.* 22, 4349–4362. doi: 10.1111/mec.12392

Roesch, L. F. W., Fulthorpe, R. R., Pereira, A. B., Pereira, C. K., Lemos, L. N., Barbosa, A. D., et al. (2012). Soil bacterial community abundance and diversity in ice-free areas of Keller peninsula, antarctica. *Appl. Soil Ecol.* 61, 7–15. doi: 10.1016/j.apsoil.2012.04.009

Rudolf, V. H., and Rasmussen, N. L. (2013). Population structure determines functional differences among species and ecosystem processes. *Nat. Commun.* 4, 1–7. doi: 10.1038/ncomms3318

Schloss, P. D., Westcott, S. L., Ryabin, T., Hall, J. R., Hartmann, M., Hollister, E. B., et al. (2009). Introducing mothur: open-source, platform-independent, community-supported software for describing and comparing microbial communities. *Appl. Environ. Microbiol.* 75, 7537–7541. doi: 10.1128/AEM. 01541-09

Steele, J. A., Countway, P. D., Xia, L., Vigil, P. D., Beman, J. M., Kim, D. Y., et al. (2011). Marine bacterial, archaeal and protistan association networks reveal ecological linkages. *ISME J.* 5, 1414–1425. doi: 10.1038/ismej.2011.24

Suleiman, A. K. A., Lupatini, M., Boldo, J. T., Pereira, M. G., and Wurdig Roesch, L. F. (2013). Shifts in soil bacterial community after eight years of land-use change. *Syst. Appl. Microbiol.* 36, 137–144. doi: 10.1016/j.syapm.2012.10.007

Taylor, R. (1990). Interpretation of the correlation-coefficient - a basic review. *J. Diagn. Med. Sonogr.* 6, 35–39. doi: 10.1177/875647939000600106

Vick-Majors, T. J., Priscu, J. C. A, and Amaral-Zettler, L. (2014). Modular community structure suggests metabolic plasticity during the transition to polar night in ice-covered Antarctic lakes. *ISME J.* 8, 778–789. doi: 10.1038/ismej.2013.190

Zhang, X., Liu, W., Schloter, M., Zhang, G., Chen, Q., Huang, J., et al. (2013). Response of the abundance of key soil microbial nitrogen-cycling genes to multi-factorial global changes. *PLoS ONE* 8:e76500. doi: 10.1371/journal.pone.0076500

Zhou, J., Deng, Y., Luo, F., He, Z., and Yang, Y. (2011). Phylogenetic molecular ecological network of soil microbial communities in response to elevated CO_2. *Mbio* 2, e00122–e00111. doi: 10.1128/mBio.00122-11

Conflict of Interest Statement: The authors declare that the research was conducted in the absence of any commercial or financial relationships that could be construed as a potential conflict of interest.

9

Future rainfall variations reduce abundances of aboveground arthropods in model agroecosystems with different soil types

Johann G. Zaller[1]*, Laura Simmer[1], Nadja Santer[1], James Tabi Tataw[1], Herbert Formayer[2], Erwin Murer[3], Johannes Hösch[4] and Andreas Baumgarten[4]

[1] Department of Integrative Biology and Biodiversity Research, Institute of Zoology, University of Natural Resources and Life Sciences Vienna, Austria
[2] Department of Water-Atmosphere-Environment, Institute of Meteorology, University of Natural Resources and Life Sciences Vienna, Austria
[3] Institute of Land and Water Management Research, Federal Agency for Water Management, Petzenkirchen, Austria
[4] Division for Food Security, Institute of Soil Health and Plant Nutrition, Austrian Agency for Health and Food Safety (AGES), Vienna, Austria

Edited by:
Vimala Nair, University of Florida, USA

Reviewed by:
Astrid Rita Taylor, Swedish University of Agricultural Sciences, Sweden
Holger Hoffmann, Leibniz Universität Hannover, Germany
Todd Z. Osborne, University of Florida, USA

*Correspondence:
Johann G. Zaller, Department of Integrative Biology and Biodiversity Research, Institute of Zoology, University of Natural Resources and Life Sciences Vienna, Gregor Mendel Straße 33, A-1180 Vienna, Austria
e-mail: johann.zaller@boku.ac.at

Climate change scenarios for Central Europe predict less frequent but heavier rainfalls and longer drought periods during the growing season. This is expected to alter arthropods in agroecosystems that are important as biocontrol agents, herbivores or food for predators (e.g., farmland birds). In a lysimeter facility (totally 18 3-m^2-plots), we experimentally tested the effects of long-term current vs. prognosticated future rainfall variations (15% increased rainfall per event, 25% more dry days) according to regionalized climate change models from the Intergovernmental Panel on Climate Change (IPCC) on aboveground arthropods in winter wheat (*Triticum aestivum* L.) cultivated at three different soil types (calcaric phaeozem, calcic chernozem and gleyic phaeozem). Soil types were established 17 years and rainfall treatments 1 month before arthropod sampling; treatments were fully crossed and replicated three times. Aboveground arthropods were assessed by suction sampling, their mean abundances ($\pm$ SD) differed between April, May and June with 20 ± 3 m^{-2}, 90 ± 35 m^{-2}, and 289 ± 93 individuals m^{-2}, respectively. Averaged across sampling dates, future rainfall reduced the abundance of spiders (Araneae, −47%), cicadas and leafhoppers (Auchenorrhyncha, −39%), beetles (Coleoptera, −52%), ground beetles (Carabidae, −41%), leaf beetles (Chrysomelidae, −64%), spring tails (Collembola, −58%), flies (Diptera, −73%) and lacewings (Neuroptera, −73%) but increased the abundance of snails (Gastropoda, +69%). Across sampling dates, soil types had no effects on arthropod abundances. Arthropod diversity was neither affected by rainfall nor soil types. Arthropod abundance was positively correlated with weed biomass for almost all taxa; abundance of Hemiptera and of total arthropods was positively correlated with weed density. These detrimental effects of future rainfall variations on arthropod taxa in wheat fields can potentially alter arthropod-related agroecosystem services.

Keywords: agroecology, climate change ecology, precipitation patterns, soil types, aboveground invertebrates, lysimeter, winter wheat, animal-plant interactions

INTRODUCTION

Climate change will very likely cause a seasonal shift in precipitation in Central Europe resulting in less frequent but more extreme rainfall events during summer but increased precipitation during winter (IPCC, 2007, 2013). Regionalisations of these climate models for eastern parts of Central Europe prognosticate little changes or even slight decreases in annual rainfall amounts until 2100 (Eitzinger et al., 2001; Kromp-Kolb et al., 2008). Indeed, so far for eastern Austria no change in total yearly precipitation was measured during the last decades (Formayer and Kromp-Kolb, 2009). The direction, magnitude and variability of such changes in precipitation events and their effects on ecosystem functioning will depend on how much the change deviates from the existing variability and the ability of ecosystems and inhabiting organisms to adapt to the new conditions (Beier et al., 2012).

In many natural and agriculturally managed ecosystems arthropods are the most abundant and diverse group of animals (Altieri, 1999; Speight et al., 2008). Abundances of epigeic arthropods in an arable field can reach thousands of individuals m^{-2} comprising hundreds of species (Romanowsky and Tobias, 1999; Östman et al., 2001; Pfiffner and Luka, 2003; Batary et al., 2012; Frank et al., 2012; Querner et al., 2013). These arthropods play important ecological roles as herbivores and detritivores (Seastedt and Crossley, 1984), are valued for pollination, seed dispersal and predation (Steffan-Dewenter et al., 2001), are important predators and parasitoids (Thies et al., 2003; Drapela et al., 2008; Zaller et al., 2008a, 2009) and are a food source for many vertebrates and invertebrates (Price, 1997; Brantley and Ford, 2012; Hallmann et al., 2014). As arthropods can have a strong influence on nutrient cycling processes (Seastedt and Crossley, 1984), they are also very important for ecosystem net primary production (Abbas and Parwez, 2012). Predicted longer drought intervals between rainfall events will increase drought stress for crops while changes in the amount and timing of rainfall will affect yields and the biomass production of crops (Eitzinger et al., 2001; Alexandrov et al., 2002; Thaler et al., 2008). These changes in vegetation structure and quality will also affect associated arthropods (Andow, 1991). Moreover, it has also been shown that changes in the magnitude and variability of rainfall events is likely to be more important for arthropods than changes in annual amounts of rainfall (Curry, 1994; Speight et al., 2008; Singer and Parmesan, 2010). Most studies investigating potential effects of climate change on arthropods have focused on the effects of changes in atmospheric CO_2 concentrations or temperature rather than precipitation (e.g., Cannon, 1998; Andrew and Hughes, 2004; Hegland et al., 2009; Hamilton et al., 2012). However, changes in variations of rainfall are likely to have a greater effect on species' distributions than are changes in temperature, especially among rare species (Elmes and Free, 1994).

Surprisingly, very few studies investigated the effects of different rainfall variations on aboveground arthropod abundance in arable agroecosystems, although arable land is ecologically important in terms of its diverse arthropod fauna (Frampton et al., 2000; Tscharntke et al., 2005; Drapela et al., 2008) and its interaction with natural ecosystems in a landscape matrix (Tscharntke and Brandl, 2004; Frank et al., 2012; Balmer et al., 2013; Coudrain et al., 2014). Results from studies investigating the effects of rainfall variations on arthropods are not consistent ranging from increased spider activity to a reduced activity of Collembola under reduced rainfall (Lensing et al., 2005) while others showed little influence of rainfall on spiders (Buchholz, 2010). It also appears that even short rainfall events in spring can influence various groups of farmland arthropods for the following months (Frampton et al., 2000).

To the best of our knowledge, no study assessed the effects of rainfall variations on arthropods in wheat, one of the most important cereal crops worldwide. Moreover, experiments studying the effects of precipitation on ecosystems are usually conducted at different locations with different soil types, thus confounding location with soil types and making it impossible to test to what extent soil types can potentially buffer rainfall variations on ecosystem processes (Beier et al., 2012). The few studies investigating arthropod abundance in different soil types found a significant difference in soil fauna abundance and diversity (Loranger-Merciris et al., 2007) or invertebrate community composition between different soil types (Ivask et al., 2008; Tabi Tataw et al., 2014).

Hence, the objectives of the current study were: (1) To examine effects of different rainfall variations on the abundance of aboveground arthropods in winter wheat, (2) to assess to what extent different soil types alter potential responses of aboveground arthropods to rainfall variations. The investigations were based on the hypotheses that differences in the amount and variability of rainfall alter the structure of winter wheat stands by either affecting growth of crops and/or weeds (Porter and Semenov, 2005) and consequently affecting the abundance and diversity of arthropods (Duelli and Obrist, 2003; Menalled et al., 2007). As

the composition of arthropod communities changes during the season we expected that different arthropod taxa would be differently affected by rainfall variations (Price et al., 2011). Moreover, different moisture sensitivities/drought tolerances of arthropod taxa (Finch et al., 2008) will be affected by soil types with different water holding capacities and soil types will also modify the growth and structure of vegetation that will interact with rainfall variations in affecting arthropods.

MATERIALS AND METHODS
STUDY SITE

The experiment was carried out in the lysimeter experimental facility of the Austrian Agency for Health and Food Safety (AGES), in Vienna, Austria (northern latitude $48°15'11''$, eastern longitude $16°28'47''$) at an altitude of 160 m above sea level. The facility is located in a transition area of the Western European oceanic (mild winters, wet, cool summer) and the Eastern European continental climatic area (cold winters, hot summers) ecologically referred to as the Pannonium region. Long-term mean annual precipitation at this site is 550–600 mm at a mean air temperature of $9.5°C$ (Danneberg et al., 2001).

The lysimeter facility was established in 1995 and consists of 18 cylindrical vessels made of stainless steel each with a surface area of $3.02\ m^2$ and a depth of 2.45 m (**Figure 1**). The lysimeters are arranged in two parallel rows with nine lysimeter plots in each row; one row was subjected to current rainfall the other row to prognosticated rainfall. Within each row three soil types were randomized to ensure replicates of each soil type in each row (see below for more details on treatment factors); each treatment was replicated three times ($n = 3$).

EXPERIMENTAL TREATMENTS
SOIL TYPES

In 1995, the lysimeters were filled with three different soil types representing around 80% of the agriculturally most productive

FIGURE 1 | Experimental winter wheat plots containing three different soil types (calcaric phaeozem, calcic chernozem, gleyic phaeozem) subjected to long-time current prognosticated rainfall variations according to regionalized climate change models.

area in Austria (region Marchfeld; east of Vienna, Austria): cal-caric phaeozem (S), calcic chernozem (T), and gleyic phaeozem (F; soil nomenclature after World Soil Classification, FAO, 2002). The soil material was carefully excavated from their native sites in 10 cm layers and filled into the lysimeter vessels retaining their original bulk densities of 1.4 g cm^{-3} (Danneberg et al., 2001). See Tabi Tataw et al. (2014) for further details on the soil charac-teristics. Briefly, the calcic chernozem and the calcaric phaeozem have a fully developed AC-profile, emerging from carbonate-fine siliceous material. The thickness of the A horizon is at least 30 cm, the humus form is mull with both 4.9% humus content (Nestroy et al., 2011). The calcic chernozem is moderately dry, the calcaric phaeozem is dry; both soil types consist of fine sediment to silt fine sand (Danneberg et al., 2001). The gleyic phaeozem is a soil of former hydromorphic sites with 2.1% humus content as mull; the fully developed AC-profile and the thickness of the A-horizon is at least 30 cm thick (Nestroy et al., 2011). This gleyic phaeozem is well supplied with water and consists of fine sediment to silt fine sand; its high lime content, gives this soil type neutral to slightly alkaline pH. Mean profile water contents are 375, 595, and 550 mm for S, T, and F soil, respectively.

RAINFALL SCENARIOS

Starting in 2011, the lysimeters were subjected to two rainfall regimes, one based on past local observations ("curr. rainfall") and one based on a regionalization of the IPCC 2007 climate change scenario A2 for the period 2071–2100 ("progn. rain-fall"; IPCC, 2007). Both the current and the future precipita-tion variations were calculated using the software LARS-WG (Version 3.0; Semenov and Barrow, 2002). In contrast to clas-sic approaches using directly the projected climate time series as model input our approach with LARS-WG used only the delta values (Hoffmann and Rath, 2012, 2013). The current long-term rainfall variations was based on the precipitation amount and frequency for a location in about 10 km distance from the study site (village of Großenzersdorf) between the years 1971–2000. The future rainfall scenario for the year 2071–2100 is based on the local climatology and the climate change signal from the mean of the regional climate model scenarios from the EU-project ENSEMBLES (Christensen and Christensen, 2007). This stochastic weather generator LARS-WG was used to trans-fer the derived local climate change signals to daily precipitation rates. To exclude natural rainfall the lysimeters were covered with a 5 m high roof of transparent plastic foil from March until December in each year, all sidewalls were open allowing ventilation and free movement of animals (**Figure 1**). During winter the facility was uncovered and all lysimeters received nat-ural precipitation. Rainfall amounts (tap water) according to the model calculations were applied to nine lysimeters in a row using an automatic sprinkler system. Rainfall treatments started on 22 March 2012. Until the last arthropod sampling on 18 June 2012 the curr. rainfall plots received 156.4 mm and the progn. rainfall plots 136.3 mm irrigation water (−13% less amount of rain). Averaged over the study period, the curr. rainfall plots received 3.7 mm per rain event vs. 3.2 mm per event for the progn. rainfall plots (13% difference); progn. rainfall had 25% more dry days than curr. rainfall treatments (**Figure 2A**). Irrigation

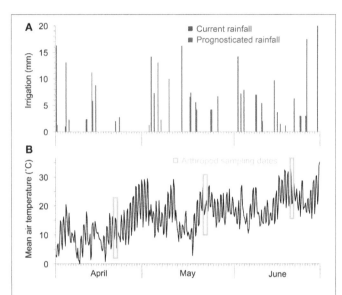

FIGURE 2 | Amount of applied rainfall applied onto treatment plots (A) and mean air temperature (B) during the course of the experiment.

was always performed in early morning at low sunlight; side walls of the transparent cover were automatically closed dur-ing irrigation. Weather stations (Delta-T Devices, Cambridge, UK) were installed between and outside of the lysimeters for monitoring air temperature (**Figure 2B**), wind speed and direc-tion, global radiation and rainfall. Soil matric potential (ψm, also called soil water potential) was measured using three pF sensors per lysimeter installed in 10 cm depth (ecoTech Umwelt-Messsysteme GmbH, Bonn, Germany). The soil matric potential was automatically measured every 15 min and represents the pres-sure it takes to pull water out of soil and increases as the soil gets drier. Technically the pF sensor measure heat capacity in a porous ceramic tip that contains a heating element and temper-ature sensors. The correlations of pF values and measured heat capacity is achieved by a sensor-specific calibration curve (www.ecotech-bonn.de/en/produkte/Bodenkunde/pF-meter.html). The matric potential changes with the soil water content and com-monly varies between different soil types. Soil matric potential is usually expressed in pF units which is the log of the soil tension in hPa (e.g., log of 10,000 hPa is equal to $pF = 4$). Daily pF val-ues were calculated by averaging the individual readings of each lysimeter. Field capacity of soil types was $pF = 1.8$, permanent wilting point for crops $pF = 4.2$.

CROP WHEAT

Winter wheat (*Triticum aestivum* L. cv. Capo) was sown at a den-sity of 400 seeds m^{-2} on 11 October 2011 after the precrop white mustard. Weeds in the treatment plots were controlled by spray-ing a mixture of the herbicides Express-SW (active ingredient: tribenuronmethyl; Kwizda Agro, Vienna, Austria) at 25 g ha^{-1}, Starane XL (a.i.: fluroxypyr and florasulam; Dow AgroSciences, Indianapolis, IN, USA) at 750 ml ha^{-1} and water at 300 l ha^{-1} on 30 March 2012. Fertilization was applied according to recommen-dations for farmers after soil analyses (**Table 1**).

Table 1 | Fertilization of winter wheat crops in lysimeter plots with different soil types (S —calcaric phaeozem, F—gleyic phaeozem, T—calcic chernozem).

Fertilizer type	Fertilizer amount per soil type (kg ha^{-1})			Date
	S	**F**	**T**	
P$_2$O$_5$–Triplesuperphosphate	0	55	55	11 October 2011
K$_2$O–Kali 60	40	0	0	11 October 2011
N–NAC (Nitramoncal 27%)	25	40	40	08 March 2012
N–NAC (Nitramoncal 27%)	30	40	40	12 April 2012
N–NAC (Nitramoncal 27%)	35	50	50	16 May 2012

Wheat growth was measured from the soil surface to the tip of the spike on 10–15 marked crop plants per lysimeter around the arthropod sampling dates (see below). Additionally, the number of weed individuals per lysimeter (weed density) was counted during these dates. Lysimeters were harvested on 5 July 2012 by cutting all vegetation (winter wheat and weeds) by hand at 5 cm above surface. Crop and weed plants were separated, crop plants devided in straw and spikes and everything was weighed after drying at 50°C for 48 h. In order to avoid boundary effects all measurements on crops were conducted in the central area of each lysimeter up to 20 cm distance from the edge of each lysimeter.

ARTHROPOD SAMPLING

All arthropods dwelling on the soil surface and on the vegetation in each of the 18 lysimeters were collected using a commercial garden vacuum (Stihl SH 56-D, Dieburg, Germany) equipped with an insect sampling net. For sampling, the suction tube was carefully moved between the crop plants across the lysimeter area in order to avoid that the sampling efficiency is too much influenced by vegetation structure, height and density (Southwood, 1978; Brook et al., 2008). To impede the escape of the arthropods, a 1 m high barrier made of plastic film was attached to the borders of the lysimeter vessels. Suction sampling was performed for 5 min in each lysimeter; afterwards, each plot was thoroughly inspected for another 20 min for remaining arthropods. This sampling procedure was performed on April 24–25, May 22–23, and June 19, 2012. Air temperature during arthropod sampling dates was on average 18.2°C on the first sampling event, 23.3°C on the second, and 30.4°C on the third sampling event (**Figure 2B**). Sampling was carried out only when the vegetation and soil surface was dry. After collection, the arthropods were sorted out, cleaned from attached soil, preserved in 80% ethylene alcohol and identified at the level of taxonomic order or families (Bellmann, 1999; Bährmann and Müller, 2005). Taxa with less than 0.3 individuals m^{-2} were lumped together in a group of rare individuals. Arthropod abundance was expressed in individuals m^{-2} and relative abundance of the identified groups to the arthropod community present in each lysimeter was calculated in percentage based on the m^{-2} values.

STATISTICAL ANALYSES

First, all measured parameters were tested for normal distribution and variance homogeneity using the Kolmogorov-Smirnov-Test and Levene-Test, respectively. The two parameters that did not meet the requirements of parametric statistics, Hemiptera and total individuals from the May sampling, were Boxcox transformed. Secondly, for all arthropod abundance parameters, repeated measurement analysis of variance (ANOVA) with the factors Rainfall (two levels: longtime current rainfall variations vs. prognosticated rainfall variations), Soil type (three levels: F, S, and T soils) and Sampling date (three dates: April, May, June sampling) were conducted. Additionally, to test for treatment effects at each sampling date separately, two-factorial ANOVAs with the factors Rainfall and Soil type and their interactions were conducted for arthropod taxa and for soil pF values. As a measure of community diversity the Simpson and the Shannon index were calculated and also tested with a two-factorial ANOVA for each sampling date separately (Rosenzweig, 1995). Pearson correlations were performed between arthropod abundance, crop height, crop and weed biomass and weed abundance. All statistical analyses were performed using the freely available software "R" (version 3.0.2; R Core Team, 2013). Statistical results with $P > 0.50 < 0.10$ were considered marginally significant. Values within the text are means ± SD.

RESULTS

Soil matrix potential was significantly affected by rainfall ($P < 0.001$) and soil types ($P < 0.001$; rainfall × soil type interaction: $P < 0.001$) with sandy soils showing the lowest and F and T soil the highest pF values under both rainfall treatments (**Figure 3**).

Arthropod abundances differed highly significantly between sampling dates; rainfall variations significantly affected arthropod abundances at different sampling dates (i.e., rainfall × sampling date interaction; repeated measures ANOVA, **Table 2, Figure 4**). Averaged across rainfall variations and soil types total arthropod abundance in April was 20.38 ± 3.24 m^{-2}, in May 89.62 ± 34.74 m^{-2} and in June 289.23 ± 92.84 m^{-2} (**Figure 4**). Overall, Hymenoptera was the dominant order in April; Hemiptera, Hymenoptera and Acari were dominant in May and Hemiptera were the most dominant group in June; especially the abundance of Hemiptera, Collembola and Acari increased from April to June.

When analyzing the arthropod abundances separately for each sampling date using Two-Way ANOVAs, prognosticated rainfall in April significantly reduced abundances of Gastropoda by 69% and of Auchenorrhyncha by 61% (**Table 3, Figure 4**). In May, prognosticated rainfall significantly reduced Collembola by 53%, Diptera by 59%, Neuroptera by 73%, and Saltatoria by 70% (**Table 3, Figure 4**). In April and May, soil types had no effect of the abundance of arthropods (except for the group of not determinable arthropods; **Table 3**). In June, prognosticated rainfall significantly reduced Araneae by 56%, Auchenorrhyncha by 47%, Coleoptera, and Collembola each by 62%, Chrysomelidae by 66%, Diptera by 77%, and total individuals by 61% (**Table 3, Figure 4**). All other arthropod taxa were not affected by rainfall. In June, soil types had no effect on arthropod abundance except for Auchenorrhyncha (**Table 3**).

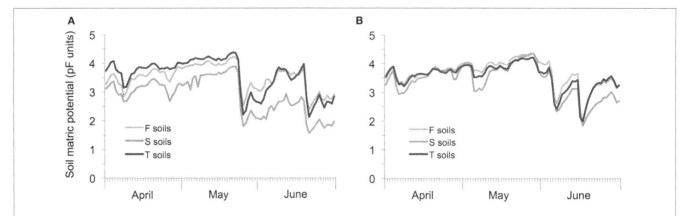

FIGURE 3 | Mean soil matric potential in pF units in winter wheat cultivated under current (A) and prognosticated rainfall variations (B) at the different soil types calcaric phaeozem (S), gleyic phaeozem (F) and calcic chernozem (T).

Table 2 | Summary of repeated measurement ANOVA results of the influence of rainfall patterns (current and prognosticated rainfall), soil types (calcaric phaeozem, calcic chernozem, and gleyic phaeozem) and sampling date (April, May, June 2012) on total abundance of arthropods in winter wheat.

Factor	F	P
Rainfall	4.36	0.059
Soil type	0.04	0.961
Sampling date	20.87	**<0.001**
Rainfall × Soil type	1.00	0.398
Rainfall × Sampling date	6.33	**0.006**
Soil type × Sampling date	0.39	0.815
Rainfall × Soil type × Sampling date	0.44	0.776

Significant effects are in bold.

Considering the relative abundance (i.e., percentage contribution to arthropod community) of the identified arthropod groups for each sampling date, rainfall variations significantly affected Collembola ($P = 0.036$) and Neuroptera ($P = 0.041$) in May and Diptera ($P = 0.041$) in June; with the exception of the relative abundance of rare individuals in April ($P = 0.027$) the composition of arthropod communities was not affected by soil types (**Figure 4**).

Across sampling dates, absolute abundance of Araneae (-43%), Coleoptera (-48%), Carabidae (-41%), Chrysomelidae (-64%), Collembola (-58%), Diptera (-75%), Auchenorrhyncha (-39%), and Neuroptera (-73%) were significantly reduced under prognosticated rainfall, also total arthropod abundance were marginally significantly lower under prognosticated rainfall than under current rainfall (**Table 3, Figure 5**). Only the abundance of Gastropoda increased by 69% in the prognosticated rainfall compared to current rainfall (**Figure 6**). There was no effect of soil types on any of the identified arthropod groups across sampling dates (**Table 3, Figure 6**). Considering the relative abundances across sampling dates, only the relative abundance of Diptera ($P = 0.027$) and Gastropoda ($P = 0.031$) were significantly affected by rainfall; soil types only

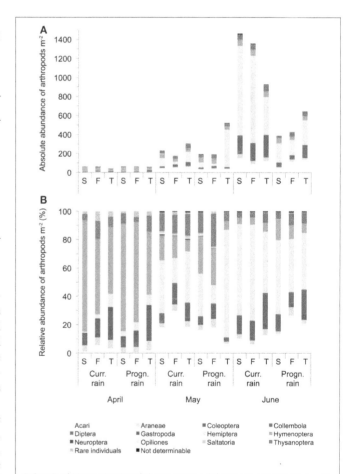

FIGURE 4 | Mean absolute (A) and relative (B) abundance of arthropods per m² at the three sampling dates in winter wheat cultivated under current and prognosticated rainfall variations at the different soil types calcaric phaeozem (S), gleyic phaeozem (F) and calcic chernozem (T).

significantly affected the relative abundance of rare individuals ($P = 0.010$). Hemiptera showed the highest relative abundance in all fields (**Figure 6**). Rainfall variations and soil types had no effect on the diversity indices of arthropod communities (data not shown).

Table 3 | Summary of ANOVA results of the influence of rainfall patterns (long time current vs. prognosticated rainfall patterns) and different soil types (calcaric phaeozem, calcic chemozem and gleyic phaeozem) on abundance of arthropods in winter wheat at different sampling dates.

| Arthropod | April sampling | | | | May sampling | | | | June sampling | | | | Across dates | | | |
| | Rainfall | | Soil types | | Rainfall | | Soil types | | Rainfall | | Soil types | | Rainfall | | Soil types | |
Taxa	F	P	F	P	F	P	F	P	F	P	F	P	F	P	F	P
Acari	1.274	0.281	0.673	0.529	2.043	0.178	0.228	0.800	0.298	0.595	0.234	0.795	0.992	0.339	0.347	0.714
Araneae	0.000	1.000	0.072	0.931	0.000	1.000	3.276	0.073	11.927	**0.005**	0.056	0.946	12.844	**0.004**	0.367	0.700
Coleoptera	1.960	0.187	0.270	0.768	0.041	0.843	0.985	0.402	21.519	**0.001**	3.227	0.075	14.757	**0.002**	2.999	0.088
Carabidae	4.050	0.067	1.050	0.380	3.559	0.084	0.912	0.428	0.966	0.345	0.490	0.624	6.169	**0.029**	1.518	0.258
Chrysomelidae	0.333	0.574	1.000	0.397	0.125	0.712	1.000	0.397	15.591	**0.002**	2.187	0.155	16.248	**0.002**	2.624	0.113
Collembola	0.865	0.371	2.758	0.103	6.019	**0.030**	2.480	0.125	10.219	**0.008**	1.750	0.215	13.750	**0.003**	2.919	0.093
Diptera	2.286	0.156	1.000	0.397	5.831	**0.032**	1.241	0.324	11.794	**0.005**	0.240	0.791	13.945	**0.003**	0.426	0.663
Gastropoda	6.750	**0.023**	2.583	0.117	0.152	0.704	4.581	**0.033**	1.333	0.271	2.083	0.167	0.107	0.749	3.663	0.057
Hemiptera	1.768	0.208	2.268	0.146	0.639	0.440	1.680	0.227	4.403	0.058	0.342	0.717	2.495	0.140	0.026	0.975
Heteroptera	0.000	1.000	0.250	0.783	0.305	0.591	1.622	0.238	0.000	1.000	0.906	0.430	0.152	0.704	1.570	0.248
Auchenorrhyncha	5.042	**0.044**	3.792	0.053	3.018	0.108	0.031	0.969	10.670	**0.007**	4.227	**0.041**	8.119	**0.015**	2.016	0.176
Hymenoptera	0.211	0.655	1.609	0.240	1.267	0.282	0.616	0.556	1.727	0.213	0.190	0.830	0.019	0.894	1.111	0.361
Not determinable	0.133	0.721	1.233	0.326	0.563	0.468	7.000	**0.010**	3.879	0.072	0.010	0.990	4.196	0.063	0.036	0.965
Neuroptera	–	–	–	–	7.111	**0.021**	0.778	0.481	0.143	0.712	1.000	0.397	–	–	–	–
Opiliones	1.333	0.271	2.333	0.139	2.128	0.170	1.340	0.298	4.500	0.055	0.722	0.506	3.641	0.081	0.323	0.730
Rare individuals	0.590	0.457	2.967	0.090	1.393	0.261	1.877	0.195	4.440	0.057	1.908	0.191	2.701	0.126	3.665	0.057
Saltatoria	1.000	0.337	2.250	0.148	12.893	**0.004**	1.750	0.215	0.098	0.760	1.390	0.286	2.691	0.127	2.161	0.158
Thysanoptera	0.000	1.000	0.247	0.785	0.111	0.744	1.192	0.337	3.499	0.086	1.319	0.304	2.626	0.131	3.031	0.086
Total individuals	0.268	0.614	0.552	0.590	0.237	0.636	0.600	0.565	6.612	**0.024**	0.076	0.930	Detailed results in **Table 2**			

No interaction between rainfall patterns and soil type were detected (except *Saltatoria* in May $P = 0.030$). Data for individual sampling dates were analyzed using two ANOVAs, data across dates were analyzed using repeated measurement ANOVAs. Significant effects are in bold.

– No data available for this date.

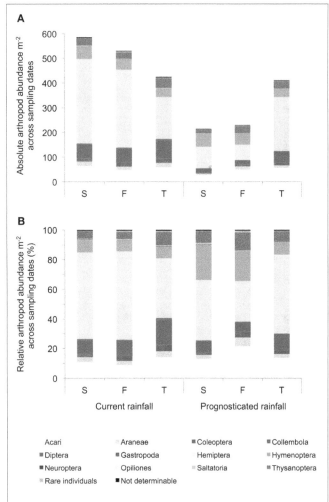

FIGURE 5 | Mean absolute (A) and relative (B) abundance of arthropods per m² across the three sampling dates (April, May, June) in winter wheat cultivated under current and prognosticated rainfall variations at the different soil types calcaric phaeozem (S), gleyic phaeozem (F) and calcic chernozem (T).

Wheat height was across sampling dates not affected by rainfall but significantly affected by soil types with lowest heights in the S soils and similarly high wheat plants in F and T soils (significant rainfall × soil type interaction; **Table 4**). Wheat straw biomass across sampling dates was significantly affected by rainfall and soil types (significant rainfall × soil type interaction; **Table 4**). Weed abundance across sampling dates was marginally significantly affected by rainfall variations and highly significantly affected by soil types (no rainfall × soil types interaction; **Table 4**). Weed biomass across sampling dates was only significantly affected by soil types with lowest weed biomass values in F soils and highest weed biomass in S soils (**Table 4**). Arthropod abundance was unrelated to winter wheat straw biomass (**Table 5**) wheat height or weed abundance (data not shown). However, abundances of Acari, Araneae, Collembola, Diptera, the group of not determinable arthropods and Thysanoptera was positively correlated with weed biomass (**Table 5**).

DISCUSSION

Results of this study show substantial reductions in the abundances of various arthropod groups but no changes on the diversity of arthropod communities under rainfall variations prognosticated for the years 2071–2100. Given the average 45% reduction of total arthropod abundance under prognosticated rainfall means that instead of $86\,m^{-2}$ only $48\,m^{-2}$ arthropod individuals would be inhabiting these wheat agroecosystems. Arthropod abundance data from the current study fit well with those from a conventional cereal field in Denmark also assessed with suction sampling in late June over 2 years (Reddersen, 1997): Araneae $(5.4–17.8\,m^{-2})$, Collembola (0.65–155.9), Hemiptera $(14.1–2146\,m^{-2})$, Hymenoptera $(13.5–23.9\,m^{-2})$, however much more Coleoptera $(51.5–110.4\,m^{-2})$, Diptera $(66.3–104.1\,m^{-2})$, and Lepidoptera $(0.43\,m^{-2})$ were reported. Similar to our study, Moreby and Sotherton (1997) also found low abundances of Diptera $(5.4\,m^{-2})$, Carabidae $(0.82\,m^{-2})$, and Chrysomelidae $(1.36\,m^{-2})$ in conventional winter wheat fields in southern England with suction samplings in June and July. Reasons for differences in arthropod abundances in different studies reflect climatic differences, effects of surrounding landscape structure, influence of different insecticide usage or differences in wheat varieties. The finding that mainly abundances but not diversity was reduced suggests that the size of arthropod populations seem to be the sensitive parameter responding to rainfall variations. Whether effects of rainfall variations on arthropod abundances have consequences on how fast arthropod populations can react to environmental changes remains to be investigated by a specific experiment. We also found great differences in arthropod abundances between sampling dates from April to June reflecting the natural fluctuations due to different seasonal development of the various arthropod taxa (Frampton et al., 2000; Afonina et al., 2001; Abbas and Parwez, 2012).

ARTHROPOD ABUNDANCES AS INFLUENCED BY RAINFALL

Predicted rainfall variations reduced arthropod abundances mainly in June but had little influence in April and May. We explain this by the fact that rainfall treatments were established only 1 month before the first arthropod sampling and by the relatively small difference between the rainfall scenarios in April and May that may have been insufficient to cause shifts in arthropod abundance. Moreover, until the first arthropod sampling in April the prognosticated rainfall plots (38 mm) received even more rainfall than the current rainfall plots (33 mm rainfall). Until the second sampling date in May the current rainfall plots received 91 mm and the prognosticated rainfall plots 81 mm. Even, until the June sampling the difference between the two rainfall treatments was only 20 mm, however rainfall amount combined with extended dry periods was obviously enough to lead to several significant differences in arthropod abundances. Moreover, the increased soil matric potential in the prognosticated rainfall plots showed that soil water was less available than under the current rainfall treatment affecting wheat biomass production and weed abundance. Further, rainfall showed different effects on the availability of water in different soil types as indicated by a significant interaction between rainfall and soil types for soil matric potential.

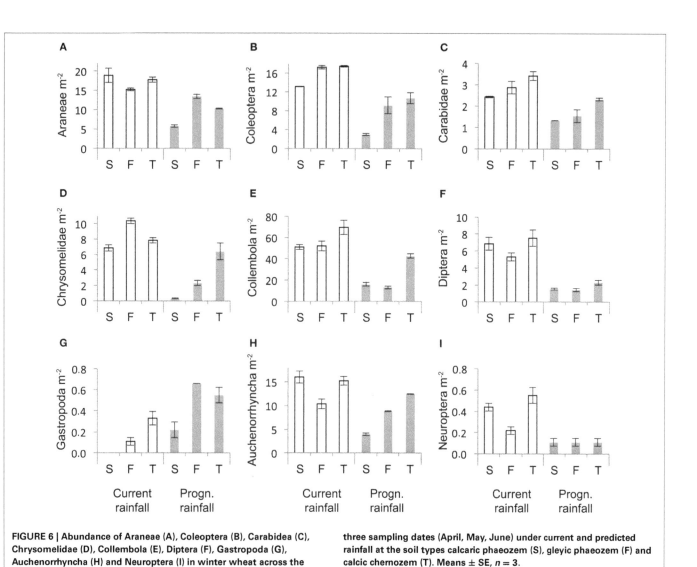

FIGURE 6 | Abundance of Araneae (A), Coleoptera (B), Carabidea (C), Chrysomelidae (D), Collembola (E), Diptera (F), Gastropoda (G), Auchenorrhyncha (H) and Neuroptera (I) in winter wheat across the three sampling dates (April, May, June) under current and predicted rainfall at the soil types calcaric phaeozem (S), gleyic phaeozem (F) and calcic chernozem (T). Means ± SE, $n = 3$.

Despite the small differences in rainfall it was interesting to see significant differences in abundances of Gastropoda and Auchenorrhyncha in April. However, given the small abundances of these taxa ($0.31\,m^{-2}$ for Gastropoda and $0.46\,m^{-2}$ for Auchenorrhyncha) results should be interpreted with caution. On the other hand, the predicted rainfall plots received more precipitation than the current plots until April and Gastropoda are known to be very sensitive to rainfall (Choi et al., 2004) and might thus be sensitive indicators for changes in moisture. In our experiment Auchenorrhyncha (e.g., cicadas) also seemed to be sensitive to rainfall, although others found no differences in the abundance in summer drought plots compared to plots under ambient climate condition (Masters et al., 1998). Collembola, Diptera, Neuroptera, and Saltatoria responded to rainfall scenarios in May. This can be explained by a higher sensitivity to changes of these four orders, so that small differences in rainfall amounts (9.8 mm) and variation were effective, whereas the other orders appear to be more tolerant against changes in rainfall. Others also found that mites were not responsive to precipitation treatments, but Collembola were (Kardol et al., 2011).

In June 11 of the 18 arthropod groups investigated were affected by rainfall treatments suggesting that 20 mm difference in the amount of rainfall and 25% more dry days were enough for these taxa to respond. Finding that certain arthropod taxa were affected by rainfall treatments in 1 month but not in the other (e.g., Gastropoda, Saltatoria) can be explained by spatial and temporal variations of arthropod distribution between agroecosystems and the surrounding landscape (Afonina et al., 2001; Tscharntke et al., 2002; Zaller et al., 2008b). Clearly, to better understand the mechanisms underlying the relationship between rainfall amounts/variations and arthropod abundances an analysis at the species level would be desirable. However, it can be concluded from the current study that changes in rainfall variations with a slightly decreased amount of rainfall, more dry days and more intensive rainfall events will most likely decrease the abundance of aboveground arthropods in winter wheat crops.

Vegetation structural complexity, including crop biomass and weed abundance which differed between the rainfall treatments, is an important determinant of arthropod abundance and diversity in agroecosystems (Honek, 1988; Lagerlöf and Wallin, 1993;

Table 4 | Wheat height, wheat straw mass, weed abundance and biomass (all averaged across several sampling dates) in lysimeters cultivated with wheat in response to current vs. prognosticated rainfall variations and different soil types (S—calcaric phaeozem, F—gleyic phaeozem, T—calcic chernozem).

Parameter/Soil type	Treatments Current Rainfall	Progn. Rainfall
WHEAT HEIGHT (cm)		
S soil	42.8 ± 0.7	46.3 ± 0.9
F soil	51.3 ± 0.3	48.9 ± 2.5
T soil	48.7 ± 0.7	50.9 ± 1.9
ANOVA RESULTS FOR WHEAT HEIGHT		
Rainfall	$P = 0.121$	
Soil types	$P < 0.001$	
Rainfall × Soil types	$P = 0.009$	
WHEAT STRAW BIOMASS (g m^{-2})		
S soil	49.2 ± 1.1	54.8 ± 3.0
F soil	59.3 ± 0.5	57.2 ± 3.6
T soil	55.8 ± 1.1	61.4 ± 3.0
ANOVA RESULTS FOR WHEAT STRAW BIOMASS		
Rainfall	$P = 0.018$	
Soil types	$P < 0.001$	
Rainfall × Soil types	$P = 0.021$	
WEED ABUNDANCE (ind. m^{-2})		
S soil	345.8 ± 104.8	239.6 ± 68.1
F soil	118.1 ± 48.9	87.5 ± 11.6
T soil	191.7 ± 50.6	156.9 ± 40.7
ANOVA RESULTS FOR WEED ABUNDANCE		
Rainfall	$P = 0.070$	
Soil types	$P < 0.001$	
Rainfall × Soil types	$P = 0.503$	
WEED BIOMASS (g m^{-2})		
S soil	15.1 ± 3.2	18.9 ± 2.8
F soil	8.5 ± 5.8	8.8 ± 3.3
T soil	12.3 ± 3.7	9.6 ± 3.7
ANOVA RESULTS FOR WEED BIOMASS		
Rainfall	$P = 0.807$	
Soil types	$P = 0.008$	
Rainfall × Soil types	$P = 0.382$	

Means ± SD. Statistical results from Two-Way ANOVAs, significant effects are in bold.

Table 5 | Correlation between arthropod abundance (June sampling) and straw and weed biomass (Pearson's product-moment correlation).

	Straw biomass		Weed biomass	
	R	**P**	**R**	**P**
Acari	−0.292	0.240	0.576	**0.012**
Araneae	0.120	0.635	0.517	**0.028**
Coleoptera	0.338	0.170	0.396	0.103
Collembola	0.168	0.505	0.542	**0.020**
Diptera	−0.002	0.995	0.687	**0.002**
Hemiptera	0.048	0.849	0.360	0.143
Hymenoptera	−0.181	0.471	−0.355	0.148
Not determinable	−0.013	0.960	0.607	**0.008**
Rare individuals	0.381	0.119	0.340	0.167
Saltatoria	−0.226	0.368	0.091	0.720
Thysanoptera	0.135	0.593	0.745	**<0.001**
Total individuals	0.029	0.908	0.451	0.060

Significant correlations in bold.

Frank and Nentwig, 1995; Kromp, 1999). Correlations between arthropod abundance and crop and weed biomass suggest that the rainfall effects indirectly affect arthropods by changes on crops and weeds. Many studies describe the interrelation between weeds and arthropods, in which greater weed density and diversity is associated with higher numbers of arthropods (Moreby and Sotherton, 1997; Moreby and Southway, 1999; Marshall et al., 2003). In the current study, 45% less weed biomass were found in the predicted rainfall plots than in current plots and thus the significant correlations for the abundance of arthropods (Acari, Araneae, Collembola, Diptera, and Thysanoptera) and weed biomass are not surprising. However, it is somewhat counterintuitive, that there was no correlation between numbers of individuals of weeds and abundance of arthropods, except

for Hemiptera and total individuals in May. Also in contrast to other studies is the lack of a correlation between arthropod abundance and crop height (Frampton et al., 2000; Perner et al., 2005) indicating that our treatment factors rainfall and soil types influenced relationships between arthropods and plants. For example, the observed increased soil matric potential under progn. rainfall suggests that crop and weed plants in these treatments had soil water less easily available than plants in curr. rainfall treatments which could have affected the nutritional quality and structure of the crop-weed communities for arthropods (Masters et al., 1998). Plant responses to soil water availability can influence herbivore population dynamics with implications for multitrophic arthropod-plant interactions (Masters et al., 1993; Gange and Brown, 1997). Plant-mediated indirect effects of rainfall on arthropods have been described in detail for aphids where the performance of aphids on drought-stressed relative to healthy plants was increased, decreased or unchanged depending on the aphid species, host-plant, timing and severity of the drought stress (Pons and Tatchell, 1995). Whatever the causal mechanisms are, the decrease in arthropod abundance can have potential consequences for ecosystem function such as biological control, nutrient cycling, pollination, seed dispersal, plant decomposition, and soil alteration (Price, 1997; Bokhorst et al., 2008; Brantley and Ford, 2012). Arthropods control populations of other organisms and provide a major food source for other taxa, like birds or amphibia. Many farmland bird species are declining in Europe, and one reason could be a decreasing availability of arthropods (Moreby and Southway, 1999; Wilson et al., 1999; Hallmann et al., 2014). Insects are also an important supplementary human food source in many regions of the world, but as arthropods can also cause damage through feeding injury or transmission of plant-diseases, natural biological control in form of antagonistic arthropods are crucial for agricultural systems worldwide (Foottit and Adler, 2009). Our study also indicates that prognosticated rainfall variations might have little influence on biological control within the wheat agroecosystem as both important antagonists for pests (Araneae, Carabidae) and pests themselves (Chrysomelidae, Auchenorrhyncha) are reduced. However, the influence of rainfall on these pest-antagonist interactions demand more detailed investigations.

When interpreting our data one has to keep in mind that in climate change models temperature and precipitation are closely linked. Since we only investigated rainfall effects while leaving temperature unchanged, different impacts that the ones reported here could occur when both factors, temperature and precipitation, are studied simultaneously.

ARTHROPOD ABUNDANCE LITTLE INFLUENCED BY SOIL TYPES

Unlike expected, the soil types had no effect on arthropod abundances despite of clear differences in the availability of soil water as measured by the soil matric potential. Surprisingly, also orders which live in soil for most of its life cycle such as Collembola did not respond to soil types and the availability of soil water indicating that these taxa are rather tolerant to environmental conditions. As the factor soil type was rarely considered in studies on arthropods there is little literature to compare with. Differences in soil matric potential could also influence communities of soil bacteria and fungi and indirectly affect mycophagous and detritivorous arthropod species; however this remains to be investigated. Our current results of little influence of soil types on arthropods are in contrast with those who found significant differences in the abundance of spiders, carabides and Heteroptera in three different types of Estonian cultivated field soils; but there was also no difference between soil types regarding the number of Coleoptera (Ivask et al., 2008). When comparing those data one has to keep in mind that in the former study pitfall traps were used as opposed to suction sampling in the current study; moreover different times of the year in very different climatic regions were studied. In our study soil types influenced wheat height and weed abundance and the finding that some arthropod taxa were correlated with vegetation density suggests some relationship (Chapman et al., 1999). However, other factors, including competition between arthropod taxa from different trophic levels (Perner et al., 2005) might have overruled possible effects of soil types. In order to interpret these data in more detail, further studies investigating interactions between crop species and soil types would be necessary.

CONCLUSION

Taken together, this study suggests that future rainfall variations with less rainfall and longer drought periods during the vegetation period will significantly reduce the abundance of aboveground arthropods in winter wheat fields. The lack of significant effects of soil types suggests that rainfall variations most likely will have similar effects on different soil types. Weeds associated with winter wheat were shown to play an important role in promoting arthropod abundance while effects of rainfall on crop growth seemed to be of minor importance. The strong response of arthropod abundances to only small differences in rainfall amounts demands more appreciation of the effects of rainfall variations when studying climate change effects on ecological interactions in agroecosystems. As this is among the first studies investigating the combined effects of rainfall variations and soil types on the abundance of aboveground arthropods, more research is needed to get a better understanding of their consequences on ecosystem functioning and services.

AUTHOR CONTRIBUTIONS

Laura Simmer, James Tabi Tataw, Johannes Hösch, Erwin Murer, Johann G. Zaller conducted field work; Laura Simmer, Johann G. Zaller, Nadja Santer, Erwin Murer analyzed the data; Johann G. Zaller, Andreas Baumgarten, Herbert Formayer, Johannes Hösch, Johann G. Zaller conceived and designed the experiment; all authors wrote on the manuscript.

ACKNOWLEDGMENTS

We are grateful to Helene Berthold for providing logistical support during this project. Thanks to Karl Moder for statistical advice. This research was funded by the Austrian Climate and Energy Fund as part of the program ACRP2.

REFERENCES

Abbas, M. J., and Parwez, H. (2012). Impact of edaphic factors on the diversity of soil microarthropods in an agricultural ecosystem at Aligarh. *Indian J. Fundam. Appl. Life Sci.* 2, 185–191.

Afonina, V. M., Tshernyshev, W. B., Soboleva-Dokuchaeva, I. I., Timokhov, A. V., Timokhova, O. V., and Seifulina, R. R. (2001). Arthropod complex of winter wheat crops and its seasonal dynamics. *IOBC Wprs Bull.* 24, 153–164.

Alexandrov, V., Eitzinger, J., Cajic, V., and Oberforster, M. (2002). Potential impact of climate change on selected agricultural crops in north-eastern Austria. *Glob. Change Biol.* 8, 372–389. doi: 10.1046/j.1354-1013.2002.00484.x

Altieri, M. A. (1999). The ecological role of biodiversity in agroecosystems. *Agric. Ecosyst. Environ.* 74, 19–31. doi: 10.1016/S0167-8809(99)00028-6

Andow, D. A. (1991). Vegetational diversity and arthropod population response. *Annu. Rev. Entomol.* 36, 561–586. doi: 10.1146/annurev.en.36.010191.003021

Andrew, N. R., and Hughes, L. (2004). Species diversity and structure of phytophagous beetle assemblages along a latitudinal gradient: predicting the potential impacts of climate change. *Ecol. Entomol.* 29, 527–542. doi: 10.1111/j.0307-6946.2004.00639.x

Bährmann, R., and Müller, H. J. (2005). *Bestimmung Wirbelloser Tiere.* Heidelberg: Spektrum Akademischer Verlag.

Balmer, O., Pfiffner, L., Schied, J., Willareth, M., Leimgruber, A., Luka, H., et al. (2013). Noncrop flowering plants restore top-down herbivore control in agricultural fields. *Ecol. Evol.* 3, 2634–2646. doi: 10.1002/ece3.658

Batary, P., Holzschuh, A., Orci, K. M., Samu, F., and Tscharntke, T. (2012). Responses of plant, insect and spider biodiversity to local and landscape scale management intensity in cereal crops and grasslands. *Agric. Ecosyst. Environ.* 146, 130–136. doi: 10.1016/j.agee.2011.10.018

Beier, C., Beierkuhnlein, C., Wohlgemuth, T., Penuelas, J., Emmett, B., Körner, C., et al. (2012). Precipitation manipulation experiments—challenges and recommendations for the future. *Ecol. Lett.* 15, 899–911. doi: 10.1111/j.1461-0248.2012.01793.x

Bellmann, H. (1999). *Der Neue Kosmos-Insektenführer.* Stuttgart: Franckh-Kosmos.

Bokhorst, S., Huiskes, A., Convey, P., Van Bodegom, P., and Aerts, R. (2008). Climate change effects on soil arthropod communities from the Falkland Islands and the Maritime Antarctic. *Soil Biol. Biochem.* 40, 1547–1556. doi: 10.1016/j.soilbio.2008.01.017

Brantley, S. L., and Ford, P. L. (2012). "Climate change and arthropods: pollinators, herbivores, and others," in *Climate Change in Grasslands, Shrublands, and Deserts of the Interior American West: a Review and Needs Assessment,* ed D. M. Finch (Fort Collins, CO: US Department of Agriculture, Forest Service, Rocky Mountain Research Station), 35–47.

Brook, A., Woodcock, B., Sinka, M., and Vanbergen, A. (2008). Experimental verification of suction sampler capture efficiency in grasslands of differing vegetation height and structure. *J. Appl. Ecol.* 45, 1357–1363. doi: 10.1111/j.1365-2664.2008.01530.x

Buchholz, S. (2010). Simulated climate change in dry habitats: do spiders respond to experimental small-scale drought? *J. Arachnol.* 38, 280–284. doi: 10.1636/P09-91.1

Cannon, R. J. C. (1998). The implications of predicted cli- mate change for insect pests in the UK, with emphasis on non-indigenous species. *Glob. Change Biol.* 4, 785–796. doi: 10.1046/j.1365-2486.1998.00190.x

Chapman, P. A., Armstrong, G., and Mckinlay, R. G. (1999). Daily movements of *Pterostichus melanarius* between areas of contrasting vegetation density within crops. *Entomol. Exp. Appl.* 91, 477–480. doi: 10.1046/j.1570-7458.1999.00516.x

Choi, Y., Bohan, D., Powers, S., Wiltshire, C., Glen, D., and Semenov, M. (2004). Modelling *Deroceras reticulatum* (Gastropoda) population dynamics based on daily temperature and rainfall. *Agric. Ecosyst. Environ.* 103, 519–525. doi: 10.1016/j.agee.2003.11.012

Christensen, J. H., and Christensen, O. B. (2007). A summary of the PRUDENCE model projections of changes in European climate by the end of this century. *Clim. Change* 81, 7–30. doi: 10.1007/s10584-006-9210-7

Coudrain, V., Schüepp, C., Herzog, F., Albrecht, M., and Entling, M. (2014). Habitat amount modulates the effect of patch isolation on host-parasitoid interactions. *Front. Environ. Sci.* 2:27. doi: 10.3389/fenvs.2014.00027

Curry, J. P. (1994). *Grassland Invertebrates: Ecology, Influence on Soil Fertility and Effects on Plant Growth.* London: Chapman & Hall.

Danneberg, O., Baumgarten, A., Murer, E., Krenn, A., and Gerzabek, M. H. (2001). Stofftransport im System Boden—Wasser—Pflanze: Lysimeterversuche (Exkursion P2). *Mitt. Österr. Bodenkundlichen Ges.* 63, 193–208.

Drapela, T., Moser, D., Zaller, J. G., and Frank, T. (2008). Spider assemblages in winter oilseed rape affected by landscape and site factors. *Ecography* 31, 254–262. doi: 10.1111/j.0906-7590.2008.5250.x

Duelli, P., and Obrist, M. K. (2003). Regional biodiversity in an agricultural landscape: the contribution of seminatural habitat islands. *Basic Appl. Ecol.* 4, 129–138. doi: 10.1078/1439-1791-00140

Eitzinger, J., Zalud, Z., Alexandrov, V., Van Diepen, C. A., Trnka, M., Dubrovsky, M., et al. (2001). A local simulation study on the impact of climate change on winter wheat production in north-eastern Austria. *Bodenkultur* 52, 279–292.

Elmes, G. W., and Free, A. (1994). *Climate Change and Rare Species.* London, UK: HMSO.

FAO. (2002). "FAO/UNESCO Digital Soil Map of the World and derived soil properties. Land and Water Digital Map Series #1 rev. 1." (Rome, Italy: FAO).

Finch, O. D., Loffler, J., and Pape, R. (2008). Assessing the sensitivity of *Melanoplus frigidus* (Orthoptera: Acrididae) to different weather conditions: a modeling approach focussing on development times. *Insect Sci.* 15, 167–178. doi: 10.1111/j.1744-7917.2008.00198.x

Foottit, R. G., and Adler, P. H. (2009). *Insect Biodiversity: Science and Society.* Chichester: Wiley-Blackwell.

Formayer, H., and Kromp-Kolb, H. (2009). *Hochwasser und Klimawandel. Auswirkungen des Klimawandels auf Hochwasserereignisse in Österreich (Endbericht WWF 2006). BOKU-Met Report 7.* Wien.

Frampton, G. K., Van Den Brink, P. J., and Gould, P. J. (2000). Effects of spring drought and irrigation on farmland arthropods in southern Britain. *J. Appl. Ecol.* 37, 865–883. doi: 10.1046/j.1365-2664.2000.00541.x

Frank, T., Aeschbacher, S., and Zaller, J. G. (2012). Habitat age affects beetle diversity in wildflower areas. *Agric. Ecosyst. Environ.* 152, 21–26. doi: 10.1016/j.agee.2012.01.027

Frank, T., and Nentwig, W. (1995). Ground dwelling spiders (Araneae) in sown weed strips and adjacent fields. *Acta Oecol.* 16, 179–193.

Gange, A. C., and Brown, V. K. (1997). *Multitrophic Interactions in Terrestrial Systems.* Oxford, UK: Blackwell Science.

Hallmann, C. A., Foppen, R. P. B., van Turnhout, C. A. M., de Kroon, H., and Jongejans, E. (2014). Declines in insectivorous birds are associated with high neonicotinoid concentrations. *Nature* 511, 341–343. doi: 10.1038/nature13531

Hamilton, J., Zangerl, A. R., Berenbaum, M. R., Sparks, J. P., Elich, L., Eisenstein, A., et al. (2012). Elevated atmospheric CO_2 alters the arthropod community in a forest understory. *Acta Oecologica* 43, 80–85. doi: 10.1016/j.actao.2012.05.004

Hegland, S. J., Nielsen, A., Lázaro, A., Bjerknes, A.-L., and Totland, O. (2009). How does climate warming affect plant-pollinator interactions? *Ecol. Lett.* 12, 184–195. doi: 10.1111/j.1461-0248.2008.01269.x

Hoffmann, H., and Rath, T. (2012). Meteorologically consistent bias correction of climate time series for agricultural models. *Theor. Appl. Climatol.* 110, 129–141. doi: 10.1007/s00704-012-0618-x

Hoffmann, H., and Rath, T. (2013). Future bloom and blossom frost risk for *Malus domestica* considering climate model and impact model uncertainties. *PLoS ONE* 8:e75033. doi: 10.1371/journal.pone.0075033

Honek, A. (1988). The effect crop density and microclimate on pitfall trap catches of Carabidae, Staphylinidae (Coleoptera) and Lycosidae (Araneae) in cereal fields. *Pedobiologia* 32, 233–242.

IPCC. (2007). *Climate Change 2007: The Physical Science Basis. Fourth Assessment Report of the Intergovernmental Panel on Climate Change.* Cambridge: Cambridge University Press.

IPCC. (2013). *Climate Change 2013: The Physical Science Basis. Contribution to the Fifth Assessment Report of the Intergovernmental Panel on Climate Change* (Cambridge: Cambridge University Press).

Ivask, M., Kuu, A., Meriste, M., Truu, J., Truu, M., and Vaater, V. (2008). Invertebrate communities (Annelida and epigeic fauna) in three types of Estonian cultivated soils. *Eur. J. Soil Biol.* 44, 532–540. doi: 10.1016/j.ejsobi.2008.09.005

Kardol, P., Reynolds, W. N., Norby, R. J., and Classen, A. T. (2011). Climate change effects on soil microarthropod abundance and community structure. *Appl. Soil Ecol.* 47, 37–44. doi: 10.1016/j.apsoil.2010.11.001

Kromp, B. (1999). Carabid beetles in sustainable agriculture: a review on pest control efficacy, cultivation impacts and enhancement. *Agric. Ecosyst. Environ.* 74, 187–228. doi: 10.1016/S0167-8809(99)00037-7

Kromp-Kolb, H., Formayer, H., Eitzinger, J., Thaler, S., Kubu, G., and Rischbeck, P. (2008). "Potentielle Auswirkungen und Anpassungsmaßnahmen der Landwirtschaft an den Klimawandel im Nordosten Österreichs (Weinviertel-Marchfeld Region)," in *Auswirkungen des Klimawandels in Niederösterreich*, ed H. Formayer (St. Pölten: Niederösterreichische Landesregierung), 96–140.

Lagerlöf, J., and Wallin, H. (1993). The abundance of arthropods along two field margins with different types of vegetation composition: an experimental study. *Agric. Ecosyst. Environ.* 43, 141–154. doi: 10.1016/0167-8809(93)90116-7

Lensing, J. R., Todd, S., and Wise, D. H. (2005). The impact of altered precipitation on spatial stratification and activity-densities of springtails (Collembola) and spiders (Araneae). *Ecol. Entomol.* 30, 194–200. doi: 10.1111/j.0307-6946.2005.00669.x

Loranger-Merciris, G., Imbert, D., Bernhard-Reversat, F., Ponge, J.-F., and Lavelle, P. (2007). Soil fauna abundance and diversity in a secondary semi-evergreen forest in Guadeloupe (Lesser Antilles): influence of soil type and dominant tree species. *Biol. Fertil. Soils* 44, 269–276. doi: 10.1007/s00374-007-0199-5

Marshall, E. J. P., Brown, V. K., Boatman, N. D., Lutman, P. J. W., Squire, G. R., and Ward, L. K. (2003). The role of weeds in supporting biological diversity within crop fields. *Weed Res.* 43, 77–89. doi: 10.1046/j.1365-3180.2003.00326.x

Masters, G., Brown, V., Clarke, I., Whittaker, J., and Hollier, J. (1998). Direct and indirect effects of climate change on insect herbivores: auchenorrhyncha (Homoptera). *Ecol. Entomol.* 23, 45–52. doi: 10.1046/j.1365-2311.1998.00119.x

Masters, G. J., Brown, V. K., and Gange, A. C. (1993). Plant mediated interactions between above- and below-ground insect herbivores. *Oikos* 66, 148–151. doi: 10.2307/3545209

Menalled, F. D., Smith, R. G., Dauer, J. T., and Fox, T. B. (2007). Impact of agricultural management on carabid communities and weed seed predation. *Agric. Ecosyst. Environ.* 118, 49–54. doi: 10.1016/j.agee.2006.04.011

Moreby, S., and Sotherton, N. (1997). A comparison of some important chick-food insect groups found in organic and conventionally-grown winter wheat fields in southern England. *Biol. Agric. Horticult.* 15, 51–60. doi: 10.1080/01448765.1997.9755181

Moreby, S., and Southway, S. (1999). Influence of autumn applied herbicides on summer and autumn food available to birds in winter wheat fields in southern England. *Agric. Ecosyst. Environ.* 72, 285–297. doi: 10.1016/S0167-8809(99)00007-9

Nestroy, O., Aust, G., Blum, W. E. H., Englisch, M., Hager, H., Herzberger, E., et al. (2011). *Systematische Gliederung der Böden Österreichs. Österreichische Bodensystematik 2000 in der revidierten Fassung von 2011.* Wien: Österreichische Bodenkundliche Gesellschaft.

Östman, Ö., Ekbom, B., Bengtsson, J., and Weibull, A.-C. (2001). Landscape complexity and farming practice influence the condition of polyphagous carabid beetles. *Ecol. Appl.* 11, 480–488. doi: 10.1890/1051-0761(2001)011[0480:LCAFPI]2.0.CO;2

Perner, J., Wytrykush, C., Kahmen, A., Buchmann, N., Egerer, I., Creutzburg, S., et al. (2005). Effects of plant diversity, plant productivity and habitat parameters on arthropod abundance in montane European grasslands. *Ecography* 28, 429–442. doi: 10.1111/j.0906-7590.2005.04119.x

Pfiffner, L., and Luka, H. (2003). Effects of low-input farming systems on carabids and epigeal spiders–a paired farm approach. *Basic Appl. Ecol.* 4, 117–127. doi: 10.1078/1439-1791-00121

Pons, X., and Tatchell, G. M. (1995). Drought stress and cereal aphid performance. *Ann. Appl. Biol.* 126, 19–31. doi: 10.1111/j.1744-7348.1995.tb05000.x

Porter, J. R., and Semenov, M. A. (2005). Crop responses to climatic variation. *Philos. Trans. R. Soc. B Biol. Sci.* 360, 2021–2035. doi: 10.1098/rstb.2005.1752

Price, P. W. (1997). *Insect Ecology.* New York, NY: John Wiley & Sons.

Price, P. W., Denno, R. F., Eubanks, M. D., Finke, D. L., and Kaplan, I. (2011). *Insect Ecology: Behavior, Populations and Communities.* Cambridge: Cambridge University Press.

Querner, P., Bruckner, A., Drapela, T., Moser, D., Zaller, J. G., and Frank, T. (2013). Landscape and site effects on Collembola diversity and abundance in winter oilseed rape fields in eastern Austria. *Agric. Ecosyst. Environ.* 164, 145–154. doi: 10.1016/j.agee.2012.09.016

R Core Team. (2013). *R: A Language and Environment for Statistical Computing.* Vienna: R Foundation for Statistical Computing. Available online at: http://www.R-project.org/

Reddersen, J. (1997). The arthropod fauna of organic versus conventional cereal fields in Denmark. *Biol. Agricult. Horticul.* 15, 61–71. doi: 10.1080/01448765.1997.9755182

Romanowsky, T., and Tobias, M. (1999). Vergleich der Aktivitätsdichten von Bodenarthropoden (insbesondere Laufkäfern, Carabidae) in zwei agrarisch geprägten Lebensräumen – Untersuchung zum Nahrungspotential einer Population der Knoblauchkröte (*Pelobates fuscus* Laurenti, 1768). *Rana Sonderh.* 3, 49–57.

Rosenzweig, M. L. (1995). *Species Diversity in Space and Time.* Cambridge: Cambridge University Press.

Seastedt, T. R., and Crossley, D. A. (1984). The influence of arthropods on ecosystems. *BioScience* 34, 157–161. doi: 10.2307/1309750

Semenov, M. A., and Barrow, E. M. (2002). "LARS-WG. A stochastic weather generator for use in climate impact studies," in *User Manual* (Harpenden: Rothamsted Research), 27.

Singer, M. C., and Parmesan, C. (2010). Phenological asynchrony between herbivorous insects and their hosts: signal of climate change or pre-existing adaptive strategy? *Phil. Trans. R. Soc. B* 365, 3161–3176. doi: 10.1098/rstb.2010.0144

Southwood, T. R. E. (1978). *Ecological Methods.* London: Methuen.

Speight, M. R., Hunter, M. D., and Watt, A. D. (2008). *Ecology of Insects: Concepts and Applications.* Oxford, UK: Wiley-Blackwell.

Steffan-Dewenter, I., Münzenberg, U., and Tscharntke, T. (2001). Pollination, seed set and seed predation on a landscape scale. *Proc. R. Soc. Lond. B* 268, 1685–1690. doi: 10.1098/rspb.2001.1737

Tabi Tataw, J., Hall, R., Ziss, E., Schwarz, T., Von Hohberg Und Buchwald, C., Formayer, H., et al. (2014). Soil types will alter the response of arable agroecosystems to future rainfall patterns. *Ann. Appl. Biol.* 164, 35–45. doi: 10.1111/aab.12072

Thaler, S., Eitzinger, J., Dubrovsky, M., and Trnka, M. (2008). "Climate change impacts on selected crops in Marchfeld, Eastern Austria. Paper 10.7," in *28th Conference on Agricultural and Forest Meteorology, 28 April - 2 May 2008*, ed American Meteorological Society (Orlando).

Thies, C., Steffan-Dewenter, I., and Tscharntke, T. (2003). Effects of landscape context on herbivory and parasitism at different spatial scales. *Oikos* 101, 18–25. doi: 10.1034/j.1600-0706.2003.12567.x

Tscharntke, T., and Brandl, R. (2004). Plant-insect interactions in fragmented landscapes. *Annu. Rev. Entomol.* 49, 405–430. doi: 10.1146/annurev.ento.49.061802.123339

Tscharntke, T., Klein, A. M., Kruess, A., Steffan-Dewenter, I., and Thies, C. (2005). Landscape perspectives on agricultural intensification and biodiversity—ecosystem service management. *Ecol. Lett.* 8, 857–874. doi: 10.1111/j.1461-0248.2005.00782.x

Tscharntke, T., Steffan-Dewenter, I., Kruess, A., and Thies, C. (2002). Contribution of small habitat fragments to conservation of insect communities of grassland-cropland landscapes. *Ecol. Appl.* 12, 354–363. doi: 10.1890/1051-0761(2002)012[0354:COSHFT]2.0.CO;2

Wilson, J. D., Morris, A. J., Arroyo, B. E., Clark, S. C., and Bradbury, R. B. (1999). A review of the abundance and diversity of invertebrate and plant foods of granivorous birds in northern Europe in relation to agricultural change. *Agric. Ecosyst. Environ.* 75, 13–30. doi: 10.1016/S0167-8809(99)00064-X

Zaller, J. G., Moser, D., Drapela, T., Schmoger, C., and Frank, T. (2008a). Insect pests in winter oilseed rape affected by field and landscape characteristics. *Basic Appl. Ecol.* 9, 682–690. doi: 10.1016/j.baae.2007.10.004

Zaller, J. G., Moser, D., Drapela, T., Schmöger, C., and Frank, T. (2008b). Effect of within-field and landscape factors on insect damage in winter oilseed rape. *Agric. Ecosyst. Environ.* 123, 233–238. doi: 10.1016/j.agee.2007.07.002

Zaller, J., Moser, D., Drapela, T., and Frank, T. (2009). Ground-dwelling predators can affect within-field pest insect emergence in winter oilseed rape fields. *Biocontrol* 54, 247–253. doi: 10.1007/s10526-008-9167-8

Conflict of Interest Statement: The authors declare that the research was conducted in the absence of any commercial or financial relationships that could be construed as a potential conflict of interest.

10

Biochar increases soil N_2O emissions produced by nitrification-mediated pathways

*María Sánchez-García, Asunción Roig, Miguel A. Sánchez-Monedero and María L. Cayuela **

Department of Soil and Water Conservation and Waste Management, CEBAS-CSIC, Campus Universitario de Espinardo, Murcia, Spain

Edited by:
Christophe Darnault, Clemson University, USA

Reviewed by:
Anniet M. Laverman, Universite Pierre et Marie Curie, France
Lukas Van Zwieten, New South Wales Department of Primary Industries, Australia
Meihua Deng, Tsinghua University, China
Bing-Jie Ni, The University of Queensland, Australia

*Correspondence:
María L. Cayuela, CEBAS-CSIC, Campus Universitario de Espinardo, 30100 Murcia, Spain
e-mail: mlcayuela@cebas.csic.es

In spite of the numerous studies reporting a decrease in soil nitrous oxide (N_2O) emissions after biochar amendment, there is still a lack of understanding of the processes involved. Hence the subject remains controversial, with a number of studies showing no changes or even an increase in N_2O emissions after biochar soil application. Unraveling the exact causes of these changes, and in which circumstances biochar decreases or increases emissions, is vital to developing and applying successful mitigation strategies. With this objective, we studied two soils [Haplic Phaeozem (HP) and Haplic Calcisol (HC)], which showed opposed responses to biochar amendment. Under the same experimental conditions, the addition of biochar to soil HP decreased N_2O emissions by 76%; whereas it increased emissions by 54% in soil HC. We combined microcosm experiments adding different nitrogen fertilizers, stable isotope techniques and the use of a nitrification inhibitor (dicyciandiamide) with the aim of improving our understanding of the mechanisms involved in the formation of N_2O in these two soils. Evidence suggests that denitrification is the main pathway leading to N_2O emissions in soil HP, and ammonia oxidation and nitrifier-denitrification being the major processes generating N_2O in soil HC. Biochar systematically stimulated nitrification in soil HC, which was probably the cause of the increased N_2O emissions. Here we demonstrate that the effectiveness of using biochar for reducing N_2O emissions from a particular soil is linked to its dominant N_2O formation pathway.

Keywords: nitrous oxide, charcoal, nitrification, DCD, codenitrification, nitrogen fertilizers

INTRODUCTION

Biochar, a carbonaceous material produced during the pyrolysis of biomass, has been found to decrease N_2O emissions from soils (Spokas and Reikosky, 2009; Cayuela et al., 2010; Van Zwieten et al., 2010). A recent meta-analysis of 30 papers (published from 2007 to 2013) revealed a statistically significant reduction of 54% in N_2O emissions when soils were amended with biochar (Cayuela et al., 2014). However, a substantial number of studies contradict this result, they reporting no difference or even an increase in soil N_2O emissions after biochar application (Clough et al., 2010; Saarnio et al., 2013; Suddick and Six, 2013). A remarkable finding was that the same biochar could lead to opposite effects (increasing or decreasing N_2O emissions) depending on the soil to which the biochar was applied (Yoo and Kang, 2012; Malghani et al., 2013).

Soils are a major source of N_2O, which is a potent greenhouse gas and contributor to ozone layer destruction. N_2O is produced during several soil processes and its release to the atmosphere is almost entirely controlled by microbial activities. Current knowledge suggests five N_2O-genic soil microbial sources (Baggs, 2011; Spott et al., 2011). These are the nitrate or nitrite reducing processes of denitrification and dissimilatory nitrate reduction to ammonium (DNRA), and ammonia oxidation (the first step in nitrification, facilitated by ammonia oxidizing bacteria). Nitrifier denitrification, the ability of ammonia oxidizing bacteria to

denitrify, is often also seen as a separate process. Finally, codenitrification has also been identified as a relevant N_2O formation pathway in soils (Spott et al., 2011). Understanding the mechanisms of the interactions of biochar with soil N_2O formation pathways represents a difficult challenge. No evidence has been reported that would serve to unambiguously define the cause for the observed variations (increase or decline) in soil N_2O fluxes. This is due to the extremely complex set of reactions leading to N_2O formation and consumption in soils and also to the fact that the number of studies which analyze how biochar influences specific N_2O formation pathways is still very limited.

In a recent study using the ^{15}N gas flux method, Cayuela et al. (2013) observed a consistent decrease in the N_2O/N_2 ratio after biochar amendment in 15 agricultural soils, pointing to denitrification as the N_2O formation pathway that biochar might be altering. According to this, biochar would enhance the last step of denitrification (i.e., the reduction of $N_2O–N_2$). Subsequently, Harter et al. (2014) found that soil biochar amendment increased the relative gene and transcript copy numbers of the nosZ-encoded bacterial N_2O reductase, a result which could explain the previous mechanistic findings. Nevertheless, Cayuela et al. (2013) also found contrasting results for the flux of total denitrified N ($N_2O + N_2$), which was significantly reduced in the majority of soils (10 out of 15), but highly amplified in others. No conclusive explanation was found for this paradoxical finding.

In this study we aimed to look more closely at the reasons for these contrasting results. Our hypothesis was that, besides denitrification, other microbial processes (e.g., nitrifier-denitrification, dissimilatory nitrate reduction to ammonia, codenitrification) could have led to N_2O and N_2 formation in these soils, mechanisms that had not been addressed in previous studies. Hence, we studied two soils that, under identical experimental conditions, showed opposite responses to biochar amendment, i.e., whereas biochar addition decreased N_2O emissions in one soil, it increased emissions in the other. The main objective was to investigate by ^{15}N gas measurements and the use of nitrification inhibitors, the main pathways leading to N_2O formation in these two soils, with the aim of understanding why biochar might be influencing N_2O emissions differently.

MATERIALS AND METHODS
SOILS AND BIOCHAR SELECTED FOR THE EXPERIMENTS
Two agricultural soils were selected for the experiments (**Table 1**). Soil HP was used as a reference soil, since it had been previously used in numerous studies that proved that denitrification was the major process responsible for N_2O emissions (Čuhel et al., 2010). Soil HC was selected from a series of agricultural soils because it was the only one where (under identical optimal denitrifying conditions) the addition of greenwaste biochar increased N_2O emissions. The soils were sampled from a depth of 0–0.25 m, air-dried and sieved (<2 mm).

We used a biochar produced by continuous slow pyrolysis of greenwaste at 550°C provided by Pacific Pyrolysis Pty. Ltd. (Australia) (**Table 1**). Herbaceous and woody biochars have been found to be the most promising for mitigating N_2O emissions from soil (Cayuela et al., 2014). Therefore, this biochar was selected for its mitigation potential and as a representative standard biochar commonly used in other studies. The biochar was ground to a particle size <1 mm before soil application.

MICROCOSMS EXPERIMENTS
The incubation experiments were performed in 250 ml polypropylene jars at optimum conditions for denitrification: 25°C and moisture content of 90% water filled pore space (WFPS). The control treatments consisted of 100 g dry soil and the biochar treatments of 98 g dry soil and 2 g biochar (2% w:w). The biochar was thoroughly mixed with the dry soil to obtain a completely homogeneous mixture. Subsequently deionized water (or a solution containing the appropriate concentration of N fertilizer) was added to reach 90% WFPS (and the required N concentration in the fertilized treatments). The jars were incubated aerobically, covered with a polyethylene sheet that allows gas exchange but minimizes evaporation. Moisture was gravimetrically adjusted every other day with the addition of deionised water for each individual jar. The experiments were laid out as randomized block designs with four replicates per treatment.

Experiment 1. Impact of biochar on soil N_2O emissions and mineral N after the addition of different N fertilizers
A set of 48 jars [2 soils (HP/HC) × 2 management treatments (biochar/control) × 3 fertilization treatments (no

Table 1 | Physical and chemical characteristics of soil and biochar samples used in the experiments.

	Soil HP	Soil HC	Biochar
Management	Pasture	Olive orchard (organic farm)	–
Location	48°52′ N, 14°13′ E	38°23′ N 1°22′ W	–
Cassification (WRB)	Haplic phaeozem	Haplic calcisol	–
Texture	Loamy sand	Sandy loam	–
Sand (%)	78	57	–
Clay (%)	6	16	–
Volatile matter (%)	–	–	26.8
Ash (%)	–	–	7.0
H:C$_{org}$	–	–	0.534
pH (in water, 1:20 w:w 25°C)	6.89	8.01	7.87
EC (μS cm^{-1})	140	518	166
Ca CO$_3$ (%)	–	30	–
TOC (g kg^{-1})	11.6	16.8	701.7
Total N (g kg^{-1})	2.0	2.4	2.7
DC (mg kg^{-1})	439.5	694.0	285.1
DOC (mg kg^{-1})	315.7	356.9	113.2
DN (mg kg^{-1})	34.7	74.0	8.6
DON (mg kg^{-1})	10.2	35.9	7.1
NH$_4^+$- N (mg kg^{-1})	19.3	5.0	1.3
NO$_2^-$ -N (mg kg^{-1})	<0.2	16.2	<0.2
NO$_3^-$ -N (mg kg^{-1})	5.3	16.9	<0.2

TOC, total organic carbon; DN, dissolved nitrogen; DON, dissolved organic nitrogen; DC, dissolved carbon; DOC, dissolved organic carbon.

fertilizer/KNO$_3$/CO(NH$_2$)$_2$] × 4 replicates] was set up for the first experiment. The fertilizers were homogeneously distributed in the soil at a rate of 200 kg N Ha^{-1} (corresponding to 55 mg N kg^{-1} based on a plough layer of 25 cm). N_2O samples were taken twice a day during the first 2 days decreasing subsequently to daily measurements, then every other day, then three times per week, etc. (see **Figure 1**). At the end of the incubation (14 days) mineral N (NH$_4^+$, NO$_3^-$, and NO$_2^-$) was extracted and determined in all jars.

Experiment 2. Isotopic composition of N_2O and N_2 emitted after application of labeled ^{15}N fertilizers
The following ^{15}N-tracer experiments were performed:

(i) Soil HP + $^{15}NO_3^-$, vs. soil HP + $^{15}NO_3^-$ + biochar,
(ii) Soil HC + $^{15}NO_3^-$ vs. soil HC + $^{15}NO_3^-$ + biochar,
(iii) Soil HC + CO($^{15}NH_2$)$_2$ vs. Soil HC + CO($^{15}NH_2$)$_2$ + biochar

Moisture was adjusted to 90% WFPS in each jar by adding the required volume of a solution containing K$^{15}NO_3$ or CO($^{15}NH_2$)$_2$ (>99% ^{15}N enrichment) at the appropriate concentration to obtain 90% WFPS and exactly 5.5 mg of ^{15}N-per jar. Rewetting the soils in this way guaranteed a homogenous ^{15}N pool. Gas samples for isotopic analysis were taken daily during the

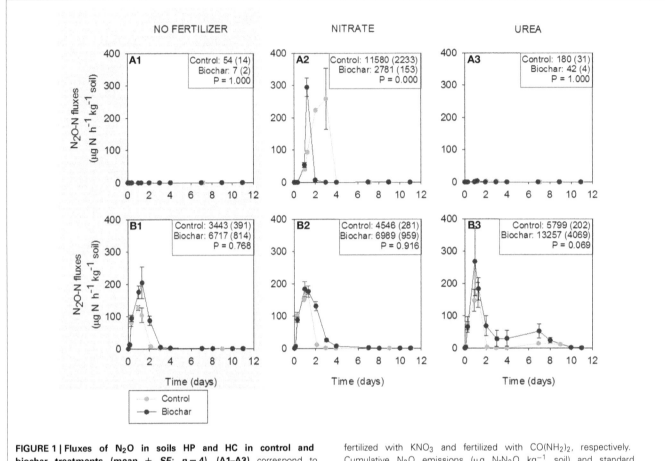

FIGURE 1 | Fluxes of N₂O in soils HP and HC in control and biochar treatments (mean ± *SE*; *n* = 4). (A1–A3) correspond to soil HP unfertilized, fertilized with KNO₃ and fertilized with CO(NH₂)₂, respectively. **(B1–B3)** correspond to soil HC unfertilized, fertilized with KNO₃ and fertilized with CO(NH₂)₂, respectively. Cumulative N₂O emissions (μg N-N₂O kg⁻¹ soil) and standard errors by the end of the incubation are reported in the right-above box for each treatment.

first 3 days and on day 10. For each treatment, two gas samples were collected using a 12-ml syringe and needle: one immediately after the screw cap was fitted to the jar ($t = 0$) and the second after 60 min ($t = 60$). The gas samples were transferred to 12-ml vials (Labco) previously purged with He and evacuated. Selected samples (a total of 192 samples) were analyzed for the isotope ratios of N_2 [29/28 (29R) and 30/28 (30R)] and N_2O [45/44 (45R) and 46/44 (46R)] by automated isotope ratio mass spectroscopy [ThermoFinnigan GasBench and PreCon trace gas concentration system interfaced to a ThermoScientific Delta V Plus isotope-ratio mass spectrometer (Bremen, Germany)].

Experiment 3. N_2O emissions, mineral N, and N_2O isotopic composition after addition of NO_2^- in soil HC

Experiments 1 and 2 were reproduced in soil HC with a different source of nitrogen: $NaNO_2$ was added to a set of 8 jars [4 replicates × 2 management treatments (biochar/control)] and homogeneously distributed in the soil at a rate of 200 kg N Ha⁻¹. N_2O and final concentrations of mineral N were determined as for Experiment 1 (see **Figure 5**).

Subsequently, the following ^{15}N tracer experiment was performed: Soil HC +$Na^{15}NO_2$ vs. Soil HC + $Na^{15}NO^2$ + biochar (as for Experiment 2).

Moisture was adjusted to 90% WFPS in each jar by adding the required volume of a solution containing $NaNO_2$ (>98% ^{15}N enrichment) at the appropriate concentration to obtain 90% WFPS and exactly 5.5 mg of ^{15}N-per jar. Gas samples for isotopic analysis were taken daily during the first 3 days and on the 10th day of incubation in the same way as in Experiment 2. A total of 64 gas samples [2 management treatments (biochar/control) × 4 replicates × 4 days (1/2/3/10) × 2 times per day ($t = 0/t = 60$)] were analyzed.

Experiment 4. N_2O emissions and mineral N after addition of dicyandiamide to soil HC

The nitrification inhibitor dicyandiamide (DCD) was applied in combination with N fertilizers in soil HC. DCD inhibits the first stage of nitrification, the oxidation of NH_4^+ to NH_2OH, by rendering the enzyme ammonia monooxygenase (AMO) ineffective. It is not a bactericide, and does not affect other heterotrophs responsible of the soil biological activity (Zacherl and Amberger, 1990).

A set of 24 jars [2 management treatments (biochar/control) × 3 fertilization treatments (no fertilizer/KNO₃/CO(NH₂)₂) × 4 replicates] was set up for the experiment. DCD was applied at a rate of 30 mg kg⁻¹ soil to ensure its persistence over the entire incubation period (Rajbanshi et al., 1992). The fertilizers were homogeneously distributed in the soil at the same rate as in the previous experiments (200 mg N Ha^{-1}) in the solution including the DCD. N_2O samples were taken following the same intervals as in Experiment 1. Mineral N (NH_4^+, NO_3^-, and NO_2^-) was also extracted and determined in all jars at the end of the incubation period.

N_2O SAMPLING AND MEASUREMENTS

For N_2O sampling each unit was sealed with gas-tight polypropylene screw caps for an accumulation period of 60 min. The headspace gas was then sampled directly with a membrane air pump (Optimal 250, Schego, Offenbach am Main, Germany), attached to a gas chromatograph (VARIAN CP-4900 Micro-GC, Palo Alto, CA, USA) (Mondini et al., 2010).

N_2O fluxes were calculated assuming a linear increase during the accumulation (closed) period, an approach which was verified prior to the experiments. Cumulative N_2O was calculated assuming linear changes in fluxes between adjacent measurement points (Velthof et al., 2003).

CHEMICAL-PHYSICAL ANALYSES OF BIOCHAR AND SOILS
Biochar

Proximate analysis was conducted using ASTM D1762-84 Chemical Analysis of Wood Charcoal. Total N and C were analyzed by automatic elemental analysis (FlashEA 1112 Series, Thermo scientific, Madrid, Spain). Water soluble C and N were determined in 1:10 (w/v) water extracts using a Photometer Nanocolor 500 D MACHEREY-NAGEL. Electrical conductivity (EC) and pH were determined in a 1:10 (w/v) water-soluble extract. NH_4^+ was extracted with 2.0 M KCl at 1:10 (w/v) and determined by a colorimetric method based on Berthelot's reaction. NO_3^- and NO_2^- were extracted with water at 1:10 (w/v) and determined by ion chromatography (HPLC, model 861, Metrohm AG, Herisau, Switzerland).

Soil

Soil texture was determined using the pipette method according to Kettler et al. (2001). Soils were extracted by shaking four replicates of moist soil (1/10, w/v dry weight basis) with 2.0 M KCl (for NH_4^+) or water (for NO_3^- and NO_2^-) for 2 h. Extracts were centrifuged (2509 G) and filtered (0.45 μm) before analysis. NH_4^+ was determined by a colorimetric method based on Berthelot's reaction. NO_3^- and NO_2^- were determined by ion chromatography (HPLC, model 861, Metrohm AG, Herisau, Switzerland).

^{15}N CALCULATIONS

The ^{15}N atomic fraction in N_2O was calculated from the 45/44 and 46/44 ratios of N_2O. The ^{15}N gas-flux method (Mulvaney and Boast, 1986; Stevens et al., 1993; Stevens and Laughlin, 2001) was used to quantify N_2O and N_2 emissions from denitrification in soil HP. The molar fraction of ^{15}N-NO_3^- ($^{15}X_N$) in the soil pool was calculated from $\Delta 45R$ and $\Delta 46R$ according to Stevens and Laughlin (2001). The flux of N_2 and N_2O was then calculated by the equations given by Mulvaney and Boast (1986). The presence of hybrid nitrous oxide ($^{45}N_2O$) co-metabolically introduced into the reaction pathway of denitrification was tested by the model developed by Spott and

Florian Stange (2011). This model considers two different N sources, where each source generates non-hybrid N_2O ($^{46}N_2O$ and $^{44}N_2O$) and, simultaneously, both N sources can be combined to form hybrid N_2O ($^{45}N_2O$). According to this model, the contribution of each pathway to the total N_2O formation can be calculated from the mass distribution of the released N_2O and the ^{15}N mole fraction of the labeled N source (Spott and Florian Stange, 2011).

STATISTICAL ANALYSIS

Univariate analysis of variance was used to investigate the significant differences in N_2O emissions and mineral N concentrations between biochar and control treatments with IBM SPSS Statistics 21, Sommers, USA.

RESULTS
EXPERIMENT 1. CUMULATIVE N_2O EMISSIONS AND MINERAL N IN SOILS A AND B

Soil HP emitted N_2O when NO_3^- was added but not in the absence of fertilizer or after the addition of urea. In this soil, biochar significantly reduced N_2O emissions, by an average of 76% (Figures 1A1–A3).

Soil HC emitted N_2O in all treatments: without N fertilization, after the addition of NO_3^- and urea. In this soil, biochar consistently increased total cumulative N_2O emissions and the average increase was larger in the non-fertilized (95%) and urea (129%) treatments than in the NO_3^- treatment (54%), (Figures 1B1–B3).

Comparing treatments without biochar, the addition of NO_3^- increased total N_2O emissions in soil HP (from 54 to 11580 μg N_2O-N kg^{-1} soil), whereas it increased N_2O emissions slightly in soil HC (from 3443 to 4546 μg N_2O-N kg^{-1} soil). The addition of urea had no impact on soil HP, and increased emissions in soil HC (from 3443 to 5799 μg N_2O-N kg^{-1} soil).

Figure 2 shows NH_4^+-N concentration in soils HP and HC at the end of the experiment. The original concentration of NH_4^+ in soil HP was 19.3 mg N kg^{-1} soil. After 14 days of incubation, soil HP underwent a significant increase in NH_4^+ content for all fertilization treatments (74.5–110.4 mg N kg^{-1} soil). The highest increase was observed when soil HP was fertilized with urea. Biochar addition did not have a significant impact on the final NH_4^+ concentration in this soil.

Soil HC similarly increased its NH_4^+ concentration throughout the incubation (initial concentration: 2.8 mg kg^{-1} soil), excluding the KNO3 treatment. In this soil biochar significantly decreased the amount of NH_4^+ by the end of the incubation for the non-fertilized soil. Biochar also decreased mean NH_4^+ concentration in the urea treatment, although not significantly due to the high variability in the biochar samples.

Figure 3 shows (NO_3^- + NO_2^-)-N concentrations in soils HP and HC. The concentrations of (NO_3^- + NO_2^-)-N in soil HP were very low (<2.0 mg kg^{-1}) for all fertilization treatments and biochar did not have a significant impact. However, NO_2^- was detected in biochar amended soils and not in the control. Soil HC had low (NO_3^- + NO_2^-)-N concentrations when no fertilizer was added or after the addition of urea. In contrast, 33.3 mg of NO_3^--N kg^{-1} were found in the KNO3 treatment irrespective of the biochar addition.

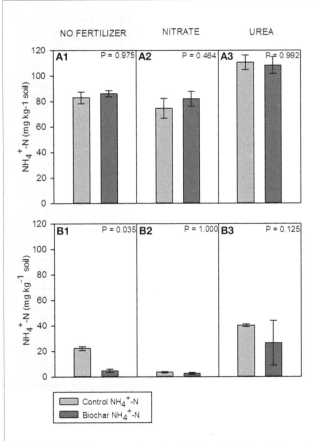

FIGURE 2 | NH$_4^+$-N concentrations in soils HP and HC after 14 days of incubation (mean ± *SE*; *n* = 4). (A1–A3) correspond to soil HP unfertilized, fertilized with KNO$_3$ and fertilized with CO(NH$_2$)$_2$ respectively. **(B1–B3)** correspond to soil HC unfertilized, fertilized with KNO$_3$ and fertilized with CO(NH$_2$)$_2$ respectively.

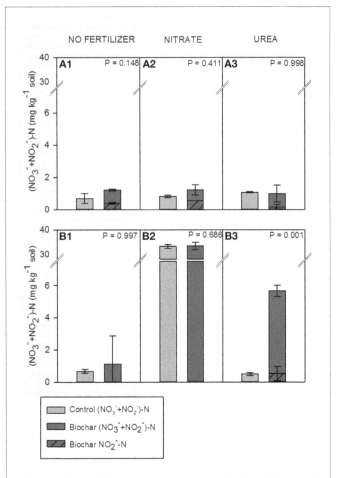

FIGURE 3 | (NO$_3^-$ + NO$_2^-$)-N concentrations in soils HP and HC after 14 days of incubation (mean ± *SE*; *n* = 4). (A1–A3) correspond to soil HP unfertilized, fertilized with KNO$_3$ and fertilized with CO(NH$_2$)$_2$, respectively. **(B1–B3)** correspond to soil HC unfertilized, fertilized with KNO$_3$ and fertilized with CO(NH$_2$)$_2$, respectively.

EXPERIMENT 2. ISOTOPIC COMPOSITION OF N$_2$O EMITTED FROM SOILS A AND B

Figure 4 shows the ^{15}N atomic fraction in N$_2$O emitted from soils HP and HC in Experiment 2. When ^{15}NO$_3^-$ was added, the initial ^{15}N atomic fraction in N$_2$O emitted from soil HP was 0.74, decreasing gradually to reach 0.04 at day 10 (**Figure 4A**). In contrast, the ^{15}N isotopic composition in soil HC followed totally different dynamics: the initial ^{15}N atomic fraction in N$_2$O was only 0.18; it increased slightly to 0.33 by day three, and reached a final value of 0.10 by day 10 (**Figure 4B1**). Biochar altered the isotopic composition of N$_2$O emitted in both soils.

When urea was added, soil HP did not emit N$_2$O (**Figure 1A3**). In soil HC (even when emissions were high) the initial ^{15}N atomic fraction in N$_2$O was zero (**Figure 4B2**), it successively increased, but always remained beneath 0.15. The biochar and control treatments showed identical ^{15}N-N$_2$O concentration dynamics.

Table 2 shows the molar fraction of ^{15}N-NO$_3^-$ and the ratio N$_2$O/(N$_2$+N$_2$O) calculated by the ^{15}N gas flux method (Mulvaney and Boast, 1986) and the contribution of codenitrification to N$_2$O formation according to Spott and Florian Stange (2011) in soil HP. The ratio N$_2$O/(N$_2$+N$_2$O) was very high during the first 3 days, which demonstrates that most N was lost

as N$_2$O. Biochar decreased the N$_2$O/N$_2$ ratio, particularly at day three (the peak of emissions in the control soil). The contribution of codenitrification was zero (see C in **Table 2**). This method of calculation could not be applied to soil HC, since other mechanisms than denitrification were operating in this soil and we could not calculate the enrichment of the source [^{15}NO$_3^-$ in soil ($^{15}X_N$)] (Mulvaney and Boast, 1986). Nonetheless, we found a high proportion of N$_2$O with a hybrid bond (^{45}N$_2$O) in soil HC.

EXPERIMENT 3. N$_2$O EMISSIONS, ^{15}N ISOTOPIC COMPOSITION AND MINERAL N AFTER FERTILIZATION OF SOIL HC WITH NO$_2^-$

Addition of NO$_2^-$ to soil HC produced the highest N$_2$O emissions peak monitored in this soil (**Figure 5B1**); fourfold higher than that of the non-fertilized soil (**Figure 1B1**). Under these conditions, the biochar amendment did not modify cumulative N$_2$O emissions.

The ^{15}N atomic fraction in N$_2$O (**Figure 5B2**) followed a different pattern than with ^{15}NO$_3^-$ (Experiment 2; **Figure 4B1**). The initial ^{15}N atomic fraction in the N$_2$O emitted was 0.30,

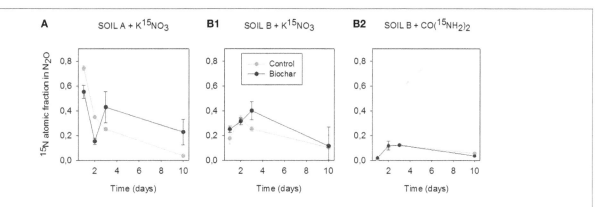

FIGURE 4 | ^{15}N atomic fraction in N_2O emitted from soils HP and HC in control and biochar treatments after 1, 2, 3, and 10 days of incubation (mean $\pm$ SE, $n = 4$). (A) corresponds to soil HP fertilized with $K^{15}NO3$ (>99% enrichment). **(B1)** and **(B2)** correspond to soil HC amended with $K^{15}NO_3$ and $CO(^{15}NH_2)_2$ respectively (both at >99% enrichment).

Table 2 | Means and standard deviations ($n = 4$) of $^{15}X_N$, the ratio $N_2O/(N_2+N_2O)$ and the three fractions (A, B, C) of hybrid and non-hybrid N_2O (Spott and Florian Stange, 2011) in soil HP.

Parameter	Treatment	Time (days)			
		1	2	3	10
$^{15}X_N$	Control	0.98 (0.00)	0.99 (0.00)	0.99 (0.00)	0.84 (0.01)
(molar fraction of $^{15}N-NO_3^-$ in soil, calculated by the ^{15}N gas flux method)	Biochar	0.99 (0.00)	0.99 (0.00)	0.92 (0.06)	0.91 (0.09)
$N_2O/(N_2+N_2O)$	Control	1.01 (0.12)	0.99 (0.01)	0.99 (0.00)	0.14 (0.27)
(calculated by the ^{15}N gas flux method)	Biochar	0.93 (0.05)	0.89 (0.08)	0.04 (0.05)	0.05 (0.08)
A	Control	0.19 (0.05)	0.02 (0.02)	0.00 (0.00)	0.95 (0.00)
(fraction of non-hybrid N_2O from the unlabeled source)	Biochar	0.03 (0.03)	0.01 (0.01)	0.48 (0.28)	0.75 (0.22)
B	Control	0.81 (0.05)	0.98 (0.02)	1.00 (0.00)	0.05 (0.00)
(fraction of non-hybrid N_2O from the labeled source)	Biochar	0.99 (0.03)	1.00 (0.01)	0.49 (0.31)	0.24 (0.22)
C	Control	0.00 (0.00)	0.00 (0.00)	0.00 (0.00)	0.00 (0.00)
(fraction of hybrid N_2O formed by a 1:1 linkage of labeled and unlabeled sources)	Biochar	−0.02 (0.00)	−0.02 (0.00)	0.04 (0.03)	0.01 (0.00)

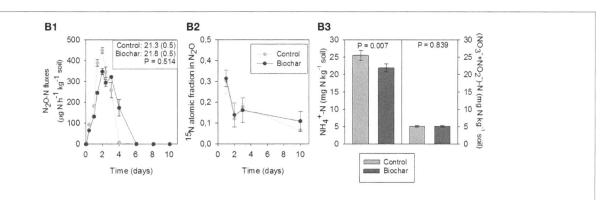

FIGURE 5 | Fluxes of N_2O (B1) ^{15}N atomic fraction in N_2O at days 1, 2, 3, and 10 (B2) and NH_4^+-N and NO_3^--N concentrations after 14 days of incubation (B3) in soil HC (mean $\pm$ SE, $n = 4$). Soil HC had been fertilized with $Na^{15}NO_2$ (>98% enrichment). Cumulative N_2O emission (mg $N-N_2O$ kg^{-1} soil) and standard error by the end of the incubation is reported in the right-above box in **B1**.

decreasing gradually to reach 0.06 at day 10 (**Figure 5B2**). Biochar did not significantly modify this pattern.

The biochar amended soil had a significantly lower concentration of NH_4^+ at the end of the incubation (**Figure 5B3**). The concentration of NO_3^- was low (below 5 mg kg^{-1} soil) and not affected by biochar addition.

EXPERIMENT 4. IMPACT OF THE NITRIFICATION INHIBITOR DICYCIANDIAMIDE (DCD) ON N_2O EMISSIONS AND MINERAL N CONCENTRATION IN SOIL HC

N_2O emissions almost ceased when DCD was added to soil HC (**Figure 6**). The highest emissions were observed when the soil was fertilized with NO_3^- (**Figure 6B2**), but still represented less

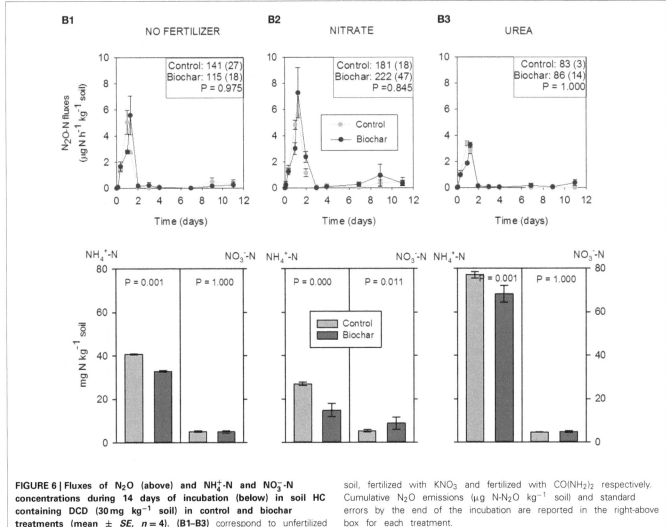

FIGURE 6 | Fluxes of N₂O (above) and NH₄⁺-N and NO₃⁻-N concentrations during 14 days of incubation (below) in soil HC containing DCD (30 mg kg⁻¹ soil) in control and biochar treatments (mean ± SE, n = 4). (B1–B3) correspond to unfertilized soil, fertilized with KNO₃ and fertilized with CO(NH₂)₂ respectively. Cumulative N₂O emissions (μg N-N₂O kg⁻¹ soil) and standard errors by the end of the incubation are reported in the right-above box for each treatment.

than 0.4% of the added N (compared to 12.7% without DCD (**Figure 1B2**).

The highest NH_4^+ concentrations were found in the soil amended with urea, followed by the non-fertilized soil and the soil amended with KNO_3. Biochar (compared to the control) systematically decreased the concentration of NH_4^+ by the end of the incubation for all treatments (non-fertilized soil, KNO_3, and urea). NO_3^- concentration was lower than the original in soil (16.9 mg NO_3^--N kg⁻¹ soil).

DISCUSSION

PRE-DOMINANT N₂O FORMATION PATHWAYS IN SOIL HP AND HC

Nitrous oxide emissions patterns and their response to the addition of different N fertilizers were different in soils HP and HC, which clearly reflected the different N_2O production pathways involved.

Figure 7 illustrates the main pathways for N_2O formation in soil. Ammonia oxidation takes place in two steps: first NH_3 is oxidized to NH_2OH, which is then oxidized to NO_2^-. N_2O may be directly released as a by-product of ammonia oxidation (nitrifier-nitrification) (Hooper and Terry, 1979) or it can be produced

through a denitrification pathway where NO_2^- is reduced to N_2O (nitrifier-denitrification) (Kool et al., 2011). The ability to denitrify is a widespread, if not ubiquitous, attribute in ammonia oxidizers (Shaw et al., 2006). Classically, denitrification (from NO_3^-) has been considered the main N_2O formation pathway in soils. However, other pathways that have been systematically overlooked in soil studies could play a more important role than originally estimated (Baggs, 2011; Spott et al., 2011). This is the case for codenitrification, which is potentially a widespread pathway of microbial N transformation in terrestrial environments (Spott et al., 2011) and dissimilatory nitrate reduction to ammonia (DNRA) (Giles et al., 2012). Although our knowledge of microbial N transformation in soil has evolved significantly over the last decades, recent findings show that, even today, our understanding of N_2O formation and consumption in soil is still very limited (Sanford et al., 2012; Long et al., 2013).

In the nearly water-saturated soil conditions used in our experiments (90% WFPS), N_2O production is expected to be dominated by denitrification of NO_3^-. This was the case in soil HP, where emissions were clearly controlled by the conventional denitrification pathway. This can be deduced from the following

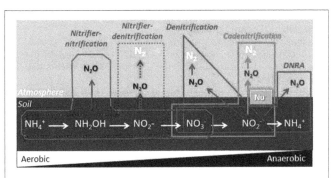

FIGURE 7 | Microbial sources of N$_2$O during transformations of mineral nitrogen in soil. Nu$^-$: nuclophile (e.g., R-NH$_2$, NH$_4^+$, amino acids or other organic N compounds). During codenitrification, nitrous acid reacts with a nucleophile in soil through nitrosation reactions forming a hybrid N-N bond (Spott et al., 2011); DNRA, dissimilatory nitrate reduction to ammonium.

facts: (i) This soil only emitted N$_2$O after the addition of NO$_3^-$ (**Figure 1A2**); (ii) the ^{15}N atomic fraction of the N$_2$O emitted at day one was 0.74 (**Figure 4A**), which shows that N$_2$O was primarily produced from the added ^{15}NO$_3^-$. The ^{15}N atomic fraction decreased over time, showing the depletion of the labeled source; (iii) given the limited nitrification activity detected in this soil, addition of NO$_3^-$ did not increase the final NH$_4^+$ concentration (with respect to the non-fertilized soil), which suggests that DNRA was not a relevant pathway, and (iv) applying the equations developed by Spott and Florian Stange (2011), codenitrification was found to be null (**Table 2**).

As previously found in other soils under analogous optimal denitrifying conditions (Cayuela et al., 2013), biochar significantly decreased total N$_2$O emissions in this soil.

In soil HC, the weak response of N$_2$O emissions to NO$_3^-$ addition pointed out to a low contribution of denitrification or DNRA in this soil. Given that the original NO$_3^-$ concentration in the soil was 16.9 mg N kg^{-1} at a natural abundance of 0.364% ^{15}N, and that we added 55 mg N kg^{-1} of ^{15}NO$_3^-$ (>99% enrichment), the ^{15}N-NO$_3^-$ enrichment in the soil at the beginning of the incubation was 75.8%. Yet, the ^{15}N atomic fraction in the N$_2$O emitted at day one (**Figure 4B1**) was only 0.18, which demonstrates that some N$_2$O originated from denitrification, but also that NO$_3^-$ was not the only source of N$_2$O. Moreover, the low C:N ratio of this soil and the NH$_4^+$ concentration at the end of the incubation in the KNO$_3$ treatment (**Figure 2B2**) indicates that DNRA was not a major N$_2$O formation route in this soil (Giles et al., 2012). Instead, we hypothesize that N$_2$O formation in soil HC was mainly the result of nitrification-mediated processes. The results supporting this hypothesis can be summarized: (i) The addition of extra NO$_3^-$ did not increase N$_2$O emissions in this soil, whereas the addition of extra urea did; (ii) the ^{15}N atomic fraction of the N$_2$O emitted at day one was 17.7% (**Figure 4B1**), which shows that N$_2$O was not pre-dominantly formed from the added ^{15}NO$_3^-$. (iii) The concentration of dissolved organic N in this soil was very high (35.9 mg N kg^{-1}soil), which can explain the low contribution of the labeled urea to the emitted ^{15}N$_2$O (**Figure 4B2**). However, significant hybrid N$_2$O (^{45}N$_2$O) was produced (data not shown) and we cannot

discard the contribution of codenitrification to N$_2$O formation in soil HC.

To better understand which processes (within nitrification-mediated pathways) biochar might be modifying we performed Experiments 3 and 4.

IMPACT OF BIOCHAR IN N$_2$O BY NITRIFICATION-MEDIATED PATHWAYS

In Experiment 3 the addition of NO$_2^-$ to soil HC showed that, under high moisture conditions, this soil was able to rapidly reduce NO$_2^-$ to N$_2$O, which was emitted in large quantities (38% of added NO$_2^-$-N). It is very unlikely that the N$_2$O emitted was just the product of the chemical decomposition of NO$_2^-$ (chemodenitrification), since this process, largely controlled by soil pH, only occurs in neutral and acidic soils (Bremner, 1997). Instead, NO$_2^-$ was most probably used as electron acceptor for microbial respiration (nitrifier-denitrification). The high N$_2$O production in Experiment 3 (21.3 mg N kg^{-1}compared to 3.4 mg N kg^{-1} in Experiment 1) may be related to enhanced nitrifier-denitrification for detoxifying NO$_2^-$ (Jung et al., 2014).

The subsequent tracer experiment with application of ^{15}NO$_2^-$, demonstrated that significant nitrite reduction to N$_2$O occurs (the N$_2$O originating from the added ^{15}NO$_2^-$ at day one was 31.5%, see **Figure 5B2**), but also that it could not be the only process leading to N$_2$O emissions. This experiment demonstrated that biochar was not increasing N$_2$O emissions through the nitrifier-denitrification pathway, since N$_2$O emissions in the biochar and control treatments were not statistically different.

In our final experiment (Experiment 4), the high NH$_4^+$ and low NO$_3^-$ concentrations by the end of the experiment demonstrate the effectiveness of the DCD treatment to inhibit ammonia oxidation, which correlated with a large decrease in N$_2$O emissions for all treatments. We assumed that DCD did not inhibit other possible N$_2$O formation pathways. Although the impacts of DCD on other aspects of microbial N transformation in soil are largely unknown, Bremner and Yeomans (1986) demonstrated that DCD does not inhibit N$_2$O and N$_2$ emissions by denitrification when applied at similar rates to those used in this study. More recently, Wakelin et al. (2013) also demonstrated in a field study that the application of DCD had a minor impact on denitrifying bacteria activity (*nir*S).

Addition of biochar significantly and consistently decreased the NH$_4^+$ concentration in soil HC. These results reinforce our conclusion that the production of N$_2$O in soil HC must be the consequence of nitrification processes (nitrifier-nitrification and associated nitrifier-denitrification). It seems that biochar does not promote the denitrification from NO$_2^-$ (as was deduced from Experiment 3), but it does promote the oxidation of ammonia and concomitantly the formation of N$_2$O through nitrifier-nitrification. Clearly, if biochar raises the production of NO$_2^-$ in soil, it will intrinsically enhance its denitrification (nitrifier-denitrification) when the soil is under low oxygen conditions (as in our experiments).

Our results are in agreement with recent findings by Prommer et al. (2014), who showed that biochar promotes soil ammonia-oxidizer populations and accelerates gross nitrification rates in a calcareous arable soil. The importance of nitrifier-nitrification and nitrifier-denitrification for N_2O production in calcareous soils has been recently documented by Huang et al. (2014), who demonstrated that these processes accounted for 35–53% and 44–58% of total N_2O emissions, respectively.

Here we present preliminary evidence that explains how biochar might affect N_2O emissions differently depending on the N_2O formation pathway operating in the soil. When denitrification was the main N_2O formation pathway (soil HP), biochar was found to decrease the $N_2O/(N_2 + N_2O)$ ratio (**Table 2**), which is in agreement with previous findings (Cayuela et al., 2013). Recent studies have reported that biochar promotes an increase in the abundance of nitrous oxide reductase (nosZ) in soil (Harter et al., 2014), an enzyme that enhances the reduction of N_2O to N_2 (the last step in denitrification). In contrast, when N_2O was produced by nitrification (soil HC), biochar addition might have increased emissions by promoting gross nitrification. To our knowledge, there are not published studies explicitly relating to biochar and nitrification-N_2O production.

Another question that arises from this study is: why these two soils under identical experimental conditions follow different N_2O formation pathways, which we hypothesize might be linked to different soil microbial communities. In conclusion, predicting which N_2O formation pathway pre-dominates in a certain kind of soil will be necessary for guaranteeing the success of biochar as a N_2O mitigation strategy.

ACKNOWLEDGMENTS

We are very grateful to Prof. Miloslav Šimek (University of South Bohemia, Czech Republic) for supplying soil HP and to Pacific Pyrolysis Pty. Ltd (Australia) for providing the biochar used in the experiments. This work was possible thanks to a European Community Marie Curie Fellowship (FP7 PEOPLE-2010-MC-European Reintegration Grants (ERG) #277069). The European Social Fund is acknowledged for co-financing María Luz Cayuela's JAE-Doc contract at CSIC. The authors thank Dr. Sarah K. Wexler, expert in English editing and scientific writing, for her kind help revising this manuscript.

REFERENCES

Baggs, E. M. (2011). Soil microbial sources of nitrous oxide: recent advances in knowledge, emerging challenges and future direction. *Curr. Opin. Environ. Sust.* 3, 321–327. doi: 10.1016/j.cosust.2011.08.011

Bremner, J. M. (1997). Sources of nitrous oxide in soils. *Nutr. Cycl. Agroecosys.* 49, 7–16. doi: 10.1023/A:1009798022569

Bremner, J. M., and Yeomans, J. C. (1986). Effects of nitrification inhibitors on denitrification of nitrate in soil. *Biol. Fert. Soils* 2, 173–179. doi: 10.1007/BF00260840

Cayuela, M. L., Oenema, O., Kuikman, P. J., Bakker, R. R., and Van Groenigen, J. W. (2010). Bioenergy by-products as soil amendments? Implications for carbon sequestration and greenhouse gas emissions. *GCB Bioenergy* 2, 201–213. doi: 10.1111/j.1757-1707.2010.01055.x

Cayuela, M. L., Sánchez-Monedero, M. A., Roig, A., Hanley, K., Enders, A., and Lehmann, J. (2013). Biochar and denitrification in soils: when, how much and why does biochar reduce N_2O emissions? *Sci. Rep.* 3:1732. doi: 10.1038/srep01732

Cayuela, M. L., Van Zwieten, L., Singh, B. P., Jeffery, S., Roig, A., and Sánchez-Monedero, M. A. (2014). Biochar's role in mitigating soil nitrous oxide emissions: a review and meta-analysis. *Agr. Ecosyst. Environ.* 191, 5–16. doi: 10.1016/j.agee.2013.10.009

Clough, T. J., Bertram, J. E., Ray, J. L., Condron, L. M., O'Callaghan, M., Sherlock, R. R., et al. (2010). Unweathered wood biochar impact on nitrous oxide emissions from a bovine-urine-amended pasture soil. *Soil Sci. Soc. Am. J.* 74, 852–860. doi: 10.2136/sssaj2009.0185

Čuhel, J., Šimek, M., Laughlin, R. J., Bru, D., Chèneby, D., Watson, C. J., et al. (2010). Insights into the effect of soil pH on N_2O and N_2 emissions and denitrifier community size and activity. *Appl. Environ. Microbiol.* 76, 1870–1878. doi: 10.1128/AEM.02484-09

Giles, M. E., Morley, N. J., Baggs, E. M., and Daniell, T. J. (2012). Soil nitrate reducing processes – drivers, mechanisms for spatial variation and significance for nitrous oxide production. *Front. Microbiol.* 3:407. doi: 10.3389/fmicb.2012.00407

Harter, J., Krause, H.-M., Schuettler, S., Ruser, R., Fromme, M., Scholten, T., et al. (2014). Linking N2O emissions from biochar-amended soil to the structure and function of the N-cycling microbial community. *ISME J.* 8, 660–674. doi: 10.1038/ismej.2013.160

Hooper, A. B., and Terry, K. R. (1979). Hydroxylamine oxidoreductase of nitrosomonas: production of nitric oxide from hydroxylamine. *Biochim. Biophy. Acta* 571, 12–20. doi: 10.1016/0005-2744(79)90220-1

Huang, T., Gao, B., Hu, X.-K., Lu, X., Well, R., Christie, P., et al. (2014). Ammonia-oxidation as an engine to generate nitrous oxide in an intensively managed calcareous Fluvo-aquic soil. *Sci. Rep.* 4:3950. doi: 10.1038/srep03950

Jung, M.-Y., Well, R., Min, D., Giesemann, A., Park, S.-J., Kim, J.-G., et al. (2014). Isotopic signatures of N2O produced by ammonia-oxidizing archaea from soils. *ISME J.* 8, 1115–1125. doi: 10.1038/ismej.2013.205

Kettler, T. A., Doran, J. W., and Gilbert, T. L. (2001). Simplified method for soil particle-size determination to accompany soil-quality analyses. *Soil Sci. Soc. Am. J.* 65, 849–852. doi: 10.2136/sssaj2001.653849x

Kool, D. M., Dolfing, J., Wrage, N., and Van Groenigen, J. W. (2011). Nitrifier denitrification as a distinct and significant source of nitrous oxide from soil. *Soil Biol. Biochem.* 43, 174–178. doi: 10.1016/j.soilbio.2010.09.030

Long, A., Heitman, J., Tobias, C., Philips, R., and Song, B. (2013). Co-occurring anammox, denitrification, and codenitrification in agricultural soils. *Appl. Environ. Microb.* 79, 168–176. doi: 10.1128/AEM.02520-12

Malghani, S., Gleixner, G., and Trumbore, S. E. (2013). Chars produced by slow pyrolysis and hydrothermal carbonization vary in carbon sequestration potential and greenhouse gases emissions. *Soil Biol. Biochem.* 62, 137–146. doi: 10.1016/j.soilbio.2013.03.013

Mondini, C., Sinicco, T., Cayuela, M. L., and Sanchez-Monedero, M. A. (2010). A simple automated system for measuring soil respiration by gas chromatography. *Talanta* 81, 849–855. doi: 10.1016/j.talanta.2010.01.026

Mulvaney, R. L., and Boast, C. W. (1986). Equations for determination of nitrogen-15 labeled dinitrogen and nitrous oxide by mass spectrometry. *Soil Sci. Soc. Am. J.* 50, 360–363. doi: 10.2136/sssaj1986.03615995005000020021x

Prommer, J., Wanek, W., Hofhansl, F., Trojan, D., Offre, P., Urich, T., et al. (2014). Biochar decelerates soil organic nitrogen cycling but stimulates soil nitrification in a temperate arable field trial. *PLoS ONE* 9:e86388. doi: 10.1371/journal.pone.0086388

Rajbanshi, S. S., Benckiser, G., and Ottow, J. C. G. (1992). Effects of concentration, incubation temperature, and repeated applications on degradation kinetics of dicyandiamide (DCD) in model experiments with a silt loam soil. *Biol. Fert. Soils* 13, 61–64. doi: 10.1007/BF00337336

Saarnio, S., Heimonen, K., and Kettunen, R. (2013). Biochar addition indirectly affects N_2O emissions via soil moisture and plant N uptake. *Soil Biol. Biochem.* 58, 99–106. doi: 10.1016/j.soilbio.2012.10.035

Sanford, R. A., Wagner, D. D., Wu, Q., Chee-Sanford, J. C., Thomas, S. H., Cruz-García, C., et al. (2012). Unexpected nondenitrifier nitrous oxide reductase gene diversity and abundance in soils. *Proc. Natl. Acad. Sci. U.S.A.* 109, 19709–19714. doi: 10.1073/pnas.1211238109

Shaw, L. J., Nicol, G. W., Smith, Z., Fear, J., Prosser, J. I., and Baggs, E. M. (2006). Nitrosospira spp. can produce nitrous oxide via a nitrifier denitrification pathway. *Environ. Microbiol.* 8, 214–222. doi: 10.1111/j.1462-2920.2005.00882.x

Spokas, K. A., and Reikosky, D. C. (2009). Impacts of sixteen different biochars on soil greenhouse gas production. *Ann. Environ. Sci.* 3, 179–193. Available online at: http://hdl.handle.net/2047/d10019583

Spott, O., and Florian Stange, C. (2011). Formation of hybrid N_2O in a suspended soil due to co-denitrification of NH_2OH. *J. Plant Nutr. Soil Sci.* 174, 554–567. doi: 10.1002/jpln.201000200

Spott, O., Russow, R., and Stange, C. F. (2011). Formation of hybrid N_2O and hybrid N_2 due to codenitrification: First review of a barely considered process of microbially mediated N-nitrosation. *Soil Biol. Biochem.* 43, 1995–2011. doi: 10.1016/j.soilbio.2011.06.014

Stevens, R. J., and Laughlin, R. J. (2001). Lowering the detection limit for dinitrogen using the enrichment of nitrous oxide. *Soil Biol. Biochem.* 33, 1287–1289. doi: 10.1016/S0038-0717(01)00036-0

Stevens, R. J., Laughlin, R. J., Atkins, G. J., and Prosser, S. J. (1993). Automated determination of nitrogen-15-labeled dinitrogen and nitrous oxide by mass spectrometry. *Soil Sci. Soc. Am. J.* 57, 981–988. doi: 10.2136/sssaj1993.03615995005700040017x

Suddick, E. C., and Six, J. (2013). An estimation of annual nitrous oxide emissions and soil quality following the amendment of high temperature walnut shell biochar and compost to a small scale vegetable crop rotation. *Sci. Total Environ.* 465, 298–307. doi: 10.1016/j.scitotenv.2013.01.094

Van Zwieten, L., Kimber, S., Morris, S., Downie, A., Berger, E., Rust, J., et al. (2010). Influence of biochars on flux of N_2O and CO_2 from Ferrosol. *Aust. J. Soil Res.* 48, 555–568. doi: 10.1071/SR10004

Velthof, G., Kuikman, P., and Oenema, O. (2003). Nitrous oxide emission from animal manures applied to soil under controlled conditions. *Biol. Fert. Soils* 37, 221–230. doi: 10.1007/s00374-003-0589-2

Wakelin, S. A., Clough, T. J., Gerard, E. M., and O'Callaghan, M. (2013). Impact of short-interval, repeat application of dicyandiamide on soil N transformation in urine patches. *Agric. Ecosyst. Environ.* 167, 60–70. doi: 10.1016/j.agee.2013.01.007

Yoo, G., and Kang, H. (2012). Effects of biochar addition on greenhouse gas emissions and microbial responses in a short-term laboratory experiment. *J. Environ. Qual.* 41, 1193–1202. doi: 10.2134/jeq2011.0157

Zacherl, B., and Amberger, A. (1990). Effect of the nitrification inhibitors dicyandiamide, nitrapyrin and thiourea onNitrosomonas europaea. *Fert. Res.* 22, 37–44. doi: 10.1007/BF01054805

Conflict of Interest Statement: The authors declare that the research was conducted in the absence of any commercial or financial relationships that could be construed as a potential conflict of interest.

11

Soil phosphorus saturation ratio for risk assessment in land use systems

Vimala D. Nair *

Soil and Water Science Department, University of Florida, Gainesville, FL, USA

Edited by:
Pankaj Kumar Arora, Yeungnam University, South Korea

Reviewed by:
Gurbir S. Bhullar, Research Institute of Organic Agriculture (FiBL), Switzerland
Gustavo Habermann, São Paulo State University (UNESP), Brazil
Bing-Jie Ni, The University of Queensland, Australia

***Correspondence:**

Vimala D. Nair, Soil and Water Science Department, University of Florida, PO Box 110290, 2181 McCarty Hall A, Gainesville, FL 32611, USA
e-mail: vdn@ufl.edu

The risk of phosphorus loss from agricultural soils can have serious implications for water quality. This problem has been noted particularly in sandy soils in several parts of the world including Europe (e.g., the Netherlands, Italy, and UK) and the southeastern USA. However, the capacity of a soil to retain P is limited and even non-sandy soils have the potential to eventually release P when inorganic or organic fertilizer is added over a period of time. A threshold phosphorus saturation ratio (PSR), calculated from P, Fe, and Al in an oxalate or a soil test solution such as Mehlich 1 or Mehlich 3, has been recognized as a practical means of determining when a soil has reached a level of P loading that constitutes an environmental risk. When soils are below a threshold PSR value, the equilibrium P concentration (EPC_0) is minimal. Further, the soil P storage capacity calculated from the same data is directly linked to the strength of P bonding (K_L) as determined from Langmuir isotherms, and K_D, the distribution coefficient related to the strength of sorption. While the PSR is occasionally used as a predictor of the onset of environmentally significant P loss from a soil, the procedure might be adopted as a routine soil test.

Keywords: Langmuir, linear isotherm, Mehlich 1, Mehlich 3, oxalate, phosphorus risk assessment, soil P storage capacity, threshold phosphorus saturation ratio

INTRODUCTION

Non-point source pollution from agricultural and other anthropogenic sources have been identified as the major cause of degradation of water bodies (USEPA, 2002). Excess application of inorganic or organic phosphorus (P) fertilizers beyond that required for plant uptake would result in the loss of P from the soil to adjacent water bodies through surface or subsurface movement (Sims et al., 1998; Hooda et al., 2000; Sharpley and Tunney, 2000). Phosphorus loss from agricultural lands has an adverse impact on water quality and therefore affects human and animal health, biodiversity, nutrient cycling, and ecosystem functioning in addition to recreational facilities such as swimming and fishing. Eutrophication of surface waters has been noted in various parts of the world including the Netherlands and Italy (Breeuwsma and Silva, 1992), Florida (Nair et al., 1995), and other parts of the southeastern USA where the soils are predominantly sandy (Sims et al., 1998). However, all soils have a finite capacity to retain P and continued application of fertilizers even to soils that have high P retention capabilities, will ultimately reach the environmental limit for safe storage of P. This review is on the P saturation ratio (PSR) concept where the threshold value is an indicator for risk assessment in land use systems.

Since the ability of soils to retain P varies substantially, P loss from a soil is heavily dependent on the soil components that retain P. Determining when a soil has reached a level of P loading in agricultural and related land-use systems that constitutes an environmental risk is indeed challenging. The degree of phosphorus saturation (DPS) was first introduced in the Netherlands as a tool to predict environmental limits for soil P in sandy soils (van der Zee et al., 1987; Breeuwsma and Silva, 1992) but has since been extended to other parts of the world. The DPS is normally expressed as a percentage and calculated as the ratio of acid ammonium oxalate-extractable [P] to [Al + Fe] (van der Zee and van Riemsdijk, 1988).

$$DPS_{OX} = \frac{Oxalate - extractable\, P}{\alpha\, Oxalate - extractable\, [Fe + Al]} \times 100 \qquad (1)$$

While the original method of calculation of the *DPS* specified oxalate-extractable *P*, *Fe*, and *Al*, modifications were based on soil test phosphorus parameters used in various parts of the USA. Mehlich 1 extracts (DPS_{M1}) (Nair and Graetz, 2002; Beck et al., 2004; Nair et al., 2004) and Mehlich 3 extracts (DPS_{M3}) (Maguire and Sims, 2002; Sims et al., 2002; Nair et al., 2004) have been shown to be suitable to calculate *DPS* for sandy soils of the southeastern USA. Included in the *DPS* calculation is an empirical α factor in the denominator to account for the fraction of *Fe* and *Al* responsible for *P* sorption for soils of a given region. The corrective factor α may be omitted and a simple ratio of molar *P* to molar [*Fe* + *Al*], referred to as the PSR, used for soils with similar properties (Maguire and Sims, 2002; Nair and Harris, 2004).

Many researchers use an arbitrary value of 0.5 for α (Nair et al., 2004) though others such as Paulter and Sims (2000) specified a value of 0.68 for their soils.

The PSR (or *DPS*) is related to soil solution *P* concentration, and allows an establishment of threshold values corresponding approximately to a set critical solution concentration (Breeuwsma and Silva, 1992; Nair et al., 2004). A plot of water soluble *P* (WSP) against the PSR (or *DPS*) illustrates this concept using *P*, *Fe*, and *Al* in an oxalate solution in the calculations (**Figure 1**). Sharpley et al. (2013) have pointed out that recently attempts have been made "to quantify the concept both from parameterized models

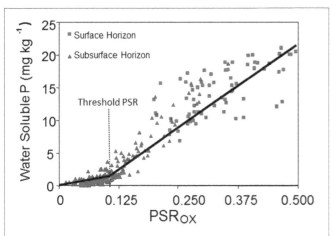

FIGURE 1 | Relationship between the concentration of water soluble phosphorus (WSP) and the phosphorus saturation ratio (PSR) for manure-impacted surface and subsurface horizons. The threshold PSR = 0.1. Source: Modified from Nair et al. (2004), replacing the degree of phosphorus saturation (DPS) with PSR.

describing P sorption isotherms to models describing P sorption saturation." In a groundwater field monitoring study in Delaware, Andres and Sims (2013) showed that the use of the PSR concept and the calculated soil P storage capacity (SPSC; see below) were effective in risk assessment of P loss from the site.

APPLICATION OF THE PSR CONCEPT TO NON-SANDY SOILS

The PSR concept was originally developed for sandy soils where leaching is a primary mode of P transport. However, subsurface horizons of major soil orders such as the Bt horizon of Ultisols in Florida can have loamy to clayey textures, and Chakraborty et al. (2012) identified a threshold PSR for these horizons as well. Their results also showed that even when there is an abundance of crystalline components that contribute to P sorption above the threshold PSR, the most tenaciously-bound P is associated with non-crystalline metal oxides such as those extracted by an oxalate or a soil test solution such as Mehlich 1 or Mehlich 3. Further, the authors identified a threshold PSR for Bh horizons, which have organically-complexed Al. The Bh horizon occurs in Spodosols, the most extensively-occurring soil order in Florida, USA. Therefore, the PSR concept appears applicable to a wide range of soil types and not just sandy soils or soils with inorganically-complexed Al.

SOIL PSR FOR WETLAND SOILS

Preliminary data collected on wetland soils (Mukherjee et al., 2009) suggested that the threshold PSR might be a practical indicator to assess nutrient enrichment in wetland soils where P solubility is regulated by Fe and Al. More recent work (Nair and Reddy, 2013) confirmed the applicability of the PSR concept to inundated soils. When occurring on the same landscape, the threshold PSR value is identical to that of adjacent upland soils. This threshold PSR has been identified as 0.1 for a range of upland (Nair et al., 2010) and wetland (Nair and Reddy, 2013) soils in Florida, USA. The organic matter in the wetland soil does not

contribute to P sorption below the threshold PSR. In the Nair and Reddy (2013) study, organic matter ranged from 0 to 95% for wetland soils both above and below the threshold value, confirming that below the 0.1 threshold, P retention is dictated only by Fe and Al.

THRESHOLD PSR VALUES FOR A RANGE OF LOCATIONS AND SOIL TYPES

A literature review of various studies on DPS/PSR by researchers in different parts of the world showed a remarkable similarity in the threshold PSR value (**Table 1**) for all soils—located on Alfisols, Entisols, Inceptisols, Mollisols, Oxisols, Spodosols Ultisols, or Vertisols. While the WSP-PSR relationship is often used in determining the threshold PSR value, some researchers (Sims et al., 2002) identified threshold PSR values using runoff in rainfall simulation experiments as a function of soil PSR. Some other procedures used are identified in **Table 1**. Further, the PSR itself is determined using different extractants such Mehlich 1, Mehlich 3, or oxalate.

When P sorption maximum (P fixation maximum) is used instead of [Fe + Al], the α value is not specified and it is not clear how this parameter can be related to [Fe + Al] in PSR calculations. Despite these variations in methodology for the threshold PSR calculations, the value appears to be in the 0.10–0.15 range (**Table 1**). The PSR determination appears to be operationally dependent; however, it could eventually be possible to obtain a single threshold PSR value for a range of soils if the method of PSR determination is maintained the same.

The PSR values are soil-specific and not system-specific, i.e., two soils can have the same PSR, but the environmental risk would be different because the P retentive properties of the soils are different. However, once a threshold PSR is determined for a group of soils with similar properties (such as soils with inorganically-complexed Al or organo-Al complexed horizons such as the Bh horizon), then a more quantitative measure, the soil P storage capacity (SPSC) can be calculated (see the next Section).

OBTAINING ISOTHERM PARAMETERS FROM SOIL TEST DATA

Since the strength of P bonding in the Langmuir adsorption model is an indicator of how firmly P is held to soil components such as Fe and Al, it follows that below the threshold PSR Langmuir K_L, the P bonding constant should be high and the P would be released from the soil to the water once the threshold PSR is reached (Dari et al., 2012). Therefore the PSR value obtained from a soil testing solution such as Mehlich 1 or Mehlich 3 can be used to predict when a soil will begin to become an environmental P loss risk. Below the threshold PSR, the equilibrium P concentration (EPC$_0$) of a soil will be a minimum, but will increase once the soil is above the threshold value (Chakraborty et al., 2012).

The amount of P that can be safely stored within a soil prior to the soil becoming an environmental risk can be calculated from the threshold PSR (Nair and Harris, 2004). The soil P storage capacity (SPSC) is calculated using the following generalized equation:

Table 1 | Threshold P saturation ratio (PSR) and the corresponding degree of P saturation (DPS) for surface soils from various locations and soil orders.

Location	Soil order	Procedure[†]	PSR	DPS (%)	References
Arkansas, USA	Various	Ox-PSR[‡] vs. rainfall simulation runoff; visually determined	0.12	12	Vadas et al., 2005
Delaware, USA	Entisols/Ultisols/Inceptisols	M3-PSR[‡] vs. rainfall simulation runoff	0.14	N/A	Sims et al., 2002
		M3-PSR vs. column leachate	0.21		
Florida, USA (uplands)	Entisols/Ultisols	Ox-PSR vs. WSP	0.10	20	Nair et al., 2004
		M1-PSR[‡] vs. WSP	0.10	20	
		M3-PSR vs. WSP	0.08	16	
Florida, USA (wetlands)	Spodosols	M1-PSR vs. WSP	0.10	N/A	Nair and Reddy, 2013
Minnesota, USA	Alfisols/Mollisols	M3-PSR vs. WSP	0.11	22	Laboski and Lamb, 2004
Brazil	Ultisols	M3-P/P sorption maximum vs. WSP	0.23[a]	23	Abdala et al., 2012
Canada	N/A	M3-P/M3-Al	0.08	15	Khiari et al., 2000
Italy	Alfisols/Vertisols	Olsen-P vs. P-fixation maximum	0.18[a]	18	Indiati and Sequi, 2004
Netherlands and Italy	N/A	Ox-PSR	0.13	25	Breeuwsma and Silva, 1992
Switzerland	N/A	Ox-PSR vs. WSP	0.12	24	Roger et al., 2014
UK	Various	Ox-PSR vs. desorbed P	0.10	10	Hooda et al., 2000
Uganda	Oxisols	M3-PSR vs. WSP	0.10	20	Nkedi-Kizza and Nair, unpublished

[†]Procedures used in the threshold PSR determinations vary including replacing water soluble P (WSP) with runoff P from rainfall simulation experiments or leachate from column experiments.

[‡]Ox-PSR, M3-PSR, and M1-PSR; Threshold P saturation ratio determined from P, Fe, and Al in oxalate, Mehlich 3 and Mehlich 1 solutions, respectively.

[a]α considered to be 1 in the PSR calculations.

N/A Not available.

$$SPSC = (\text{Threshold PSR} - \text{Soil PSR})$$
$$* [Fe + Al] * 31 \, mg \, kg^{-1} \quad (2)$$

where P, Fe, and Al can be determined in either an oxalate or a soil test solution. Below the threshold PSR (e.g., 0.1), SPSC is positive (the soil is a P sink) while SPSC becomes negative (soil is a P source) above the threshold value. This concept has been shown to be applicable to subsurface soil horizons (Chakraborty et al., 2011) and wetland soils (Reddy et al., 2012; Nair and Reddy, 2013) as well. The SPSC has an additional advantage in that it is able to provide a more meaningful and valued P loss risk indicator (Nair and Harris, 2004) since it takes into account previous P loading and enables a prediction of the amount of P that can be added to a soil prior to the soil becoming an environmental risk. Below a threshold PSR (i.e., when SPSC is positive), SPSC is related to Langmuir K_L (Dari et al., 2012), allowing prediction of K_L values from soil test data.

SUMMARY AND CONCLUSIONS

The threshold PSR has enormous power in predicting P stability in a soil with values below the threshold value indicating that P release from the soil is minimal. Once the threshold PSR is reached, the soil becomes a P source. While the PSR concept was originally developed for sandy surface soils, recent research indicated its validity for subsurface horizons including soils that have loamy to clayey textures. Further, the approach is applicable to

wetland soils where the organic matter does not contribute to P retention below the threshold value. Therefore, determination of the PSR of soils (surface and/or subsurface) affords a procedure to predict when P loss from a site via runoff or leaching would begin to become an environmental concern. Despite methodology differences in obtaining a threshold PSR value, most soils tend to have a PSR value in the 0.1–0.15 range. Since SPSC can be calculated from the same data that is required for PSR calculations; it follows that both the PSR and the SPSC can be obtained by sending a soil sample to a routine soil testing lab. The PSR/SPSC concept can be easily adopted by farmers and others who are interested in management practices that minimize the risk of P loss from soils.

ACKNOWLEDGMENTS

The author thanks Willie Harris for comments and suggestions on an earlier version of the review.

REFERENCES

Abdala, D. B., Ghosh, A. K., da Silva, I. R., de Novias, R. F., and Venegas, V. H. A. (2012). Phosphorus saturation of a tropical soil and related P leaching caused by poultry litter addition. *Agric. Ecosyst. Environ.* 162, 15–23. doi: 10.1016/j.agee.2012.08.004

Andres, A. S., and Sims, J. T. (2013). Assessing potential impacts of a wastewater rapid infiltration basin system on groundwater quality: a Delaware case study. *J. Environ. Qual.* 42, 391–404. doi: 10.2134/jeq2012.0273

Beck, M. A., Zelazny, L. W., Daniels, W. L., and Mullins, G. L. (2004). Using the Mehlich 1 extract to measure soil phosphorus saturation for environmental risk assessment. *Soil Sci. Soc. Am. J.* 68, 1762–1771. doi: 10.2136/sssaj 2004.1762

Breeuwsma, A., and Silva, S. (1992). "Phosphorus fertilisation and environmental effects in The Netherlands and the Po Region (Italy)," in *Report 57. Agricultural Research Department* (Wageningen: The Winand Staring Centre for Integrated Land, Soil, and Water Research).

Chakraborty, D., Nair, V. D., Chrysostome, M., and Harris, W. G. (2011). Soil phosphorus storage capacity in manure-impacted Alaquods: implications for water table management. *Agric. Ecosyst. Environ.* 142, 167–175. doi: 10.1016/j.agee.2011.04.019

Chakraborty, D., Nair, V. D., Harris, W. G., and Rhue, R. D. (2012). Environmentally-relevant phosphorus retention capacity of sandy coastal plain soils. *Soil Sci.* 177, 701–707. doi: 10.1097/SS.0b013e31827d8685

Dari, B., Nair, V. D., Rhue, R. D., and Mylavarapu, R. (2012). "Relationship of Langmuir parameters to the soil phosphorus saturation ratio," in *ASA/CSSA/SSSA 2012 International Annual Meetings* (Cincinnati, OH: CD Rom Publication).

Hooda, P. S., Rendell, A. R., Edwards, A. C., Withers, P. J. A., Aitken, M. N., and Truesdale, V. W. (2000). Relating soil phosphorus indices to potential phosphorus release to water. *J. Environ. Qual.* 29, 1166–1171. doi: 10.2134/jeq2000.00472425002900040018x

Indiati, R., and Sequi, P. (2004). Phosphorus intensity-quantity relationships in soils highly contrasting in phosphorus adsorption properties. *Commun. Soil Sci. Plant Anal.* 35, 131–143. doi: 10.1081/CSS-120027639

Khiari, L., Parent, L. E., Pellerin, A., Alimi, A. R. A., Tremblay, C., Simard, R. R., et al. (2000). An agri-environmental phosphorus saturation index acid coarse-textured soils. *J. Environ. Qual.* 29, 1561–1567. doi: 10.2134/jeq2000.00472425002900050024x

Laboski, C. A. M. and Lamb, L. A. (2004). Impact of manure application on soil phosphorus sorption characteristics and subsequent water quality implications. *Soil Sci.* 169, 440–448. doi: 10.1097/01.ss.0000131229.58849.0f

Maguire, R. O., and Sims, J. T. (2002). Soil testing to predict phosphorus leaching. *J. Environ. Qual.* 31, 1601–1609. doi: 10.2134/jeq2002.1601

Mukherjee, A., Nair, V. D., Clark, M. W., and Reddy, K. R. (2009). Development of indices to predict phosphorus release from wetland soils. *J. Environ. Qual.* 38, 878–886. doi: 10.2134/jeq2008.0230

Nair, V. D., and Graetz, D. A. (2002). Phosphorus saturation in spodosols impacted by manure. *J. Environ. Qual.* 31, 1279–1285. doi: 10.2134/jeq2002.1279

Nair, V. D., Graetz, D. A., and Portier, K. M. (1995). Forms of phosphorus in soil profiles from dairies of South Florida. *Soil Sci. Soc. Am. J.* 59, 1244–1249. doi: 10.2136/sssaj1995.03615995005900050006x

Nair, V. D., and Harris, W. G. (2004). A capacity factor as an alternative to soil test phosphorus in phosphorus risk assessment. *N.Z. J. Agric. Res.* 47, 491–497. doi: 10.1080/00288233.2004.9513616

Nair, V. D., Harris, W. G., Chakraborty, D., and Chrysostome, M. (2010). *Understanding Soil Phosphorus Storage Capacity.* SL 336. Available online at: http://edis.ifas.ufl.edu/pdffiles/SS/SS54100.pdf

Nair, V. D., Portier, K. M., Graetz, D. A., and Walker, M. L. (2004). An environmental threshold for degree of phosphorus saturation in sandy soils. *J. Environ. Qual.* 33, 107–113. doi: 10.2134/jeq2004.1070

Nair, V. D., and Reddy, K. R. (2013)."Phosphorus sorption and desorption in wetland soils," in *Methods on Biogeochemistry of Wetlands*, eds R. DeLaune, K. R. Reddy, C. J. Richardson, and P. Megonigal (Madison, WI: SSSA, Inc.), 667–678.

Paulter, M. C., and Sims, J. T. (2000). Relationships between soil test phosphorus, soluble phosphorus, and phosphorus saturation in delaware soils. *Soil Sci. Soc. Am. J.* 64, 765–773. doi: 10.2136/sssaj2000.642765x

Reddy, K. R., Clark, M. W., and Nair, V. D. (2012). *Legacy Phosphorus in Agricultural Watersheds: Implications for Restoration and Management of Wetlands and Aquatic Systems.* International Atomic Energy Agency (IAEA) Newsletter. Available online at: http://www-naweb.iaea.org/nafa/swmn/public/SNL-34-2.pdf.

Roger, A., Sinaj, S., Libohova, Z., and Frossard, E. (2014). "Regional investigation of soil phosphorus saturation degree, a study case in Switzerland," in *GlobalSoilMap: Basis of the Global Spatial Soil Information System*, eds D. Arrouays, N. McKenzie, J. Hempel, A. Richer de Forges, and A. B. McBratney (London: Taylor and Francis Group), 79–83. doi: 10.1201/b16500-18

Sharpley, A., Jarvie, H. P., Buda, A., May, L., Spears, B., and Kleinman, P. (2013). Phosphorus legacy: practices to mitigate future water quality impairment. *J. Environ. Qual.* 42, 1308–1326. doi: 10.2134/jeq2013.03.0098

Sharpley, A., and Tunney, H. (2000). Phosphorus research strategies to meet agricultural and environmental challenges in the 21st century. *J. Environ. Qual.* 29, 176–181. doi: 10.2134/jeq2000.00472425002900 010022x

Sims, J. T., Maguire, R. O., Leytem, A. B., Gartley, K. L., and Paulter, M. C. (2002). Evaluation of Mehlich 3 as an agri-environment soil phosphorus test for the Mid-Atlantic United States of America. *Soil Sci. Soc. Am. J.* 66, 2016–2032. doi: 10.2136/sssaj2002.2016

Sims, J. T., Simard, R. R., and Joern, B. C. (1998). Phosphorus loss in agricultural drainage: historical perspective and current research. *J. Environ. Qual.* 27, 277–293. doi: 10.2134/jeq1998.00472425002700 20006x

U. S. Environmental Protection Agency (USEPA). (2002). *Environmental Indicators of Water Quality in the United States. EPA 841-R-02-001.* Washington, DC: USEPA Office of Water Quality.

Vadas, P. A., Kleinman, P. J. A., Sharpley, A. N., and Turner, B. L. (2005). Relating soil phosphorus to dissolved phosphorus in runoff: a single extraction coefficient for water quality modeling. *J. Environ. Qual.* 34, 572–580. doi: 10.2134/jeq2005.0572

van der Zee, S. E. A. T. M., Fokkink, L. G. J., and van Riemsdijk, W. H. (1987). A new technique for assessment of reversibly adsorbed phosphate. *Soil Sci. Soc. Am. J.* 51, 599–604. doi: 10.2136/sssaj1987.036159950051000 30009x

van der Zee, S. E. A. T. M., and van Riemsdijk, W. H. (1988). Model for long-term phosphate reaction kinetics in soil. *J. Environ. Qual.* 17, 35–41. doi: 10.2134/jeq1988.00472425001700010005x

Conflict of Interest Statement: The author declares that the research was conducted in the absence of any commercial or financial relationships that could be construed as a potential conflict of interest.

12

Managing the pools of cellular redox buffers and the control of oxidative stress during the ontogeny of drought-exposed mungbean (*Vigna radiata* L.)—role of sulfur nutrition

Naser A. Anjum[1,2], Shahid Umar[1], Ibrahim M. Aref[3] and Muhammad Iqbal[1]*

[1] Department of Botany, Faculty of Science, Hamdard University, New Delhi, India
[2] Department of Chemistry, CESAM-Centre for Environmental and Marine Studies, University of Aveiro, Aveiro, Portugal
[3] Plant Production Department, College of Food and Agricultural Sciences, King Saud University, Riyadh, Saudi Arabia

Edited by:
Adriano Sofo, Università degli Studi della Basilicata, Italy

Reviewed by:
Yogesh Abrol, Bhagalpur University, India
Kumar Ajit, University of South Australia, Australia

*Correspondence:
Muhammad Iqbal, Department of Botany, Faculty of Science, Hamdard University, New Delhi, 110062, India
e-mail: iqbalg5@yahoo.co.in

Impacts of increasing environmental stresses (such as drought) on crop productivity can be sustainably minimized by using plant-beneficial mineral nutrients (such as sulfur, S). This study, based on a pot-culture experiment conducted in greenhouse condition, investigates S-mediated influence of drought stress (imposed at pre-flowering, flowering, and pod-filling stages) on growth, photosynthesis and tolerance of mungbean (*Vigna radiata* L.) plants. Drought stress alone hampered photosynthesis functions, enhanced oxidative stress [measured in terms of H_2O_2; lipid peroxidation (LPO); electrolyte leakage (EL)] and decreased the pools of cellular redox buffers (namely ascorbate (AsA); glutathione (GSH)], and the overall plant growth (measured as leaf area and plant dry mass), maximally at flowering stage, followed by pre-flowering and pod-filling stages. Contrarily, S-supplementation to drought-affected plants (particularly at flowering stage) improved the growth- and photosynthesis-related parameters considerably. This may be ascribed to S-induced enhancements in the pools of reduced AsA and GSH, which jointly manage the balance between the production and scavenging of H_2O_2 and stabilize cell membrane by decreasing LPO and EL. It is inferred that alleviation of drought-caused oxidative stress depends largely on the status of AsA and GSH via S-supplementation to drought-stressed *V. radiata* at an appropriate stage of plant growth, when this nutrient is maximally or efficiently utilized.

Keywords: cellular buffers, drought stress, mungbean ontogeny, oxidative stress, sulfur, *Vigna radiata*

INTRODUCTION

Recognized as one of the major environmental stress factors, and as a main constraint for crop production worldwide, drought affects virtually every aspect of plant growth, physiology and metabolism (Harb et al., 2010). In particular, at the whole-plant level, drought stress affects mainly the plant photosynthetic functions, causing imbalance in "CO_2 fixation and electron transport." This facilitates transfer of electrons to reactive oxygen species (ROS), including H_2O_2, as a result of over-reduction of the electron-transport-chain components (Anjum et al., 2008a; Lawlor and Tezara, 2009). Additionally, high concentration of ROS causes oxidative damage to photosynthetic pigments, biomolecules such as lipids, proteins and nucleic acids, and leakage of electrolytes via lipid peroxidation (LPO), leading to cessation of normal plant cellular metabolism (Anjum et al., 2012a). The ascorbate-glutathione (AsA-GSH) pathway constitutes the major part of antioxidant defense system in plants where a number of ROS are effectively metabolized and/or detoxified by a network of reactions involving enzymes and metabolites with redox properties. Both AsA and GSH (tripeptide GSH; γ-glutamate-cysteine-glycine) are cellular redox buffers closely linked in major physiological functions. Nevertheless, in conjunction with other components of AsA-GSH pathway, both AsA and GSH together determine the lifetime of varied ROS and their reaction products within the cellular environment and provide crucial protection against oxidative damage (Anjum et al., 2010, 2012a, 2013). In recent studies, exogenous application of AsA or GSH was reported to help plants to withstand consequences caused by a range of abiotic stresses including Cd (Cai et al., 2011; Son et al., 2014), salinity (Wang et al., 2014) and high temperature (Nahar et al., 2015).

Maintenance of the status of mineral nutrients in plants is important for increasing the crop productivity and plant resistance to environmental stress (Cakmak, 2005; Anjum et al., 2008b, 2012b). The cumulative role of mineral nutrients in modulating cellular levels of AsA and GSH, and in strengthening the plant antioxidant defense system has been discussed extensively (Anjum et al., 2010, 2012a, 2013; Gill et al., 2011). Sulfur (S) is the fourth essential macronutrient for plants, after N, P and K, and plays a vital role in the regulation of plant growth, development and productivity (Hawkesford, 2000), via affecting leaf chlorophyll, N content and photosynthetic enzymes. Sulfur is required for protein synthesis, incorporated into organic molecules in plants, and is located in thiol ($-SH$) groups in proteins (cysteine-residues) or non-protein thiols (glutathione, GSH), the potential modulators of stress response (Anjum et al., 2008b; Lunde et al., 2008). Significance of plant ontogeny in the modulation of plant responses to abiotic stress factors such as drought (Anjum et al., 2008a) and heavy metals (Anjum et al., 2008c) has been reported. Also, considering a single plant growth-stage, the role of S nutrition in the improvement of plant growth, development and yield (Ahmad et al., 2005), and tolerance to stresses (such as Cd; Anjum et al., 2008b) has been evidenced. However, information is meager on the S-mediated control of plant responses to drought stress during plant ontogeny.

Given the paucity of information on drought sensitivity of legume crops, and on the physiological basis of mineral-nutrient-(like S)-assisted management of crop growth and productivity, the current study was undertaken (i) to screen the drought-sensitive stage(s) during plant ontogeny, (ii) to identify the plant-growth stage when S helps plants maximally to improve the pools of both cellular redox buffers (AsA, GSH) and mineral-nutrients (K, S, and Mg) in order to counteract the drought-accrued oxidative stress (measured as electrolyte leakage (EL), membrane lipid peroxidation and H_2O_2 levels). Mungbean (*Vigna radiata* L. Wilczek) was chosen as a model plant system for the current study, because it is a potential pulse crop in the Indian sub-continent due to its ready market, N_2-fixation capability, early maturity and the ability to fit well in crop-rotation program (Anjum, 2006). Additionally, S-requirement of the pulse crops, for maintaining their normal growth and development, stands just second to that of the oil-yielding crops.

MATERIALS AND METHODS
EXPERIMENTAL MATERIALS, PROCEDURE AND SOIL CHARACTERISTICS
Seeds of mungbean (*Vigna radiata* L. Wilczek) cultivar Pusa 9531 were sown in 30-cm-diameter earthen pots filled with 8 kg soil. The soil was sandy loam in texture, with 7.8 pH 7.8, 0.38 dsm electrical conductivity, 0.43% organic carbon, 70 mg kg^{-1} soil available K and 5 mg kg^{-1} soil available S. Nitrogen (N; 120 mg kg^{-1} soil) and phosphorus (P; 30 mg kg^{-1} soil) were applied at the time of sowing. S was applied to *V. radiata* plants at the rate of 40 mg kg^{-1} soil, in the form of solution, 5 days before drought-stress imposition at various growth stages. The sources of N, P, K, and S were urea, single super phosphate, gypsum, and muriate of potash, respectively. After germination, three plants per pot were maintained until harvest. The pots were kept in green house under semi-controlled condition. A polythene plastic film was used to thwart the effects of rainfall, which allowed transmittance of 90% of visible wavelength (400–700 nm) under natural day and night conditions with a day/night temperature 25/20 ± 4°C and relative humidity of 70 ± 5%. All experiments were performed using completely expanded leaves from the second youngest nodes from the top of the plants.

DROUGHT STRESS IMPOSITION AND SULFUR (S) APPLICATION SCHEDULE
Drought was imposed at pre-flowering (15 d after sowing) (group 1), flowering (30 d after sowing) (group 2) and at pod filling (50 d after sowing) (group 3) by withholding water for 5 days; this was followed by normal watering (without S). Other three groups (4–6) as well as the controls were supplied with an equal amount of S solution (40 mg S kg^{-1}). All these (1–6) plant groups, and the control group, were maintained until harvest, and watered on alternate days. Soil moisture content was measured gravimetrically on dry weight basis at the time of pre-flowering, flowering, and post-flowering (pod-filling) stages (**Table 1**). Samplings were done after re-watering the drought-exposed plants for 5 days at the given growth stage i.e., at 25, 40, and 60 days after sowing. The treatments were arranged in a randomized complete block design, and each treatment was replicated five times.

PLANT GROWTH, PHOTOSYNTHESIS AND BIOCHEMICAL ESTIMATIONS
Leaf area was measured with a leaf area meter (LI-3000A, LI-COR, Lincoln, NE). Plant dry mass was determined after drying the plant at 80°C to a constant weight with the help of an electronic balance (SD-300). Net photosynthetic rate (Pn), stomatal conductance (Gs) and intercellular CO_2 concentration (Ci) were recorded in fully expanded leaves of second youngest nodes, using infra-red gas analyzer (IRGA, LI-COR, 6400, Lincoln, NE) on a sunny day between 10:00 and 11:00 h. Chlorophyll content was estimated in fully expanded young leaves at each stage using the method given by Hiscox and Israelstam (1979). Estimation of soluble protein content was done according to Bradford (1976) using bovine serum albumin as standard.

OXIDATIVE STRESS TRAITS
We considered electrolyte leakage (EL), membrane lipid peroxidation and H_2O_2 levels as the biomarkers of oxidative stress. Cellular membrane integrity in leaves was assayed by measuring the EL according to Anjum et al. (2013). In brief, fresh leaves (1.0 g) were kept in glass vials containing 10 ml deionized water.

Table 1 | Soil moisture content (%) measured in the control and drought-stressed conditions [with and without sulfur (S) supply], at pre-flowering, flowering and pod-filling stages of mungbean (*Vigna radiata*) plants. Values are the means of five replicates ± standard deviation.

Growth stage treatment	Soil moisture content (%)
Control	20.7 ± 1.0
Pre-flowering Drought	15.3 ± 0.7[a]
Drought + S	15.5 ± 0.9[b]
Flowering Drought	13.2 ± 0.6[ab]
Drought + S	16.1 ± 0.8[ab]
Pod-filling Drought	10.0 ± 0.5[abc]
Drought + S	11.03 ± 0.6[ac]

Significant differences are: [a]*vs. Control;* [b]*vs. Drought (pre-flowering);* [c]*vs. Drought (flowering).*

The vials, covered with plastic caps, were placed in a shaking incubator at a constant temperature of 25°C for 6 h and the electrical conductivity (EC) of the solution was measured (EC1) using an electrical conductivity meter (WTW Cond 330i/SET, Weilheim, Germany). Subsequently, the same vials were kept in water bath shaker at 90°C for 2 h, cooled and EC2 was measured. EL was expressed following the formula EL = EC1/EC2 × 100.

Membrane lipid peroxidation was estimated in terms of thiobarbituric acid reactive substances (TBARS) contents adopting the method of Dhindsa et al. (1981) as described by Anjum et al. (2013). Briefly, fresh leaves (1.0 g) were ground in liquid nitrogen, mixed with 0.73% 2-thiobarbituric acid in 12% trichloroacetic acid, incubated for 30 min in boiling water, ice-cooled, centrifuged at 1000×g for 10 min at 4°C and the absorbance measured in the supernatant at 532 nm. The rate of lipid peroxidation was expressed as nmoles of TBARS formed per gram of fresh weight, using a molar extinction coefficient of $1.55 \times 105\,M^{-1}$ cm^{-1}. Leaf-H_2O_2 content was determined following the method of Loreto and Velikova (2001) as adopted and described by Dipierro et al. (2005). In brief, leaf tissues (1.0 g) were homogenized in 2 ml of 0.1% (w/v) TCA. The homogenate was centrifuged at 12,000×g for 15 min and 0.5 ml of the supernatant were mixed with 0.5 ml of 10 mM K-phosphate buffer pH 7.0 and 1 ml of 1 M KI. The H_2O_2 content of the supernatant was evaluated by comparing its absorbance at 390 nm with a standard calibration curve.

DETERMINATION OF CELLULAR BUFFERS

Both reduced GSH and AsA were considered as representative cellular redox buffers. The content of reduced glutathione (GSH) was estimated following the method of Anderson (1985). Fresh leaf tissues (1.0 g) were homogenized in 2 ml of 5% (w/v) sulphosalicylic acid at 4°C. The homogenate was centrifuged at 10,000×g for 10 min. To a 0.5 ml of supernatant, 0.6 ml of K-phosphate buffer (100 mM, pH 7.0) and 40 μl of 5′5′-dithiobis-2-nitrobenzoic acid (DTNB) were added, and absorbance was recorded after 2 min at 412 nm on a UV-VIS spectrophotometer (Lambda Bio 20, Perkin Elmer, MA, USA). The method of Law et al. (1983) was followed for estimation of reduced ascorbate (AsA). In brief, fresh leaf (0.5 g) was homogenized in 2.0 ml of K-phosphate buffer (100 mM, pH 7.0) containing 1 mM EDTA and centrifuged at 10,000×g for 10 min. To a 1.0 ml of supernatant, 0.5 ml of 10% (w/v) trichloroactetic acid (TCA) was added, thoroughly mixed and incubated for 5 min at 4°C. Then, 0.5 ml of NaOH (0.1 M) was mixed with 1.5 ml of the above solution and centrifuged at 5000×g for 10 min at 20°C. The aliquot thus obtained was equally distributed into two separate microfuge tubes (750 μl each). For estimation of AsA, 200 μl of K-phosphate buffer (150 mM, pH 7.4) was added to 750 μl of aliquot. For DHA estimation, 750 μl of aliquot was added to 100 μl of 1,4-dithiothreitol (DTT), followed by vortex-mixing, incubation for 15 min at 20°C, and addition of 100 μl of 0.5% (w/v) NEM. Both the microfuge tubes were then incubated for 30 s at room temperature. To each sample tube, 400 μl of 10% (w/v) TCA, 400 μl of H_3PO_4, 400 μl of 4% (w/v) bipyridyl dye (N'N-dimethyl bipyridyl) and 200 μl of 3% (w/v) $FeCl_3$ were added and thoroughly mixed.

Absorbance was recorded at 525 nm after incubation for 1 h at 37°C.

K, S, AND Mg CONTENT DETERMINATIONS

The method of Lindner (1944) was followed to estimate K content in digested samples using flame photometry; whereas, for S determination, 100 mg of dried fine powder of leaf was digested in a mixture of concentrated HNO_3 and 60% $HClO_4$ (85:1, v/v) and the content of sulfate was estimated using the turbidimetric method of Chesnin and Yien (1950). Leaf Mg content was determined by digesting samples in 5 ml of 96% H_2SO_4 and 3 ml of 30% H_2O_2 at 270°C; thereafter, Mg content was assayed by atomic absorption spectrometry at 285.2 nm wavelengths.

DATA ANALYSIS

SPSS (PASW statistics 18, Chicago, IL, USA) for Windows was used for statistical analysis. One-Way analysis of variance (ANOVA) was performed, followed by all pairwise multiple comparison procedures (Tukey test). Mann-Whitney U-test and Levene's test were performed in order to check the normal distribution and the homogeneity of variances, respectively. The data are expressed as mean values ± SD of five independent experiments with at least five replicates for each. The significance level was set at $P \leq 0.05$.

RESULTS

Significant results related to plant growth, photosynthetic functions, soluble-protein content, oxidative stress, cellular reducing buffers, plant mineral nutrients, and yield attributes are presented here, highlighting the significant changes observed at different (pre-flowering, flowering, and pod-filling) stages of plant growth.

PLANT GROWTH AND PHOTOSYNTHETIC FUNCTIONS

Under drought stress, plant growth, in terms of leaf area and plant dry mass, decreased significantly at pre-flowering stage (vs. control, C). On application of S, significant change was noted in the drought-induced reduction in leaf area only. Drought stress imposed during pre-flowering stage also caused significant decrease in photosynthetic functions (viz., photosynthetic rate, Pn; stomatal conductance, Gs; intercellular CO_2, Ci; chlorophyll content), as compared with the control. Supplementation of S significantly increased the drought-induced reductions in Pn, Gs, and Ci (**Tables 2, 3**).

During the flowering/reproductive stage, significant decrease in leaf area and plant dry mass was perceptible under drought stress alone (vs. C), whereas S application significantly increased the drought-induced reductions in these parameters. Pn, Gs, Ci and the chlorophyll content displayed significant decreases due to drought stress (vs. C); whereas, supplementation of S improved these traits (vs. drought at flowering). During the pod-filling stage, leaf area and plant dry mass decreased significantly due to imposition of drought stress (vs. C) and S application deepened the decline in leaf area and plant dry mass. Likewise, Pn, Gs and Ci and chlorophyll content displayed significant decreases due to drought stress imposed at the pod-filling stage (vs. C). The decrease in Pn, Ci and chlorophyll was significantly ameliorated with S supplementation (**Tables 2, 3**).

OXIDATIVE STRESS AND MODULATION OF THE POOLS OF CELLULAR REDOX BUFFERS

With-holding water for 5-days during pre-flowering stage led to significant increases in electrolyte leakage (EL) and in the contents of thiobarbituric-acid-reactive substances (TBARS) and H_2O_2 (vs. C). However, S-application significantly decreased the impact of drought stress-impact at pre-flowering by reducing EL and the contents of TBARS and H_2O_2 (**Figure 1**). The pools of cellular redox buffers namely reduced ascorbate (AsA) and glutathione (GSH) declined significantly due to pre-flowering drought stress (vs. C); S-supplementation was insignificant to mitigate these declines. Oxidative stress traits such as EL, and the contents of TBARS and H_2O_2 significantly increased due to drought stress created at flowering stage (vs. C); however, S-application significantly decreased this drought-caused oxidative stress. In contrast,

the contents of reduced AsA and GSH contents declined significantly (vs. C) due to drought stress at this stage, and supplementation of S significantly ameliorated these declines. Drought stress imposed at pod-filling stage significantly increased EL and the contents of TBARS and H_2O_2 (vs. C); whereas S-application significantly reduced the levels of H_2O_2 and EL elevated by drought at this stage. The reductions in AsA and GSH contents due to the drought stress imposed at pod-filling stage were insignificant (vs. C), and the effect of S supplementation in mitigating the impact of drought stress was also insignificant (**Figure 2**).

MINERAL NUTRIENTS

Plant nutrients, such as K, S, and Mg, displayed significant reductions due to drought stress imposed at pre-flowering stage (vs. C); however, no significant difference was observed when drought-stressed plants were supplemented with S. Drought imposition during flowering stage caused significant reduction in K, S, and Mg levels in the leaf tissue (vs. C); whereas their contents significantly increased when plants facing drought at flowering stage were supplemented with S. Among the plant nutrients studied, only K content displayed a significant reduction due to drought at pod-filling stage (vs. C); whereas S and Mg contents did not differ significantly under the stress of drought alone or drought + S imposed at pod-filling stage (vs. C). Moreover, the K content significantly increased when drought-stressed plants were supplemented with S (**Table 4**).

DISCUSSION

PLANT GROWTH AND PHOTOSYNTHETIC FUNCTIONS

Plant growth is the outcome of coordination of major physiological/biochemical processes in plants. In the present study, plant dry mass and leaf area showed a significant relationship with the severity of water deficit stress, irrespective of the phase of plant ontogeny. Earlier, plant growth in terms of dry mass accumulation and leaf area has been used as a tool for the assessment of crop productivity (Sundaravalli et al., 2005; Anjum et al., 2008a). Cell division, enlargement and differentiation and also the plant

Table 2 | Leaf area (cm^2 plant^{-1}) and plant dry mass (g plant^{-1}) in mungbean (*Vigna radiata*) as influenced by drought stress and by drought + sulfur (S) application (mg kg^{-1} soil) at pre-flowering, flowering, and pod-filling stages.

Growth stage	Treatment	Leaf area	Plant dry mass
Pre-flowering	Control	147.0 ± 10.1	1.8 ± 0.3
	Drought	87.02 ± 9.1[a]	1.0 ± 0.2[a]
	Drought + S	140.6 ± 9.3[b]	1.3 ± 0.4
Flowering	Control	362.5 ± 14.5	9.2 ± 0.6
	Drought	291.8 ± 13.2[a]	6.2 ± 0.3[a]
	Drought + S	360.6 ± 14.4[c]	8.02 ± 0.6[ac]
Pod-filling	Control	303.1 ± 10.6	16.6 ± 1.02
	Drought	192.2 ± 8.5[a]	10.6 ± 0.8[a]
	Drought + S	185.2 ± 7.9[ad]	16.1 ± 0.8[d]

Values are the means of five replicates ± standard deviation. Significant differences within the same growth stage are: [a]vs. Control; [b]vs. Drought (pre-flowering); [c]vs. Drought (flowering); [d]vs. Drought (pod-filling).

Table 3 | Net photosynthetic rate (Pn; μmol CO_2 m^{-1} s^{-1}), stomatal conductance (Gs; mol m^{-2} s^{-1}), intercellular CO_2 concentration (Ci; μmol mol^{-1}), chlorophyll (Chl) content (mg g^{-1} fresh weight, f.w.) and soluble protein content (mg g^{-1} f.w.) in mungbean (*Vigna radiata*), as influenced by drought stress and by drought + sulfur (S) application (mg kg^{-1} soil) at pre-flowering, flowering, and pod-filling stages.

Growth stage	Treatment	Pn	Gs	Ci	Chl
Pre-flowering	Control	13.2 ± 1.0	0.38 ± 0.07	192.6 ± 36.6	1.17 ± 0.2
	Drought	9.6 ± 1.3[a]	0.25 ± 0.05[a]	174.0 ± 33.1[a]	0.76 ± 0.1[a]
	Drought + S	10.5 ± 2.2[ab]	0.50 ± 0.09[ab]	185.0 ± 35.2[ab]	0.9 ± 0.1[a]
Flowering	Control	20.0 ± 3.4	0.70 ± 0.2	206.9 ± 39.3	1.4 ± 0.2
	Drought	10.7 ± 3.2[a]	0.48 ± 0.1[a]	154.9 ± 29.4[a]	0.8 ± 0.07[a]
	Drought + S	13.3 ± 2.8[ac]	1.0 ± 0.17[ac]	198.4 ± 37.7[ac]	1.2 ± 0.13[ac]
Pod-filling	Control	14.8 ± 4.4	0.47 ± 0.1	189.4 ± 36.0	1.3 ± 0.2
	Drought	10.2 ± 3.5[a]	0.4 ± 0.1	178.3 ± 34.0[a]	0.7 ± 0.06[a]
	Drought + S	11.6 ± 4.3[ad]	0.6 ± 0.1[ad]	195.2 ± 37.0[ad]	1.03 ± 0.1[d]

Values are the means of five replicates ± standard deviation. Significant differences within the same growth stage are: [a]vs. Control; [b]vs. Drought (pre-flowering); [c]vs. Drought (flowering); [d]vs. Drought (pod-filling).

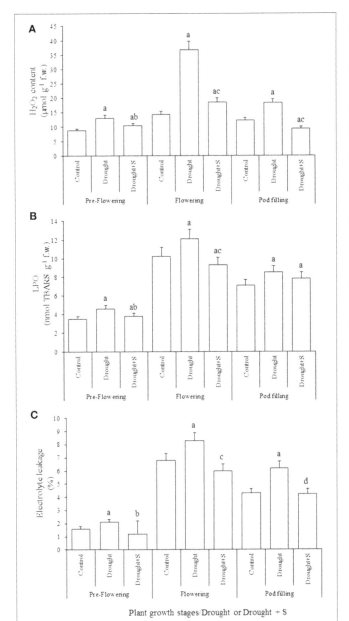

FIGURE 1 | Levels of H₂O₂ (nmol g⁻¹ fresh weight, f.w.) (A), lipid peroxidation (LPO; nmol thiobarbituric acid reactive substances, TBARS g⁻¹ f.w.) (B) and electrolyte leakage (%) (C) in the mungbean (*Vigna radiata*) leaf as influenced by drought stress and by drought + sulfur (S) application (mg kg⁻¹ soil) at pre-flowering, flowering, and pod-filling stages of plant growth. Values are the means of five replicates ± standard deviation. Significant differences within the same growth stage are: ᵃvs. Control; ᵇvs. Drought (pre-flowering); ᶜvs. Drought (flowering); and ᵈvs. Drought (pod-filling).

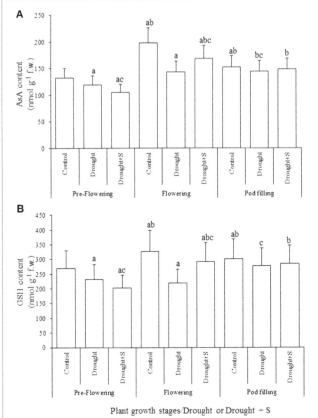

FIGURE 2 | The reduced ascorbate (AsA) (A) and reduced glutathione (GSH) (B) contents (nmol g⁻¹ fresh weight) in the mungbean (*Vigna radiata*) leaf as influenced by drought stress and sulfur (S) application (mg kg⁻¹ soil) at pre-flowering, flowering, and pod-filling stages of plant growth. Values are the means of five replicates ± standard deviation. Significant differences within same growth stage are: ᵃvs. Control; ᵇvs. Drought (pre-flowering); ᶜvs. Drought (flowering); and ᵈvs. Drought (pod-filling).

genetic make-up are significantly influenced by water-deficit stress, which in turn affects plant growth (Aref et al., 2013). In the present study, previously mentioned processes might be impacted by drought tress severely during vegetative/flowering stage which coincides with drought-mediated considerable decreases in leaf area and photosynthesis, as observed earlier also (Sundaravalli et al., 2005; Anjum et al., 2008a; Husen et al., 2014). However, the

drought-induced huge reduction in leaf area (a major component of plant growth) may be a strategy that plants adopt to adjust with water-deficit stress. Earlier, the reduced leaf-expansion/area was evidenced to conserve the internal water/moisture through the reduced rate of transpiration (reviewed by Mahajan and Tuteja, 2005).

Photosynthesis (P*n*) and its related variables (G*s*, C*i*, chlorophyll content) are highly regulated multi-step processes and exhibit great sensitivity to drought stress (Zlatev et al., 2006; Lawlor and Tezara, 2009; Husen et al., 2014). In the current study, drought stress alone significantly decreased P*n*, G*s*, C*i* and chlorophyll content irrespective of the plant ontogenetic stages. In fact, photosynthesis and its related variables are tightly interwoven and hence changes in one component significantly affect the performance of others (Lawlor and Tezara, 2009). Our findings on the drought stress-accrued reductions in G*s* and leaf C*i* coincide with those of Zlatev et al. (2006) and Meyer and Genty (1998), who considered G*s* as the major factor for controlling C*i* and hence the P*n*. Additionally, unavailability of chlorophyll

Table 4 | Potassium (K), sulfur (S), and magnesium (Mg) contents (μmol g^{-1} dry weight) in mungbean (*Vigna radiata*) leaves as influenced by drought stress and by drought +S application (mg kg^{-1} soil) at pre-flowering, flowering, and pod-filling stages.

Growth stage	Treatment	K	S	Mg
Pre-Flowering	Control	3.3 ± 0.3	1.6 ± 0.2	1.6 ± 0.2
	Drought	2.2 ± 0.2^a	1.2 ± 0.1^a	0.7 ± 0.1^a
	Drought + S	2.4 ± 0.2^a	1.6 ± 0.2	0.9 ± 0.1^a
Flowering	Control	5.6 ± 0.5	2.45 ± 0.2	2.1 ± 0.2
	Drought	3.0 ± 0.3^a	0.2 ± 0.02^a	0.8 ± 0.1^a
	Drought + S	5.2 ± 0.4^c	1.7 ± 0.13^c	2.0 ± 0.2^c
Pod-filling	Control	3.1 ± 0.3	1.3 ± 0.1	1.8 ± 0.2
	Drought	2.4 ± 0.2^a	1.1 ± 0.1	1.4 ± 0.2
	Drought + S	3.0 ± 0.3^d	1.3 ± 0.1	1.8 ± 0.2

Values are the means of five replicates ± standard deviation. Significant differences within same growth stage are: [a] *vs. Control;* [b] *vs. Drought (pre-flowering);* [c] *vs. Drought (flowering);* [d] *vs. Drought (pod-filling).*

also contributes to drought-induced decrease in Pn (Lawlor and Tezara, 2009). The drought-induced decrease in chlorophyll content has been reported earlier also due to reduction in the lamellar content of the light-harvesting chlorophyll a/b protein, inhibition in biosynthesis of chlorophyll-precursors and/or degradation of chlorophyll (Khanna-Chopra et al., 1980). Our findings on drought-mediated decrease in Pn, Gs, Ci and the content of chlorophyll confirm some earlier reports (Khanna-Chopra et al., 1980; Anjum et al., 2008a; Husen et al., 2014).

Regardless of irrigation treatments, our results also revealed that S-application significantly increased the growth and chlorophyll content and Pn. It was more effective when applied at flowering stage of the plant. The adequate and balanced supply of mineral nutrients has been shown to play a vital role in sustaining food security (Cakmak, 2005). S is involved in the light reaction of photosynthesis as an integral part of ferredoxin, a non-haem iron-sulfur protein (Marschner, 1995). Additionally, it plays essential roles in mechanisms like vitamin co-factors, GSH in redox homeostasis, and detoxification of xenobiotics (Anjum et al., 2012b). The S requirement by plants varies with growth stage and with species, varying normally between 0.1 and 1.5% of dry weight. Anjum et al. (2008b) suggested that adequate S supply may improve the pools of these compounds in plants to a great extent that may lead to increased photosynthetic efficiency, dry mass and crop yield. Sufficient S supplies improved photosynthesis and growth of *Brassica juncea* through regulating N assimilation (Khan et al., 2005). The maximum utilization of S in *Brassica campestris* crop takes place when applied at flowering stage (Ahmad et al., 2005; Anjum, 2006). Application of S increased the seed yield and attributing characters in other crops also (Anjum et al., 2012b).

OXIDATIVE STRESS AND MODULATION OF THE POOLS OF CELLULAR REDOX BUFFERS AND MINERAL NUTRIENTS

Production of ROS, such as H$_2$O$_2$, is mediated by O$_2$ reduction and subsequent oxidative damages in drought-exposed plants

(Khanna-Chopra and Selote, 2007; Anjum et al., 2012a). Plant membrane is regarded as the first target of many plant stresses due to increase in its permeability and loss of integrity under environmental stresses including the drought stress (Candan and Tarhan, 2003). In the present study, the drought-stress sensitivity of the reproductive phase of drought-exposed *V. radiata* was evidenced by significantly high levels of H$_2$O$_2$, the content of TBARS (the cytotoxic products of lipid peroxidation and indicator of extent of stress-led ROS-mediated high oxidative stress) and the EL (the measure of stress-mediated changes in membrane leakage and injury) at flowering stage, followed by pre-flowering and post-flowering stages. These results are in close agreement with the findings of Qureshi et al. (2007). Earlier, the least peroxidation of membrane lipids and the ability of cell membranes to tightly control the rate of ion movement in and out of cells have been used as tests of damage to a great range of tissues (Candan and Tarhan, 2003). However, the drought-stressed plants exhibited least contents of H$_2$O$_2$, TBARS and percent EL, when supplemented with S at their flowering and pod-filling stages. These results suggested that the S-mediated decrease in contents of H$_2$O$_2$, TBARS and percent EL depends on application of S to drought-stressed plants at appropriate growth stage when S is efficiently and differentially utilized to strengthen plants to withstand the enhanced lipid peroxidation and subsequent leakage of electrolytes due to elevated levels of H$_2$O$_2$. Thus, S-application protected differentially the drought-stressed plants against H$_2$O$_2$-mediated localized oxidative damage, disruption of metabolic functions, LPO and leakage of electrolytes (Zlatev et al., 2006). Our observations on drought alone-mediated significant increases in H$_2$O$_2$ content, lipid peroxidation (in terms of TBARS content) and percent EL in *V. radiata* plants coincide with the findings of Sreenivasulu et al. (2000) and Selote and Khanna-Chopra (2006) on different crop plants.

Plant resistance to stresses is closely associated with the efficiency of the antioxidant defense system (comprising both enzymatic and non-enzymatic components of AsA-GSH pathway) in the maintenance of the balance between the basal production of ROS and their elimination (Anjum et al., 2010, 2012c). In this perspective, AsA and GSH are important water-soluble non-enzymatic antioxidants and major cellular redox buffers in plants (Anjum et al., 2010, 2012c, 2014). Both are interlinked in terms of their physiological role in AsA-GSH pathway for effective elimination of ROS (such as H$_2$O$_2$) in plant cells (Anjum et al., 2010, 2012a,c). Contrary to an earlier report (Shehab et al., 2010) on drought-induced increase in AsA and GSH levels in different plant species, our study revealed a significant decrease in the contents of both AsA and GSH in *V. radiata*, irrespective of the growth stage at which the drought stress was imposed. However, our findings are in conformity with those of Khanna-Chopra and Selote (2007) on drought-exposed *Triticum aestivum*. The application of S improved the AsA and GSH contents and was thus beneficial when applied to drought-stressed plants at their flowering or post-flowering stages. It was, therefore, significant for protection of *V. radiata* against ROS-mediated oxidative stress. This substantiates our earlier report suggesting improved AsA and GSH contents in Cd-stressed *Brassica campestris* plants by S supplementation (Anjum et al., 2008b). However, it is imperative to mention here that exhibition of higher levels of AsA and GSH in plants receiving drought stress + S supply at pod-filling stage may be due to S-mediated maintenance of elevated activities of AsA-GSH-regenerating enzymes such as dehydroascorbate reductase, monodehydroascorbate reductase, and GSH reductase (Eltayeb et al., 2007; Anjum et al., 2008b). Moreover, as AsA and GSH are key players in cellular redox homeostasis; the S-mediated improvement in their reduced pools must help plants to run normally the ascorbate peroxide-dependent H_2O_2 metabolism under drought-stress conditions. Therefore, S application mitigated, although partially, the drought-induced decrease in AsA content by maintaining elevated activities of dehydroascorbate reductase and monodehydroascorbate reductase (data not shown)—the key components in maintaining the reduced pool of AsA and hence the plant tolerance to oxidative stress (Eltayeb et al., 2007). Considering K, S, and Mg responses to drought and S, the uptake of the available nutrient ions dissolved in the soil solution by

plants depends upon water flow through the soil-root-shoot pathway. It also depends on root growth and nutrient mobility in the soil (Fageria et al., 2002). In this study, drought stress significantly impacted the contents of K, S, and Mg in leaves contingent upon the plant-growth stage exposed. However, as reported also in earlier studies (Abdin et al., 2003; Malvi, 2011), a synergistic interaction of S with K and Mg was revealed herein, where S-application ameliorated drought-induced reductions in the leaf K, S, and Mg contents, maximally when applied at reproductive stage.

CONCLUSIONS

Drought stress in isolation enhanced ROS generation and decreased the cellular redox buffers (AsA and GSH) and eventually hampered photosynthetic functions. These results were significant at flowering stage, followed by the pre-flowering and post-flowering (pod-filling) stages (**Figure 3**). However, improvements in these parameters due to S application was apparent (at the flowering/reproductive stage), which enhanced the pools of cellular redox buffers (AsA and GSH), which in turn managed a balance between the production and scavenging of H_2O_2 and stabilized the cell membrane by decreasing LPO (**Figure 3**). Overall, the study inferred that supplementation of S to drought-exposed plants at their flowering stage can improve their growth, photosynthesis and related variables via efficiently being utilized, and in turn managing the pools of AsA and GSH, and subsequently controlling the drought-accrued oxidative stress.

ACKNOWLEDGMENTS

Partial financial support received from the Hamdard National Foundation (HNF), New Delhi, India (DSW/HNF-18/2006) and the Portuguese Foundation for Science and Technology (FCT) (SFRH/BPD/64690/2009; SFRH/BPD/84671/2012) is gratefully acknowledged.

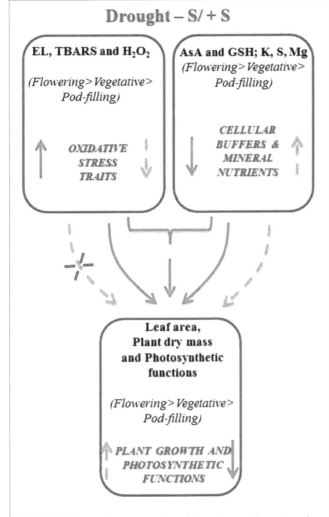

FIGURE 3 | Schematic representation of drought stress impacts and the role of sulfur (S) in mungbean (*Vigna radiata*) during its ontogeny. Complete and broken lines indicate respectively drought alone and drought + S conditions; whereas, increase and decrease have been indicated by the up and down arrows respectively.

REFERENCES

Abdin, M. Z., Ahmad, A., Khan, N., Khan, I., Jamal, A., and Iqbal, M. (2003). "Sulphur interaction with other nutrients," in *Sulphur in Plants*, eds Y. Abrol and A. Ahmad (Dordrecht: Kluwer Academic Publishers), 359–374. doi: 10.1007/978-94-017-0289-8_20

Ahmad, A., Khan, I., Anjum, N. A., Diva, I., Abdin, M. Z., and Iqbal, M. (2005). Effect of timing of sulfur fertilizer application on growth and yield of rapeseed. *J. Plant Nutr.* 28, 1049–1059. doi: 10.1081/PLN-200058905

Anderson, M. E. (1985). Determination of glutathione and glutathione disulfides in biological samples. *Methods Enzymol.* 113, 548–570. doi: 10.1016/S0076-6879(85)13073-9

Anjum, N. A. (2006). *Effect of Abiotic Stresses on Growth and Yield of Brassica campestris L. and Vigna radiata (L.) Wilczek under Different Sulfur Regimes.* Ph.D. thesis, New Delhi: Jamia Hamdard.

Anjum, N. A., Ahmad, I., Mohmood, I., Pacheco, M., Duarte, A. C., Pereira, E., et al. (2012c). Modulation of glutathione and its related enzymes in plants' responses to toxic metals and metalloids - a review. *Environ. Exp. Bot.* 75, 307–324. doi: 10.1016/j.envexpbot.2011.07.002

Anjum, N. A., Ahmad, I., Rodrigues, S., Henriques, B., Cruz, N., Coelho, C., et al. (2013). *Eriophorum angustifolium* and *Lolium perenne* metabolic adaptations to metals- and metalloids-induced anomalies in the vicinity of a chemical industrial complex. *Environ. Sci. Pollut. Res.* 20, 568–581. doi: 10.1007/s11356-012-1062-2

Anjum, N. A., Gill, S. S., Gill, R., Hasanuzzaman, M., Duarte, A. C., Pereira, E., et al. (2014). Metal/metalloid stress tolerance in plants: role of ascorbate, its redox couple and associated enzymes. *Protoplasma* 251, 1265–1283. doi: 10.1007/s00709-014-0636-x

Anjum, N. A., Gill, S. S., Umar, S., Ahmad, I., Duarte, A. C., and Pereira, E. (2012b). Improving growth and productivity of oleiferous *Brassicas* under changing environment: significance of nitrogen and sulphur nutrition, and underlying mechanisms. *Sci. World J.* 2012:657808. doi: 10.1100/2012/657808

Anjum, N. A., Umar, S., and Ahmad, A. (2012a). *Oxidative Stress in Plants: Causes, Consequences and Tolerance.* New Delhi: IK International Publishing House Pvt. Ltd.

Anjum, N. A., Umar, S., Ahmad, A., Iqbal, M., and Khan, N. A. (2008b). Sulphur protects mustard (*Brassica campestris* L.) from cadmium toxicity by improving leaf ascorbate and glutathione. *Plant Growth Regul.* 54, 271–279. doi: 10.1007/s10725-007-9251-6

Anjum, N. A., Umar, S., Ahmad, A., Iqbal, M., and Khan, N. A. (2008c). Ontogenic variation in response of *Brassica campestris* L. to cadmium toxicity. *J. Plant Interact.* 3, 189–198. doi: 10.1080/17429140701823164

Anjum, N. A., Umar, S., and Chan, M. T. (2010). *Ascorbate-Glutathione Pathway and Stress Tolerance in Plants.* Dordrecht: Springer. doi: 10.1007/978-90-481-9404-9

Anjum, N. A., Umar, S., Iqbal, M., and Khan, N. A. (2008a). Growth characteristics and antioxidant metabolism of moongbean genotypes differing in photosynthetic capacity subjected to water deficit stress. *J. Plant Interact.* 3, 127–136. doi: 10.1080/17429140701810732

Aref, M. I., Ahmed, A. I., Khan, P. R., El-Atta, H., and Iqbal, M. (2013). Drought-induced adaptive changes in the seedling anatomy of *Acacia ehrenbergiana* and *Acacia tortilis* subsp. raddiana. *Trees* 27, 959–971. doi: 10.1007/s00468-013-0848-2

Bradford, M. M. (1976). A rapid and sensitive method for the quantitation of microgram quantities of protein utilizing the principle of protein-dye binding. *Anal. Biochem.* 72, 248–254. doi: 10.1016/0003-2697(76)90527-3

Cai, Y., Cao, F., Wei, K., Zhang, G., and Wu, F. (2011). Genotypic dependent effect of exogenous glutathione on Cd-induced changes in proteins, ultrastructure and antioxidant defense enzymes in rice seedlings. *J. Hazard. Mater.* 192, 1056–1066. doi: 10.1016/j.jhazmat.2011.06.011

Cakmak, I. (2005). The role of potassium in alleviating detrimental effects of abiotic stresses in plants. *J. Plant Nutr. Soil Sci.* 168, 521–530. doi: 10.1002/jpln.200420485

Candan, N., and Tarhan, L. (2003). The correlation between antioxidant enzyme activities and lipid peroxidation levels in *Mentha pulegium* organs grown in Ca^{2+}, Mg^{2+}, Cu^{2+}, Zn^{2+} and Mn^{2+} stress conditions. *Plant Sci.* 163, 769–779. doi: 10.1016/S0168-9452(03)00269-3

Chesnin, L., and Yien, C. H. (1950). Turbidimetric determination of available sulphates. *Proc. Soil Sci. Soc. Am.* 51, 149–151.

Dhindsa, R. H., Plumb-Dhindsa, P., and Thorpe, T. A. (1981). Leaf senescence correlated with increased level of membrane permeability, lipid per oxidation and decreased level of SOD and CAT. *J. Exp. Bot.* 32, 93–101. doi: 10.1093/jxb/32.1.93

Dipierro, N., Mondelli, D., Paciolla, C., Brunetti, G., and Dipierro, S. (2005). Changes in the ascorbate system in the response of pumpkin (*Cucurbita pepo* L.) roots to aluminium stress. *Plant Physiol.* 162, 529–536. doi: 10.1016/j.jplph.2004.06.008

Eltayeb, A. M., Kawano, N., Badawi, G. H., Kaminaka, H., Sanekata, T., Shibahara, T., et al. (2007). Overexpression of monodehydroascorbate reductase in transgenic tobacco confers enhanced tolerance to ozone, salt and polyethylene glycol stresses. *Planta* 225, 1255–1264. doi: 10.1007/s00425-006-0417-7

Fageria, N. K., Baligar, V. C., and Clark, R. B. (2002). Micronutrients in crop production. *Adv. Agron.* 77, 185–187. doi: 10.1016/S0065-2113(02)77015-6

Gill, S. S., Khan, N. A., Anjum, N. A., and Tuteja, N. (2011). Amelioration of cadmium stress in crop plants by nutrients management: morphological, physiological and biochemical aspects. *Plant Stress* 5, 1–23.

Harb, A., Krishnan, A., Ambavaram, M. M., and Pereira, A. (2010). Molecular and physiological analysis of drought stress in Arabidopsis reveals early responses leading to acclimation in plant growth. *Plant Physiol.* 154, 1254–1271. doi: 10.1104/pp.110.161752

Hawkesford, M. J. (2000). Plant responses to sulfur deficiency and the genetic manipulation of sulfate transporters to improve S-utilization efficiency. *J. Exp. Bot.* 51, 131–138. doi: 10.1093/jexbot/51.342.131

Hiscox, J. H., and Israelstam, G. F. (1979). A method for extraction of chlorophyll from leaf tissues without maceration. *Can. J. Bot.* 57, 1332–1334. doi: 10.1139/b79-163

Husen, A., Iqbal, M., and Aref, I. M. (2014). Water status and leaf characteristics of *Brassica carinata* under drought stress and re-hydration conditions. *Braz. J. Bot.* 37, 217–227. doi: 10.1007/s40415-014-0066-1

Khan, N. A., Mobin, M., and Samiullah. (2005). The influence of gibberellic acid and sulfur fertilization rate on growth and S-use efficiency of mustard (*Brassica juncea*). *Plant Soil* 270, 269–274. doi: 10.1007/s11104-004-1606-4

Khanna-Chopra, R., Chaturverdi, G., Aggarwal, P., and Sinha, S. (1980). Effect of potassium on growth and nitrate reductase during water stress and recovery in maize. *Physiol. Plant* 49, 495–500. doi: 10.1111/j.1399-3054.1980.tb03340.x

Khanna-Chopra, R., and Selote, D. S. (2007). Acclimation to drought stress generates oxidative stress tolerance in drought resistant than susceptible wheat cultivar under field conditions. *Environ. Exp. Bot.* 60, 276–283. doi: 10.1016/j.envexpbot.2006.11.004

Law, M. Y., Charles, S. A., and Halliwell, B. (1983). Glutathione and ascorbic acid in spinach (*Spinaceu oleraceu*) chloroplasts. The effect of hydrogen peroxide and of paraquat. *Biochem. J.* 210, 899–903.

Lawlor, D. W., and Tezara, W. (2009). Causes of decreased photosynthetic rate and metabolic capacity in water-deficient leaf cells: a critical evaluation of mechanisms and integration of processes. *Ann. Bot.* 103, 561–579. doi: 10.1093/aob/mcn244

Lindner, R. C. (1944). Rapid analytical method for some of the more common organic substances of plant and soil. *Plant Physiol.* 19, 76–84. doi: 10.1104/pp.19.1.76

Loreto, F., and Velikova, V. (2001). Isoprene produced by leaves protects the photosynthetic apparatus against ozone damage, quenches ozone products, and reduces lipid peroxidation of cellular membranes. *Plant Physiol.* 127, 1781–1787. doi: 10.1104/pp.010497

Lunde, C., Zygadlo, A., Simonsen, H. T., Nielsen, P. L., Blennow, A., and Haldrup, A. (2008). Sulfur starvation in rice: the effect on photosynthesis, carbohydrate metabolism, and oxidative stress protective pathways. *Physiol. Plant* 134, 508–521. doi: 10.1111/j.1399-3054.2008.01159.x

Mahajan, S., and Tuteja, N. (2005). Cold, salinity and drought stresses: an overview. *Arch. Biochem. Biophys.* 444, 139–158. doi: 10.1016/j.abb.2005.10.018

Malvi, U. R. (2011). Interaction of micronutrients with major nutrients with special reference to potassium. *Karnataka J. Agric. Sci.* 24, 206–109.

Marschner, H. (1995). *Mineral Nutrition of Higher Plants, 1st Edn.* New York, NY: Academic Press.

Meyer, S., and Genty, B. (1998). Mapping intercellular CO_2 mole fraction (Ci) in Rosa leaves fed with ABA by using chlorophyll fluorescence imaging. Significance of Ci estimated from leaf gas exchange. *Plant Physiol.* 116, 947–958. doi: 10.1104/pp.116.3.947

Nahar, K., Hasanuzzaman, M., Alam, M. M., and Fujita, M. (2015). Exogenous glutathione confers high temperature stress tolerance in mung bean (*Vigna radiata* L.) by modulating antioxidant defense and methylglyoxal detoxification system. *Environ. Exp. Bot.* 112, 44–54. doi: 10.1016/j.envexpbot.2014.12.001

Qureshi, M. I., Qadir, S., and Zolla, L. (2007). Proteomics-based dissection of stress-responsive pathways in plants. *J. Plant Physiol.* 164, 1239–1260. doi: 10.1016/j.jplph.2007.01.013

Selote, D. S., and Khanna-Chopra, R. (2006). Drought acclimation confers oxidative stress tolerance by inducing co-ordinated antioxidant defence at cellular and subcellular level in leaves of heat seedlings. *Physiol. Plant* 127, 494–506. doi: 10.1111/j.1399-3054.2006.00678.x

Shehab, G. G., Ahmed, O. K., and El-Beltagi, H. S. (2010). Effects of various chemical agents for alleviation of drought stress in rice plants (*Oryza sativa* L.). *Not. Bot. Hort. Agrobot. Cluj-Napoca* 38, 139–148.

Son, J. A., Narayanankutty, D. P., and Roh, K. S. (2014). Influence of exogenous application of glutathione on rubisco and rubisco activase in heavy metal-stressed tobacco plant grown *in vitro*. *Saudi J. Biol. Sci.* 21, 89–97. doi: 10.1016/j.sjbs.2013.06.002

Sreenivasulu, N., Grimm, B., Wobus, U., and Weschke, W. (2000). Differential response of antioxidant compounds to salinity stress in salt-tolerant and salt sensitive seedlings of foxtail millet (*Setaria italica*). *Physiol. Plant* 109, 435–442. doi: 10.1034/j.1399-3054.2000.100410.x

Sundaravalli, V., Paliwal, K., and Ruckmani, A. (2005). Effect of water stress on photosynthesis, protein content and nitrate reductases activity of *Albizzia* seedlings.
J. Plant Biol. 32, 13–17.

Wang, R., Liu, S., Zhou, F., Ding, C., and Hua, C. (2014). Exogenous ascorbic acid and glutathione alleviate oxidative stress induced by salt stress in the
chloroplasts of *Oryza sativa* L. *Z. Naturforsch.* 69c, 226–236. doi: 10.5560/ZNC.
2013-0117

Zlatev, Z. S., Lidon, F. C., Ramalho, J. C., and Yordanov, I. T. (2006). Comparison
of resistance to drought of three bean cultivars. *Biol. Plant* 50, 389–394. doi:
10.1007/s10535-006-0054-9

Conflict of Interest Statement: The authors declare that the research was conducted in the absence of any commercial or financial relationships that could be
construed as a potential conflict of interest.

13

Evaluation of soil and water conservation practices in the north-western Ethiopian highlands using multi-criteria analysis

Akalu Teshome[1,2]*, Jan de Graaff[1] and Leo Stroosnijder[1]

[1] Soil Physics and Land Management Group, Environmental Science Department, University of Wageningen, Wageningen, Netherlands
[2] Amhara Regional Agricultural Research Institute, Bahir Dar, Ethiopia

Edited by:
Luuk Fleskens, University of Leeds, UK

Reviewed by:
Robert Zougmoré, International Crops Research Institute for the Semi-Arid Tropics, Mali
Doan Nainggolan, University of Aarhus, Denmark

*Correspondence:
Akalu Teshome, Amhara Regional Agricultural Research Institute, P.O. Box +527, Bahir Dar, Ethiopia
e-mail: akalu_firew@yahoo.com

Investments by farmers in soil and water conservation (SWC) practices are influenced by the physical effectiveness, financial efficiency, and social acceptability of these practices. The objective of this study is to evaluate different SWC practices in the north-western highlands of Ethiopia using various qualitative criteria and weightings based on ecological, economic and social impacts using Multi-Criteria Analysis (MCA). The study reveals that MCA is a useful evaluation tool that takes into account non-monetary and less quantifiable effects of SWC practices. Farmers employ a range of criteria to evaluate the performance of SWC practices. The relative importance of each criterion in their selection of SWC alternatives depends mostly on slope categories. In steeply sloping areas, farmers assigned the highest score for criteria related to ecological impacts; whilst preferring practices with stronger positive economic impacts in moderate and gentle sloping areas. Policy makers and development practitioners are encouraged to pay greater attention to both farmer preferences and slope specific circumstances when designing SWC strategies and programmes.

Keywords: multi-criteria analysis, soil and water conservation, farmer preferences, slope, Ethiopia

INTRODUCTION

Agriculture is the major source of livelihood in Ethiopia. However, land degradation in the form of soil erosion has hampered agricultural productivity and economic growth of the nation (Haileslassie et al., 2005; Hengsdijk et al., 2005; Balana et al., 2010). Land degradation, low agricultural productivity and poverty are critical and closely related problems in the Ethiopian highlands (Pender and Gebremedhin, 2007; Yitbarek et al., 2012).

Investments[1] in soil and water conservation (SWC) practices enhance crop production, food security and household income (Adgo et al., 2013). Recognizing these connections, the government of Ethiopia is promoting SWC technologies for improving agricultural productivity, household food security and rural livelihoods. Particularly in the Ethiopian highlands, different SWC technologies have been promoted among farmers to control erosion. These technologies include stone bunds, soil bunds and *Fanya juu* bunds (made by digging a trench and moving the soil uphill to form an embankment). However, the adoption rates of these SWC technologies vary considerably within the country (Kassie et al., 2009; Tefera and Sterk, 2010; Tesfaye et al., 2013; Teshome et al., 2014), largely because investments by farmers in SWC are influenced by the ecological, economic and social impacts of the SWC technologies.

The impact of SWC measures in Ethiopia and elsewhere is mostly evaluated in monetary terms (cost-benefit analysis; CBA) (Bizoza and de Graaff, 2012; Teshome et al., 2013). However, SWC measures also have ecological and social impacts that cannot be easily quantified in monetary values (Tenge, 2005). Moreover, CBA is sometimes criticized in that it does not take into account the interactions between different impacts. More rigorous evaluation methods of SWC measures are of paramount importance in quantifying the monetary and non-monetary value of SWC measures to ensure better decision-making processes of policy makers and development practitioners.

Availability of several SWC alternatives, conflicting objectives and a range of evaluation criteria of farmers hamper their decision-making and adoption of SWC measures (Amsalu, 2006). Farmers' investment objectives often differ considerably from those of researchers and extension personnel, as they have other objectives in addition to reducing soil loss and maximizing financial benefits of SWC measures (Tenge, 2005). These objectives are often conflicting, resulting in no single SWC measure that can provide the best outcome for all farmers.

Therefore, there is a need to evaluate the objectives and criteria of farm households in decision-making of SWC practices based on ecological, economic and social impacts. In order to identify and analyse multiple and conflicting objectives and goals, Multi-Criteria Analysis (MCA) represents a more suitable tool (Romero and Rehman, 2003). In addition, MCA methods are an appropriate modeling tool for addressing economic-environmental

---

[1] Investments refer to any efforts (e.g. labor, knowledge, and time) made by farmers to combat water erosion and enhance soil fertility.

evaluation issues (Munda et al., 1994; Mendoza and Martins, 2006).

The objective of this study is to evaluate different SWC practices using qualitative criteria by different stakeholders (farmers and experts) based on perceived ecological, economic and social impacts.

MULTIPLE CRITERIA ANALYSIS (MCA) FOR SOIL AND WATER CONSERVATION EVALUATION

Most SWC investment activities are evaluated using a CBA, which assumes that complex soil and water objectives can be converted into one basic objective of "maximizing profit." However, the objective function consists of a single choice criterion, yet within SWC investments, there are usually several objectives or goals (Prato, 1999). Therefore, a discrete MCA has been developed as a decision-making tool when different objectives have to be fulfilled. Recently, Fleskens et al. (2014) revealed that scenario assessments with integrated models help determine location-specific, financially viable technologies to effectively combat land degradation problems, and provide input into multilevel land management decision-making processes. Moreover, choice experiments, a stated preference valuation method, are also a tool that can assign monetary values to environmental impact assessment (Vega and Alpízar, 2011).

MCA is a decision-making tool applied to choice problems in the face of a number of different alternatives and conflicting criteria (Hajkowicz et al., 2000). CIFOR (1999) defined MCA as a decision-making tool developed for complex multi-criteria problems that include qualitative and/or quantitative aspects of the problem in the decision-making process. MCA is an evaluation method, based on sustainable development economic theory, that ranks or scores the performance of decision options against multiple criteria (Hajkowicz, 2007), ensuring the final results have clear meaning in terms of sustainability (Boggia and Cortina, 2010).

The main characteristics of MCA are: multiplicity of objectives, heterogeneity of objectives and plurality of decision makers (Seo and Sakawa, 1988). However, in terms of evaluating SWC practices, MCA has some advantages and disadvantages (de Graaff, 1996; Prato, 1999) (**Table 1**), but offers great potential in addressing the shortcomings of other SWC evaluation methods.

For evaluation of SWC investments, CBA only compares one "with" case with one "without case" (or "before" and "after" case), resulting in all effects being valued in monetary values, and focused mostly on the efficiency criterion. MCA has the disadvantage that it does not allow for an easy comparison of streams of costs and benefits over time, and relies on subjective weightings attached to several criteria by the stakeholders concerned and represented (**Table 1**). An appropriate solution to evaluate SWC is the use of CBA results as one of the criteria (efficiency) in the MCA evaluation of SWC measures (de Graaff, 1996). Therefore, MCA appears to be one of the more appropriate tools to evaluate SWC practices.

STEPS IN MULTI-CRITERIA ANALYSIS (MCA) METHODS

MCA uses a number of defining steps to identify the best alternatives on the basis of relevant criteria (Voogd, 1982; Munda et al.,

1994; Tenge, 2005; Hajkowicz and Higgins, 2008; Ananda and Herath, 2009). The major steps in the MCA are the following:

Step 1: Determination of objectives.

Step 2: Identification of alternatives/options, that contributes to achieving the objectives.

Step 3: Determination of the evaluation criteria to assess the performance of the alternatives.

Step 4: Determination of the effects (score) on alternatives.
The effects of alternatives are identified, measured (quantitative or qualitative) and determined according to the measurable criteria set, established in step 3.

Step 5: Standardization of the effects.
Making the unit of scores comparable, on a scale between 0 and 1, eliminates the effect of different dimension scoring of alternatives.

Step 6: Formulation of weights.
Weights are assigned to criteria by farmers, policymakers, or other stakeholders to represent their relative importance for the respective group.

Step 7: Aggregation and ranking.
Involves combining weighted scores for each alternative. Among the discrete MCA methods, the most important aggregation methods are the Additive Weighting and the Sequential Elimination methods.

MATERIALS AND METHODS
STUDY AREAS

The study was undertaken in three watersheds in the East and West Gojam Zones of the Amhara region of Ethiopia, i.e., the Anjeni, Dijil, and Debre-Mewi watersheds (**Figure 1**). The watersheds are part of the north-western highlands of Ethiopia. These watersheds were selected because of their specific experience with SWC activities. Moreover, the watersheds have diverse

Table 1 | Advantages and disadvantages of MCA for evaluating SWC.

Advantages of MCA	Disadvantages of MCA
Focus on several objectives and alternatives.	Non-comparability among objectives.
Considered the intangible effects of SWC.	Exposed to subjectivity problem: subjective weights attached to several criteria.
Use of both qualitative and quantitative effects.	Use of qualitative scales, where quantitative could be used.
Holistic approach: it can also incorporate CBA and other financial efficiency criteria.	Different methods give different results.
Increases the rationality of the decision process.	Difficult to incorporate the time dimension.
Identifies gaps in knowledge in SWC practices.	Pays little attention to uncertainty and to possible trade-offs among some of the objectives.
Interactive method.	Different conflicting evaluation criteria are taken into consideration.

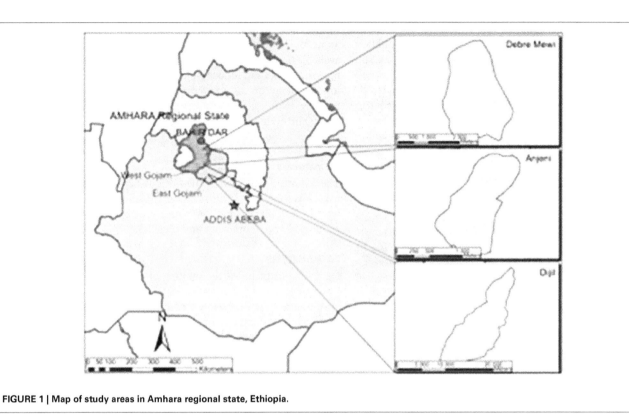

FIGURE 1 | Map of study areas in Amhara regional state, Ethiopia.

biophysical and socio-economic characteristics (**Table 2**). The dominant farming system in the watersheds is characterized as crop-livestock mixed farming.

DATA COLLECTION

Farmers and experts are the main stakeholders in SWC activities in the Ethiopian highlands. Qualitative and quantitative data were collected from these stakeholders using group discussions and a formal survey in 2013. Two group discussions were undertaken in each watershed. The number of participants in each group ranged from 9–11 farmers. Group discussions were followed by an individual survey to crosscheck the information provided. For this survey a total of 50 farm households (20 from Debre Mewi, 15 from Anjeni and 15 from Dijil) were carefully selected from an earlier much larger household survey[2]. In addition, 16 experts were interviewed from different levels of the Department of Agriculture (*kebele*[3], district, zone, and region).

SWC alternatives and evaluation criteria were identified during the two previous surveys and group discussions (Teshome et al., 2014, Forthcoming). SWC alternatives and evaluation criteria were compiled and presented for discussions with farmers. During the group discussions, some SWC alternatives and criteria were removed as they were either not very relevant, or not commonly practiced in the prevalent farming system

(alternatives and criteria were fine-tuned during group discussions). For example, for farmers in Anjeni, soil bunds were not important in their watershed. Thus, weightings were reassigned through group consensus to criteria dependent on the different slope categories (steep, moderate, and gentle). This is because farmers mainly classify their land parcels into three categories, i.e., steep (*tedafat*), moderate (*mekakelegna*), and gentle (*deledala/medama*). A fixed point scoring technique was applied in this study (Hajkowicz et al., 2000), where the decision-maker is required to distribute a fixed number of points among the criteria. Thus, a higher point score indicates that the criterion has greater importance. Fixed point scoring is the most direct means of obtaining weighting information from the decision maker.

SOIL AND WATER CONSERVATION ALTERNATIVES

SWC measures are part and parcel of the farming system evident in the study areas. Almost all farmers perceived erosion problems while many of them also believed that SWC measures are profitable (Teshome et al., 2013). Thus, different SWC measures to avert erosion problems were introduced by government and non-government organizations. Soil bunds, *Fanya juu* bunds and stone bunds are the major SWC measures that are widely implemented by farmers. Therefore, these three SWC measures and the "No measure" alternative were included in the evaluation.

Soil bunds

Soil bunds are embankments made from topsoil along the contour to control erosion (**Figure 2**). They require less labor for

---

[2]These 50 farm households are part of large survey of households. This large survey included 60, 125, and 115 households from Anjeni, Dijil, and Debre Mewi watersheds, respectively.

[3]*Kebele* is the lowest administrative body in Ethiopia and is part of the sub-district.

construction compared to stone and *Fanya juu* bunds as excavated material from the ditch is placed downhill rather than uphill, as is the case in the construction of *Fanya juu* bunds. However, soil bunds require more labor for maintenance than *Fanya juu* and stone bunds. The uphill drains of the soil bund are impacted by accumulated material (silt) and therefore require more labor to regularly excavate the ditches, as farmers need to ensure effective evacuation of excess water. Grass is grown on the riser to stabilize the bunds. Soil bunds can be easily eroded during heavy rainfall in steeply sloping areas.

Fanya juu bunds

Fanya juu bunds are made by digging a trench and moving the soil uphill to form an embankment (**Figure 2**), and are thus more labor-intensive during construction. A terrace can therefore be created in a relatively short period of time. They provide an opportunity to grow fodder or grass on the riser, but they can also experience water-logging.

Table 2 | Socio- and physical characteristics of the study watersheds.

Features	Anjeni	Dijil	Debre Mewi
Size of watershed (ha)	113	936	523
Altitude (m.a.s.l)	2450	2480	2300
Average annual rainfall (mm)	1790	1300	1260
Dominant soil types	Alisols, Nitosols, Regosols, Leptisols	Nitosols	Vertic Nitosols, Nitosols, Vertisols
Degradation	Degraded	Very degraded	Not heavily degraded
Soil pH	5.7	4.3	6.7
Slope class			
Flat to gentle (<10%) (%)	30.5	13.9	19.7
Medium (10–20%) (%)	28.6	41.4	41.9
Steep (>20%) (%)	40.9	44.7	38.4
Dominant crop in farming systems	Barley	Oats	Tef
Productivity	Low	Low	High
Number of households	95	628	324
All weather road	Poor	Good	Good
Distance to district town (km)	20	8	12

Sources: SCRP, 1991; Liu et al., 2008; Zegeye, 2009; Tesfaye, 2011.

Stone bunds

Stone bunds are usually constructed where stones are readily available on or near the field. Stone bunds are stable and durable measures. They can reduce runoff and soil erosion in steeply sloping areas, and excess water can pass more easily through stone terraces. However, construction does require a large amount of labor. Furthermore, they are not convenient for ox-plowing and can harbor rodents.

No measures

The *No measure* is one of the options available in the farmers' SWC investment decision-making. This alternative would be preferred by farmers when SWC measures have minimal impacts (ecologically, economically, and socially).

DATA ANALYSIS

Data analysis included the ranking of the most important SWC alternatives and standardization of the effects. Average weightings were used in our analysis to accommodate the different views of the farmers and experts on the relative importance of each criterion. Farmers and experts evaluated SWC measures by giving scores to each criteria on a scale of 1 for worst and 4 for best (and 3 in case of Anjeni, where only three alternatives were considered). We used the mode (most typical value) to aggregate rankings of individual farmers and experts. Scoring of the alternatives was also calculated by averaging the scales to crosscheck the results.

Regime Analysis method was used to obtain a complete ranking and further detailed information on the relative importance between the alternatives (Hinloopen and Nijkamp, 1990). The Regime Analysis method is one of the most common weighting methods, particularly in the case of qualitative data. This method is used to analyse ordinal and cardinal data. This method is based on pairwise comparison of two alternatives i according to criteria j (Hinloopen and Nijkamp, 1990).

Consider two alternatives i and i'. The pairwise comparison of these two alternatives according to criterion j ($e_{ii'j}$) is therefore:

$$e_{ii'j} = 1 \text{ if } p_{ij} > p_{i'j} \tag{1}$$

$$e_{ii'j} = -1 \text{ if } p_{ij} < p_{i'j} \tag{2}$$

Where p_{ij} and $p_{i'j}$ are the ranks of alternatives i and i' according to criteria j. The regime vector ($e_{ii'}$) for each pair of alternatives is then constructed by extending the comparison of the alternatives

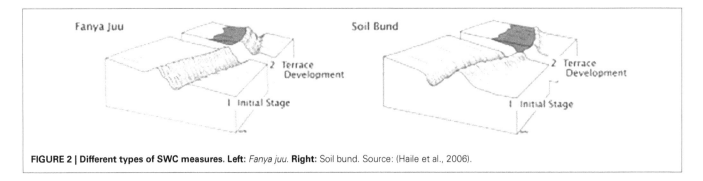

FIGURE 2 | Different types of SWC measures. Left: *Fanya juu.* **Right:** Soil bund. Source: (Haile et al., 2006).

i and i' to all criteria j = 1, 2, ... J as follows:

$$e_{ii'} = (e_{ii',1}, e_{ii',2},...e_{ii',J}) \qquad (3)$$

Positive "+" and negative "−" signs are used to indicate the relative dominance of one alternative over another, and "0" for no dominance. Based on the pairwise comparison of the alternatives obtained, the weighted dominance of alternative i with respect to i' ($p_{ii'}$) is defined as:

$$p_{ii'} = \sum_{j=1}^{j} w_j * e_{ii',j} \qquad (4)$$

where

w_j = weight relative to criterion j,
$e_{ii'}$ = pairwise comparison of alternative i and i', and
j = the criterion.

RESULTS

MAJOR ACTORS AND THEIR OBJECTIVES

Erosion has adverse impacts on ecological, economic, and social aspects of farming communities. Farmers evaluate these multiple effects of the problem in their SWC investment decisions. In our formal and informal surveys we found that the major objectives of farmers in relation to SWC investments are ecological restoration (erosion control, enhanced soil fertility and increased water retention), economic benefits (increase production and decrease costs), and diminishing socially adverse effects of erosion and SWC measures. Similarly, the major objective revealed from the experts is to improve the livelihood of the farmers through comprehensive and integrated natural resource management and development.

EVALUATION CRITERIA AND WEIGHTINGS

Farmers and experts defined and used 10 evaluation criteria to evaluate SWC measures, which were then categorized into ecological, economic, social, and other criteria (**Table 3**).

Ecological criteria

Three criteria were identified for evaluating the ecological impacts of SWC alternatives. The criteria reveal that farmers would like SWC measures that are effective in erosion control, enhance soil fertility and improve water retention.

Economic criteria

Four criteria were mentioned to evaluate the economic impact of SWC alternatives. These evaluation criteria focus on the costs and benefits aspects of SWC alternatives.

Social and other criteria

Farmers in the study areas predominantly preferred SWC measures that have social benefits as well as measures that have no adverse effects on the farming system.

Farmers and experts gave weightings for the different evaluation criteria (**Tables 4, 5**). The results show that farmers and experts gave different weights and that these vary by slope

category. The ecological impact criteria had the highest weighting within the steep slope category. On the other hand, economic impact criteria received the highest weighting in the gentle slope category. According to the farmers' views steeper slopes are more prone to erosion and that it is relatively more important to preserve them. The gentle plots, on the other hand, have higher economic potential.

Farmers gave relatively high scores to the social and other impacts criteria of SWC measures compared to the experts (**Tables 4, 5**). These criteria are: contributions of SWC measures to avoid disputes with adjacent farmers due to erosion, ox-plowing convenience of the measures and the risk of pest harboring effects of the measures. This shows that farmers pay more attention to everyday aspects of their lives during SWC investments while experts have larger scales than the field/farm, e.g., watershed level, in mind.

Anjeni farmers gave a higher weighting for maximizing crop yield, maximizing plowing convenience, and minimizing disputes with adjacent farmers as compared to other watersheds. This could be due to long term SWC activities implemented in the watershed within the last three decades and thereby farmers perceived the benefits of conservation measures over time. Most of the *Fanya juu* bunds in Anjeni have stabilized into bench terraces.

Table 3 | Farmers'/experts' evaluation criteria of SWC measures.

Objectives	Criteria	Unit of measurement
ECOLOGICAL IMPACTS		
Erosion control (**C1**)	Minimize soil loss	Rank
Enhance soil fertility (**C2**)	Minimize nutrient loss	Rank
Water retention (**C3**)	Maximize water retention	Rank
ECONOMIC IMPACTS		
Crop yields (**C4**)	Maximize crop yields	Rank
Grass production (**C5**)	Maximize grass production	Rank
Labor requirements for establishment (**C6**)	Minimize labor for establishment	Rank
Maintenance costs (**C7**)	Minimize maintenance costs	Rank
SOCIAL AND OTHER IMPACTS		
Ox-plowing convenience (**C8**)	Maximize ox-plowing convenience	Rank
Risk of pest harboring effect (**C9**)	Minimize risk of pest harboring effect	Rank
Avoid dispute with adjacent farmers (**C10**)	Minimize dispute with adjacent farmers	Rank

Source: Own surveys.

This results in the diminution of slope angles and increased topsoil depth behind the bunds, which has a positive effect on yields.

FARMERS AND EXPERTS MULTI-CRITERIA RANKING OF THE ALTERNATIVES

The results of farmers' and experts' ranking of SWC measures based on the evaluation criteria are presented in **Tables 6–8**. The scores indicate the perceived level of importance of each SWC alternative with respect to the criteria defined.

Farmers and experts ranked stone bunds first for erosion control in the steep slope category in the three watersheds, while soil bunds are preferred by experts and farmers in Debre Mewi and Dijil watershed to control erosion in the moderate slope category. Soil bunds are not common practice in Anjeni watershed, but regardless, farmers selected *Fanya juu* bunds to control erosion in the moderate slope category. In all watersheds, farmers gave priority to *Fanya juu* bunds to control erosion in gentle slope areas. Farmers' rankings of SWC alternatives for maximizing crop yield were highly correlated with the degree of erosion control of the measures, except for *Fanya juu* in Debre Mewi. Even though *Fanya juu* bunds were preferred to control erosion on gentle slopes, their contribution to increased yield was not ranked as high, probably due to the water logging effects of the measures in Debre Mewi (**Tables 6–8**).

Farmers did not prefer stone bunds due largely to high labor demands for establishment, plowing inconvenience and the risks of pest harboring effects, as evident across all watersheds and slope categories (**Tables 6, 7**). Soil bunds were next to "no measure" ranked first in minimizing labor requirements for establishment of SWC. On the other hand, it was ranked last in minimizing maintenance costs. In general, farmer preferences reflect their experiences, perceptions, and attitudes about the merits and drawbacks of SWC alternatives under different situations.

THE EVALUATION MATRIX

Pairwise comparisons refine a complex decision problem into a series of one-to-one judgments regarding the significance of each alternative relative to the criterion that it describes (Balana et al., 2010). A pairwise comparison of the SWC alternatives against the evaluation criteria is presented in **Table 9**. Each alternative under a given criterion is compared with every other alternative under that criterion to evaluate its relative importance.

RANKING OF THE ALTERNATIVES

The weighted scores (p) of the pairwise comparisons and overall rank of the alternatives for each slope category are given in **Tables 10, 11**. The higher the evaluation score, the better the perceived performance of the SWC alternative. The evaluation score of each alternative pair in descending order provides a list of SWC measure from best to worst performing.

Steep slope category

In steeply sloping areas, stone bunds are the most viable SWC alternative in all watershed areas, followed by soil bunds. Stone bunds are durable and stable in controlling high runoff in steep areas. Stone bunds are also the first alternative for experts. Other SWC measures are easily eroded by runoff. Farmers

Table 5 | Experts' weighting (%) sets of evaluation criteria for each slope category.

Criteria	Slope		
	Steep	Moderate	Gentle
Erosion control	30.4	21.4	10
Enhance fertility	15.7	14.1	9.9
Water retention	13.1	11.4	14.7
Crop yields	12.6	21.2	30.2
Grass production	9.5	8.1	5.5
Labor for establishment	3.5	7.1	10.9
Maintenance cost	3.9	6.1	7.5
Plowing convenience	2.1	4.2	5.9
Pest harboring effect	2.8	3.5	4.3
Dispute with adjacent farmers	5.9	2.9	1.1
Total (%)	100	100	100

Table 4 | Farmers' weight sets of evaluation criteria for each slope category in percentages.

Criteria	Steep slope			Moderate slope			Gentle slope		
Watershed	D. Mewi	Anjeni	Dijil	D. Mewi	Anjeni	Dijil	D. Mewi	Anjeni	Dijil
Erosion control	30.0	31.0	30.0	20.0	18.0	22.5	7.7	5.0	5.5
Enhance fertility	16.6	14.0	16.0	12.3	9.0	12.5	7.0	6.0	13.5
Water retention	11.7	10.0	9.0	11.0	13.0	10.0	15.3	14.0	13.5
Crop yields	13.3	14.0	15.0	23.3	27.6	25.0	30.0	35.0	32.5
Grass production	7.7	6.6	7.5	5.3	3.8	3.5	3.3	3.2	2.5
Labor for establishment	3.0	3.1	2.25	8.0	5.8	5.0	13	10.1	7.5
Maintenance cost	2.7	3.3	2.75	5.0	5.8	4.0	8.7	8.7	7.5
Plowing convenience	2.3	2.6	2.25	5.3	5.6	4.5	7.7	8.8	8.5
Pest harboring effect	3.0	3.4	2.75	3.7	3.8	4.5	4.3	5.6	5.5
Dispute with adjacent farmers	9.7	12.0	12.5	6.0	7.6	8.5	3.0	3.6	3.5
Total (%)	100	100	100	100	100	100	100	100	100

Table 6 | Farmers' ranking of SWC measures on the evaluation criteria for different slopes (4, Best; 1, Worst): Debre Mewi and Dijil watersheds.

Criteria	Slope	Watershed	Soil bund	*Fanya juu*	Stone bund	No measure
Minimize soil losses(erosion control)	Steep	Debre Mewi	3	2	4	1
		Dijil	3	2	4	1
	Moderate	Debre Mewi	4	2	3	1
		Dijil	4	3	2	1
	Gentle	Debre Mewi	3	4	2	1
		Dijil	3	4	2	1
Enhance soil fertility	Steep	Debre Mewi	3	2	4	1
		Dijil	3	2	4	1
	Moderate	Debre Mewi	4	3	2	1
		Dijil	3	4	2	1
	Gentle	Debre Mewi	3	4	2	1
		Dijil	3	4	2	1
Maximize water retention	Steep	Debre Mewi	4	3	2	1
		Dijil	4	2	3	1
	Moderate	Debre Mewi	3	4	2	1
		Dijil	3	4	2	1
	Gentle	Debre Mewi	3	4	2	1
		Dijil	3	4	2	1
Maximize crop yields	Steep	Debre Mewi	3	2	4	1
		Dijil	3	2	4	1
	Moderate	Debre Mewi	4	3	2	1
		Dijil	3	4	2	1
	Gentle	D. Mewi	4	3	2	1
		Dijil	3	4	2	1
Maximize fodder (grass) production	Steep	Debre Mewi	3	4	1	1
		Dijil	3	4	1	1
	Moderate	Debre Mewi	3	4	1	1
		Dijil	3	4	1	1
	Gentle	Debre Mewi	3	4	1	1
		Dijil	3	4	1	1
Minimize labor requirement for establishment	Steep	Debre Mewi	3	2	1	4
		Dijil	3	2	1	4
	Moderate	Debre Mewi	3	2	1	4
		Dijil	3	2	1	4
	Gentle	Debre Mewi	3	2	1	4
		Dijil	3	2	1	4
Minimize maintenance costs	Steep	Debre Mewi	1	2	3	4
		Dijil	1	2	3	4
	Moderate	Debre Mewi	1	2	3	4
		Dijil	1	2	3	4
	Gentle	Debre Mewi	1	2	3	4
		Dijil	1	2	3	4
Maximize ox-plowing convenience	Steep	Debre Mewi	3	2	1	4
		Dijil	3	2	1	4
	Moderate	Debre Mewi	3	2	1	4
		Dijil	3	2	1	4
	Gentle	Debre Mewi	3	2	1	4
		Dijil	3	2	1	4

(Continued)

Table 6 | Continued

Criteria	Slope	Watershed	Soil bund	*Fanya juu*	Stone bund	No measure
Minimize risks of pest harboring effect	Steep	Debre Mewi	2	3	1	4
		Dijil	3	2	1	4
	Moderate	Debre Mewi	2	3	1	4
		Dijil	3	2	1	4
	Gentle	Debre Mewi	2	3	1	4
		Dijil	3	2	1	4
Minimize dispute with adjacent farmers	Steep	Debre Mewi	3	2	4	1
		Dijil	3	2	4	1
	Moderate	Debre Mewi	4	3	3	1
		Dijil	4	3	2	1
	Gentle	Debre Mewi	4	3	2	1
		Dijil	3	4	2	1

Table 7 | Farmers' ranking of SWC measures on the evaluation criteria for different slopes (3, Best; 1, Worst): Anjeni watershed.

Criteria	Slope	*Fanya juu*	Stone bund	No measure
Minimize soil losses (erosion control)	Steep	2	3	1
	Moderate	3	2	1
	Gentle	3	2	1
Enhance soil fertility	Steep	2	3	1
	Moderate	3	2	1
	Gentle	3	2	1
Maximize water retention	Steep	3	2	1
	Moderate	3	2	1
	Gentle	3	2	1
Maximize crop yields	Steep	2	3	1
	Moderate	3	2	1
	Gentle	3	2	1
Maximize fodder (grass) production	Steep	3	1	1
	Moderate	3	1	1
	Gentle	3	1	1
Minimize labor requirement for establishment	Steep	2	1	3
	Moderate	2	1	3
	Gentle	2	1	3
Minimize maintenance costs	Steep	1	2	3
	Moderate	1	2	3
	Gentle	1	2	3
Maximize ox-plowing convenience	Steep	2	1	3
	Moderate	2	1	3
	Gentle	2	1	3
Minimize risks of pest harboring effect	Steep	2	1	3
	Moderate	2	1	3
	Gentle	2	1	3
Minimize dispute with adjacent farmers	Steep	2	3	1
	Moderate	3	2	1
	Gentle	3	2	1

Table 8 | Experts' ranking of SWC measures on the evaluation criteria for different slopes (4, Best; 1, Worst).

Criteria	Slope	Soil bund	*Fanya juu*	Stone bund	No measure
Minimize soil losses (erosion control)	Steep	3	2	4	1
	Moderate	4	3	2	1
	Gentle	3	4	2	1
Enhance soil fertility	Steep	3	2	4	1
	Moderate	4	3	2	1
	Gentle	3	3	2	1
Maximize water retention	Steep	3	3	3	1
	Moderate	3	3	2	1
	Gentle	3	4	2	1
Maximize crop yields	Steep	3	2	4	1
	Moderate	4	3	2	1
	Gentle	3	4	2	1
Maximize fodder (grass) production	Steep	4	3	1	1
	Moderate	3	4	1	1
	Gentle	3	4	1	1
Minimize labor requirement for establishment	Steep	3	2	1	4
	Moderate	3	2	1	4
	Gentle	3	2	1	4
Minimize maintenance costs	Steep	2	2	3	4
	Moderate	2	2	3	4
	Gentle	2	2	3	4
Maximize ox-plowing convenience	Steep	3	2	1	4
	Moderate	3	2	1	4
	Gentle	3	2	1	4
Minimize risks of pest harboring effect	Steep	3	2	1	4
	Moderate	3	2	1	4
	Gentle	3	2	1	4
Minimize dispute with adjacent farmers	Steep	3	2	4	1
	Moderate	4	3	2	1
	Gentle	3	3	2	1

and experts ranked *Fanya juu* as the least effective alternative next to "No measures" in steeply sloping areas, as high runoff would easily rupture these structure. Thus, the main objective of farmers in steep areas is to control erosion and consequently, farmers know that stone bunds are the most effective measure.

Moderate slope category

For moderate sloping areas, soil bunds are the best alternative for farmers (Debre Mewi) and experts; however in the Dijil and Anjeni areas, *Fanya juu* is preferred. As water erosion is not as severe on moderate slopes, soil embankments are deemed a suitable measure. Farmer weightings also indicate that increasing yields represents the main objective on moderate slopes (**Table 4**).

Gentle slope category

Fanya juu bunds were the most preferred alternative on plots with gentle slopes in Dijil and Anjeni but not in the Debre Mewi watershed. Farmers invest in SWC on gentle sloped areas to increase production and productivity to help achieve higher profitability for their practice; similar to the moderate slope category. Farmers of Debre Mewi preferred soil bunds for moderate and gentle slopes categories due to their long time experience with soil bunds (**Table 11**).

DISCUSSION

There are differences between the weight sets of farmers and those of experts. Farmers in the three watersheds give social and other criteria on all three slope categories a weighting of about 17%, while this is only 11% among experts. The latter underestimate

Table 9 | Pairwise comparison of SWC measures based on the evaluation criteria (from Table 3).

(A) FARMERS IN Debre Mewi WATERSHED

Regime vector	Steep slope										Moderate slope										Gentle slope									
	C1	C2	C3	C4	C5	C6	C7	C8	C9	C10	C1	C2	C3	C4	C5	C6	C7	C8	C9	C10	C1	C2	C3	C4	C5	C6	C7	C8	C9	C10
e-12	1	1	1	1	−1	−1	−1	1	−1	1	−1	1	−1	1	−1	1	−1	1	−1	1	1	−1	−1	1	−1	1	1	1	−1	1
e-13	−1	−1	1	−1	−1	−1	−1	1	−1	1	1	1	1	1	1	1	1	1	−1	1	1	1	−1	1	1	1	−1	1	−1	1
e-14	1	1	1	1	1	−1	−1	1	1	1	1	1	1	1	1	1	1	1	−1	1	1	1	1	1	1	1	1	1	−1	1
e-23	−1	−1	1	−1	−1	−1	−1	−1	−1	0	−1	1	1	1	1	1	1	1	−1	1	1	1	1	1	1	1	1	1	−1	1
e-24	−1	1	1	1	−1	−1	−1	−1	−1	1	1	1	1	1	1	1	1	1	−1	1	1	1	1	1	1	1	1	−1	−1	1
e-34	1	1	1	1	0	−1	1	−1	−1	1	1	1	1	1	0	1	1	−1	−1	1	1	1	1	1	0	1	1	−1	−1	1

(B) FARMERS IN Anjeni WATERSHED

Regime vector	Steep slope										Moderate slope										Gentle slope									
	C1	C2	C3	C4	C5	C6	C7	C8	C9	C10	C1	C2	C3	C4	C5	C6	C7	C8	C9	C10	C1	C2	C3	C4	C5	C6	C7	C8	C9	C10
e-12	1	1	1	1	1	1	−1	1	−1	1	1	1	1	1	1	−1	−1	1	−1	1	1	1	1	1	1	−1	−1	1	−1	1
e-13	1	1	1	1	1	−1	−1	1	−1	1	1	1	1	1	1	−1	−1	1	−1	1	1	1	1	1	1	−1	−1	1	−1	1
e-23	1	1	1	1	0	−1	−1	1	−1	1	1	1	1	1	0	−1	−1	1	−1	1	1	1	1	1	0	−1	−1	1	−1	1

(C) FARMERS IN Dijil WATERSHED

Regime vector	Steep slope										Moderate slope										Gentle slope									
	C1	C2	C3	C4	C5	C6	C7	C8	C9	C10	C1	C2	C3	C4	C5	C6	C7	C8	C9	C10	C1	C2	C3	C4	C5	C6	C7	C8	C9	C10
e-12	−1	1	1	1	1	1	1	1	−1	1	1	1	1	1	1	1	1	1	−1	1	1	1	1	1	1	1	1	1	−1	1
e-13	−1	−1	1	1	1	−1	−1	1	−1	1	1	1	1	1	1	1	−1	1	−1	1	1	1	1	1	1	1	−1	1	−1	1
e-14	−1	−1	1	1	1	1	1	1	1	1	1	1	1	1	1	1	1	1	1	1	1	1	1	1	1	1	1	1	1	1
e-23	−1	−1	1	1	1	−1	−1	1	−1	1	1	1	1	1	1	−1	−1	1	−1	1	1	1	1	1	1	−1	−1	1	−1	1
e-24	−1	−1	1	1	1	−1	−1	1	−1	1	1	1	1	1	0	−1	−1	1	−1	1	1	1	1	1	0	−1	−1	1	−1	1
e-34	1	1	1	1	0	−1	1	−1	−1	1	1	1	1	1	0	1	1	−1	−1	1	1	1	1	1	0	1	1	−1	−1	1

(D) EXPERTS

Regime vector	Steep slope										Moderate slope										Gentle slope									
	C1	C2	C3	C4	C5	C6	C7	C8	C9	C10	C1	C2	C3	C4	C5	C6	C7	C8	C9	C10	C1	C2	C3	C4	C5	C6	C7	C8	C9	C10
e-12	1	1	0	1	1	0	0	1	1	1	1	0	0	1	1	1	0	1	1	1	1	0	1	1	1	1	0	1	1	0
e-13	−1	−1	0	1	1	−1	−1	1	−1	1	1	1	−1	1	1	1	−1	1	−1	1	1	1	−1	1	1	1	−1	1	−1	1
e-14	1	1	1	1	1	1	1	1	1	1	1	1	1	1	1	1	1	1	1	1	1	1	1	1	1	1	1	1	1	1
e-23	−1	−1	0	1	1	−1	−1	1	−1	1	1	1	1	1	1	−1	−1	1	−1	1	1	1	1	1	1	−1	−1	1	−1	1
e-24	−1	1	1	1	1	−1	−1	1	−1	1	1	1	1	1	1	−1	−1	1	−1	1	1	1	1	1	1	−1	−1	1	−1	1
e-34	1	1	1	1	0	−1	1	−1	−1	1	1	1	1	1	0	1	1	−1	−1	1	1	1	1	1	0	1	1	−1	−1	1

Table 10 | The weighted scores of the pairwise comparisons and overall rank (Rk) of the alternatives for each slope category, by watershed and for farmers and experts.

	Ranking by farmers						Ranking by experts			Ranking by farmers		
R.S	Debre Mewi watershed			Dijil watershed			All watersheds			Anjeni watershed		
	S	M	G	S	M	G	S	M	G	S	M	G
p12	73.2	49.9	7.4	79.5	−10.0	−57.0	45.5	60.5	−39.3	−48.6	88.4	82.6
p13	−44.6	89.9	82.6	−52.5	92.0	85.0	−32.6	87.8	85	75.2	58.0	33.6
p14	78.0	55.9	32.6	80.0	64	42.0	74.9	58.2	42.8	68.6	54.2	30.4
p23	−44.6	43.9	82.6	−70.5	70.5	70.5	−32.6	87.8	85			
p24	78.0	55.9	32.6	80.0	64	42.0	74.9	58.2	42.8			
p34	70.3	50.6	29.3	72.5	60.5	39.5	65.4	50.1	37.3			
Rk	3>1 > 2>4	1>2 > 3>4	1>2 > 3>4	3>1 > 2>4	2>1 > 3>4	2>1 > 3>4	3>1 > 2>4	1>2 > 3>4	2>1 > 3>4	2>1 > 3	1>2 > 3	1>2 > 3

Table 11 | MCA ranking of the SWC measures by farmers and experts.

Watershed	Slope	Ranking
RANKING BY FARMERS		
Debre Mewi	Steep	Stone bunds > soil bunds > Fanya juu > no measure
	Moderate	Soil bunds > Fanya juu > stone bunds > no measure
	Gentle	Soil bunds > Fanya juu > stone bunds > no measure
Anjeni	Steep	Stone bunds > Fanya juu > no measure
	Moderate	Fanya juu > stone bunds > no measure
	Gentle	Fanya juu > stone bunds > no measure
Dijil	Steep	Stone bunds > soil bunds > Fanya juu > no measure
	Moderate	Fanya juu > soil bund > stone bund > no measure
	Gentle	Fanya juu > soil bund > stone bund > no measure
RANKING BY EXPERTS		
All watersheds	Steep	Stone bund > soil bund > Fanya juu > no measure
	Moderate	Soil bund > Fanya juu > stone bund > no measure
	Gentle	Fanya juu > soil bund > stone bund > no measure

the issues of ease of plowing, pest harboring effects, and disputes. The experts on the other hand attach higher weighting to the three ecological criteria and to earnings from grass production; aspects which they focus on in their extension messages. This is in line with the findings of Tenge (2005).

The results of the analysis also illustrate that farmers often stick to practices that they are more familiar with. In Debre Mewi, farmers on moderate and gentle slopes prefer soil bunds above *Fanya juu*, since they have become accustomed to soil bunds and not (yet) to *Fanya juu*. In Anjeni and (to a lesser extent) Digil, farmers already have lengthy (over 30 years) experience with *Fanya juu* and therefore prefer this SWC measure on

moderate and gentle slopes. It is interesting to note that among the experts, there was one favorite measure for each slope category: stone bunds for steep slopes, soil bunds for moderate slopes and *Fanya juu* for gentle slopes. But experts should still look at each particular situation and should not come up with rigid guidelines.

Farmers take into account ecological, economic, social, and other impacts of the SWC when they select SWC practices to meet multiple objectives. Thus, adoption of SWC practices by farmers is not solely based on economic or monetary values as usually demonstrated through CBA (Tenge, 2005; Amsalu, 2006). This suggests that SWC practices that fulfill both economic and other considerations of farmers can contribute to the continued adoption of SWC.

Furthermore, this study revealed that SWC practices have ecological, economic and social benefits. However, SWC practices are mostly evaluated by CBA. These SWC practices are sometimes not profitable from a private-economic point of view (Kassie et al., 2011; Adimassu et al., 2012). This is because the ecological and social benefits of SWC practices they are not quantified in monetary values. Thus, holistic evaluation methods (e.g., MCA) are important to evaluate the overall benefits of SWC practices. Moreover, MCA accommodates diverse views, interests, preferences, and expertise of the stakeholders (Balana et al., 2010). However, MCA with ordinal data does not incorporate the time dimension of costs and benefits, which is pertinent for SWC measures that need a long time for benefits to be realized. The time dimension of SWC can be incorporated within MCA through the use of an efficiency criteria of CBA. The following describes such an example.

Undertaking a CBA for a standard slope of 10%, with an assumed 20 years lifetime of the measures and a 12.5% discount rate, it was found that in Debre Mewi soil bunds (with grassed risers) and stone bunds had a Net Present Value (NPV) of 1819 and 1265 EtB ha^{-1} over 20 years, respectively (1 EtB ≈ 0.056 \$US in 2013). Since *Fanya Juu* was only recently introduced in the Debre Mewi watershed, it was not taken into account. In the Anjeni watershed, soil bunds (rare in this watershed), *Fanya juu* (both with grassed risers) and stone bunds, scored a NPV of 1902, 2718, and 2217 EtB ha^{-1}, respectively (Teshome et al., 2013). Since these detailed calculations were only made for the most common slopes of 10%, these results can only be used for the moderate slope category, for two watersheds and three SWC alternatives. This information on financial efficiency (expressed by NPV) was subsequently used for the economic impact within the Regime Analysis (replacing the four separate cost and benefit criteria). The information on the other (ecological and social/other) evaluation criteria remained the same. The results of this analysis show the same ranking as in the previous analysis for the moderate slopes in the two watersheds: soil bunds better than stone bunds in Debre Mewi watershed and *Fanya juu* better than stone bunds in Anjeni watershed. The measures were in both cases better than no measure.

CONCLUSIONS

A number of SWC practices in the north-western highlands of Ethiopia using MCA were evaluated to assess their ecological, economic and social impacts. The study revealed that MCA is an effective evaluation tool that can take into account non-monetary and less quantifiable effects of SWC measures, which is not possible with a CBA.

The results of the analysis indicate that farmers in the north-western highlands of Ethiopia have a range of criteria to evaluate the performance of SWC measures. The relative importance of each criterion in the selection of SWC alternatives depends to a large extent on slope categories. Farmers in the study areas gave the highest score for criteria related to ecological impacts in steeply sloping areas, and prefer alternatives with stronger positive economic impacts in moderate and gentle sloping areas. Furthermore, stone bunds were deemed the best SWC alternative on steep slopes in all watersheds. *Fanya juu* bunds are the most preferred alternative on plots with gentle slopes in the Dijil and Anjeni watersheds. This indicates that SWC alternatives must be promoted based on farmers' preferences and specific agro-ecological conditions such as slope. Thus, in order to facilitate the adoption of SWC practices, a blanket recommendation approach must be avoided. The extension service should deliver a range of SWC options for the needy farmers from which to select an appropriate SWC measure that is governed by their preferences and plot situation. In addition, the Research-Extension-Farmers linkage must be strengthened in order to identify and disseminate appropriate technologies based on farmer needs. To conclude, policy makers and development practitioners should pay more attention to farmer SWC preferences and particular circumstances (e.g., slope) in designing SWC strategies and programmes.

REFERENCES

Adgo, E., Teshome, A., and Mati, B. (2013). Impacts of long-term soil and water conservation on agricultural productivity: the case of Anjenie watershed, Ethiopia. *Agric. Water Manag.* 117, 55–61. doi: 10.1016/j.agwat.2012.10.026

Adimassu, Z., Mekonnen, K., Yirga, C., and Kessler, A. (2012). Effect of soil bunds on runoff, soil and nutrient losses, and crop yield in the central highlands of Ethiopia. *Land Degrad. Dev.* doi: 10.1002/ldr.2182

Amsalu, A. (2006). *Caring for the Land: Best Practices in Soil and Water Conservation in Beresa Watershed, Highlands of Ethiopia.* Ph.D. thesis, Wageningen University, Wageningen.

Ananda, J., and Herath, G. (2009). A critical review of multi-criteria decision making methods with special reference to forest management and planning. *Ecol. Econ.* 68, 2535–2548. doi: 10.1016/j.ecolecon.2009.05.010

Balana, B.B., Mathijs, E., and Muys, B. (2010). Assessing the sustainability of forest management: an application of multi-criteria decision analysis to community forests in northern Ethiopia. *J. Environ. Manage.* 91, 1294–1304. doi: 10.1016/j.jenvman.2010.02.005

Bizoza, A.R., and de Graaff, J. (2012). Financial cost-benefit analysis of bench terraces in Rwanda. *Land Degrad. Dev.* 23, 103–115. doi: 10.1002/ldr.1051

Boggia, A., and Cortina, C. (2010). Measuring sustainable development using a multi-criteria model: a case study. *J. Environ. Manage.* 91, 2301–2306. doi: 10.1016/j.jenvman.2010.06.009

CIFOR (Centre for International Forestry Research). (1999). *Guidelines for Applying Multi criteria Analysis to the Assessment of Criteria and Indicators.* Jakarta: CIFOR.

de Graaff, J. (1996). *The Price of Soil Erosion: an Economic Evaluation of Soil Conservation and Watershed Development.* Ph.D. thesis, Wageningen Agricultural University, Wageningen.

Fleskens, L., Nainggolan, D., and Stringer, L. C. (2014). An exploration of scenarios to support sustainable land management using integrated environmental socio-economic models. *Environ. Manag.* 54, 1005–1021.

Haile, M., Herweg, K., and Stillhardt, B. (2006). *Sustainable Land Management - A New Approach to Soil and Water Conservation in Ethiopia.* Land Resources Management & Environmental Protection Departement Mekelle University, Ethiopia, Centre for Development and Environment (CDE), National Centre of Competence in Research (NCCR) North-South, Mekelle.

Haileslassie, A., Priess, J., Veldkamp, E., Teketay, D., and Lesschen, J.P. (2005). Assessment of soil nutrient depletion and its spatial variability on smallholders' mixed farming systems in Ethiopia using partial versus full nutrient balances. *Agric. Ecosyst. Environ.* 108, 1–16. doi: 10.1016/j.agee.2004.12.010

Hajkowicz, S. (2007). A comparison of multiple criteria analysis and unaided approaches to environmental decision making. *Environ. Sci. Policy* 10, 177–184. doi: 10.1016/j.envsci.2006.09.003

Hajkowicz, S., and Higgins, A. (2008). A comparison of multiple criteria analysis techniques for water resource management. *Eur. J. Oper. Res.* 184, 2550–2265. doi: 10.1016/j.ejor.2006.10.045

Hajkowicz, S., McDonald, G. T., and Smith, P.H. (2000). An evaluation of multiple objective decision support weighting techniques in natural resource management. *J. Environ. Plann. Manag.* 43, 505–518 doi: 10.1080/7136 76575

Hengsdijk, H., Meijerink, G. W., and Mosugu, M. E. (2005). Modeling the effect of three soil and water conservation practices in Tigray, Ethiopia. *Agric. Ecosyst. Environ.* 105, 29–40. doi: 10.1016/j.agee.2004.06.002

Hinloopen, E., and Nijkamp, P. (1990). Qualitative multiple criteria choice analysis: the dominant regime method. *Qual. Quant.* 24, 37–56.

Kassie, M., Kohlin, G., Bluffstone, R., and Holden, S. (2011). Are soil conservation technologies "win-win"? A case study of Anjeni in the North-Western Ethiopian Highlands. *Nat. Resour. Forum* 35, 89–99 doi: 10.1111/j.1477-8947.2011.01379.x

Kassie, M., Zikhali, P., Manjur, K., and Edwards, S. (2009). Adoption of sustainable agriculture practices: Evidence from a semi-arid region of Ethiopia. *Nat. Resour. Forum* 33, 189–198. doi: 10.1111/j.1477-8947.2009.01224.x

Liu, B.M., Collick, A.S., Zeleke, G., Adgo, E., Easton, Z.M., and Steenhuis, T.S. (2008). Rainfall-discharge relationships for a monsoonal climate in the Ethiopian highlands. *Hydrol. Process.* 22, 1059–1067. doi: 10.1002/hyp.7022

Mendoza, G.A., and Martins, H. (2006). Multi-criteria decision analysis in natural resource management: a critical review of methods and new modelling paradigms. *Forest Ecol. Manag.* 230, 1–22. doi: 10.1016/j.foreco.2006.03.023

Munda, G., Nijkamp, P., and Rietveld, P. (1994). Qualitative multicriteria evaluation for environmental management. *Ecol. Econ.* 10, 97–112. doi: 10.1016/0921-8009(94)90002-7

Pender, J., and Gebremedhin, B. (2007). Determinants of agricultural and land management practices and impacts on crop production and household income in the highlands of Tigray, Ethiopia. *J. Afr. Econ.* 17, 395–450. doi: 10.1093/jae/ejm028

Prato, T. (1999). Multiple attribute decision analysis for ecosystem management. *Ecol. Econ.* 30, 207–222 doi: 10.1016/S0921-8009(99)00002-6

Romero, C., and Rehman, T. (2003). *Multiple Criteria Analyis for Agricultural Decision, 2nd Edn.* Amsterdam: Elseveir Science B. V.

SCRP (Soil Conservation Research Project). (1991). *Soil Conservation Progress Report 8, Vol. 9.* Bern: University of Bern.

Seo, F., and Sakawa, M. (1988). *Multiple Criteria Decision Analysis in Regional Planning. Concepts, Methods and Application.* Tokyo: D. Reidel Publishing Company.

Tefera, B., and Sterk, G. (2010). Land management, erosion problems and soil and water conservation in Fincha'a watershed, Western Ethiopia. *Land Use Policy* 27, 1027–1037. doi: 10.1016/j.landusepol.2010.01.005

Tenge, A.J. (2005). *Participatory Appraisal for Farm-Level Soil and Water Conservation Planning in West Usambara Highlands, Tanzania.* Ph.D. thesis, Wageningen University, Wageningen.

Tesfaye, A., Negatu, W., Brouwer, R., and van der Zaag, P. (2013). Understanding soil conservation decision of farmers in the Gedeb watershed, Ethiopia. *Land Degrad. Dev.* 25, 71–79. doi: 10.1002/ldr.2187

Tesfaye, H. (2011). *Assessment of Sustainable Watershed Management Approach Case Study Lenche Dima, Tsegur Eyesus and Dijil Watershed.* Master of Professional Studies thesis, Cornell University, Bahir Dar.

Teshome, A., de Graaff, J., and Kassie, M. (Forthcoming). Household level determinants of soil and water conservation adoption phases in the north-western Ethiopian highlands. *Environ. Manag.*

Teshome, A., de Graaff, J., Ritsema, C., and Kassie, M. (2014). Farmers' perceptions about influence of land quality, land fragmentation and tenure systems on sustainable land management investments in the North Western Ethiopian Highlands. *Land Degrad. Dev.* doi: 10.1002/ldr.2298

Teshome, A., Rolker, D., and de Graaff, J. (2013). Financial viability of soil and water conservation technologies in north-western Ethiopian highlands. *Appl. Geogr.* 37, 139–149. doi: 10.1016/j.apgeog.2012.11.007

Vega, D. C., and Alpízar, F. (2011). Choice experiments in environmental impact assessment: the case of the Toro 3 hydroelectric project and the Recreo Verde tourist center in Costa Rica. *Impact Assess. Proj. Appraisal* 29, 252–262. doi: 10.3152/146155111X12959673795804

Voogd, J.H. (1982). *Multi-Criteria Evaluation for Urban and Regional Planning.* London: Pion.

Yitbarek, T.W., Belliethathan, S., and Stringer, L.C. (2012). The onsite cost of gully erosion and cost-benefit of gully rehabilitation: a case study in Ethiopia. *Land Degrad. Dev.* 23, 157–166. doi: 10.1002/ldr.1065

Zegeye, A.D. (2009). *Assessment of Upland Erosion Processes and Farmer's Perception of Land Conservation in Debre-Mewi Watershed, Near Lake Tana, Ethiopia.* Master of Professional Studies thesis, Cornell University, Bahir Dar.

Conflict of Interest Statement: The authors declare that the research was conducted in the absence of any commercial or financial relationships that could be construed as a potential conflict of interest.

14

Brackish groundwater membrane system design for sustainable irrigation: optimal configuration selection using analytic hierarchy process and multi-dimension scaling

Beni Lew[1,2], Lolita Trachtengertz[3], Shany Ratsin[3], Gideon Oron[4,5] and Amos Bick[6]*

[1] Department of Civil Engineering, Ariel University, Ariel, Israel
[2] Agriculture Research Organization Volcani Center, Institute of Agricultural Engineering, Bet Dagan, Israel
[3] The Department of Chemical Engineering, Shenkar College of Engineering and Design, Ramat-Gan, Israel
[4] Department of Environment Water Resources, The Jacob Blaustein Institutes for Desert Research, Ben-Gurion University of the Negev, Midrashet Ben-Gurion, Israel
[5] The Department of Industrial Engineering and Management, and The Environmental Engineering Program, Ben-Gurion University of the Negev, Beer Sheva, Israel
[6] Bick & Associates, Ganey-Tikva, Israel

Edited by:
Abdel-Tawab H. Mossa, National Research Centre, Egypt

Reviewed by:
Olimpio Montero, Spanish Council for Scientific Research, Spain
Abid Ali Khan, Jamia Millia Islamia University, India

*Correspondence:
Amos Bick, Bick & Associates, 7 Harey Jerusalem, Ganey–Tikva 55900, Israel
e-mail: amosbick@gmail.com

The recent high demands for reuse of salty water for irrigation affected membrane producers to assess new potential technologies for undesirable physical, chemical, and biological contaminants removal. This paper studies the assembly options by the analytic hierarchy process (AHP) model and the multi-dimension scaling (MDS) techniques. A specialized form of MDS (CoPlot software) enables presentation of the AHP outcomes in a two dimensional space and the optimal model can be visualized clearly. Four types of 8″ membranes were selected: (i) Nanofiltration low rejection and high flux (ESNA1-LF-LD, 86% rejection, 10,500 gpd); (ii) Nanofiltration medium rejection and medium flux (ESNA1-LF2-LD, 91% rejection, 8200 gpd); (iii) Reverse Osmosis high rejection and high flux (CPA5-MAX, 99.7 rejection, 12,000 gpd); and (iv) Reverse Osmosis medium rejection and extreme high flux (ESPA4-MAX, 99.2 rejection, 13,200 gpd). The results indicate that: (i) Nanofiltration membrane (High flux and Low rejection) can produce water for irrigation with valuable levels of nutrient ions and a reduction in the sodium absorption ratio (SAR), minimizing soil salinity; this is an attractive option for agricultural irrigation and is the optimal solution; and (ii) implementing the MDS approach with reference to the variables is consequently useful to characterize membrane system design.

Keywords: analytical hierarchical process, irrigation, multi-dimension scaling, nanofiltration, reverse-osmosis

INTRODUCTION

Reuse of brackish water, mainly for agricultural irrigation, simultaneously solves water shortage problem and allows for overcoming environmental pollution problems (Qadir et al., 2007; Ghermandi and Mesalem, 2009). It is recognized that water for irrigation should be without particles with average size larger than 50–100 micron and, have a low salt content to avoid certain ion toxicities and increase of soil salinity (Ben-Gal et al., 2008; Edelstein et al., 2009). Salt concentration is generally tested by electrical conductivity (EC) or total dissolved solids (TDS) levels (Bui, 2013). In addition, the sodium to water hardness (calcium and magnesium) ratio should be managed to keep sodium absorption ratio (SAR) at values lower than 3–4, if possible. Together with that, precipitation potential of the water should be to some extent negative in order to avoid the precipitation of calcium and magnesium on the productive system. Also, iron and manganese ions need to be managed to prevent staining harms.

Other parameters, such as boron, fluoride, and heavy metals should also be low to ease the possibility of ion toxicity. Besides, irrigation water needs to be without any disease-causing microorganisms, viruses, bacteria, fungi, nematodes, cysts, etc. (Yermiyahu et al., 2007; Bustan et al., 2013).

Hence, irrigation water has to be treated properly to get rid of all the undesirable physical, chemical, and biological contaminants that are able to: (i) reduce the choice of crops (Levy and Tai, 2013); (ii) cut down crop yield (Rasouli et al., 2013); (iii) damage crop quality (Bernstein et al., 2011); (iv) injure soil appropriateness (Liu et al., 2013); and (v) harm the irrigation tools (Connor et al., 2012).

Water treatment regarding irrigation purposes is quite new to the agricultural business. The advanced technology is based on membrane technology (Oron et al., 2008; Bick et al., 2012). Mainly, NanoFiltration (NF) and Reverse-Osmosis (RO) can usually be implemented for removal of organic matter, particles, turbidity, pathogenic micro-organisms, and selective ions without the use of disinfectants (Mrayed et al., 2011; Riera et al., 2013).

Membrane system presents several advantages: modularity, automation, low maintenance costs, and low chemicals consumption (Miller, 2006; Birnhack et al., 2010). However, there are several drawbacks: high energy consumption, permeate quality (lack of vital ions for plant growth, e.g., magnesium), high costs relative to other sources of water and in particular chemical and biological fouling of the membranes (Oren et al., 2012).

For a known feed water source, the selection between membrane and any other alternative for water treatment is usually made on technical, economical, and/or political criteria. Yet, environmental concern have demonstrated its importance to integrate sustainability factors into the judgment process by comparing the "green" impacts of the different alternatives (Vince et al., 2008; Lew et al., 2009).

Downstream supervision on desalination plants, such as brine dilution, treatment, and/or removal may diminish the ecological impacts. However, reducing the upstream sources through means of optimization may be more successfully in minimizing the environmental impacts. The actual performances of the membrane course of action outcome from model choices, which usually are designed to minimize the total water cost, taking in account technical constraints and project necessities but, with no consideration for the environmental impact (Bartels et al., 2008; Garcia et al., 2013; Wei et al., 2013). The environmental performances of the membrane system could be enhanced by revaluating these previously defined outline selections (Gur-Reznik et al., 2011; Chakrabortty et al., 2013; Zheng et al., 2013; Zhou et al., 2013). It is thus anticipated the introduction of the environmental criteria directly in the near the beginning design phases, in order to classify environmental responsible process configurations (Vince et al., 2008).

Concerning agricultural applications, NF membranes showed up to be a new opportunity for salty water desalination (Hilal et al., 2005): (i) high rejection rate for divalent ions, capability of knocking down TDS significantly and may constitute a potentially cost-effective alternative for irrigation with brackish water; (ii) higher tolerance (in general) for fouling conditions, as compared to RO (Gutman et al., 2012; Sotto et al., 2013); (iii) removal of specific pollutants, so that the concentrate stream can be used without increased membrane fouling; and (iv) operation at lower applied pressures, as compared to RO, saving energy and operating costs (Nederlof et al., 2005; Al-Amoudi and Lovitt, 2007; Zhao et al., 2012).

NF-RO configuration can be a promising solution, since (i) the NF stage can separate the agriculturally advantageous ions, such as magnesium from the non-desirable ions, such as sodium, which are finally removed from the water by the following RO stage; and (ii) the advantageous ions (in the NF brine flow) can be blended with the free sodium water produced in the RO stage (RO permeate stream) to create a nutrient-enriched, low-salt water (Mrayed et al., 2011).

The Arava Valley in Israel was used for this study, and the modeling and design of a desalination plant featuring both NF and RO membranes is discussed. The objectives of this study are: (i) to examine the designs of NF and RO membranes using typical brackish water from Hatzeva-Idan aquifer in the Arava Valley (Israel), which is characterized to some extent by saline water with a typical TDS concentration of 1200 ppm; (ii) to simulate RO and NF technologies for desalination of brackish water (using IMSDesign software, Verhuelsdonk et al., 2010; Karabelas et al., 2012); (iii) to compare the design configuration using Analytic

Hierarchy Process (AHP) model (Caputo et al., 2013); and (iv) to rank the various alternatives using multi-dimension scaling (MDS) technique (Lespinats et al., 2009) in order to find the optimal design configuration for technology selection.

MATERIALS AND METHOD
MANAGEMENT MODELING AND THE ANALYTICAL HIERARCHICAL PROCESS (AHP)

The amount of data and information collected and retained by organizations and businesses is persistently growing, due to advances in data gathering, computerization of transactions, and breakthroughs in storage tools. In order to take out practical information from such large datasets, it is crucial to be able to spot patterns, trends and interactions in the data and visualize their global configuration to ease decision making. Decision making techniques used for the ranking of various options on the basis of more than one attribute are strictly dependent on the attributes setting and thus can be completely different for different settings (Saaty, 1980).

The analytical hierarchical process (AHP) is a strong and flexible decision-supporting process that helps in setting priorities and making the best decision when mutually qualitative and quantitative aspects are measured (Tzfati et al., 2011; Zhang et al., 2013). The AHP is designed for individual evaluation of a set of options based on multiple attributes, and is approved in a hierarchical building (Kayastha et al., 2013). The estimate of the options is founded on a pair-wise judgment (Bozoki et al., 2011). The pair-wise comparisons are translated from linguistic/verbal expressions in numerical values (integers 1–9) using the original Saaty's Scale (Saaty, 2003) for the comparative judgments (**Table 1**).

MULTI DIMENSION SCALING (MDS) AND COPLOT

Many research questions dealing with information require the analysis of complex multivariate data (Gomez-Silvan et al., 2013).

Table 1 | Fundamental Saaty's Scale for pair-wise comparison (Saaty, 1980).

Numerical values	Verbal term	Explanation
1	Equally important	Two elements have equal importance
3	Moderately more important	Experience or judgment slightly favors one element
5	Strongly more important	Experience or judgment strongly favors one element
7	Very strongly more important	Dominance of one element proved in practice
9	Extremely more important	The highest order dominance of one element over another
2,4,6,8	Important intermediate values	Compromise is needed

Briefly, most multivariate approaches can be roughly classified as dependence methods (e.g., multiple regression, discriminant analysis, multivariate analysis of variance), that are usually used to assess the relationship between dependent and independent variables; or as interdependence methods (e.g., principal component analysis, factor analysis, cluster analysis), that are typically used to estimate the mutual association among all variables with no difference made among the variable types (Schilli et al., 2010). Multi dimensional scaling (MDS) is an interdependence methods that facilitates the examination of multivariate data by reducing multidimensional data into a two-dimensional structure that attempts to expose the "out of sight structure" in a data set by creating a pictorial or mapping image of the data. The MDS map graphically represents the proximities (or similarities) between objects (i.e., observations or events). Similarities between the observations in the data set are transformed into distances on a map such that observations with great similarity are closer together than less similar observations.

In this way, a single picture illustrates the relationships among all the observations. MDS, initially developed in the 1960s, has been used to evaluate the relationships among observations, to identify clusters of similar observations and to locate outliers (Bick et al., 2011). However, MDS maps have two key limitations: (i) MDS does not simultaneously map the variables and the observations; and (ii) the MDS map has no orientation, thereby limiting the map's interpretability.

CoPlot, a variant of multi-dimensional scaling, addresses both these limitations and locates each decision-making unit in a two-dimensional space in which the location of each observation is strong-minded by every variables simultaneously (hence, its name) (Lipshitz and Raveh, 1994). The graphical put on view technique exhibits observations as points and variables as arrows, relative to the same center-of-gravity. CoPlot is rooted in the integration of mapping concepts, using a variant of regression analysis that superimposes two graphs sequentially. Additionally, CoPlot maps the observations and variables in a manner that preserves their relationships, allowing richer interpretations of the data. Importantly, CoPlot allows analysis of a dataset where the number of variables is greater than the number of observations. Also, CoPlot map can be used to detect outliers and errors in the data, assessment of the relationships within the data and for selection of key variables for subsequent analysis (Adler and Raveh, 2008).

Coplot's output is a visual display of its findings [Given an input matrix Yn × v of v variable values for each of n observations (see for example **Table 2**)] and it is based on two graphs that are superimposed on each other (Bravata et al., 2008). The first graph maps the n observations into a two-dimensional space: those observations that are perceived to be very similar to each other are placed near each other on the map, and those observations that are perceived to be very different from each other are placed far away from each other on the map. The second graph (**Figure 1**) consists of v arrows, representing the variables and it shows the direction of the gradient for each arrow.

The CoPlot analysis consists of four stages, two preliminary adaptations of the data matrix and two subsequent stages that compute two maps sequentially that are then superimposed to

Table 2 | Production technologies for irrigation.

Design code	Membrane type	Number of passes	Additional data
NF-L-H	ESNA1-LF-LD 86% rejection 10,500 gpd	1	Nanofiltration Low rejection High flux
NF-M-M	ESNA1-LF2-LD 91% rejection 8200 gpd	1	Nanofiltration Medium rejection Medium flux
RO-H-H	CPA5-MAX 99.7% rejection 12,000 gpd	1	Reverse-Osmosis High rejection High flux
RO-M-H	ESPA4-MAX 99.2% rejection 13,200 gpd	1	Reverse-Osmosis Medium rejection Extreme high flux
NF-RO-1P	NF-M-M at the first stage RO-M-H at the second stage	1	
NF-RO-2P	NF-M-M at pass I RO-M-H at pass II	2	Desalination of Nanofiltration permeate (the brine of the second pass is recycled to the feed)

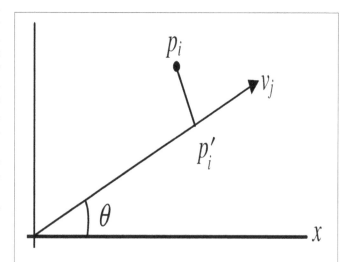

FIGURE 1 | CoPlot: Adding the variable vectors. The point p_i corresponds to the coordinates for observation i = 1,..., n. The vector v_j is for the variable j = 1,... m. The x-axis is rotated through an angle θ to give a point p_i', which is the projection of p_i onto the vector v_j. The correlation between the n new projected scores and the original n values for variable j are computed, and the choice for θ is the one that maximizes this correlation.

produce a single map. The goal of the first stage is to normalize the variables, which is needed in order to allow the variables to relate to each other, although each variable has a different unit and scale. This is done in the usual way, the difference between the elements of the matrix Y_{ij} (which are scores) and the deviations from column means $\overline{Y}_j$ is divided by their standard deviations (D_j), given the normalized Z_{ij}, according to Equation (1).

$$Z_{ij} = \left(Y_{ij} - \overline{Y}_j\right) / D_j \qquad (1)$$

In the second stage, a measure of dissimilarity ($S_{ik} \geq 0$) between each pair of observations (rows of $Zn \times v$) is chosen and a symmetric $n \times n$ matrix is produced from all the different pairs of observations. To measure S_{ik}, the sum of absolute deviations (generally defined as city-block distance) as a measure of dissimilarity is used (Equation 2).

$$S_{ik} = \sum_{j=1}^{v} |Z_{ij} - Z_{kj}| \qquad (2)$$

In the third stage the matrix S_{ik} is mapped by a MDS method. The algorithm maps the matrix S_{ik} into an Euclidean space, of two dimensions in our case, such that "similar" observations (with a small dissimilarity between them) are close to each other on the map, while "different" ones are also distant on the map. Formally, the requirement is as follows: consider two observations, i and k, which are mapped at a d_{ik} distance from each other. This distance has to reflect the dissimilarity S_{ik} (which is actually a relative measurement), taking in consideration that the important constraints are $S_{ik} < S_{lm}$ if $d_{ik} < d_{lm}$.

CoPlot procedure uses the Guttman's smallest space analysis (SSA) with the "coefficient of alienation" (Θ) as a measure of "goodness of fit" (Guttman, 1968; Raveh, 2000). The intuition for Θ comes directly from the above MDS requirement of dissimilarity measures and the map distances: A success of satisfying it implies that the product of the differences between the dissimilarity measures and the map distances are positive. In a normalized form a new variable is defined, μ_{ca} (Equation 3).

$$\mu_{ca} = \frac{\sum_{i,k,l,m} (S_{ik} - S_{lm}) (d_{ik} - d_{lm})}{\sum_{i,k,l,m} |S_{ik} - S_{lm}| |d_{ik} - d_{lm}|} \qquad (3)$$

μ_{ca} can attain a maximal value of 1 (Raveh, 2000) and is used to define Θ according to Equation (4).

$$\Theta = \sqrt{1 - \mu_{ca}^2} \qquad (4)$$

The details of the SSA algorithm are beyond the scope of this paper and were presented in the literature (Guttman, 1968). The SSA algorithm is an extensively used method in social sciences and several examples along with intuitive descriptions can be found (Raveh, 2000). The outcome of this stage is a two-dimensional map of n observations and the CoPlot user can color code observations with any definite variable that has up to 16 different values.

The map generated so far is a classical MDS map, however without orientation or meaningful axes. In the fourth stage of the CoPlot method, v arrows are drawn on the Euclidean space previously obtained. Each variable j is represented by an arrow j, emerging from the center of gravity of the n points. The graphical display technique plots observations as points and variables as arrows, relative to the same arbitrary center-of-gravity.

Observations are mapped such that similar entities are closely located on the plot, signifying that they belong to a group possessing comparable characteristics and behavior. The location of the center of gravity is in the middle of the plot in order to introduce all the observations and it does not affect the arrows direction. The direction of each arrow is selected so that the correlation between the actual values of the variable j and their projections on the arrow is maximal (the arrows' length is undefined). Therefore, observations with a high j value will be located in the part of the space which the arrow points to, while observations with a low j value will be located at the other side of the map. The magnitude of the j maximal correlations measures the "goodness of fit" of the R_j regressions. Higher is the correlation better is the arrow representation of the variables and, those having low correlations should be eliminated.

Moreover, arrows linked with highly correlated variables will point in about the same direction and vice versa. As a result, the cosines of angles between these arrows are approximately proportional to the correlations between their associated variables (Raveh, 2000).

The "goodness of fit" measured for each correlation (R_j) is obtained as follows: For each possible variable vector, CoPlot projects the points onto the vector, thereby yielding n projected values. These projected values can now be compared with the observed values. The axis that is chosen is the one that maximizes the correlation between the projected values and the observed values. **Figure 1** depicts how this is performed. The point p_i corresponds to the coordinates for observation i = 1,..., n. The vector v_j is for the variable j = 1,..., m. The x-axis is rotated through an angle θ to give a point p_i', which is the projection of p_i onto the vector v_j. The correlation between the n new projected scores and the original n values for variable j are computed and, the choice for "goodness of fit" is the one that maximizes this correlation. This maximization can be achieved numerically by calculating all $360°$ possibilities for θ, which is performed separately for each variable vector.

These variable vectors have four useful properties; first, vectors for highly correlated variables point in the same direction, vectors for highly negatively correlated variables are oriented along the same axis but in opposing directions and, vectors for variables that are not correlated are orthogonal to each other. Second, each vector emanates from the center of gravity, which serves as the origin. An observation located at or near the origin is an average observation (it has an average value in all variables). Third, the length of each vector is proportional to the correlation (namely R_j) between the original data for that variable and the projections of the observations onto the vector. Finally, the cosines of angles between the arrows are approximately proportional to the correlations between their associated Variables. Therefore, the correlational structure among the variables can be studied in a single graphical output (Raveh, 2000).

In practical terms, the user imports data, selects variables and observations for inclusion in the analysis, creates the CoPlot map, evaluates Θ parameters, selects the map to view (observations only, variables only, or both observations and variables) and, then selects variables for color coding the observations for greater interpretation. Qualitative variables can be selected for color coding and may either be included in the computation of the map or can be excluded from the computation of the map but still used for color coding. For example, if a variable was found to have low R_j, it might be excluded from the computation of the map but could still be used to color code variables to facilitate the interpretation of the data.

CoPlot produces two "goodness of fit" measurements, one that describes how well the CoPlot map represents the observations and another that describes how well the CoPlot map represents the variables. The first measure is a "coefficient of alienation" (Θ), which indicates the relative loss of information that arises when the multidimensional data are transformed into two dimensions. The lower the Θ value, the smaller the information loss in the process of reducing the original data set to a two-dimensional map. In other words, the lower the Θ, the more precise the representation of the MDS model to the proximities. A general rule-of-thumb states that the map is statistically significant if $\Theta \leq 0.15$ (Guttman, 1968).

In general, as the number of variables increases, Θ also increases. In such case, Θ measures the discrepancy between every pair of points and the original matrix of "similarities" that comprises distances between points, so that this index provides a comparison between two matrices, the matrix similarities (which are inputs) and the matrix of the distances on the map (which are outputs) obtained by the algorithm. When these two matrices (inputs and outputs) are identical, Θ is zero (precise).

The second "goodness of fit" measure is obtained at the stage of calculating the correlation between the original data, for each variable and, the projection of each observation onto that vector in the CoPlot map. In general, the methodology maximizes the correlations (actually the normalized cross-products) of the vector of inputs (the actual distances from each point to every other point) and the outputs (the coordinates of the vectors that go into the map); in other words, the R_j measurements are the correlational measure that relates the input with the output. The closer these are, in a correlational sense, higher the fit. Individual correlations are obtained for each of the variables separately. If a vector has a correlation of 1 it means that there is a perfect fit with the original variable data. In general, as the number of (poor) variables decreases the average correlation increases and, average of correlations with values of 0.7 or greater provides maps that fit the data (Bravata et al., 2008).

CASE STUDIES

The Arava Valley in Southern Israel is an example of highly efficient agriculture and greenhouse technologies (Villarreal-Guerrero et al., 2012) in a region of extreme water scarcity (Hillel et al., 2013). The water quality of the Arava Valley (Hatzeva wells) is as follows: Total Dissolved Solids-1577 ppm; Barium-0.2 ppm; Calcium-150 pm; Potassium-12.5 ppm; Magnesium-82.5 ppm; Sodium-225 ppm; Chloride-359 ppm; Bicarbonates-208 ppm; Nitrate-9.6 ppm; Sulfates-505 ppm (Ghermandi and Mesalem, 2009).

Simulation of Arava feed water treatment can be done by RO process design software's that were developed by membrane constructors such as ROSA from Filmtec or IMSDesign from Hydranautics (Penate and García-Rodríguez, 2011). Besides

defining constructor good practices for membrane operation and shortcuts methods for pressure vessel modeling, the simulation allowed to verify flexible RO and NF configurations for different commercial membranes (Alghoul et al., 2009).

In this work, four types of $8''$ commercial membranes were selected: (i) Nanofiltration low rejection and high flux (ESNA1-LF-LD, 86% rejection, 10,500 gpd); (ii) Nanofiltration medium rejection and medium flux (ESNA1-LF2-LD, 91% rejection, 8200 gpd); (iii) Reverse Osmosis high rejection and high flux (CPA5-MAX, 99.7 rejection, 12,000 gpd); and (iv) Reverse Osmosis medium rejection and extreme high flux (ESPA4-MAX, 99.2 rejection, 13,200 gpd). The simulation procedure includes six configurations (**Table 2**): (i) one pass with ESNA1-LF-LD membranes (code NF-L-H); (ii) one pass with ESNA1-LF2-LD membranes (code NF-M-M); (iii) one pass with CPA5-MAX membranes (code RO-H-H); (iv) one pass with ESPA4-MAX membranes (code RO-M-H); (v) one pass with ESNA1-LF2-LD membranes at the first stage and ESPA4-MAX membranes at the second stage (code NF-RO-1P); and (vi) two pass with ESNA1-LF2-LD membranes at the first pass and ESPA4-MAX membranes at the second pass (code NF-RO-2P).

The output of the simulation using IMSDesign software of the six different membrane configurations from **Table 2** is presented in **Table 3** (the design is based on six elements per vessel, production of 960 m³/day, and operation at a permeate flux of 20 l/m²-h). The pilot plant flow (960 m³/day) can be increases without limitations, like any other membrane plants based on NF membranes. The 20 l/m²-h permeated flux is based on pilot plant best performance as indicated by the model (increasing the flux in the system may cause blocking and fouling problems) The comparative performance of configuration technologies by the various alternative methods is very complex and is highly depending on various site-specific operational and economic attributes. General attributes must be fulfilled by an objective function (Gursoy et al., 2013): (i) minimization of energy consumption; (ii) minimization of brine production; (iii) minimization of brine concentration; and (iv) minimization of membrane types and passes. In order to support adequate selection the decision-maker has to define a utility function (Z_T) (Equation 5) that takes in account all the previous attributes.

$$Z_T = \begin{pmatrix} \text{Min} \\ \text{Energy} \\ \text{Consumption} \end{pmatrix} + \begin{pmatrix} \text{Min} \\ \text{Brine} \\ \text{Production} \end{pmatrix}$$
$$+ \begin{pmatrix} \text{Min} \\ \text{Brine} \\ \text{Concentration} \end{pmatrix} + \begin{pmatrix} \text{Min} \\ \text{Number} \\ \text{of Passes} \end{pmatrix} \quad (5)$$

To use the AHP, the decision-maker must specify his requirements (based on previous experience) for achieving the overall goal.

MODEL IMPLEMENTATION AND DISCUSSION

From the matrix obtained in **Table 3**, the geometric mean (w_i, approximately the product of the elements in each row regarding to a matrix of n rows and n columns) and the normalized geometric mean (p_i) are determined according to Equations (6) and (7), respectively.

$$w_i = \left(\prod_{j=1}^{n} a_{ij} \right)^{1/n} \quad i = 1, \ldots, n \qquad (6)$$

$$p_i = \left(\prod_{i=1}^{n} a_{ij} \right)^{1/n} / \sum_{i=1}^{n} \left(\prod_{j=1}^{n} a_{ij} \right)^{1/n} \quad i = 1, \ldots, n \qquad (7)$$

where, a_{ij} is an evaluated value, i is the row index (alternative, i = 1,..., n) and j is the column index (quality attribute, j = 1,...,

n). In **Table 4** its shown that for each decision attribute chosen, the importance of the design options. It designates how the alternatives are preferred over others with respect to each objective as well as to the whole objective (Cay and Uyan, 2013).

Table 5 compares minimization of energy consumption related to each of the options. One can see that according to the comparison the configurations NF-M-M is favorable (0.381). A comparable comparison was run for each of the attributes. The resultant set of weights for each of the option (in this case technology) with respect to each attributes is presented in **Table 6**.

Discrepancies in the response assessments can occur and are related to human errors along the process. Let's assume that assumption A is preferred over assumption B and, assumption B is preferred over assumption C. Then, it can be assumed that assumption A should be preferred over assumption C by a wide margin. If throughout the pair-wise comparison A is vaguely preferred over C, then a contradiction is taken place. In this work, the statistics and implementation of a judgment matrix (Yang et al., 2013) was implemented.

Based on the results of the correlations between variables, the CoPlot diagram presented in **Figure 2** shows "min brine salinity" and "min energy" relatively close to each other. "Min brine production"is pointing to the opposite direction and "min passes" is lying in-between. Consequently "min brine salinity"(or "min energy") is identified as offering diminutive information and can be taken away from the analysis with no effect (or little effect) on the results. The results show that: NF-RO-2P is disconnected from the other units because it has the highest "min passes" value; RO-H-H is separated from the other units because it has the highest "min brine production" value; and NF-M-M is separated from the other units because it has the highest "min brine salinity"and "min energy" values. NF-RO-2P, RO-H-H and NF-M-M could be considered outliers (see **Figure 2**). Finally, NF-L-H is average in all variables and in conclusion appears quite near to the center-of-gravity, the point from which all arrows begin. It is being claimed in this paper that MDS can be used to present the result graphically, the lower value (min objective function) a unit gets, the more effective the design is assumed over that specific attribute and it is clearly shown (**Figure 2**) that the best configuration is achieved with NF-L-H.

Table 3 | Output technologies for irrigation (IMSDesign software).

Design code	Energy kwhr/m^3	Brine percent	Brine salinity ppm	Number of passes
NF-L-H	0.26	18.6	5377	1
NF-M-M	0.20	22.5	4524	1
RO-H-H	0.52	16.1	6111	1
RO-M-H	0.33	17.0	5817	1
NF-RO-1P	0.31	19.5	5145	1
NF-RO-2P	0.28	16.3	5280	2

Table 4 | Variability in importance across design options.

Attribute	Design option
Min energy consumprion kwhr/m^3	NF-M-M>NF-L-H>NF-RO-2P>NF-RO-1P>RO-M-H>RO-H-H
Min brine production percent	RO-H-H>NF-RO-2P>RO-M-H>NF-L-H>NF-RO-1P>NF-M-M
Min brine salinity ppm	NF-M-M>NF-RO-1P>NF-RO-2P>NF-L-H>RO-M-H>RO-H-H

The notations > and ≅ symbolize the option preceding the sign in "preferable to" and "equal to" the one after the sign, respectively.

Table 5 | AHP pair-wise evaluation of min energy consumption (Number are based on Saaty's Scale and an expert subjective point of view).

Design	NF-L-H	NF-M-M	RO-H-H	RO-M-H	NF-RO-1P	NF-RO-2P	Geometric mean (*)	Normalization (**)
NF-L-H	1	1/2	5	4	3	2	1.98	0.252
NF-M-M	2	1	6	5	4	3	2.99	0.381
RO-H-H	1/5	1/6	1	1/2	1/3	1/4	0.33	0.042
RO-M-H	1/4	1/5	2	1	1/2	1/3	0.51	0.064
NF-RO-1P	1/3	1/4	3	2	1	1/2	0.79	0.101
NF-RO-2P	1/2	1/3	4	3	2	1	1.26	0.160
Total							7.87	

*For example, the geometric mean of NF-M-M is $(2 \cdot 1 \cdot 6 \cdot 5 \cdot 4 \cdot 3)^{1/6} = 2.99$

**For example, the normalization of NF-M-M is 2.99/7.87 = 0.38.

The "goodness of fit" in the Coplot system is calculated by two types of measures: the "coefficient of alienation" (Θ) and the extend of the v maximal correlations that measure the "goodness of fit" of the j regressions. Smaller "coefficient of alienation" indicates better the output and, values lower than 0.15 are assumed precise. In this study (**Figure 2**) a coefficient of 0.038 is considered as an excellent figure. The "goodness of fit" of the j regressions helps to decide whether one eliminates or adds variables: Variables that do not fit into the graphical display, meaning, have low correlations, should be removed. Higher the variable's correlation, the better the variable's arrow represents common direction and order of the projections of the n points along the axis.

Based on this test case, the correlations of MDS data are high: average 0.905 ("min energy" = 0.93; "min brine production" = 0.86; "min brine salinity" = 0.89; "min passes" = 0.94). The two "goodness of fit" measurements ("coefficient of alienation" for the first step and four correlations for each one of the variables for the second step) allow the researchers to point out that according to this map the optimal solution is Nanofiltration technology (low rejection and high flux).

There are some limitations of these approaches: (i) Some problems are too complex resulting in a very complicated pair wise process that cannot be conducted by AHP and MDS, (ii) Not all individuals are capable in using and understanding the process; there is a need for experts, and (iii) The focus is on immediate technology impacts and not on the gain in substantial knowledge expertise and future improvements.

Table 6 | AHP pair-wise results of attribute weights.

Design	Min energy	Min brine production	Min brine salinity	Min number of passes
NF-L-H	0.252	0.101	0.101	0.196
NF-M-M	0.381	0.042	0.381	0.196
RO-H-H	0.042	0.381	0.042	0.196
RO-M-H	0.064	0.160	0.064	0.196
NF-RO-1P	0.101	0.064	0.252	0.196
NF-RO-2P	0.160	0.252	0.160	0.022

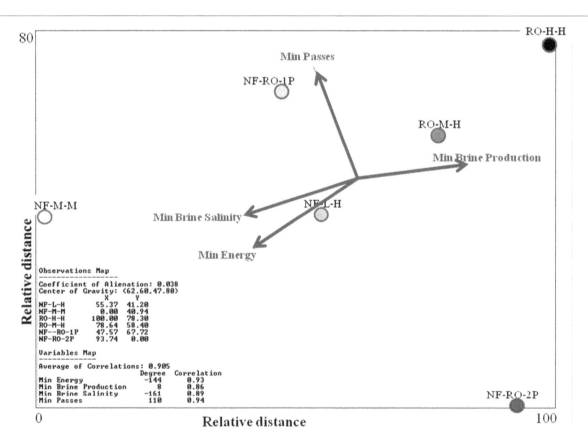

FIGURE 2 | MDS map, generated by the proposed method (CoPlot): NF and RO configuration performance (at a permeate flux of 20 l/m²-h, six elements per vessel). Design: (i) one pass with ESNA1-LF-LD membranes (code NF-L-H); (ii) one pass with ESNA1-LF2-LD membranes (code NF-M-M); (iii) one pass with CPA5-MAX membranes (code RO-H-H); (iv) one pass with ESPA4-MAX membranes (code RO-M-H); (v) one pass with ESNA1-LF2-LD membranes at the first stage and ESPA4-MAX membranes at the second stage (code NF-RO-1P); and (vi) two pass with ESNA1-LF2-LD membranes at the first pass and ESPA4-MAX membranes at the second pass (code NF-RO-2P). Each variable is represented by an arrow, emerging from the center of gravity of the n points. The direction of each arrow is chosen so that the correlation between the actual values of the variable and their projections on the arrow is maximal. (MDS Statistics: "Coefficient of alienation" = 0.038, "Average of Correlations" = 0.905). The "goodness of fit" measures (concerning the "Coefficient of alienation" and the variables correlations) is satisfied both for the configurations and the variables.

For the use of NF technology a reform of water pricing may be needed, which is most often driven by budget government pressure, rise of the costs of the services provided and, government aspiration to reduce subsidies. The World Bank has encouraged governments to employ a policy of cost recovery for many years, on the belief that users should cover O&M and some of the capital costs (Bustan et al., 2013).

The Public-Private Partnerships (PPP) can be viewed as the governance policy for minimizing transaction costs and; coordinating and enforcing relations between partners engaged in agricultural goods and services production. The Built Own and Operate (BOO) projects promotes an optimal policy tactic to enable social and economic development, bringing together competence, plasticity, and competence for the private sector with the responsibility, long-term outlook and social interest of the public sector (Johannessen et al., 2014).

The use of membrane technology in agriculture can lead to additional gains: (i) understanding the entire supply-chain instead of the bottlenecks, (ii) improved quality and quantity of sales' product, (iii) new market penetration, and (iv) positive social effects. Agricultural research concerning NF technology responds to multiple objectives. Clearness of objectives in determining water charges is vital and there are limits of valuing as a practical tool for irrigation demand. The authorities must take into consideration the environmental benefits by including incentives for appropriate water allocation and demand management.

CONCLUSIONS

Membrane technology selection for water treatment is often a subjective task: Combining quantitative methods into the evaluation procedure permits the decision makers to recognize the most suitable option in an objective and efficient way. In this study two methodologies were used and discussed to analyse multiple variable data:AHP and MDS (CoPlot).

The AHP method establishes an evaluation model for design of membrane technology based on IMSDesign software. Analysis results using Saaty's Scale indicates that the decision-makers may select the most appropriate option based on the following attributes: minimization of energy consumption, minimization of brine production, minimization of brine salinity, and minimization of design passes (investment cost).

Using the AHP method for assortment of optimal technology provides a systematically decision making a framework with several characteristics: (i) the performance of different technologies can be evaluated using multiple attributes—both quantitative and qualitative—rather than profitability alone; (ii) ratings allow to evaluate the use of the technologies for the end user; (iii) AHP method provides an effective way for managing process documentation; and (iv) the proposed method forms the basis for a intermittent process of planning and managing technology, so that the priorities of the technologies can easily be modified and updated.

AHP database is graphically introduced by the CoPlot technique, a multivariate statistical method that is remarkably robust and provides new insights about membrane configurations and design, giving a strong picture of what needs to be done next. The CoPlot technique provides a powerful analytic tool for brackish groundwater membrane system: the optimal configuration using AHP and MDS design is based on NF membrane (low rejection and high flux).

ACKNOWLEDGMENT

The research was funded by the Chief Scientist, Israeli Ministry of Agriculture and Rural Development (Project n. 459-4437-12).

REFERENCES

Adler, N., and Raveh, A. (2008). Presenting DEA graphically. *Omega* 36, 715–729. doi: 10.1016/j.omega.2006.02.006

Al-Amoudi, A., and Lovitt, R. W. (2007). Fouling strategies and the cleaning system of NF membranes and factors affecting cleaning efficiency. *J. Membr. Sci.* 303, 4–28. doi: 10.1016/j.memsci.2007.06.002

Alghoul, M. A., Poovanaesvaran, P., Sopian, K., and Sulaiman, M. Y. (2009). Review of brackish water reverse osmosis (BWRO) system designs. *Renew. Sust. Energ. Rev.* 13, 2661–2667. doi: 10.1016/j.rser.2009.03.013

Bartels, C., Hirose, M., and Fujioka, H. (2008). Performance advancement in the spiral wound RO/NF element design. *Desalination* 221, 207–214. doi: 10.1016/j.desal.2007.01.077

Ben-Gal, A., Ityel, E., Dudley, L., Cohen, S., Yermiyahu, U., Presnov, E., et al. (2008). Effect of irrigation water salinity on transpiration and on leaching requirements: a case study for bell peppers. *Agric. Water Manage.* 95, 587–597. doi: 10.1016/j.agwat.2007.12.008

Bernstein, N., Ioffe, M., Luria, G., Bruner, M., Nishri, Y., Philosoph-Hadad, S., et al. (2011). Effects of K and N nutrition on function and production of *Ranunculus asiaticus*. *Pedosphere* 21, 288–301. doi: 10.1016/S1002-0160(11)60129-X

Bick, A., Gillerman, L., Manor, Y., and Oron, G. (2012). Economic assessment of an integrated membrane system for secondary effluent polishing for unrestricted reuse. *Water* 4, 219–236. doi: 10.3390/w4010219

Bick, A., Yang, F., Shandalov, S., Raveh, A., and Oron, G. (2011). Multidimension scaling as an exploratory tool in the analysis of an immersed membrane bioreactor. *Membr. Water Treat.* 2, 105–119. doi: 10.12989/mwt.2011.2.2.105

Birnhack, L., Shlesinger, N., and Lahav, O. (2010). A cost effective method for improving the quality of inland desalinated brackish water destined for agricultural irrigation. *Desalination* 262, 152–160. doi: 10.1016/j.desal.2010.05.061

Bozoki, S., Fulop, J., and Koczkodaj, W. W. (2011). An LP-based inconsistency monitoring of pairwise comparison matrices. *Math. Comput. Model.* 54, 789–793. doi: 10.1016/j.mcm.2011.03.026

Bravata, D. M., Shojania, K. G., Olkin, I., and Raveh, A. (2008). CoPlot: a tool for visualizing multivariate data in medicine. *Stat. Med.* 27, 2234–2247. doi: 10.1002/sim.3078

Bui, E. N. (2013). Soil salinity: a neglected factor in plant ecology and biogeography. *J. Arid Environ.* 92, 14–25. doi: 10.1016/j.jaridenv.2012.12.014

Bustan, A., Avni, A., Yermiyahu, U., Ben-Gal, A., Riov, J., Erel, R., et al. (2013). Interactions between fruit load and macroelement concentrations in fertigated olive (Olea europaea L.). trees under arid saline conditions. *Sci. Hortic.* 152, 44–55. doi: 10.1016/j.scienta.2013.01.013

Caputo, A. C., Pelagagge, P. M., and Salini, P. (2013). AHP-based methodology for selecting safety devices of industrial machinery. *Saf. Sci.* 53, 202–218. doi: 10.1016/j.ssci.2012.10.006

Cay, T., and Uyan, M. (2013). Evaluation of reallocation criteria in land consolidation studies using the Analytic Hierarchy Process (AHP). *Land Use Policy* 30, 541–548. doi: 10.1016/j.landusepol.2012.04.023

Chakrabortty, S., Roy, M., and Pal, P. (2013). Removal of fluoride from contaminated groundwater by cross flow nanofiltration: transport modelling and economic evaluation. *Desalination* 313, 115–124. doi: 10.1016/j.desal.2012.12.021

Connor, J. D., Schwabe, K., King, D., and Knapp, K. (2012). Irrigated agriculture and climate change: the influence of water supply variability and salinity on adaptation. *Ecol. Econ.* 77, 149–157. doi: 10.1016/j.ecolecon.2012.02.021

Edelstein, M., Plaut, Z., Dudai, N., and Ben-Hur, M. (2009). Vetiver (*Vetiveria zizanioides*) responses to fertilization and salinity under irrigation conditions. *J. Environ. Manage.* 91, 215–221. doi: 10.1016/j.jenvman.2009.08.006

Garcia, N., Moreno, J., Cartmell, E., Rodriguez-Roda, I., and Judd, S. (2013). The cost and performance of an MF-RO/NF plant for trace metal removal. *Desalination* 309, 181–186. doi: 10.1016/j.desal.2012.10.017

Ghermandi, A., and Mesalem, R. (2009). The advantages of NF desalination of brackish water for sustainable irrigation: the case of the Arava Valley in Israel. *Desalination Water Treat.* 10, 101–107. doi: 10.5004/dwt.2009.824

Gomez-Silvan, C., Arevalo, J., Perez, J., Gonzalez-Lopez, J., and Rodelas, B. (2013). Linking hydrolytic activities to variables influencing a submerged membrane bioreactor (MBR) treating urban wastewater under real operating conditions. *Water Res.* 47, 66–78. doi: 10.1016/j.watres.2012.09.032

Gur-Reznik, S., Koren-Menashe, I., Heller-Grossman, L., Rufel, O., and Dosoretz, C. G. (2011). Influence of seasonal and operating conditions on the rejection of pharmaceutical active compounds by RO and NF membranes. *Desalination* 277, 250–256. doi: 10.1016/j.desal.2011.04.029

Gursoy, B. B., Mason, O., and Sergeev, S. (2013). The analytic hierarchy process, max algebra and multi-objective optimisation. *Linear Algebra Appl.* 438, 2911–2928. doi: 10.1016/j.laa.2012.11.020

Gutman, J., Fox, S., and Gilron, J. (2012). Interactions between biofilms and NF/RO flux and their implications for control-a review of recent developments. *J. Membr. Sci.* 421–422, 1–7. doi: 10.1016/j.memsci.2012.06.032

Guttman, L. (1968). A general non-metric technique for finding the smallest space for a configuration of points. *Psychometrica* 33, 479–506.

Hilal, N., Al-Zoubi, H., Darwish, N. A., and Mohammad, A. W. (2005). Characterisation of nanofiltration membranes using atomic force microscopy. *Desalination* 177, 187–199. doi: 10.1016/j.desal.2004.12.008

Hillel, D., Belhassen, Y., and Shani, A. (2013). What makes a gastronomic destination attractive? Evidence from the Israeli Negev. *Tourism Manage.* 36, 200–209. doi: 10.1016/j.tourman.2012.12.006

Johannessen, A., Rosemarin, A., Thomalla, F., Swartling, A. G., Stenström, T. A., Vulturius, G., et al. (2014). Strategies for building resilience to hazards in water, sanitation and hygiene (WASH) systems: the role of public private partnerships. *Int. J. Disaster Risk Reduct.* 10, 102–115. doi: 10.1016/j.ijdrr.2014.07.002

Karabelas, A. J., Koutsou, C. P., Gragopoulos, J., Isaias, N. P., and Al Rammah, A. S. (2012). A novel system for continuous monitoring of salt rejection characteristics of individual membrane elements in desalination plants. *Sep. Pur. Technol.* 88, 29–38. doi: 10.1016/j.seppur.2011.12.002

Kayastha, P., Dhital, M. R., and De Smedt, F. (2013). Application of the analytical hierarchy process (AHP). for landslide susceptibility mapping: a case study from the Tinau watershed, west Nepal. *Comput. Geosci.* 52, 398–408. doi: 10.1016/j.cageo.2012.11.003

Lespinats, S., Fertil, B., Villemain, P., and Herault, J. (2009). RankVisu: mapping from the neighborhood network. *Neurocomputing* 72, 2964–2978. doi: 10.1016/j.neucom.2009.04.008

Levy, D., and Tai, G. C. C. (2013). Differential response of potatoes (Solanum tuberosum L.) to salinity in an arid environment and field performance of the seed tubers grown with fresh water in the following season. *Agric. Water Manage.* 116, 122–127. doi: 10.1016/j.agwat.2012.06.022

Lew, B., Cochva, M., and Lahav, O. (2009). Potential effects of desalinated water quality on the operation stability of wastewater treatment plants. *Sci. Total Environ.* 407, 2404–2410. doi: 10.1016/j.scitotenv.2008.12.023

Lipshitz, G., Raveh, A. (1994). Applications of the CoPlot method in the study of socioeconomic differences among cities: a basis for a differential development policy. *Urban Stud.* 31, 123–135.

Liu, Y., Li, X., Xing, Z., Zhao, X., and Pan, Y. (2013). Responses of soil microbial biomass and community composition to biological soil crusts in the revegetated areas of the Tengger Desert, *Appl. Soil Ecol.* 65, 52–59. doi: 10.1016/j.apsoil.2013.01.005

Miller, G. W. (2006). Integrated concepts in water reuse: managing global water needs. *Desalination* 187, 65–75. doi: 10.1016/j.desal.2005.04.068

Mrayed, S. M., Sanciolo, P., Zou, I., and Leslie, G. (2011). An alternative membrane treatment process to produce low-salt and high-nutrient recycled water suitable for irrigation purposes. *Desalination* 274, 144–149. doi: 10.1016/j.desal.2011.02.003

Nederlof, M. M. and van Paassen, J. A. M., Jong, R. (2005). Nanofiltration concentrate disposal: experiences in the Netherlands. *Desalination* 178, 303–312. doi: 10.1016/j.desal.2004.11.041

Oren, S., Birnhack, L., Lehmann, O., and Lahav, O. (2012). A different approach for brackish-water desalination, comprising acidification of the feed-water and CO_2(aq). reuse for alkalinity, Ca^{2+} and Mg^{2+} supply in the post treatment stage. *Sep. Purif. Technol.* 89, 252–260. doi: 10.1016/j.seppur.2012.01.027

Oron, G., Gillerman, L., Bick, A., Manor, Y., Buriakovsky, N., and Hagin, J. (2008). Membrane technology for sustainable treated wastewater reuse: agricultural, environmental and hydrological considerations. *Water Sci. Technol.* 57, 1383–1388. doi: 10.2166/wst.2008.243

Penate, B., and García-Rodríguez, L. (2011). Reverse osmosis hybrid membrane inter-stage design: a comparative performance assessment. *Desalination* 281, 354–363. doi: 10.1016/j.desal.2011.08.010

Qadir, M., Sharma, B. R., Bruggeman, A., Choukr-Allah, R., Karajeh, F. (2007). Non-conventional water resources and opportunities for water augmentation to achieve food security in water scarce countries. *Agric. Water Manage.* 87, 2–22. doi: 10.1016/j.agwat.2006.03.018

Rasouli, F., Pouya, A. K., and Karimian, N. (2013). Wheat yield and physico-chemical properties of a sodic soil from semi-arid area of Iran as affected by applied gypsum. *Geoderma* 193–194, 246–255. doi: 10.1016/j.geoderma.2012.10.001

Raveh, A. (2000). CoPlot: a Graphic display method for geometrical representations of MCDM. *Eur. J. Oper. Res.* 125, 670–678. doi: 10.1016/S0377-2217(99)00276-3

Riera, F. A., and Suarez, A., Muro, C. (2013). Nanofiltration of UHT flash cooler condensates from a dairy factory: Characterisation and water reuse potential. *Desalination* 309, 52–63. doi: 10.1016/j.desal.2012.09.016

Saaty, T. L. (1980). *The Analytic Hierarchy Process.* New York, NY: McGraw-Hill.

Saaty, T. L. (2003). Decision-making with the AHP: why is the principal eigenvector necessary. *Eur. J. Oper. Res.* 145, 85–91. doi: 10.1016/S0377-2217(02)00227-8

Schilli, C., Lischeid, G., and Rinklebe, J. (2010). Which processes prevail?: analyzing long-term soil solution monitoring data using nonlinear statistics. *Geoderma* 158, 412–420. doi: 10.1016/j.geoderma.2010.06.014

Sotto, A., Arsuaga, J. M., and Van der Bruggen, B. (2013). Sorption of phenolic compounds on NF/RO membrane surfaces: Influence on membrane performance. *Desalination* 309, 64–73. doi: 10.1016/j.desal.2012.09.023

Tzfati, E., Sein, M., Rubinov, A., Raveh, A., and Bick, A. (2011). Pre-treatment of wastewater: optimal coagulant selection using partial order scaling analysis (POSA). *J. Hazard Mater.* 190, 51–59. doi: 10.1016/j.jhazmat.2011.02.023

Yermiyahu, U., Tal, A., Ben-Gal, A., Bar-Tal, J., and Tarchitzky, O. (2007). Lahav, rethinking desalinated water quality and agriculture. *Science* 318, 920–921. doi: 10.1126/science.1146339

Verhuelsdonk, M., Attenborough, T., Lex, O., and Altmann, T. (2010). Design and optimization of seawater reverse osmosis desalination plants using special simulation software. *Desalination* 250, 729–733. doi: 10.1016/j.desal.2008.11.031

Villarreal-Guerrero, F., Kacira, M., Fitz-Rodríguez, E., Kubota, C., Giacomelli, G. A., Linker, R., et al. (2012). Comparison of three evapotranspiration models for a greenhouse cooling strategy with natural ventilation and variable high pressure fogging. *Sci. Hortic.* 134, 210–221. doi: 10.1016/j.scienta.2011.10.016

Vince, F., Marechal, F., Aoustin, E., and Breant, P. (2008). Multi-objective optimization of RO desalination plants. *Desalination* 222, 96–118. doi: 10.1016/j.desal.2007.02.064

Wei, J., Qiu, C., Wang, Y. N., Wang, R., and Tang, C. Y. (2013). Comparison of NF-like and RO-like thin film composite osmotically-driven membranes-Implications for membrane selection and process optimization, J. *Membrane Sci.* 427, 460–471. doi: 10.1016/j.memsci.2012.08.053

Yang, X., Yan, L., and Zeng, L. (2013). How to handle uncertainties in AHP: The cloud delphi hierarchical analysis. *Inform. Sci.* 222, 384–404. doi: 10.1016/j.ins.2012.08.019

Zhang, R., Zhang, X., Yang, J., and Yuan, H. (2013). Wetland ecosystem stability evaluation by using analytical hierarchy process (AHP) approach in Yinchuan Plain, China. *Math. Comput. Model.* 57, 366–374. doi: 10.1016/j.mcm.2012.06.014

Zhao, S., Zou, L., and Mulcahy, D. (2012). Brackish water desalination by a hybrid forward osmosis–nanofiltration system using divalent draw solute. *Desalination* 284, 175–181. doi: 10.1016/j.desal.2011.08.053

Zheng, Y., Yu, S., Shuai, S., Zhou, Q., Cheng, Q., Liu, M., et al. (2013). Color removal and COD reduction of biologically treated textile effluent through

Brackish groundwater membrane system design for sustainable irrigation: optimal configuration selection...

191

submerged filtration using hollow fiber nanofiltration membrane. *Desalination* 314, 89–95. doi: 10.1016/j.desal.2013.01.004

Zhou, F., Wang, C., and Wei, J. (2013). Separation of acetic acid from monosaccharides by NF and RO membranes: performance comparison. *J. Membr. Sci.* 429, 243–251. doi: 10.1016/j.memsci.2012.11.043

Conflict of Interest Statement: The authors declare that the research was conducted in the absence of any commercial or financial relationships that could be construed as a potential conflict of interest.

15

Microbial regulation of nitrogen dynamics along the hillslope of a natural forest

Kazuo Isobe[1]*, Nobuhito Ohte[1], Tomoki Oda[1], Sho Murabayashi[1], Wei Wei[1], Keishi Senoo[1], Naoko Tokuchi[2] and Ryunosuke Tateno[2]

[1] *Graduate School of Agricultural and Life Sciences, The University of Tokyo, Tokyo, Japan*
[2] *Field Science Education and Research Center, Kyoto University, Kyoto, Japan*

Edited by:
Wilfred Otten, Abertay University, UK

Reviewed by:
Peter S. Hooda, Kingston University London, UK
Miguel Ángel Sánchez-Monedero, Spanish National Research Council, Spain

*Correspondence:
Kazuo Isobe, Department of Applied Biological Chemistry, Graduate School of Agricultural and Life Sciences, The University of Tokyo, 1-1-1 Yayoi, Bunkyo-ku, Tokyo 113-8657, Japan
e-mail: akisobe@ mail.ecc.u-tokyo.ac.jp

Topography affects the soil physicochemistry, soil N dynamics, and plant distribution and growth in forests. In Japan, many forests are found in mountainous areas and these traits are often highly variable along steep slopes. In this study, we investigated how the microbial population dynamics reflected the bioavailable N dynamics with the physicochemical gradient along the slope in soils of a natural forest in Japan. We measured the gross rates of NH_4^+ production, nitrification, and NH_4^+/ NO_3^- immobilization using the N isotope dilution method to analyze the N dynamics in the soils. We also determined the abundance of the bacterial 16S rRNA gene and bacterial and archaeal ammonia monooxygenase gene (*amoA*) using qPCR to assess the populations of total bacteria and nitrifiers. We found that gross rates of NH_4^+ production and nitrification were higher in the lower part of the slope, they were positively correlated with the abundance of the bacterial 16S rRNA gene and archaeal *amoA*, respectively; and the availability of N, particularly NO_3^-, for plants was higher in the lower part of the slope because of the higher microbial nitrification activity and low microbial NO_3^- immobilization activity. In addition, path analysis indicated that gross rates of NH_4^+ production and nitrification were regulated mainly by the substrate (dissolved organic N and NH_4^+) concentrations and population sizes of total bacteria and nitrifiers, respectively, and their population sizes were strongly affected by the soil physicochemistry such as pH and water content. Our results suggested that the soil physicochemical gradient along the slope caused the spatial gradient of gross rates of NH_4^+ production and nitrification by altering the communities of ammonifiers and nitrifiers in the forest slope, which also affected plant distribution and growth via the supply of bioavailable N to plants.

Keywords: ammonia-oxidizing archaea, ammonification, hillslope, nitrification, nitrogen dynamics

INTRODUCTION

In forests, the topography affects the soil environment such as the soil depth, soil formation processes, moisture content, and nutrient status, as well as the local climate (Hook and Burke, 2000; Tromp-van Meerveld and McDonnell, 2006; Penna et al., 2009). It may enhance the plant diversity in a forest, because spatial heterogeneity of environmental conditions allows plants with various physiologies to coexist. In Japan, many forests are found in mountainous areas; thus, the soil environments and plant diversity are highly variable along a steep slope (Enoki, 2003; Tateno and Takeda, 2003; Koyama et al., 2013). In general, a hydrological gradient is observed along a slope, from dry soil at the top part to wet soil at the bottom part. The soil chemistry such as pH and nutrient availability as well as the light intensity also exhibit gradients from the upper to lower parts. These gradients could affect the transient vegetation dynamics directly or indirectly (i.e., growth and distribution). The spatial differentiation of plants (e.g., plants adapted only to upper or lower slope conditions, or ubiquitously distributed plants) has often been observed and has been explained according to the availabilities of water,

nutrients, or light for plants (Hook and Burke, 2000; Tromp-van Meerveld and McDonnell, 2006; Engelbrecht et al., 2007). Nitrogen is considered to be a limiting nutrient for plant growth in most temperate forests (Vitousek and Howarth, 1991). The N availability for plants can be strongly affected by the hydrological factors on a slope (Hill and Kemp, 1999; Band et al., 2001). Therefore, many studies have focused on the patterns of soil N dynamics and NO_3^- production as well as their controlling factors in forest slopes (Hirobe et al., 1998, 2003; Tokuchi et al., 2000; Nishina et al., 2009a,b; Koyama et al., 2013).

Microorganisms are responsible for the production of NH_4^+ and NO_3^-, which are the main forms of bioavailable N to plants, via the degradation of organic N and nitrification, respectively (Isobe et al., 2011a; Isobe and Ohte, 2014). Given that microbial activity levels reflect the dynamics of the bioavailable N in soil, we can expect that microorganisms will also have direct or indirect associations with the growth and distribution of plants via the supply of bioavailable N to plants. This association can also be found in the competition for N between microorganisms and plants. This competition can be harder for plants in N-limiting environments, because microorganisms generally have a higher affinity for bioavailable N than plants (Kuzyakov and Xu, 2013). Microbial communities and their activities can be influenced by environmental conditions. Therefore, an environmental gradient may cause changes in the bioavailable N supply for plants via alterations in the microbial activity levels. In particular, many recent studies have shown that the rates of dissimilatory N transformation, such as nitrification and denitrification, could be explained by the sizes of nitrifier or denitrifier populations (Hawkes et al., 2005; Philippot et al., 2009; Isobe et al., 2012). Therefore, although few previous studies on forest slopes have focused on microbial population dynamics, we can expect that soil environmental gradients affect the transient vegetation dynamics on forest slopes by altering the population size of microorganisms responsible for the supply of bioavailable N to plants. With this, we can obtain a more mechanistic understanding of how soil environmental gradients affect the dynamics of bioavailable N and the plant diversity on forest slopes by considering microbial population dynamics and their associated activities.

The research site in the present study, Ashiu Forest Research Station, is a broadleaved deciduous forest located in Kyoto, Japan. Previous studies have found the gradient of soil physicochemistry and transient vegetation dynamics on a hill slope in this forest (Tateno and Takeda, 2003, 2010; Tateno et al., 2004, 2005). These studies showed that the soil properties, i.e., the water content, pH, and N content, increased from the upper part to the lower part of the slope. The dominant plant species also differed between the upper and lower parts of the slope. *Fagus crenata*, a dominant and ubiquitously distributed plant on the slope, exhibited different growth rates and N utilization patterns. In the lower part of the slope, *F. crenata* grew faster, and its biomass distribution was more likely to be weighted to the aboveground sections; moreover, it utilized more NO_3^- than the upper slope. Previous studies did not analyze the N dynamics on the basis of gross rates and microbial populations responsible for the N dynamics on the forest slope. However, we hypothesized that the population dynamics of ammonifiers and nitrifiers in the soil environment along the slope affects the production and consumption of NH_4^+ and NO_3^- which can influence plant distribution or growth rate and the N utilization pattern of *F. crenata*.

In the present study, we investigated how the microbial population dynamics reflected the bioavailable N dynamics with the physicochemical gradient along a slope in the Ashiu forest. We specifically analyzed how the soil environmental gradient affected the NH_4^+ and NO_3^- production rates by altering populations of ammonifiers and nitrifiers. We also addressed how these relationships affected N uptake and growth of plants. We hypothesized that we could consider almost all bacteria to be ammonifiers because NH_4^+ production can occur via the assimilation of small organic N compounds such as amino acids, amino sugars, and nucleotides which all bacteria can be involved in (Schimel and Bennett, 2004; Myrold and Bottomley, 2008; Bottomley et al., 2012; Isobe and Ohte, 2014). We also analyzed ammonia-oxidizing bacteria and archaea as nitrifiers, which are responsible for the rate-limiting step of nitrification, ammonia oxidation (Isobe et al., 2011a).

MATERIALS AND METHODS

STUDY SITE

The study was conducted in a cool-temperate broadleaved deciduous forest in the Kyoto University Ashiu Forest Research Station, Kyoto Prefecture, Japan (35°18′N, 135°43′E). The forest is located in a mountainous area at elevations of 680–720 masl. In this area, forests are dominated by broadleaved deciduous tree species, including *F. crenata* Blume and *Quercus crispula* Blume, and they have remained intact since 1898 or earlier. The mean annual temperature and precipitation over a 56-year period at a weather station (640 masl) located approximately 1 km from the study site were approximately 10°C and 2495 mm, respectively. More detailed information about the site was reported by Tateno and Takeda (2003).

SOIL SAMPLING

Soil sampling was conducted in June 2013. A 30–200-m transect (0.6 ha) from the valley bottom to the ridge top on a northwest-facing slope was established in a previous study. We sampled approximately 500 g of soil from the surface 0–10 cm in the mineral layer at 11 points every 20 m from the top to the bottom along the center line of the transect. The soil was sieved through a 2-mm mesh.

SOIL CHEMISTRY ANALYSIS

The soil water content was measured by drying 20 g of soil at 105°C for 24 h in a ventilated oven. The soil pH was measured using a pH meter (Horiba, Kyoto, Japan) after extracting 5 g of soil in 25 mL of water. The NH_4^+ and NO_3^- concentrations in soil samples were determined using the indophenol and denitrifier methods (Isobe et al., 2011b), respectively, after extracting 7 g of soil with 35 mL of 2-M KCl solution. The concentration of dissolved organic nitrogen (DON) in the soil extracted with KCl solution was determined using a TOC/TN analyzer (TOC-V; Shimadzu, Kyoto, Japan).

MEASUREMENT OF THE GROSS AND NET RATES OF NH_4^+ PRODUCTION AND NITRIFICATION

The gross rates of NH_4^+ production and nitrification in soils were determined using the isotope dilution method (Hart et al., 1994). Two subsamples (7 g each, equivalent to about 3.5 g of dry soil) from each soil sample were used for the analysis of NH_4^+ production or nitrification during 24-h incubation. The detailed procedures have been reported previously (Isobe et al., 2011b,c; Urakawa et al., 2014). The concentrations and N isotope ratios of NH_4^+ or NO_3^- in the 2M KCl extracts were determined according to the method of Isobe et al. (2011b). The gross soil NH_4^+ production and nitrification rates were calculated according to the equations of Kirkham and Bartholomew (1954) using the concentrations and N isotope ratios for NH_4^+ and NO_3^-, respectively. The gross NH_4^+ and NO_3^- consumption rates were calculated in the same manner. The gross NH_4^+ immobilization rate was calculated by substituting the gross nitrification rate with the gross NH_4^+ consumption rate. The gross NO_3^- immobilization rate was considered to be the same as the NO_3^- consumption rate because denitrification can be minimized during aerobic incubation. Immobilization includes microbial assimilation and any other consumption process, but we assumed that the immobilization process during the 24-h incubation was the microbial assimilation process because abiotic consumption processes are more likely to occur in a short period after the ^{15}N addition (i.e., <2 h). The percentage of nitrification was calculated by gross NH_4^+ production rate divided by gross nitrification rate which presented the proportion of the produced NH_4^+ that is being converted to NO_3^- via nitrification. The net rates of NH_4^+ production and nitrification were calculated as the concentration changes in NH_4^+ and NO_3^-, respectively, during the 24-h incubation.

QUANTIFICATION OF THE BACTERIAL 16S rRNA GENE AND BACTERIAL AND ARCHAEAL AmoA GENES

The bacterial 16S rRNA gene and bacterial and archaeal AmoA genes (*amoA*) were quantified to estimate the sizes of the populations of ammonifiers and nitrifiers. The bacterial 16S rRNA gene was determined by qPCR using the primers, 357f-520r, and the StepOne real-time PCR system (Applied Biosystems, Tokyo, Japan). The bacterial and archaeal *amoA* genes were also determined by qPCR using the primers amoA1f and amoA2r and primers CrenamoA23f and CrenamoA616r (Nicol et al., 2008), respectively. Each reaction mixture (20 μL) contained the KOD SYBR green PCR master mixture (Toyobo, Tokyo, Japan), 0.2 μM of each primer, 0.5 μg mL^{-1} of bovine serum albumin, and 10 ng of DNA template. To generate standard curves (10^1–10^7 copies per reaction mixture), we used the bacterial 16S rRNA gene fragment of *Pseudomonas stuzeri*, the bacterial *amoA* fragment of *Nitrosospira multiformis* ATCC 25196, and the archaeal *amoA* fragment of the soil clone obtained in a previous study (Isobe et al., 2012). The reactions were performed in the following conditions: initial annealing at 98°C for 2 min, followed by 45 cycles at 98°C for 10 s, 58°C for 30 s for the bacterial 16S rRNA gene or 55°C for 30 s for bacterial/archaeal *amoA*, and 72°C for 30 s. The amplification efficiency of all the genes during standard curve generation was >90% and the standard curves had high correlation coefficients ($R^2 > 0.95$). The amplification of DNA fragments of the correct size was confirmed by dissociation curve analysis and agarose gel electrophoresis.

DATA ANALYSIS

To highlight the differences in the soil properties, N dynamics, and gene abundances between the upper and lower parts of the slope, we divided the data obtained into two categories [data derived from soils sampled at points 0–120 m from the top as the upper part ($N = 5$), and from 120 to 200 m as the lower part ($N = 6$)], which we compared using an unequal variance t-test. Plant distribution was clearly different between upper and lower parts of the slope (Tateno and Takeda, 2003). The relationship between the two variables in all samples was assessed by correlation analysis to identify possible factors that affected the gross rate of N transformation and gene abundances ($N = 11$). An $\alpha < 0.05$ was considered statistically significant for both tests. Path analysis was also used to determine the factors that affected the gross rates and abundances. In the path analysis, an experimentally supported theory was used to formulate a simple model of the causal and noncausal relationships between the variables (**Figure 1A**). Specifically, the gross N transformation rate can be affected by the

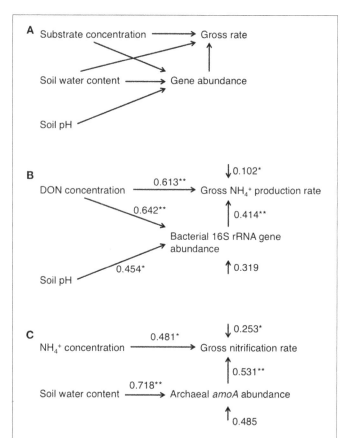

FIGURE 1 | Path diagrams representing the full model (A) and the final models used to describe the patterns observed in the gross NH_4^+ production rate (B) and gross nitrification rate (C). The numbers associated with the arrows between two variables are the partial regression coefficients derived from multiple regressions. The numbers associated with the arrows of single variables represent the unexplained variation ($1-R^2$), which represent the effects of unmeasured variables and measurement error. All pathways are significant in the final model (*$p < 0.05$; **$p < 0.01$).

N-substrate supply, the population size of microorganisms that utilize the substrate for its transformation, and the water content. The population size is also affected by the N-substrate supply, soil pH, and soil water content. The substrate and microorganisms should be the minimum requirements for N transformation. Water mediates all physiological reactions and microbial growth, and pH also has a great impact on growth. In the analysis of the gross NH_4^+ production rate, we used the DON concentration as the substrate supply and the abundance of the bacterial 16S rRNA gene as the population size. In the analysis of the gross nitrification rate, we used the NH_4^+ concentration and the gross NH_4^+ production rate as the substrate supply, and the abundance of the bacterial and archaeal *amoA* as the population size. Reduced models were created by eliminating the paths with the highest probability values in a stepwise manner until all paths between two variables were significant ($p < 0.05$). Because of the small number of samples and variables, we performed these processes manually and only one reduced model maintained all paths between two variables with significance (**Figures 1B,C**). To account for differences in the magnitudes of data, the measurement scales were changed by standardizing all data to a similar numeric magnitude before calculating the variance–covariance matrices. All statistical analyses were performed using the R software (version 3.1.1, R Development Core Team, 2007) and path analysis was performed using the structural equation modeling function in the sem package in R.

RESULTS

Table 1 shows the soil properties (water content, pH, and DON/NH_4^+/NO_3^- concentrations), N transformation rates (gross rates of NH_4^+ production/consumption/immobilization, nitrification, and NO_3^- immobilization, and net rates of NH_4^+ production and nitrification), and gene abundances (abundances of the bacterial 16S rRNA gene and archaeal *amoA*) in the upper and lower parts of the slope. Bacterial *amoA* was detected only in the sampling points at the lowest positions (7.1×10^6 copy/g-dry soil). With respect to the soil properties, the water content, soil pH, and NH_4^+ concentration did not differ significantly between the upper and lower parts, whereas the DON and NO_3^- concentrations were higher in the lower part. The net NH_4^+ production rate was higher in the upper part, whereas the net nitrification rate was higher in the lower part. The gross rates of N transformation were higher in the lower part, except for NH_4^+ and NO_3^- immobilization. The percentage of nitrification was also higher in the lower part. The abundances of the bacterial 16S rRNA gene and archaeal *amoA* were higher in the lower part.

To estimate the factors that affected the gross rates of NH_4^+ production and nitrification, we performed correlation analyses of the gross rates and soil physicochemical properties or gene abundances and the results are shown in **Table 2**. In particular, the gross rate of NH_4^+ production correlated with the abundance of the bacterial 16S rRNA gene (**Figure 2**). The gross rate of nitrification correlated with the abundance of archaeal *amoA* (**Figure 2**).

The correlation analysis (**Table 2**) showed that the soil water content, DON concentration, and the abundance of the bacterial 16S rRNA gene were possible factors that affected the gross NH_4^+ production rate, whereas the gross NH_4^+ production rate, NH_4^+ concentration, and the abundance of archaeal *amoA* were possible factors that affected the gross nitrification rate. These variables do not always have direct effects on the gross rates of NH_4^+ production or nitrification; thus, we performed path analysis based on the assumptions, described in **Figure 1A**, to consider the indirect influences on the gross rates. The results of the analysis (**Figures 1B,C**) suggest that the gross NH_4^+ production rate was affected by both the DON concentration and the abundance of the bacterial 16S rRNA gene, whereas the abundance of the bacterial 16S rRNA gene was affected by both the DON concentration and soil pH. The gross nitrification rate was affected by both the NH_4^+ concentration and the abundance of archaeal *amoA*, whereas the abundance of archaeal *amoA* was affected by the soil water content. We did not use the gross NH_4^+ production rate as the resource parameter (**Figure 1A**) instead of or in addition to the NH_4^+ concentration in the analysis of the gross nitrification rate, because there was high multicollinearity between the gross NH_4^+ production rate and the abundance of archaeal *amoA*, which interfered with the statistical evaluation of the influence of the gross NH_4^+ production rate.

DISCUSSION

SOIL CHEMISTRY AND N DYNAMICS ON THE SLOPE

The soil water content and pH did not differ significantly between the upper and lower parts of the slope because of the large standard error, but they increased gradually down the slope from the top to the bottom (see "vs. sampling position" in **Table 2**), which was consistent with the results of a previous study conducted in 1997, as reported previously (Tateno and Takeda, 2003). The net rates of NH_4^+ production and nitrification suggest that

Table 1 | Soil properties, N transformation rates, and gene abundances in upper and lower parts on the forest slope.

		Upper part	Lower part	*p*-Value
Water content	(%)	33.8 (2.56)	41.4 (4.16)	0.08
Soil pH		4.1 (0.12)	4.5 (0.21)	0.06
Soil DON concentration	(mg-N/kg-dry soil)	**28.7 (5.45)**	**58.5 (17.2)**	0.05
Soil NH_4^+ concentration	(mg-N/kg-dry soil)	8.9 (2.1)	13.2 (1.8)	0.08
Soil NO_3^- concentration	(mg-N/kg-dry soil)	**13.2 (1.8)**	**21.5 (5.3)**	0.05
Net NH_4^+ production rate	(mg-N/kg-dry soil/day)	**2.06 (1.25)**	**0.13 (0.51)**	<0.01
Net nitrification rate	(mg-N/kg-dry soil/day)	**0.08 (0.26)**	**6.31 (1.62)**	<0.01
Gross NH_4^+ production rate	(mg-N/kg-dry soil/day)	**2.71 (0.46)**	**12.36 (3.09)**	<0.01
Gross NH_4^+ consumption rate	(mg-N/kg-dry soil/day)	**0.36 (1.22)**	**12.23 (2.78)**	<0.01
Gross NH_4^+ immobilization rate	(mg-N/kg-dry soil/day)	0.42 (1.46)	5.38 (2.39)	0.06
Gross nitrification rate	(mg-N/kg-dry soil/day)	**0.48 (0.16)**	**6.85 (1.15)**	<0.01
Gross NO_3^- immobilization rate	(mg-N/kg-dry soil/day)	0.6 (0.42)	0.53 (0.63)	0.46
Percent nitrification	(%)	**17.2 (4.2)**	**63.3 (9.7)**	<0.01
Bacterial 16S rRNA gene abundance	(copy/g-dry soil)	**4.3×10^{10} (1.8×10^9)**	**5.6×10^{10} (2.8×10^9)**	<0.01
Archaeal *amoA* abundance	(copy/g-dry soil)	**5.1×10^7 (1.1×10^7)**	**2.0×10^8 (2.7×10^7)**	<0.01

N = 6 for upper part and N = 5 for lower part. Average and standard error in parenthesis. Bold values denote the significant difference between upper and lower parts (p < 0.05).

Table 2 | Correlation with gross rates of NH_4^+ production and nitrification, abundance of bacterial 16S rRNA gene and archaeal *amoA*, or sampling position.

	vs. Gross NH_4^+ production rate		vs. Gross nitrification rate		vs. Bacterial 16S rRNA gene abundance		vs. Archaeal *amoA* abundance		vs. Sampling position	
	R	*p*-Value	R	*p*-Value	R	*p*-Value	R	*p*-Value	R	*p*-Value
Water content	**0.82**	<0.01	0.66	0.055	**0.70**	0.024	**0.66**	0.040	**0.69**	0.028
Soil pH	0.47	0.174	0.62	0.073	0.52	0.121	0.46	0.181	**0.76**	0.011
Soil DON concentration	**0.90**	<0.01	0.57	0.110	**0.69**	0.027	**0.66**	0.036	0.56	0.096
Soil NH_4^+ concentration	0.51	0.128	**0.75**	0.020	0.32	0.372	0.41	0.238	0.41	0.240
Soil NO_3^- concentration	**0.98**	<0.01	**0.88**	<0.01	**0.81**	<0.01	**0.82**	<0.01	**0.80**	<0.01
Gross NH_4^+ production rate	–	–	**0.83**	<0.01	**0.84**	<0.01	**0.83**	<0.01	**0.80**	<0.01
Gross NH_4^+ consumption rate	**0.97**	<0.01	**0.82**	<0.01	**0.83**	<0.01	**0.81**	0.008	**0.87**	<0.01
Gross NH_4^+ immobilization rate	**0.82**	<0.01	0.43	0.243	**0.71**	0.032	0.62	0.076	0.64	0.061
Gross nitrification rate	**0.83**	<0.01	–	–	**0.70**	0.034	**0.77**	0.014	**0.84**	<0.01
Gross NO_3^- immobilization rate	−0.05	0.891	−0.03	0.946	−0.08	0.845	0.03	0.935	−0.17	0.662
Percent nitrification	0.29	0.457	**0.71**	0.033	0.34	0.376	0.53	0.145	0.52	0.156
Bacterial 16S rRNA gene abundance	**0.84**	<0.01	**0.70**	0.034	–	–	**0.92**	<0.01	**0.84**	<0.01
Archaeal *amoA* abundance	**0.83**	<0.01	**0.77**	0.014	**0.92**	<0.01	–	–	**0.84**	<0.01

N = 11. The strength of the linear relationship between two variables is described by the Pearson Correlation Coefficient, R, and negative numbers correspond to negative correlations. Bold values denote the significant correlation (p < 0.05)."vs. Sampling position" indicate whether gradiation of each variable along the slope is significant or not.

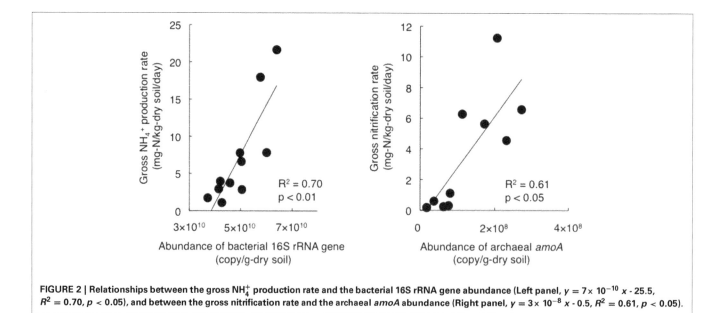

FIGURE 2 | Relationships between the gross NH_4^+ production rate and the bacterial 16S rRNA gene abundance (Left panel, $y = 7 \times 10^{-10} x - 25.5$, $R^2 = 0.70$, $p < 0.05$), and between the gross nitrification rate and the archaeal *amoA* abundance (Right panel, $y = 3 \times 10^{-8} x - 0.5$, $R^2 = 0.61$, $p < 0.05$).

NH_4^+ tended to accumulate in the upper part of the slope whereas NO_3^- tended to accumulate in the lower part (**Table 1**). This indicates that the available N for plants in the surface 0–10 cm mineral soil was likely to be NH_4^+ in the upper part and NO_3^- in the lower part. NH_4^+ accumulation was not observed in the lower part, but the gross NH_4^+ production rates in the lower part were much larger (**Table 1**), thereby suggesting that the concurrent production and consumption of larger amounts of NH_4^+ occurred in the lower part. Nitrification and NH_4^+ immobilization were the dominant pathways of NH_4^+ consumption, with similar rates (**Table 1**). The higher gross nitrification rates in the

lower part showed that larger amounts of NO_3^- were produced, and the higher percentage of nitrification in the lower part suggests that more of the NH_4^+ produced via DON degradation was oxidized to NO_3^- via nitrification. In addition, the gross rates of NH_4^+ immobilization and nitrification were much higher than the gross NO_3^- immobilization rate in the lower part. This suggested that soil microorganisms were likely to utilize NH_4^+ rather than NO_3^- for N assimilation, and a large amount of the NO_3^- produced via nitrification was not utilized. These results suggest that the N dynamics related to NH_4^+ and NO_3^- were more prominent in the lower part because of the greater microbial NH_4^+

production and consumption activities and the N availability, particularly the NO_3^- availability, for plants was greater in the lower part because of the higher microbial nitrification activity but lower NO_3^- immobilization activity.

MICROBIAL POPULATION RESPONSIBLE FOR N TRANSFORMATION

Similar to the gross rates of NH_4^+ production and nitrification, the populations of total bacteria and NH_4^+-oxidizing archaea were higher in the lower part (**Table 1**). We hypothesized that almost all bacteria were involved in NH_4^+ production, and the positive correlation between the gross NH_4^+ production rates and abundances of the bacterial 16S rRNA gene supported our hypothesis. In general, N mineralization via DON degradation is measured in the form of NH_4^+ production as a single process, but it is actually the sum of multiple distinct physiological processes (Isobe and Ohte, 2014). Therefore, it is difficult to identify the microorganisms responsible for NH_4^+ production solely based on genetic information such as amoA for ammonia oxidizers or nirK/S for denitrifiers (Hawkes et al., 2005; Philippot et al., 2009; Isobe et al., 2012). Alternatively, NH_4^+ can potentially be produced via the direct enzymatic cleavage of a free amino group, amine-N, or amide-N (R-NH_2) inside cells that take up R-NH_2, and deaminase and deamidase enzymes can be produced by most bacteria (Schimel and Bennett, 2004; Myrold and Bottomley, 2008; Bottomley et al., 2012; Isobe and Ohte, 2014); thus, we hypothesized that almost all bacteria are involved in NH_4^+ production. There are few studies on microbial ecology associating with NH_4^+ production; however, our study presents an alternative method for analyzing the mechanisms that regulate NH_4^+ production rate in soils. The positive correlation between the gross nitrification rates and the abundance of archaeal amoA (**Figure 1**) suggests that NH_3-oxidizing archaea were mainly responsible for NH_3 oxidation. Because NH_3-oxidizing bacteria were detected only at the lowest point, their contribution to NH_3 oxidation on the slope could be limited or site-specific. Previous studies have shown that gross or net nitrification rates correlate with the abundance of bacterial amoA in grassland or agricultural fields (Hawkes et al., 2005; Di et al., 2009; Jia and Conrad, 2009), but with the abundance of archaeal amoA in forests (Isobe et al., 2012). This supports the results of the present study, thereby suggesting that NH_3-oxidizing archaea are likely to predominate in forests.

REGULATION OF N DYNAMICS ON THE SLOPE

The results of the path analysis of the regulation of the gross rates of NH_4^+ production and nitrification appeared to be reasonable because a substrate supply and microorganisms that utilize the substrate for transformation are the primary requirements for N transformation. The soil pH was suggested to be a major factor that affected the total bacterial population size. Previous study has also shown that the bacterial population size can be strongly affected by the soil pH (Rousk et al., 2010). The soil water content was suggested to be a strong factor that affected the population size of NH_4^+-oxidizing archaea. Recently, Bustamante et al. (2012) also showed that the population size of NH_4^+-oxidizing archaea responded positively to water availability. The gross NH_4^+ production rate was indicated to be a factor that affected the gross nitrification rate, although we did not use it in the analysis.

A positive correlation between the gross rates of NH_4^+ production and nitrification is observed in many forests (Booth et al., 2005; Kuroiwa et al., 2011). Petersen et al. (2012) demonstrated the mutual correlations among the gross NH_4^+ production rate, abundance of bacterial amoA, and the net nitrification rate in various ecosystems in Alaska. Therefore, multicollinearity between the gross NH_4^+ production rate and the abundance of amoA during the analysis of gross nitrification may be observed in many forests. However, the gross NH_4^+ production rate is not always a dominant factor that affects the gross nitrification rate. Hawkes et al. (2005) showed that the gross NH_4^+ production rate was not considered to be a factor that affected either the gross nitrification rate or the abundance of bacterial amoA in a grassland soil, despite a strong correlation between them. We also have found that the gross nitrification rates positively correlated with the abundance of archaeal amoA in subtropical forest soils with different gross NH_4^+ production rates (Isobe, 2011). In the present study, the lack of statistical evaluation of the effect of the gross NH_4^+ production rate on the gross nitrification rate and the abundance of archaeal amoA was attributable to the low number of samples examined. Microbial NH_4^+ immobilization was one of main N transformation processes on the slope. As demonstrated by the correlation between the gross NH_4^+ immobilization rate and the abundance of the bacterial 16S rRNA gene (**Table 2**), gross NH_4^+ immobilization rate could be affected strongly by the size of the bacterial population because NH_4^+, but not NO_3^-, was the main form of N assimilated by microorganisms in the lower part of the slope. The results of the present study suggest that the gradients of soil environmental properties such as the pH and water content along the forest slope affected N transformation rates by altering the sizes of the microbial populations responsible for the N transformations. **Figure 3** shows a conceptual diagram of the possible regulation of N transformation on the slope.

INTERACTIONS BETWEEN MICROORGANISMS AND PLANTS VIA THE REGULATION OF N DYNAMICS

The N availability for plants was higher in the lower part of the slope. In particular, the larger population size of NH_3-oxidizing archaea was related to the higher NO_3^- availability for plants. In addition, the microbial communities preferred NH_4^+ for N assimilation and supplied free NO_3^- via nitrification. These results indicate that partitioning the bioavailable N can occur between microbial communities and plants. This could facilitated the higher growth and more active utilization of NO_3^- by F. crenata on the lower slope, as shown in a previous study (Tateno et al., 2005). The higher water availability in the lower part, which is one of the general characteristics of forest slopes, also led to increases in the population size of NH_3-oxidizing archaea and acceleration of the diffusion of the NO_3^- produced, which could relieved the plants from water- and N-limiting conditions. However, the lower part could presented severe conditions for plants that selectively uptake NH_4^+ for N assimilation. The increased concentration of DON of which supply originates from the degradation of the plant litter could facilitate increases in the population size of total bacteria as the ammonifiers and the NH_4^+ production rate.

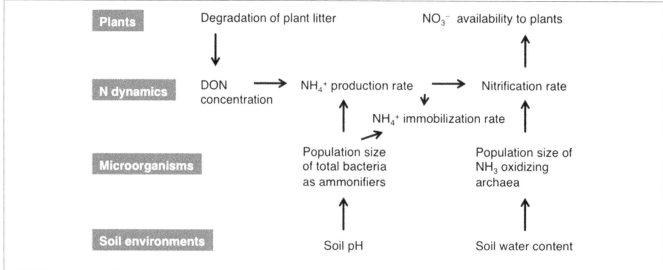

FIGURE 3 | Conceptual diagram showing the regulation of N transformation on the examined slope. The effects of the gradients in the soil environmental properties on the bioavailable N dynamics are mediated via the microbial population dynamics. The relationships among the three categories are influenced by plant litter fall, which affects the plant N availability. The gross NH_4^+ production rate is affected by the concentration of DON and the total bacterial population size as ammonifiers. The total bacterial population size is affected by the soil water content. The gross nitrification rate is affected by the NH_4^+ concentration and the NH_3-oxidizing archaea population size as nitrifiers. The NH_3-oxidizing archaea population size is affected by the soil pH. The NO_3^- produced is available to plants because soil microorganisms prefer NH_4^+ for N assimilation.

In summary, the results of this study suggest that the soil physicochemical gradient along the slope caused the spatial gradient of gross rates of NH_4^+ production and nitrification by altering the communities of ammonifiers and nitrifiers in the forest slope, which also affected plant growth via the supply of bioavailable N to plants. Many studies have investigated microorganism–plant interactions in forests by focusing on direct interactions such as ectomycorrhizal or endomycorrhizal symbiosis (Kohzu et al., 1999; Toljander et al., 2006; Hobbie and Hobbie, 2008). However, our study suggests that microorganism–plant interactions occur indirectly via microbial regulation of supply of the bioavailable N. Let us be cautioned, however, that this study was performed in a single forest slope. Obtaining a definitive picture of the microbial regulation of bioavailable N dynamics in forest soils will require studies in forests with different vegetation types (e.g., planted forests) or different pattern of N dynamics, which should be the objective of future work.

ACKNOWLEDGMENTS

This work was supported by Grants-in-Aid for Scientific Research from the Japanese Society for the Promotion of Science (Nos. 25252026, 25550009, 26292085, and 26712015) and the GRENE/Ecoinformatics project from the Ministry of Education, Culture, Sports, Science and Technology, Japan.

REFERENCES

Band, L. E., Tague, C. L., Groffman, P., and Belt, K. (2001). Forest ecosystem processes at the watershed scale: hydrological and ecological controls of nitrogen export. *Hydrol. Process.* 15, 2013–2028. doi: 10.1002/hyp.253

Booth, M. S., Stark, J. M., and Rastetter, E. (2005). Controls on nitrogen cycling in terrestrial ecosystems: a synthetic analysis of literature data. *Ecol. Monogr.* 75, 139–157. doi: 10.1890/04-0988

Bottomley, P. J., Taylor, A. E., and Myrold, D. D. (2012). A consideration of the relative contributions of different microbial subpopulations

to the soil N cycle. *Front. Microbiol.* 3:373. doi: 10.3389/fmicb.2012.00373

Bustamante, M., Verdejo, V., Zúñiga, C., Espinosa, F., Orlando, J., and Carú, M. (2012). Comparison of water availability effect on ammonia-oxidizing bacteria and archaea in microcosms of a Chilean semiarid soil. *Front. Microbiol.* 3:282. doi: 10.3389/fmicb.2012.00282

Di, H. J., Cameron, K. C., Shen, J. P., Winefield, C. S., O'Callaghan, M., Bowatte, S., et al. (2009). Nitrification driven by bacteria and not archaea in nitrogen-rich grassland soils. *Nat. Geosci.* 2, 621–624. doi: 10.1038/ngeo613

Engelbrecht, B. M. J., Comita, L. S., Condit, R., Kursar, T. A., Tyree, M. T., Turner, B. L., et al. (2007). Drought sensitivity shapes species distribution patterns in tropical forests. *Nature* 447, 80–82. doi: 10.1038/nature05747

Enoki, T. (2003). Microtopography and distribution of canopy trees in a subtropical evergreen broad-leaved forest in the northern part of Okinawa Island, Japan. *Ecol. Res.* 18, 103–113. doi: 10.1046/j.1440-1703.2003.00549.x

Hart, S. C., Stark, J. M., Davidson, E. A., and Firestone, M. K. (1994). "Nitrogen mineralization, immobilization, and nitrification," in *Methods of Soil Analysis. Part 2, Biochemical and Microbiological Properties,* eds R. Weaver, S. Angle, P. Bottomley, D. Bezdicek, S. Smith, A. Tabatabai, and A. Wollum (Madison: Soil Science Society of America), 985–1018.

Hawkes, C. V., Wren, I. F., Herman, D. J., and Firestone, M. K. (2005). Plant invasion alters nitrogen cycling by modifying the soil nitrifying community. *Ecol. Lett.* 8, 976–985. doi: 10.1111/j.1461-0248.2005.00802.x

Hill, A., and Kemp, W. (1999). Nitrogen chemistry of subsurface storm runoff on forested Canadian Shield hillslopes. *Water Resour. Res.* 35, 811–821. doi: 10.1029/1998WR900083

Hirobe, M., Koba, K., and Tokuchi, N. (2003). Dynamics of the internal soil nitrogen cycles under moder and mull forest floor types on a slope in a *Cryptomeria japonica* D. Don plantation. *Ecol. Res.* 18, 53–64. doi: 10.1046/j.1440-1703.2003.00532.x

Hirobe, M., Tokuchi, N., and Iwatsubo, G. (1998). Spatial variability of soil nitrogen transformation patterns along a forest slope in a *Cryptomeria japonica* D. Don plantation. *Eur. J. Soil Biol.* 34, 123–131. doi: 10.1016/S1164-5563(00)88649-5

Hobbie, E. A., and Hobbie, J. E. (2008). Natural abundance of 15N in nitrogen-limited forests and tundra can estimate nitrogen cycling through mycorrhizal fungi: a review. *Ecosystems* 11, 815–830. doi: 10.1007/s10021-008-9159-7

Hook, P., and Burke, I. (2000). Biogeochemistry in a shortgrass landscape: control by topography, soil texture, and microclimate. *Ecology* 81, 2686–2703. doi: 10.1890/0012-9658(2000)081[2686:BIASLC]2.0.CO;2

Isobe, K. (2011). *Nitrogen Flow and Nitrifying Microbial Communities in Subtropical Forest Soils Receiving High N Deposition in CHINA.* Doctoral dissertation of The University of Tokyo, Tokyo.

Isobe, K., Koba, K., Otsuka, S., and Senoo, K. (2011a). Nitrification and nitrifying microbial communities in forest soils. *J. For. Res.* 16, 351–362. doi: 10.1007/s10310-011-0266-5

Isobe, K., Koba, K., Suwa, Y., Ikutani, J., Fang, Y., Yoh, M., et al. (2012). High abundance of ammonia-oxidizing archaea in acidified subtropical forest soils in southern China after long-term N deposition. *FEMS Microbiol. Ecol.* 80, 193–203. doi: 10.1111/j.1574-6941.2011.01294.x

Isobe, K., Koba, K., Ueda, S., Senoo, K., Harayama, S., and Suwa, Y. (2011c). A simple and rapid GC/MS method for the simultaneous determination of gaseous metabolites. *J. Microbiol. Methods* 84, 46–51. doi: 10.1016/j.mimet.2010.10.009

Isobe, K., and Ohte, N. (2014). Ecological perspectives on microbes involved in N-cycling. *Microbes Environ.* 29, 4–16. doi: 10.1264/jsme2.ME13159

Isobe, K., Suwa, Y., Ikutani, J., Kuroiwa, M., Makita, T., Takebayashi, Y., et al. (2011b). Analytical techniques for quantifying 15N/14N of nitrate, nitrite, total dissolved nitrogen and ammonium in environmental samples using a gas chromatograph equipped with a quadrupole mass spectrometer. *Microbes Environ.* 26, 46–53. doi: 10.1264/jsme2.ME10159

Jia, Z., and Conrad, R. (2009). Bacteria rather than Archaea dominate microbial ammonia oxidation in an agricultural soil. *Environ. Microbiol.* 11, 1658–1671. doi: 10.1111/j.1462-2920.2009.01891.x

Kirkham, D., and Bartholomew, W. V. (1954). Equations for following nutrient transformations in soil, utilizing tracer data1. *Soil Sci. Soc. Am. J.* 18, 33. doi: 10.2136/sssaj1954.03615995001800010009x

Kohzu, A., Yoshioka, T., Ando, T., Takahashi, M., Koba, K., and Wada, E. (1999). Natural 13 C and 15 N abundance of field-collected fungi and their ecological implications. *New Phytol.* 144, 323–330. doi: 10.1046/j.1469-8137.1999.00508.x

Koyama, L., Hirobe, M., Koba, K., and Tokuchi, N. (2013). Nitrate-use traits of understory plants as potential regulators of vegetation distribution on a slope in a Japanese cedar plantation. *Plant Soil* 362, 119–134. doi: 10.1007/s11104-012-1257-9

Kuroiwa, M., Koba, K., Isobe, K., Tateno, R., Nakanishi, A., Inagaki, Y., et al. (2011). Gross nitrification rates in four Japanese forest soils: heterotrophic versus autotrophic and the regulation factors for the nitrification. *J. For. Res.* 16, 363–373. doi: 10.1007/s10310-011-0287-0

Kuzyakov, Y., and Xu, X. (2013). Competition between roots and microorganisms for nitrogen?: mechanisms and ecological relevance. *New Phytol.* 198, 656–669. doi: 10.1111/nph.12235

Myrold, D. D., and Bottomley, P. J. (2008). "Nitrogen mineralization and immobilization," in *Nitrogen in Agricultural Systems*, eds J. S. Schepers and W. R. Raun (Madison, WI: ASA-CSSA-SSSA), 153–168.

Nicol, G. W., Leininger, S., Schleper, C., and Prosser, J. I. (2008). The influence of soil pH on the diversity, abundance and transcriptional activity of ammonia oxidizing archaea and bacteria. *Environ. Microbiol.* 10, 2966–2978. doi: 10.1111/j.1462-2920.2008.01701.x

Nishina, K., Takenaka, C., and Ishizuka, S. (2009a). Spatial variations in nitrous oxide and nitric oxide emission potential on a slope of Japanese cedar (*Cryptomeria japonica*) forest. *Soil Sci. Plant Nutr.* 55, 179–189. doi: 10.1111/j.1747-0765.2007.00315.x

Nishina, K., Takenaka, C., and Ishizuka, S. (2009b). Spatiotemporal variation in N_2O flux within a slope in a Japanese cedar (*Cryptomeria japonica*) forest. *Biogeochemistry* 96, 163–175. doi: 10.1007/s10533-009-9356-2

Penna, D., Borga, M., Norbiato, D., and Dalla Fontana, G. (2009). Hillslope scale soil moisture variability in a steep alpine terrain. *J. Hydrol.* 364, 311–327. doi: 10.1016/j.jhydrol.2008.11.009

Petersen, D. G., Blazewicz, S. J., Firestone, M., Herman, D. J., Turetsky, M., and Waldrop, M. (2012). Abundance of microbial genes associated with nitrogen cycling as indices of biogeochemical process rates across a vegetation gradient in Alaska. *Environ. Microbiol.* 14, 993–1008. doi: 10.1111/j.1462-2920.2011.02679.x

Philippot, L., Cuhel, J., Saby, N. P. A., Chèneby, D., Chronáková, A., Bru, D., et al. (2009). Mapping field-scale spatial patterns of size and activity of the denitrifier community. *Environ. Microbiol.* 11, 1518–1526. doi: 10.1111/j.1462-2920.2009.01879.x

Rousk, J., Bååth, E., Brookes, P. C., Lauber, C. L., Lozupone, C., Caporaso, J. G., et al. (2010). Soil bacterial and fungal communities across a pH gradient in an arable soil. *ISME J.* 4, 1340–1351. doi: 10.1038/ismej.2010.58

Schimel, J. P., and Bennett, J. (2004). Nitrogen mineralization: challenges of a changing paradigm. *Ecology* 85, 591–602. doi: 10.1890/03-8002

Tateno, R., Hishi, T., and Takeda, H. (2004). Above- and belowground biomass and net primary production in a cool-temperate deciduous forest in relation to topographical changes in soil nitrogen. *For. Ecol. Manag.* 193, 297–306. doi: 10.1016/j.foreco.2003.11.011

Tateno, R., Osada, N., Terai, M., Tokuchi, N., and Takeda, H. (2005). Inorganic nitrogen source utilization byFagus crenata on different soil types. *Trees* 19, 477–481. doi: 10.1007/s00468-005-0409-4

Tateno, R., and Takeda, H. (2003). Forest structure and tree species distribution in relation to topography-mediated heterogeneity of soil nitrogen and light at the forest floor. *Ecol. Res.* 18, 559–571. doi: 10.1046/j.1440-1703.2003.00578.x

Tateno, R., and Takeda, H. (2010). Nitrogen uptake and nitrogen use efficiency above and below ground along a topographic gradient of soil nitrogen availability. *Oecologia* 163, 793–804. doi: 10.1007/s00442-009-1561-0

Tokuchi, N., Hirobe, M., and Koba, K. (2000). Topographical differences in soil N transformation using15N dilution method along a slope in a conifer plantation forest in Japan. *J. For. Res.* 5, 13–19. doi: 10.1007/BF02762758

Toljander, J. F., Eberhardt, U., Toljander, Y. K., Paul, L. R., and Taylor, A. F. S. (2006). Species composition of an ectomycorrhizal fungal community along a local nutrient gradient in a boreal forest. *New Phytol.* 170, 873–883. doi: 10.1111/j.1469-8137.2006.01718.x

Tromp-van Meerveld, H. J., and McDonnell, J. J. (2006). On the interrelations between topography, soil property, soil depth, soil moisture, transpiration rates and species distribution at the hillslope scale. *Adv. Water Resour.* 29, 293–310. doi: 10.1016/j.advwatres.2005.02.016

Urakawa, R., Ohte, N., Shibata, H., Tateno, R., Hishi, T., Fukushima, K., et al. (2014). Biogeochemical nitrogen properties of forest soils in the Japanese archipelago. *Ecol Res.* doi: 10.1007/s11284-014-1212-8. (in press)

Vitousek, P., and Howarth, R. (1991). Nitrogen limitation on land and in the sea: how can it occur? *Biogeochemistry* 13, 87–115. doi: 10.1007/BF00002772

Conflict of Interest Statement: The authors declare that the research was conducted in the absence of any commercial or financial relationships that could be construed as a potential conflict of interest.

Permissions

List of Contributors

Edward Jones and Balwant Singh
Department of Environmental Sciences, Faculty of Agriculture and Environment, The University of Sydney, Sydney, NSW, Australia

Urs Feller
Institute of Plant Sciences and Oeschger Centre for Climate Change Research, University of Bern, Bern, Switzerland

IrinaI.Vaseva
Institute of Plant Sciences and Oeschger Centre for Climate Change Research, University of Bern, Bern, Switzerland
Plant Stress Molecular Biology Department, Institute of Plant Physiology and Genetics, Bulgarian Academy of Sciences, Sofia, Bulgaria

María JesúsI. Briones
Departamento de Ecología y Biología Animal, Facultad de Biología, Universidad de Vigo, Vigo, Spain

Erik Braudeau
Qatar Foundation, Qatar Environment and Energy Research Institute, Doha, Qatar
Institutde Recherche pour le Développement, Bondy, France

Amjad T. Assi
Qatar Foundation, Qatar Environment and Energy Research Institute, Doha, Qatar
Department of Agricultural and Biological Engineering, Purdue University, West Lafayette, IN, USA

Hassan Boukcim
Valorhiz SAS, Parc Scientifique Agropolis II Bat6, Montferrier-sur-Lez, France

Rabi H. Mohtar
Biological and Agricultural Engineering Department, and Zachry Department of Civil Engineering, Texas A&M University, College Station, TX, USA

Amjad T. Assi
Qatar Foundation, Qatar Environment and Energy Research Institute, Doha, Qatar
Department of Agricultural and Biological Engineering, Purdue University, West Lafayette, IN, USA

Joshua Accola
Qatar Foundation, Qatar Environment and Energy Research Institute, Doha, Qatar
Department of Biological Systems Engineering, University of Wisconsin-Madison, Madison, WI, USA

Gaghik Hovhannissian
Institutde Recherche pourle Développement (IRD),Pédologie Hydrostructurale, Bondy, France

Rabi H. Mohtar
Department of Agricultural and Biological Engineering, Purdue University, West Lafayette, IN, USA
Biological and Agricultural Engineering Department and Zachry Department of Civil Engineering, Texas A&M University, College Station, TX, USA

Erik Braudeau
Qatar Foundation, Qatar Environment and Energy Research Institute, Doha, Qatar
Institutde Recherche pourle Développement (IRD),Pédologie Hydrostructurale, Bondy, France

Erik F. Braudeau
Hydrostructural Pedology, Centre de Recherche Ile de France, IRD France Nord, Institutde Recherche pourle Développement, Bondy, France

Rabi H. Mohtar
Department of Biological and Agricultural Engineering, Zachry Department of Civil Engineering, Texas A&M University, College Station, TX, USA

Nathaniel K. Newlands
Science and Technology Branch, Agriculture and Agri-Food Canada, Lethbridge Research Centre, Lethbridge, AB, Canada
Department of Statistics, University of British Columbia, Vancouver, BC, Canada

David S. Zamar
Department of Chemical and Biological Engineering, University of British Columbia, Vancouver, BC, Canada

Louis A. Kouadio
Science and Technology Branch, Agriculture and Agri-Food Canada, Lethbridge Research Centre, Lethbridge, AB, Canada

Yinsuo Zhang4,
4 Science and Technology Branch, National Agroclimate Information Service, Agriculture and Agri-Food Canada, Ottawa, ON, Canada

Aston Chipanshi
Science and Technology Branch, National Agroclimate Information Service, Agriculture and Agri-Food Canada, Regina, SK, Canada

Andries Potgieter
Queensland Alliance for Agriculture and Food Innovation, University of Queensland, Toowoomba, QLD, Australia

Souleymane Toure
Habitat Conservation Management Division, Canadian Wildlife Service, Environment Canada, Montreal, QC, Canada

Harvey S. J. Hill
Science and Technology Branch, National Agroclimate Information Service, Agriculture and Agri-Food Canada, Saskatoon, SK, Canada

Manoeli Lupatini
Departamento de Solos, Universida de Federal de Santa Maria, Santa Maria, Brazil
Department of Microbial Ecology, Netherlands Institute of Ecology (NIOO-KNAW), Wageningen, Netherlands

A fnan K. A. Suleiman, Rodrigo J. S. Jacques, Zaidal. Antoniolli
Departamento de Solos, Universida de Federal de Santa Maria, Santa Maria, Brazil

Adãode Siqueira Ferreira
Institutode Ciências Agrárias, Universida de Federal de Uberlândia, Uberlândia, Brazil

Eiko E. Kuramae
Department of Microbial Ecology, Netherlands Institute of Ecology (NIOO-KNAW), Wageningen, Netherlands

Luiz F. W. Roesch
Universida de Federaldo Pampa, São Gabriel, Brazil

Johann G. Zaller, Laura Simmer, Nadja Santer and James Tabi Tataw
Department of Integrative Biology and Biodiversity Research, Institute of Zoology, University of Natural Resources and Life Sciences Vienna, Austria

Herbert Formayer
Department of Water-Atmosphere-Environment, Institute of Meteorology, University of Natural Resources and Life Sciences Vienna, Austria

Erwin Murer
Institute of Land and Water Management Research, Federal Agency for Water Management, Petzenkirchen, Austria

Johannes Hösch and Andreas Baumgarten
Division for Food Security, Institute of Soil Health and Plant Nutrition, Austrian Agency for Health and Food Safety (AGES), Vienna, Austria

María Sánchez-García, Asunción Roig, Miguel A. Sánchez-Monedero and María L. Cayuela
Department of Soil and Water Conservation and Waste Management, CEBAS-CSIC, Campus Universitario de Espinardo, Murcia, Spain

Vimala D. Nair
Soil and Water Science Department, University of Florida, Gainesville, FL, USA

Naser A. Anjum
Department of Botany, Faculty of Science, Hamdard University, New Delhi, India
Department of Chemistry, CESAM-Centre for Environmental and Marine Studies, University of Aveiro, Aveiro, Portugal

Shahid Umar and Muhammad Iqbal
Department of Botany, Faculty of Science, Hamdard University, New Delhi, India

Ibrahim M. Aref
Plant Production Department, College of Food and Agricultural Sciences, King Saud University, Riyadh, Saudi Arabia

Akalu Teshome
Soil Physics and Land Management Group, Environmental Science Department, University of Wageningen, Wageningen, Netherlands
Amhara Regional Agricultural Research Institute, Bahir Dar, Ethiopia

Jande Graaff and Leo Stroosnijder
Soil Physics and Land Management Group, Environmental Science Department, University of Wageningen, Wageningen, Netherlands

Beni Lew
Department of Civil Engineering, Ariel University, Ariel, Israel
Agriculture Research Organization Volcani Center, Institute of Agricultural Engineering, Bet Dagan, Israel

Lolita Trachtengertz and Shany Ratsin
The Department of Chemical Engineering, Shenkar College of Engineering and Design, Ramat-Gan, Israel

Gideon Oron
Department of Environment Water Resources, The Jacob Blauste in Institutes for Desert Research, Ben-Gurion University of the Negev, Midrashet Ben-Gurion, Israel
The Department of Industrial Engineering and Management, and The Environmental Engineering Program, Ben-Gurion University of the Negev, Beer Sheva, Israel

Amos Bick
Bick & Associates, Ganey-Tikva, Israel

Kazuo Isobe, Nobuhito Ohte, Tomoki Oda, Sho Murabayashi, Wei Wei and Keishi Senoo
Graduate School of Agricultural and Life Sciences, The University of Tokyo, Tokyo, Japan

Naoko Tokuchi and Ryunosuke Tateno
Field Science Education and Research Center, Kyoto University, Kyoto, Japan

Printed in the USA
CPSIA information can be obtained
at www.ICGtesting.com
JSHW051441221024
72173JS00006B/1544